MW01677247

1999 NATIONAL PLUMBING & HVAC ESTIMATOR

By James A. Thomson

Includes inside the back cover:
An estimating CD with all the costs in the book plus an interactive multimedia video that shows how to use this disk to compile estimates, and a billing program that can export your estimates to *QuickBooks Pro* for progressive billing and job costing.

Craftsman Book Company
6058 Corte del Cedro / P.O. Box 6500 / Carlsbad, CA 92018

Acknowledgments

The sample "Standard Form Subcontract" and "Subcontract Change Order" forms used in the final section of this book are reprinted with the permission of the publisher, the Associated General Contractors of America (National Office), 1957 E Street NW, Washington, District of Columbia 20006.

Looking for Other Construction Reference Manuals?

Craftsman has the books to fill your needs. Call 1-800-829-8123 or write to:
Craftsman Book Company, P.O. Box 6500, Carlsbad, CA 92018
for a Free Catalog of over 100 books, including how-to manuals,
annual cost books, and estimating software.
Visit our Web site: http://www.craftsman-book.com

Cover photos by: *Tom Page Photography,* www.redrom.com/tom

© 1998 Craftsman Book Company

ISBN 1-57218-063-3
Published October 1998 for the year 1999.

Contents

The National Estimator (the disk in the envelope inside the back cover) is an electronic version of this manual. Use the disk to transfer the National Estimator program and all the information in this book to the hard disk of your computer. Instructions for using National Estimator begin on page 365.

How to Use This Book

This 1999 National Plumbing & HVAC Estimator is a guide to estimating labor and material costs for plumbing, heating, ventilating and air conditioning systems in residential, commercial and industrial buildings.

Inside the back cover of this book you'll find an envelope with a compact disk that you can install on your computer's hard drive. The disk has *National Estimator*, an easy-to-use estimating program, all the cost data in this book and *Job Cost Wizard* (turns estimates into invoices and exports to *QuickBooks*). An interactive, multimedia video shows how to use the software to create estimates for your jobs. Instructions for using *National Estimator* and *Job Cost Wizard* begin on page 365.

Costs in This Manual will apply within a few percent on a wide variety of projects. Using the information given on the pages that follow will explain how to use these costs and suggest procedures to follow when compiling estimates. Reading the remainder of this section will help you produce more reliable estimates for plumbing and HVAC work.

Manhour Estimates in This Book will be accurate for some jobs and inaccurate for others. No manhour estimate fits all jobs because every construction project is unique. Expect installation times to vary widely from job to job, from crew to crew, and even for the same crew from day to day.

There's no way to eliminate all errors when making manhour estimates. But you can minimize the risk of a major error by:

1. Understanding what's included in the manhour estimates in this book, and
2. Adjusting the manhour estimates in this book for unusual job conditions.

The Craft@Hrs Column. Manhour estimates in this book are listed in the column headed *Craft@Hrs*. For example, on page 15 you'll see an estimate for installing a 6 gallon hot water heater. In the *Craft@Hrs* column opposite 6 gallon you'll see:

P1@.500

To the left of the @ symbol you see an abbreviation for the recommended work crew.

Page 7 shows the wage rates and craft codes used in this book.

To the right of the @ symbol you see a number. The number is the estimated manhours (not crew hours) required to install each unit of material listed. In the case of a 6 gallon hot water heater, P1@.500 means that .500 manhours are required to install 1 hot water heater.

Costs in the Labor $ Column are based on manhour estimates in the Craft@Hrs column. Multiply the manhour estimate by the assumed hourly labor cost to find the installation cost in the Labor $ column. For example, .500 manhours times $24.44 (the average wage for crew P1) is $12.22.

Manhour Estimates include all productive labor normally associated with installing the materials described. These estimates assume normal conditions: experienced craftsmen working on reasonably well planned and managed new construction with fair to good productivity. Labor estimates also assume that materials are standard grade, appropriate tools are on hand, work done by other crafts is adequate, layout and installation are relatively uncomplicated, and working conditions don't slow progress.

All manhour estimates include tasks such as:

- Unloading and storing construction materials, tools and equipment on site.
- Working no more than two floors above or below ground level.
- Working no more than 10 feet above an uncluttered floor.
- Normal time lost due to work breaks.
- Moving tools and equipment from a storage area or truck not more than 200 feet from the work area.
- Returning tools and equipment to the storage area or truck at the end of the day.
- Planning and discussing the work to be performed.
- Normal handling, measuring, cutting and fitting.
- Regular cleanup of construction debris.
- Infrequent correction or repairs required because of faulty installation.

If the work you're estimating won't be done under these conditions, you need to apply a correction factor to adjust the manhour estimates in this book to fit your job.

Applying Correction Factors. Analyze your job carefully to determine whether a labor correction factor is needed. Failure to consider job conditions is probably the most common reason for inaccurate estimates.

Use one or more of the recommended correction factors in Table 1 to adjust for unusual job conditions. To make the adjustment, multiply the manhour estimate by the appropriate conversion factor. On some jobs, several correction factors may be needed. A correction factor less than 1.00 means that favorable working conditions will reduce the manhours required.

Supervision Expense to the installing contractor is not included in the labor cost. The cost of supervision and non-productive labor varies widely from job to job. Calculate the cost of supervision and non-productive labor and add this to the estimate.

Hourly Labor Costs also vary from job to job. This book assumes an average manhour labor cost of $28.34 for plumbers and $27.76 for sheet metal workers. If these hourly labor costs are not accurate for your jobs, adjust the labor costs up or down by an appropriate percentage. Instructions on the next page explains how to make these adjustments. If you're using the National Estimator disk, it's easy to set your own wage rates. See page 372.

Hourly labor costs in this book include the basic wage, fringe benefits, the employer's contribution to welfare, pension, vacation and apprentice funds, and all tax and insurance charges based on wages. Table 2 at the top of the next page shows how hourly labor costs in this book were calculated. It's important that you understand what's included in the figures in each of the six columns in Table 2. Here's an explanation:

Column 1, the base wage per hour, is the craftsman's hourly wage. These figures are representative of what many contractors are paying plumbers, sheet metal workers and helpers in 1999.

Column 2, taxable fringe benefits, includes vacation pay, sick leave and other taxable benefits. These fringe benefits average about 5.15% of the base wage for many plumbing and HVAC contractors. This benefit is in addition to the base wage.

Condition	Correction Factor
Work in large open areas,no partitions	.85
Prefabrication under ideal conditions, bench work	.90
Large quantities of repetitive work	.90
Very capable tradesmen	.95
Work 300' from storage area	1.03
Work 400' from storage area	1.05
Work 500' from storage area	1.07
Work on 3rd through 5th floors	1.05
Work on 6th through 9th floors	1.10
Work on 10th through 13th floors	1.15
Work on 14th through 17th floors	1.20
Work on 18th through 21st floors	1.25
Work over 21 floors	1.35
Work in cramped shafts	1.30
Work in commercial kitchens	1.10
Work above a sloped floor	1.25
Work in attic space	1.50
Work in crawl space	1.20
Work in a congested equipment room	1.20
Work 15' above floor level	1.10
Work 20' above floor level	1.20
Work 25' above floor level	1.30
Work 30' above floor level	1.40
Work 35' to 40' above floor level	1.50

Table 1 Recommended Correction Factors

Column 3, insurance and employer-paid taxes in percent, shows the insurance and tax rate for the craft workers. The cost of insurance in this column includes workers' compensation and contractor's casualty and liability coverage. Insurance rates vary widely from state to state and depend on a contractor's loss experience. Note that taxes and insurance increase the hourly labor cost by approximately 30%. There is no legal way to avoid these costs.

Column 4, insurance and employer taxes in dollars, shows the hourly cost of taxes and insurance. Insurance and taxes are paid on the costs in both columns 1 and 2.

Column 5, non-taxable fringe benefits, includes employer paid non-taxable benefits such as medical coverage and tax-deferred pension and profit sharing plans. These fringe benefits average 4.55% of the base wage for many plumbing and HVAC contractors.

Column Number	1	2	3	4	5	6
Craft	Base wage per hour	Taxable fringe benefits (at 5.15% of base wage)	Insurance and employer taxes (%)	Insurance and employer taxes ($)	Non-taxable fringe benefits (at 4.55% of base wage)	Total hourly cost used in this book
Laborer	14.22	0.73	33.05%	4.94	0.65	20.54
Plumber	20.67	1.06	26.08%	5.67	0.94	28.34
Sheet Metal Worker	19.97	1.03	27.88%	5.85	0.91	27.76
Operating Engineer	20.52	1.06	26.10%	5.63	0.93	28.14
Sprinkler Fitter	20.32	1.05	26.99%	5.77	0.92	28.06

Craft Code	Crew Composition	Average Hourly Cost per Manhour
ER	4 building plumbers, 2 building laborers, 1 operating engineer	26.09
SN	4 building sheet metal workers, 2 building laborer, 1 operating engineer	25.75
P1	1 building plumber and 1 building laborer	24.44
SK	4 sprinkler fitters, 2 building laborers, 1 operating engineer	25.92
SL	1 sprinkler fitter and 1 laborer	24.30
S2	1 building sheet metal worker, 1 building laborer	24.15

Table 2 Labor Costs Used in This Book

The employer pays no taxes or insurance on these benefits.

Column 6, the total hourly cost in dollars, is the sum of columns 1, 2, 4, and 5. The labor costs in Column 6 were used to compute costs in the Labor $ column of this book.

Adjusting Costs in the Labor $ Column. The hourly labor costs used in this book may apply within a few percent on many of your jobs. But wage rates may be much higher or lower in some areas. If the hourly costs shown in Column 6 of Table 2 are not accurate for your work, adjust labor costs to fit your jobs.

For example, suppose your hourly labor costs are as follows:

Plumber	$14.00
Laborer	$10.00
Total hourly crew cost	$24.00

Your average cost per manhour would be $12.00 ($24.00 per crew hour divided by 2 because this is a crew of two).

A labor cost of $12.00 is about 50% of the $24.44 labor cost used for crew P1. Multiply costs in the Labor $ column by .50 to find your estimated cost.

For example, notice on page 15 that the labor cost for installing a 6 gallon hot water heater is $18.20 per linear foot. If installed by your plumbing crew working at an average cost of $12.00 per manhour, your estimated cost would be 50% of $18.20 or $9.10 per heater.

Adjusting the labor costs in this book will make your estimates much more accurate. Making adjustments to labor costs is both quick and easy if you use the National Estimator disk.

Equipment Cost will vary according to need and application. It typically is $35 per hour for a 10-ton hydraulic truck-mounted crane.

Material Costs in this manual are intended to reflect what medium- to low-volume contractors will be paying in 1999 after applying normal discounts. These costs include charges for delivery to within 25 to 30 miles of the supplier.

Overhead and Profit for the installing contractor are not included in the costs in this manual unless specifically identified in the text. Markup can vary widely with local economic conditions, competition and the installing contractor's operating expenses. Add the markup that's appropriate for your company, the job and the competitive environment.

How Accurate Are These Figures? As accurate as possible considering that the editors don't know your material suppliers, haven't seen the plans or specifications, don't know what building code applies or where the job is, had to project material costs at least six months into the future, and had no record of how much work the crew that will be assigned to the job can handle.

You wouldn't bid a job under those conditions. And I don't claim that all plumbing and HVAC work is done at these prices.

Estimating Is an Art, not a science. There is no one price that applies on all jobs. On many jobs the range between high and low bid will be 10% or more. There's room for legitimate disagreement on what the correct costs are, even when complete plans and specifications are available, the date and site are established, and labor and material costs are identical for all bidders.

No estimate fits all jobs. Good estimates are custom made for a particular project and a single contractor through judgment, analysis and experience. This book is not intended as a substitute for judgment, analysis and sound estimating practice. It's an aid in developing an informed opinion of cost, not an answer book.

Additional Costs to Consider

Here's a checklist of additional costs to consider before submitting any bid.

1. Sales taxes
2. Mobilization costs
3. Payment and performance bond costs
4. Permits and fees
5. Storage container rental costs
6. Utility costs
7. Tool costs
8. Callback costs during warranty period
9. Demobilization costs

Exclusions and Clarifications

Neither the job specifications nor the contract may identify exactly what work should be included in the plumbing and HVAC bid. Obviously, you have to identify what work is included in the job.

The most efficient way to define the scope of the work is to prepare a list of tasks not normally performed by your company and attach that list to each bid submitted. Here's a good list of work that should be excluded from your bid:

Your Bid Should Exclude

Final cleaning of plumbing fixtures
Backings for plumbing fixtures
Toilet room accessories
Electrical work, including motor starters
Electrical wiring and conduit over 100 volts
Temporary utilities
Painting, priming and surface preparation
Structural cutting, patching or repairing
Fire protection and landscape sprinklers
Equipment supports
Surveying and layout of control lines
Removal or stockpiling of excess soil
Concrete work, including forming and rebar
Setting of equipment furnished by others
Equipment, unless shown, and personnel hoisting
Wall and floor blockouts
Pitch pockets
The costs of performance or payment bonds
Site utilities
Asbestos removal or disposal
Contaminated soil removal or disposal
Major increases in copper material prices
Fire dampers not shown on the plans

Your Bid Should Include

Trash sweep-up only. Others haul it away.
Site utilities from building to property line only.
Piping to 5 feet outside the building only.
Plumbing & HVAC permits for your work only.

Beware of Price Changes

There's no way to be sure what prices will be in three to six months. All labor, equipment, material and subcontract prices in a bid should be based on costs anticipated when the project is expected to be built, not when the estimate is compiled. That presents a problem. Except for the installation of underground utilities, most plumbing and HVAC work is done six months to a year after the bid is submitted. When possible, get price protection in writing from your suppliers and subcontractors. If your suppliers and subs won't guarantee prices, include an escalation allowance in your bid to cover anticipated price increases.

Material Pricing Conditions

All equipment and material prices quoted by your vendors will be conditional. They usually don't include sales tax and are subject to specific payment and shipping terms. Every estimator should understand the meaning of common shipping terms. They define who pays the freight and who has responsibility for processing freight-damage claims. Here's a summary of important conditions you should understand.

F.O.B. Factory (Free On Board at the Factory): Title passes to the buyer when the goods are delivered by the seller to the freight carrier. The buyer pays the freight and is responsible for freight-damage claims.

F.O.B. Factory F.F.A. (Free On Board at the Factory, Full Freight Allowed): The title passes to the buyer when the goods are delivered by the seller to the freight carrier. The seller pays the freight charges, but the buyer is responsible for freight-damage claims.

F.O.B. (city of destination) (Free On Board to your city): The title passes to the buyer when the goods are delivered by the seller to the freight terminal in the city, or nearest city, of destination. The seller pays the freight and is responsible for freight-damage claims to the terminal. The buyer pays the freight charge and is responsible for freight-damage claims from the terminal to the final destination.

F.O.B. Job Site (Free On Board at job site, or contractor's shop): The title passes to the buyer when the goods are delivered to the job site (or shop). The seller pays the freight and is responsible for freight-damage claims.

F.A.S. Port [of a specific city] (Free Alongside Ship at the nearest port): The title passes to the buyer when goods are delivered to the ship dock or port terminal. The seller pays the freight and is responsible for freight-damage claims to the ship dock or port terminal only. The buyer pays the freight and is responsible for freight-damage claims from the ship dock or port terminal to the designated delivery point.

Obviously, it's to your advantage to instruct all vendors to quote costs F.O.B. the job site or your shop.

Reducing Costs

Most construction specifications allow the use of alternative equipment and materials. It's the estimator's responsibility to select the most cost-effective products. Research and compare your costs before making any decisions. Avoid selecting any material or equipment simply because that's what you've always done.

Don't recommend plastic products such as ABS, PVC, or polypropylene pipe or corrugated flexible ducts until you've checked local code requirements. Most building codes prohibit use of these materials inside public buildings such as schools, care centers and hospitals.

It's wise to select 100% factory-packaged equipment. Beware of equipment labeled "Some assembly required." Field labor costs for mounting loose coils, motors and similar equipment are very high.

Value Engineering

Let's suppose you've submitted a combined plumbing and HVAC bid for $233,000. Your cutthroat competitor put in a bid at $4,000 less, $229,000. Obviously there's no way you're going to get the job. Right?

Not so fast! Maybe value engineering can help you win that contract — while fattening your profit margin.

Suppose the proposal you submitted had two parts. Part I is the bid for $233,000, based entirely on job plans and specs, just the way they were written. But appended to your proposal is Part II, a list of suggestions for saving money without sacrificing any of the capacity or quality designed into the system. Here's an example of what might be in Part II:

1. Deduct for providing pipe hanger spacings per UPC in lieu of specified spacings: $1,750.00
2. Deduct for reducing heating hot water pipe sizes by using 40 degrees F Delta T in lieu of specified 20 degrees F Delta T: $4,600.00

3. Deduct for providing pressure/temperature taps at air handling units, pumps and chillers in lieu of specified thermometers and pressure gauges: $875.00
4. Deduct for eliminating water treatment in closed piping systems: $1,800.00
5. Deduct for piping chilled and heating hot water pumps in parallel in lieu of providing 100% standby pumps: $2,900.00

Total deductions: $11,925.00

Adopting these suggestions would make you low bidder by nearly $8,000. A saving like that will be tempting to most owners, especially if the owner understands that your suggestions result in a system that is every bit as good and maybe better than the system as originally designed.

You're not offering to undercut the competition. Far from it. You're using knowledge and experience to create better value for the owner. That's called value engineering and it's likely to win the respect of nearly all cost-conscious owners.

Notice that reducing costs is only part of what value engineering is all about. You don't cut costs at the expense of system quality, integrity, capacity or performance.

Don't waste your time, and your client's, by offering to substitute cheaper or lower-quality fixtures or equipment. Any cutthroat contractor with a price list can do that. Recommend the use of inferior materials and you'll be associated with the inferior goods you promote. Some owners consider even the suggestion to be insulting.

The recommendations you make (like most of those in the example) will require design changes. You can expect to be examined (or even challenged) on these points. Be ready to explain and defend each of your suggestions. Convince the client (or the design engineer) that your ideas are based on sound engineering principles and you're well on the way to winning the owner's confidence and the contract.

Now, let's go back to the list and see how we might justify the five value engineering recommendations.

1. **Pipe Hanger Spacing.** The pipe hanger spacings recommended in the Uniform Plumbing Code (UPC) are calculated by experienced, professional structural engineers. The safety factors used in these calculations are very conservative. They've been widely used for many years and have proved to be more than adequate. There's no need for more hangers than the UPC requires.
2. **Changing HHW Delta T.** In hydronic heating systems, heat measured in Btus is pumped to terminal units. The proposed change of the Delta T, from 20 degrees F to 40 degrees F, has no effect whatsoever on how many Btus the system delivers. You're not changing anything but the volume of water being pumped. At lower volume levels, the size of the pump, the pipe and the pipe insulation can all be reduced. Not one of these changes will affect the system's ability to transmit heat. Furthermore, operating costs will also drop, since less pump horsepower will be needed to run the smaller pump.
3. **Thermometers/Pressure Gauges.** Thermometers and pressure gauges installed on or near vibrating machinery have a very short life expectancy. Gauges quickly lose accuracy under harsh conditions. Readings will become less and less reliable. That's potentially dangerous. You can avoid this problem by using insertion-type pressure/temperature taps instead. Store these sensitive gauges in a desk drawer or a tool crib when not in use. Safely stored, they're protected from damage. They'll give accurate readings longer and won't need to be replaced as often. And they're simple to use. Just insert a gauge in one of the conveniently located taps. Make the reading, then remove the gauge and put it away.
4. **Water Treatment.** ITT Bell & Gossett has done studies on corrosion in closed hydronic systems that have a make-up water rate of no more than 5% per year. These studies show that corrosion virtually stops when entrained air is either removed or depleted. No water treatment is needed in this closed system.
5. **100% Standby Pumps.** Two pumps piped and operated in parallel are more economical. Even if one pump fails, the other pump can maintain delivery at 75 to 80% of the designed flow rate. That's usually adequate for emergency operation.

These cost-saving ideas are small, but could tip the balance in your favor. I hope they demonstrate the potential that value engineering has when bidding jobs. Any time you're compiling an estimate, keep an eye out for ways to save money or reduce the owner's cost. Jot a note to yourself about each potential saving you identify. Before submitting the bid, make a list of your alternate suggestions. Maybe best of all, markup on your value engineering suggestions can be higher than your normal markup. If value engineering can cut costs by $10,000, maybe as much as $4,000 of that should end up in your pocket!

The Estimating Procedure

Every plumbing and HVAC estimator works under deadline pressure. You'll seldom have the luxury of spending as much time as you would like on an estimate. Estimators who aren't organized waste valuable time and tend to make careless errors. Try to be well-organized and consistent in your approach to estimating. For most projects, I recommend that you follow the procedures listed below and in the order listed:

1. Get a second set of project drawings and specifications for use by your suppliers and subcontractors. Remember that your subs and suppliers need access to the plans and specs and time to prepare their quotes.
2. Study the plans and specs carefully. Highlight important items. Make a list of specific tasks that require labor unit correction factors. The estimate is never complete until you're totally familiar with the project and the applicable construction codes.
3. Get the general contractor or owner to identify the proposed construction schedule and subcontractor lay-down (storage) area. Work schedule and site conditions always affect your costs.
4. Contact all potential suppliers and subcontractors as early as possible. Set a time when each can come to your office to make their take-offs from the spare set of contract documents.

When this important preliminary work is done, or in progress, it's time to begin your detailed take-off.

Guidelines for Good Estimating

You can compile estimates on a legal pad, a printed estimating form or on a computer. Regardless of the method, these guidelines will apply:

List Each Cost Separately on your take-off sheet. Don't combine system estimates, even if the materials are the same type. A combined system estimate may have to be completely redone if materials for one system are changed at a later date. Use the Estimate Detail Sheet on the facing page if you don't already have a good material take-off form.

Use Engineer's Identification Numbers when listing equipment. The word pump without any other description is ambiguous when there are several pumps included in the project.

Don't Forget Labor Adjustment factors if your labor costs are significantly higher or lower than the costs used in this book. See instructions on page 7 for adjusting labor costs.

Use Colored Pencils or highlighters to mark the items you've taken off and listed. Use a different color for each piping or ducting system.

Log Telephone Quotes and other important phone conversations on a telephone quote form. See the sample on page 14.

Project Estimated Costs for labor, material and equipment to the time when the work is expected to be done, not when the job is being estimated.

The only good estimate is a complete estimate. You've probably heard this saying, "He who makes the most mistakes is likely to be low bidder, and live to regret it."

Preparing the Proposal

It's both common courtesy and good business practice to deliver an unpriced copy of your bid or proposal letter to the general contractor three or four days before the bid deadline date. This gives the contractor time to study your proposal and obtain alternate pricing for items you may have excluded. To avoid misunderstandings, make sure your proposals include, as a minimum, the following elements:

1. The complete name and address of the proposed project.
2. Specification title and issue date.
3. A complete listing of drawings and their issue or revision date.
4. A complete list of addenda and their dates of issue.
5. A list of specification section numbers covered by your proposal.
6. A list of exclusions, clarifications and assumptions.

Your final bid can be phoned in or sent by fax, but it should reach the general contractor or owner no more than five or ten minutes before the bid deadline. Prices submitted too early may have to be revised because of last-minute price changes by subcontractors or suppliers.

Estimate Detail Sheet

Data carried forward from Take-Off Quantity Survey Sheet(s)

Company/Department ______________________ Estimator ______________________ Date ____________

Project ______________________ Checked by ______________________ Date ____________

Address ______________________ Notes: ______________________

Job description ____________ Estimate # ____________

CSI Division/ Account ____________ Estimate due ____________

Item Description	Quantity	Unit	Crew @ MH/Unit	Manhours Ext.	Materials		Labor		Equipment		Subcontract		Total $
					Unit $	Ext. $	Unit $	Ext. $	Unit $	Ext. $	Unit $	Ext. $	

Totals This Sheet	Manhours	Material $	Labor $	Equipment $	Subcontract $	Total $

Carry totals forward to Estimate Summary Sheet

Estimate # ______ Estimate Details Sheet ______ of ______

Quotation Sheet

o: ____________________

pplier: ____________________

esperson: ____________________ **Phone No:** ____________________

Plans/Specs: ____________ **Freight:** ____________ **Terms:** ____________

Description	Delivery Time	Price

y: ____________________

Record of Telephone Conversation

Date:__________ Time:__________ Project: __________

Telecon with:__________

Company: __________ Phone No:__________

Subject: __________

Details of Conversation:__________

By:__________

Description	Craft@Hrs	Unit	Material $	Labor $	Equipment $	Total $

Electric domestic hot water heater, (residential). Set in place only, (floor models). Make additional allowances for pipe and electrical connections. (see below)

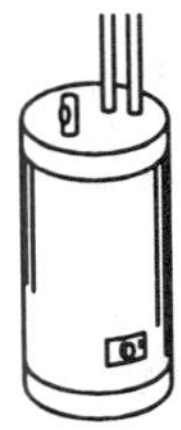

Description	Craft@Hrs	Unit	Material $	Labor $	Equipment $	Total $
6 gallon	P1@.500	Ea	120.00	12.20	—	132.20
10 gallon	P1@.500	Ea	135.00	12.20	—	147.20
15 gallon	P1@.750	Ea	140.00	18.30	—	158.30
20 gallon	P1@.750	Ea	145.00	18.30	—	163.30
30 gallon	P1@1.00	Ea	155.00	24.40	—	179.40

Electric domestic hot water heater, (commercial), 208/240 volt. Set in place only. Make additional allowances for pipe and electrical connections. (see below)

Description	Craft@Hrs	Unit	Material $	Labor $	Equipment $	Total $
96 gallon, 12 kw	P1@.500	Ea	680.00	12.20	—	692.20
96 gallon, 18 kw	P1@.500	Ea	930.00	12.20	—	942.20
96 gallon, 36 kw	P1@.750	Ea	965.00	18.30	—	983.30
120 gallon, 18 kw	P1@.750	Ea	985.00	18.30	—	1,003.30
120 gallon, 36 kw	P1@.750	Ea	1,020.00	18.30	—	1,038.30
120 gallon, 54 kw	P1@.750	Ea	1,200.00	18.30	—	1,218.30
120 gallon, 63 kw	P1@.750	Ea	1,300.00	18.30	—	1,318.30

Gas-fired domestic hot water heater, (residential). Set in place only. Make additional allowances for pipe and combustion b-vent connections. (see below)

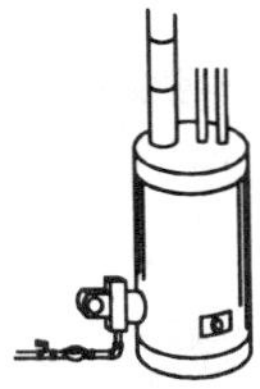

Description	Craft@Hrs	Unit	Material $	Labor $	Equipment $	Total $
30 gallon	P1@1.00	Ea	250.00	24.40	—	274.40
40 gallon	P1@1.00	Ea	405.00	24.40	—	429.40
50 gallon	P1@1.50	Ea	460.00	36.70	—	496.70

Gas-fired domestic hot water heater, (commercial), standard efficiency. Set in place only. Make additional allowances for pipe and combustion b-vent connections. (see below)

Description	Craft@Hrs	Unit	Material $	Labor $	Equipment $	Total $
50 gal./95 gph	P1@2.00	Ea	780.00	48.90	—	828.90
67 gal./106 gph	P1@2.00	Ea	925.00	48.90	—	973.90
76 gal./175 gph	P1@2.00	Ea	1,250.00	48.90	—	1,298.90
91 gal./291 gph	P1@2.00	Ea	1,480.00	48.90	—	1,528.90

Gas-fired domestic hot water heater, (commercial), energy miser. Set in place only. Make additional allowances for pipe and combustion b-vent connections. (see below)

Description	Craft@Hrs	Unit	Material $	Labor $	Equipment $	Total $
50 gal./95 gph	P1@2.00	Ea	1,130.00	48.90	—	1,178.90
67 gal./106 gph	P1@2.00	Ea	1,170.00	48.90	—	1,218.90
76 gal./175 gph	P1@2.00	Ea	1,450.00	48.90	—	1,498.90
91 gal./291 gph	P1@2.00	Ea	1,720.00	48.90	—	1,768.90

Domestic Hot Water Heater Connections

Description	Craft@Hrs	Unit	Material $	Labor $	Equipment $	Total $

Domestic hot water heater connection assembly. Includes supply, return, recirculation and relief piping and fittings, (copper) relief and isolation valves. Make additional allowances for gas and flue connections where applicable.

Description	Craft@Hrs	Unit	Material $	Labor $	Equipment $	Total $
3/4" residential	P1@1.75	Ea	150.00	42.80	—	192.80
3/4" commercial	P1@2.25	Ea	200.00	55.00	—	255.00
1" commercial	P1@2.75	Ea	265.00	67.20	—	332.20
1¼" commercial	P1@3.50	Ea	325.00	85.50	—	410.50
1½" commercial	P1@3.75	Ea	335.00	91.60	—	426.60
2" commercial	P1@4.50	Ea	360.00	110.00	—	470.00
2½" commercial	P1@5.75	Ea	750.00	141.00	—	891.00
3" commercial	P1@6.50	Ea	1,150.00	159.00	—	1,309.00

Domestic water heater flue connection

Description	Craft@Hrs	Unit	Material $	Labor $	Equipment $	Total $
2" b vent	P1@.090	LF	2.10	2.20	—	4.30
3" b vent	P1@.100	LF	2.59	2.44	—	5.03
4" b vent	P1@.110	LF	3.46	2.69	—	6.15
6" b vent	P1@.130	LF	3.95	3.18	—	7.13
power vent kit	P1@2.00	Ea	380.00	48.90	—	428.90

Description	Craft@Hrs	Unit	Material $	Labor $	Equipment $	Total $
Kitchen equipment booster heater						
1,000 watt	P1@4.00	Ea	104.00	97.70	—	201.70
Dishwasher						
Built-in	P1@5.00	Ea	635.00	122.00	—	757.00
Garbage disposal						
1/2 HP	P1@2.00	Ea	85.00	48.90	—	133.90
3/4 HP	P1@2.00	Ea	165.00	48.90	—	213.90
Grease and oil interceptor						
4 GPM	P1@4.00	Ea	326.00	97.70	—	423.70
10 GPM	P1@5.00	Ea	532.00	122.00	—	654.00
15 GPM	P1@7.00	Ea	789.00	171.00	—	960.00
20 GPM	P1@8.00	Ea	966.00	195.00	—	1,161.00
Hair and lint interceptor						
1½"	P1@.650	Ea	122.00	15.90	—	137.90
2"	P1@.750	Ea	171.00	18.30	—	189.30
All bronze 3/4" to 1½" in-line NPT pump						
1/12 HP	P1@1.50	Ea	227.00	36.70	—	263.70
1/6 HP	P1@1.50	Ea	339.00	36.70	—	375.70
1/4 HP	P1@1.50	Ea	399.00	36.70	—	435.70

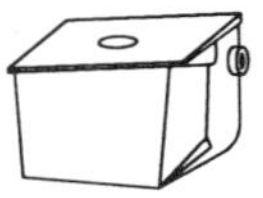

Kitchen Equipment Connections

Description	Craft@Hrs	Unit	Material $	Labor $	Equipment $	Total $
Kitchen appliance gas trim						
1/2"	P1@1.15	Ea	20.00	28.10	—	48.10
3/4"	P1@1.30	Ea	25.20	31.80	—	57.00
1"	P1@1.60	Ea	29.20	39.10	—	68.30
1¼"	P1@2.10	Ea	47.90	51.30	—	99.20
1½"	P1@2.50	Ea	61.00	61.10	—	122.10
2"	P1@3.00	Ea	81.20	73.30	—	154.50
Hot and cold water supply						
1/2"	P1@1.10	Ea	18.60	26.90	—	45.50
3/4"	P1@1.55	Ea	26.50	37.90	—	64.40
1"	P1@1.90	Ea	35.80	46.40	—	82.20
1¼"	P1@2.50	Ea	50.50	61.10	—	111.60
1½"	P1@3.00	Ea	63.70	73.30	—	137.00
Continuous waste						
2-part	P1@.250	Ea	27.10	6.11	—	33.21
3-part	P1@.350	Ea	45.70	8.55	—	54.25
4-part	P1@.450	Ea	59.20	11.00	—	70.20
Indirect waste						
1/2"	P1@1.05	Ea	3.66	25.70	—	29.36
3/4"	P1@1.50	Ea	6.14	36.70	—	42.84
1"	P1@1.90	Ea	9.82	46.40	—	56.22
1¼"	P1@2.15	Ea	14.80	52.50	—	67.30
1½"	P1@2.60	Ea	19.20	63.50	—	82.70
2"	P1@3.00	Ea	29.50	73.30	—	102.80
Kitchen fixture waste tailpiece						
1½"	P1@.100	Ea	5.95	2.44	—	8.39
Kitchen fixture trap with solder bushing						
1½"	P1@.250	Ea	18.10	6.11	—	24.21
2"	P1@.300	Ea	25.20	7.33	—	32.53

Description	Craft@Hrs	Unit	Material $	Labor $	Equipment $	Total $

Water closet, floor mounted, flush tank, white vitreous china, lined tank. Complete with trim. Make additional allowances for rough-in. Based on American Standard Cadet series. A.D.A. means American Disabilities Act compliant. [Wheelchair accessible]

Description	Craft@Hrs	Unit	Material $	Labor $	Equipment $	Total $
Round bowl	P1@2.10	Ea	140.00	51.30	—	191.30
Elongated bowl	P1@2.10	Ea	170.00	51.30	—	221.30
ADA, 18" high	P1@2.10	Ea	230.00	51.30	—	281.30

Water closet, floor mounted, flush valve, white vitreous china. Complete with trim. Make additional allowances for rough-in. Based on American Standard. A.D.A. means American Disabilities Act compliant. [Wheelchair accessible]

Description	Craft@Hrs	Unit	Material $	Labor $	Equipment $	Total $
Elongated bowl	P1@2.60	Ea	220.00	63.50	—	283.50
Elongated bowl, ADA 18" high	P1@2.60	Ea	260.00	63.50	—	323.50
Elongated bowl with a bedpan cleanser	P1@4.10	Ea	380.00	100.00	—	480.00
Elongated bowl, ADA 18" high with a bedpan cleanser	P1@4.10	Ea	410.00	100.00	—	510.00

Water closet, wall hung, flush valve, white vitreous china. Complete with fixture carrier and all trim. Make additional allowances for rough-in. Based on American Standard Afwall series.

Description	Craft@Hrs	Unit	Material $	Labor $	Equipment $	Total $
Elongated bowl	P1@3.55	Ea	350.00	86.80	—	436.80
Elongated bowl with electronic flush valve	P1@3.80	Ea	625.00	92.90	—	717.90
Elongated bowl with bedpan cleanser	P1@5.05	Ea	505.00	123.00	—	628.00
Electronic flush valve, add	P1@.600	Ea	275.00	14.70	—	289.70

Urinal, wall hung, flush valve, white vitreous. Complete with trim. Make additional allowances for rough-in.

Description	Craft@Hrs	Unit	Material $	Labor $	Equipment $	Total $
Siphon jet type	P1@3.15	Ea	350.00	77.00	—	427.00
Wash out type	P1@3.10	Ea	285.00	75.80	—	360.80
Wash down type	P1@3.00	Ea	200.00	73.30	—	273.30
Urinal carrier, add	P1@.600	Ea	65.00	14.70	—	79.70
Electronic flush valve, add	P1@.600	Ea	225.00	14.70	—	239.70

Plumbing Fixtures

Urinal, stall type, flush valve, white vitreous china. Complete with trim. Make additional allowances for rough-in.

Description	Craft@Hrs	Unit	Material $	Labor $	Equipment $	Total $
Stall urinal	P1@5.00	Ea	690.00	122.00	—	812.00

Lavatory, wall hung, white vitreous china. Complete with trim and fixture carrier. Make additional allowances for rough-in. A.D.A. means American Disabilities Act compliant. [Wheelchair accessible]

Description	Craft@Hrs	Unit	Material $	Labor $	Equipment $	Total $
Wall hung lav	P1@2.70	Ea	290.00	66.00	—	356.00
Wall hung, ADA	P1@2.70	Ea	425.00	66.00	—	491.00
Fixture carrier	P1@.500	Ea	60.00	12.20	—	72.20

Countertop lavatory, white. Complete with trim. Make additional allowances for rough-in.

Description	Craft@Hrs	Unit	Material $	Labor $	Equipment $	Total $
Vitreous china	P1@2.00	Ea	220.00	48.90	—	268.90
Enameled steel	P1@2.00	Ea	185.00	48.90	—	233.90
Acrylic	P1@2.00	Ea	135.00	48.90	—	183.90

Bathtub, white, 60" x 32". Complete with trim, including shower head. Make additional allowances for rough-in.

Description	Craft@Hrs	Unit	Material $	Labor $	Equipment $	Total $
Enameled steel	P1@2.50	Ea	305.00	61.10	—	366.10
Cast iron	P1@3.50	Ea	440.00	85.50	—	525.50
Fiberglass	P1@2.50	Ea	280.00	61.10	—	341.10
Acrylic	P1@2.50	Ea	300.00	61.10	—	361.10

Tub and shower combination, fiberglass, white. Complete with trim, Make additional allowances for rough-in.

Description	Craft@Hrs	Unit	Material $	Labor $	Equipment $	Total $
One piece	P1@4.50	Ea	625.00	110.00	—	735.00
Two piece (reno)	P1@5.50	Ea	800.00	134.00	—	934.00
Four piece (reno)	P1@6.25	Ea	850.00	153.00	—	1,003.00

Shower-stall, white, 36" x 36". Complete with trim, Make additional allowances for rough-in.

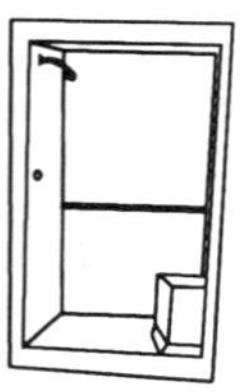

Description	Craft@Hrs	Unit	Material $	Labor $	Equipment $	Total $
Fiberglass One piece	P1@3.50	Ea	390.00	85.50	—	475.50
Fiberglass Three piece	P1@4.25	Ea	500.00	104.00	—	604.00
Acrylic One piece	P1@3.50	Ea	580.00	85.50	—	665.50
Acrylic Three piece	P1@4.25	Ea	760.00	104.00	—	864.00

Shower basin, 36" x 36". Complete with trim. (Faucet, shower head and strainer.) Make additional allowances for rough-in.

Description	Craft@Hrs	Unit	Material $	Labor $	Equipment $	Total $
Fiberglass	P1@2.50	Ea	265.00	61.10	—	326.10
Acrylic	P1@2.50	Ea	285.00	61.10	—	346.10
Molded stone	P1@2.65	Ea	275.00	64.80	—	339.80

Kitchen sink, double compartment. Complete with trim. Make additional allowances for rough-in.

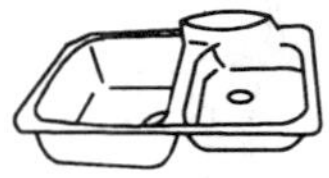

Description	Craft@Hrs	Unit	Material $	Labor $	Equipment $	Total $
Stainless steel	P1@2.15	Ea	215.00	52.50	—	267.50
Cast iron	P1@2.50	Ea	280.00	61.10	—	341.10
Acrylic	P1@2.15	Ea	255.00	52.50	—	307.50

Kitchen sink, single compartment. Complete with trim. Make additional allowances for rough-in.

Description	Craft@Hrs	Unit	Material $	Labor $	Equipment $	Total $
Stainless steel	P1@2.00	Ea	180.00	48.90	—	228.90
Cast iron	P1@2.10	Ea	210.00	51.30	—	261.30
Acrylic	P1@2.00	Ea	190.00	48.90	—	238.90

Bar sink. Complete with trim. Make additional allowances for rough-in.

Description	Craft@Hrs	Unit	Material $	Labor $	Equipment $	Total $
Stainless steel	P1@2.00	Ea	160.00	48.90	—	208.90
Acrylic	P1@2.00	Ea	105.00	48.90	—	153.90

Exam room sink. Complete with trim. Make additional allowances for rough-in.

Description	Craft@Hrs	Unit	Material $	Labor $	Equipment $	Total $
Stainless steel	P1@2.10	Ea	230.00	51.30	—	281.30
Acrylic	P1@2.10	Ea	195.00	51.30	—	246.30

Laboratory sink. Complete with trim. Make additional allowances for rough-in.

Description	Craft@Hrs	Unit	Material $	Labor $	Equipment $	Total $
Stainless steel	P1@2.25	Ea	265.00	55.00	—	320.00
Acrylic	P1@2.25	Ea	230.00	55.00	—	285.00

Laundry sink, double compartment. Complete with trim. Make additional allowances for rough-in.

Description	Craft@Hrs	Unit	Material $	Labor $	Equipment $	Total $
Cast iron	P1@3.50	Ea	310.00	85.50	—	395.50
Acrylic	P1@2.25	Ea	135.00	55.00	—	190.00

Laundry sink, single compartment. Complete with trim. Make additional allowances for rough-in.

Description	Craft@Hrs	Unit	Material $	Labor $	Equipment $	Total $
Cast iron	P1@2.75	Ea	270.00	67.20	—	337.20
Acrylic	P1@2.00	Ea	95.00	48.90	—	143.90

Mop sink, floor mounted, 36" x 24". Complete with trim. Make additional allowances for rough-in

Description	Craft@Hrs	Unit	Material $	Labor $	Equipment $	Total $
Molded stone	P1@2.65	Ea	380.00	64.80	—	444.80
Terrazzo	P1@2.65	Ea	420.00	64.80	—	484.80
Acrylic	P1@2.35	Ea	290.00	57.40	—	347.40

Description	Craft@Hrs	Unit	Material $	Labor $	Equipment $	Total $

Slop sink, enameled cast iron with P trap, standard. Complete with trim. Make additional allowances for rough-in.

Description	Craft@Hrs	Unit	Material $	Labor $	Equipment $	Total $
Slop sink with P trap, Std.	P1@3.50	Ea	620.00	85.50	—	705.50

Floor sink, recessed, enameled steel, white. Add 40% to material prices for acid-resisting finish. Complete with strainer. Make additional allowances for rough-in.

Description	Craft@Hrs	Unit	Material $	Labor $	Equipment $	Total $
9" x 9"	P1@1.00	Ea	53.00	24.40	—	77.40
12" x 12"	P1@1.00	Ea	62.00	24.40	—	86.40
15" x 15"	P1@1.15	Ea	42.00	28.10	—	70.10
18" x 18"	P1@1.25	Ea	55.00	30.50	—	85.50
24" x 24"	P1@1.50	Ea	75.00	36.70	—	111.70

Drinking fountain, refrigerated, stainless steel. Complete with trim. Make additional allowances for rough-in A.D.A. means American Disabilities Act compliant. [Wheelchair accessible]

Description	Craft@Hrs	Unit	Material $	Labor $	Equipment $	Total $
Free standing	P1@2.00	Ea	685.00	48.90	—	733.90
Semi recessed	P1@2.50	Ea	900.00	61.10	—	961.10
Fully recessed	P1@2.50	Ea	1,550.00	61.10	—	1,611.10
Wall hung	P1@2.00	Ea	635.00	48.90	—	683.90
Wall hung, ADA	P1@2.50	Ea	1,550.00	61.10	—	1,611.10

Drinking fountain, non-refrigerated. Complete with trim. Make additional allowances for rough-in. A.D.A. means American Disabilities Act compliant. [Wheelchair accessible] S.S. means stainless steel.

Description	Craft@Hrs	Unit	Material $	Labor $	Equipment $	Total $
Recessed,china	P1@2.50	Ea	520.00	61.10	—	581.10
Wall hung, china	P1@2.00	Ea	295.00	48.90	—	343.90
Recessed, S.S.	P1@2.50	Ea	590.00	61.10	—	651.10
Wall hung, S.S.	P1@2.00	Ea	315.00	48.90	—	363.90
A.D.A, S.S.	P1@2.50	Ea	540.00	61.10	—	601.10

Description	Craft@Hrs	Unit	Material $	Labor $	Equipment $	Total $

Commercial plumbing fixture rough-in. Includes type L copper supply pipe and DWV copper (to 2½") or cast iron (MJ) DWV (over 2½") drain and vent piping. Make additional allowances for plumbing fixtures and trim. Use these costs for preliminary estimates.

Description	Craft@Hrs	Unit	Material $	Labor $	Equipment $	Total $
Water closet, wall hung, flush valve, with carrier	P1@2.25	Ea	180.00	55.00	1.01	236.01
Water closet, wall hung, flush valve, no carrier	P1@1.95	Ea	65.80	47.70	0.88	114.38
Water closet, floor mounted, flush valve	P1@2.75	Ea	147.00	67.20	1.24	215.44
Water closet, floor mounted, tank type	P1@2.25	Ea	112.00	55.00	1.01	168.01
Bidet	P1@2.00	Ea	78.10	48.90	0.90	127.90
Urinal, wall hung, flush valve, with carrier	P1@3.10	Ea	197.00	75.80	1.39	274.19
Urinal, wall hung, flush valve, without carrier	P1@2.35	Ea	112.00	57.40	1.06	170.46
Lavatory, wall hung, with carrier	P1@2.40	Ea	161.00	58.60	1.53	221.13
Lavatory	P1@1.90	Ea	78.10	46.40	0.85	125.35
Sink	P1@1.90	Ea	83.90	46.40	0.85	131.15
Bath tub	P1@2.35	Ea	120.00	57.40	1.06	178.46
Shower	P1@2.60	Ea	141.00	63.50	1.17	205.67
Mop sink	P1@2.40	Ea	100.00	58.60	1.08	159.68
Slop sink	P1@2.60	Ea	71.50	63.50	1.17	136.17
Laundry tub	P1@1.95	Ea	84.80	47.70	0.88	133.38
Wash fountain	P1@2.10	Ea	91.50	51.30	0.94	143.74
Lab sink, glass drainage	P1@3.80	Ea	362.00	92.90	1.71	456.61
Lab sink acid resistant plastic drainage	P1@2.65	Ea	57.20	64.80	1.19	123.19
Drinking fountain	P1@2.20	Ea	61.90	53.80	0.99	116.69
Emergency eyewash, and shower	P1@1.75	Ea	23.80	42.80	0.79	67.39
Washing machine	P1@2.25	Ea	90.50	55.00	1.01	146.51

Description	Craft@Hrs	Unit	Material $	Labor $	Equipment $	Total $

Commercial plumbing fixture group rough-in. Includes type L copper supply pipe and DWV copper (to 2½") or cast iron (MJ) DWV (over 2½") drain and vent piping. Make additional allowances for plumbing fixtures and trim. Use these costs for preliminary estimates.

Description	Craft@Hrs	Unit	Material $	Labor $	Equipment $	Total $
3 piece washroom group	P1@5.50	Ea	215.00	134.00	1.01	350.01
3 piece washroom group back to back	P1@9.75	Ea	390.00	238.00	0.88	628.88
Kitchen sink Back to back	P1@2.15	Ea	115.00	52.50	1.24	168.74
Battery of water closets, floor mounted, tank type per water closet	P1@1.75	Ea	90.00	42.80	1.01	133.81
Battery of water closets, floor mounted, flush valve per water closet	P1@2.20	Ea	118.00	53.80	1.01	172.81
Battery of water closets, wall hung, flush valve, with carrier per water closet	P1@1.80	Ea	155.00	44.00	1.01	200.01
Battery of water closets, wall hung, flush valve, without carrier per water closet	P1@1.50	Ea	45.00	36.70	1.01	82.71
Battery of urinals, wall hung, flush valve with carrier per urinal	P1@2.45	Ea	185.00	59.90	1.01	245.91
Battery of urinals, wall hung, flush valve without carrier per urinal	P1@1.90	Ea	95.00	46.40	1.01	142.41
Battery of lav basins, wall hung, with carrier per lavatory	P1@2.00	Ea	145.00	48.90	1.01	194.91
Battery of lav basins, without carrier per lavatory	P1@1.50	Ea	65.00	36.70	1.01	102.71

Residential plumbing fixture rough-in. Includes type M copper supply pipe and ABS DWV drain and vent piping. Make additional allowances for plumbing fixtures and trim. Use these costs for preliminary estimates.

Description	Craft@Hrs	Unit	Material $	Labor $	Equipment $	Total $
Water closet, floor mounted, tank type	P1@2.00	Ea	27.60	48.90	0.90	77.40
Bidet	P1@1.85	Ea	21.00	45.20	0.83	67.03
Lavatory	P1@1.75	Ea	21.00	42.80	0.79	64.59
Counter sink	P1@1.75	Ea	22.90	42.80	0.79	66.49
Bath tub	P1@2.10	Ea	21.00	51.30	0.94	73.24
Shower	P1@2.45	Ea	30.50	59.90	1.10	91.50
Laundry tub	P1@1.75	Ea	19.10	42.80	0.79	62.69
Washing machine	P1@2.00	Ea	23.80	48.90	0.90	73.60

Type K hard-drawn copper pipe with wrought copper fittings and brazed joints is used in a wide variety of plumbing and HVAC systems such as potable water, heating hot water, chilled water, compressed air, refrigerant and A.C. condensate.

Brazed joints are those made with silver or other alloy filler metals having melting points at, or above, 1,000 degrees F. Maximum working pressure/temperature relationships for brazed joints are approximately as follows:

Maximum Working Pressure (PSIG)*			
Water Temperature (degrees F.)	**Nominal Pipe Size (inches)**		
	Up to 1	**1¼ to 2**	**2½ to 4**
Up to 350	270	190	150

*For copper pipe and solder-type fittings using brazing alloys melting at, or above 1,000 degrees F.

This section has been arranged to save the estimator's time by including all normally-used system components such as pipe, fittings, valves, hanger assemblies, riser clamps and miscellaneous items under one heading. Additional items can be found under "Plumbing and Piping Specialties." The cost estimates in this section are based on the conditions, limitations and wage rates described in the section "How to Use This Book" beginning on page 5.

Type K copper pipe with brazed joints

Description	Craft@Hrs	Unit	Material $	Labor $	Equipment $	Total $
1/2"	P1@.032	LF	.95	.78	—	1.73
3/4"	P1@.035	LF	1.75	.86	—	2.61
1"	P1@.038	LF	2.20	.93	—	3.13
1¼"	P1@.042	LF	2.95	1.03	—	3.98
1½"	P1@.046	LF	3.20	1.12	—	4.32
2"	P1@.053	LF	5.75	1.30	—	7.05
2½"	P1@.060	LF	7.45	1.47	—	8.92
3"	P1@.066	LF	9.80	1.61	—	11.41
4"	P1@.080	LF	17.55	1.95	—	19.50

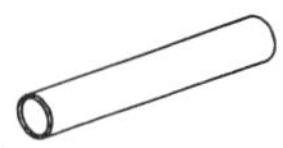

Type K copper 45 degree ell C x C with brazed joints

Description	Craft@Hrs	Unit	Material $	Labor $	Equipment $	Total $
1/2"	P1@.129	Ea	.36	3.15	—	3.51
3/4"	P1@.181	Ea	.66	4.42	—	5.08
1"	P1@.232	Ea	2.50	5.67	—	8.17
1¼"	P1@.284	Ea	3.40	6.94	—	10.34
1½"	P1@.335	Ea	6.02	8.19	—	14.21
2"	P1@.447	Ea	9.85	10.90	—	20.75
2½"	P1@.550	Ea	25.44	13.40	—	38.84
3"	P1@.654	Ea	33.45	16.00	—	49.45
4"	P1@.860	Ea	48.75	21.00	—	69.75

Copper, Type K with Brazed Joints

Description	Craft@Hrs	Unit	Material $	Labor $	Equipment $	Total $

Type K copper 90 degree ell C x C with brazed joints

Description	Craft@Hrs	Unit	Material $	Labor $	Equipment $	Total $
1/2"	P1@.129	Ea	.18	3.15	—	3.33
3/4"	P1@.181	Ea	.26	4.42	—	4.68
1"	P1@.232	Ea	2.28	5.67	—	7.95
1¼"	P1@.284	Ea	3.71	6.94	—	10.65
1½"	P1@.335	Ea	5.62	8.19	—	13.81
2"	P1@.447	Ea	10.05	10.90	—	20.95
2½"	P1@.550	Ea	20.81	13.40	—	34.21
3"	P1@.654	Ea	26.30	16.00	—	42.30
4"	P1@.860	Ea	42.25	21.00	—	63.25

Type K copper 90 degree ell Ftg x C with brazed joints

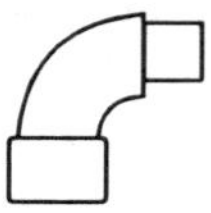

Description	Craft@Hrs	Unit	Material $	Labor $	Equipment $	Total $
1/2"	P1@.129	Ea	.49	3.15	—	3.64
3/4"	P1@.181	Ea	1.09	4.42	—	5.51
1"	P1@.232	Ea	3.25	5.67	—	8.92
1¼"	P1@.284	Ea	4.50	6.94	—	11.44
1½"	P1@.335	Ea	6.75	8.19	—	14.94
2"	P1@.447	Ea	13.50	10.90	—	24.40
2½"	P1@.550	Ea	28.50	13.40	—	41.90
3"	P1@.654	Ea	32.00	16.00	—	48.00
4"	P1@.860	Ea	64.90	21.00	—	85.90

Type K copper tee C x C x C with brazed joints

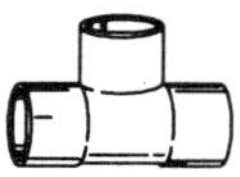

Description	Craft@Hrs	Unit	Material $	Labor $	Equipment $	Total $
1/2"	P1@.156	Ea	.32	3.81	—	4.13
3/4"	P1@.204	Ea	.79	4.99	—	5.78
1"	P1@.263	Ea	5.40	6.43	—	11.83
1¼"	P1@.322	Ea	8.60	7.87	—	16.47
1½"	P1@.380	Ea	12.48	9.29	—	21.77
2"	P1@.507	Ea	20.00	12.40	—	32.40
2½"	P1@.624	Ea	40.58	15.20	—	55.78
3"	P1@.741	Ea	62.11	18.10	—	80.21
4"	P1@.975	Ea	78.00	23.80	—	101.80

Description	Craft@Hrs	Unit	Material $	Labor $	Equipment $	Total $

Type K copper branch reducing tee C x C x C with brazed joints

Description	Craft@Hrs	Unit	Material $	Labor $	Equipment $	Total $
1/2" x 3/8"	P1@.129	Ea	2.28	3.15	—	5.43
3/4" x 1/2"	P1@.182	Ea	.79	4.45	—	5.24
1" x 1/2"	P1@.209	Ea	6.35	5.11	—	11.46
1" x 3/4"	P1@.234	Ea	6.35	5.72	—	12.07
1¼" x 1/2"	P1@.241	Ea	9.75	5.89	—	15.64
1¼" x 3/4"	P1@.264	Ea	9.75	6.45	—	16.20
1¼" x 1"	P1@.287	Ea	9.75	7.01	—	16.76
1½" x 1/2"	P1@.291	Ea	11.68	7.11	—	18.79
1½" x 3/4"	P1@.307	Ea	11.68	7.50	—	19.18
1½" x 1"	P1@.323	Ea	11.68	7.89	—	19.57
1½" x 1¼"	P1@.339	Ea	11.68	8.28	—	19.96
2" x 1/2"	P1@.348	Ea	15.77	8.50	—	24.27
2" x 3/4"	P1@.373	Ea	15.77	9.12	—	24.89
2" x 1"	P1@.399	Ea	15.77	9.75	—	25.52
2" x 1¼"	P1@.426	Ea	15.77	10.40	—	26.17
2" x 1½"	P1@.452	Ea	15.77	11.00	—	26.77
2½" x 1/2"	P1@.468	Ea	47.60	11.40	—	59.00
2½" x 3/4"	P1@.485	Ea	47.60	11.90	—	59.50
2½" x 1"	P1@.502	Ea	47.60	12.30	—	59.90
2½" x 1½"	P1@.537	Ea	47.60	13.10	—	60.70
2½" x 2"	P1@.555	Ea	47.60	13.60	—	61.20
3" x 1¼"	P1@.562	Ea	54.26	13.70	—	67.96
3" x 2"	P1@.626	Ea	56.26	15.30	—	71.56
3" x 2½"	P1@.660	Ea	56.26	16.10	—	72.36
4" x 1¼"	P1@.669	Ea	73.00	16.30	—	89.30
4" x 1½"	P1@.718	Ea	73.00	17.50	—	90.50
4" x 2"	P1@.768	Ea	73.00	18.80	—	91.80
4" x 2½"	P1@.819	Ea	73.00	20.00	—	93.00
4" x 3"	P1@.868	Ea	73.00	21.20	—	94.20

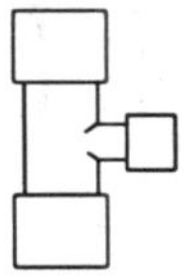

Copper, Type K with Brazed Joints

Type K copper reducer with brazed joints

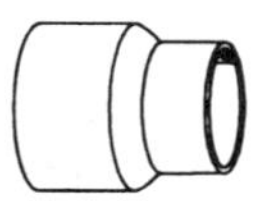

Description	Craft@Hrs	Unit	Material $	Labor $	Equipment $	Total $
1/2" x 3/8"	P1@.148	Ea	.40	3.62	—	4.02
3/4" x 1/2"	P1@.155	Ea	.72	3.79	—	4.51
1" x 3/4"	P1@.206	Ea	1.32	5.03	—	6.35
1¼" x 1/2"	P1@.206	Ea	2.20	5.03	—	7.23
1¼" x 3/4"	P1@.232	Ea	2.20	5.67	—	7.87
1¼" x 1"	P1@.258	Ea	2.20	6.30	—	8.50
1½" x 1/2"	P1@.232	Ea	3.37	5.67	—	9.04
1½" x 3/4"	P1@.257	Ea	3.37	6.28	—	9.65
1½" x 1"	P1@.283	Ea	3.37	6.92	—	10.29
1½" x 1¼"	P1@.308	Ea	3.37	7.53	—	10.90
2" x 1/2"	P1@.287	Ea	5.24	7.01	—	12.25
2" x 3/4"	P1@.277	Ea	5.24	6.77	—	12.01
2" x 1"	P1@.338	Ea	5.24	8.26	—	13.50
2" x 1¼"	P1@.365	Ea	5.24	8.92	—	14.16
2" x 1½"	P1@.390	Ea	5.24	9.53	—	14.77
2½" x 1"	P1@.390	Ea	14.30	9.53	—	23.83
2½" x 1¼"	P1@.416	Ea	14.30	10.20	—	24.50
2½" x 1½"	P1@.442	Ea	14.30	10.80	—	25.10
2½" x 2"	P1@.497	Ea	14.30	12.10	—	26.40
3" x 1¼"	P1@.468	Ea	16.70	11.40	—	28.10
3" x 1½"	P1@.493	Ea	16.70	12.00	—	28.70
3" x 2"	P1@.548	Ea	16.70	13.40	—	30.10
3" x 2½"	P1@.600	Ea	16.70	14.70	—	31.40
4" x 2"	P1@.652	Ea	32.00	15.90	—	47.90
4" x 2½"	P1@.704	Ea	32.00	17.20	—	49.20
4" x 3"	P1@.755	Ea	32.00	18.40	—	50.40

Type K copper adapter C x MPT

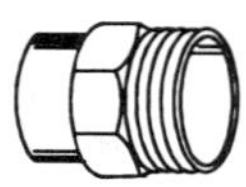

Description	Craft@Hrs	Unit	Material $	Labor $	Equipment $	Total $
1/2"	P1@.090	Ea	.97	2.20	—	3.17
3/4"	P1@.126	Ea	1.50	3.08	—	4.58
1"	P1@.162	Ea	3.69	3.96	—	7.65
1¼"	P1@.198	Ea	6.94	4.84	—	11.78
1½"	P1@.234	Ea	7.77	5.72	—	13.49
2"	P1@.312	Ea	10.01	7.62	—	17.63
2½"	P1@.384	Ea	32.60	9.38	—	41.98
3"	P1@.456	Ea	54.00	11.10	—	65.10

Type K copper adapter C x FPT

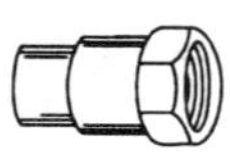

Description	Craft@Hrs	Unit	Material $	Labor $	Equipment $	Total $
1/2"	P1@.090	Ea	1.58	2.20	—	3.78
3/4"	P1@.126	Ea	2.88	3.08	—	5.96
1"	P1@.162	Ea	4.39	3.96	—	8.35
1¼"	P1@.198	Ea	9.14	4.84	—	13.98
1½"	P1@.234	Ea	11.23	5.72	—	16.95
2"	P1@.312	Ea	16.83	7.62	—	24.45
2½"	P1@.384	Ea	42.33	9.38	—	51.71
3"	P1@.456	Ea	59.19	11.10	—	70.29

Type K copper flush bushing with brazed joints

Description	Craft@Hrs	Unit	Material $	Labor $	Equipment $	Total $
1/2" x 3/8"	P1@.148	Ea	.98	3.62	—	4.60
3/4" x 1/2"	P1@.155	Ea	1.64	3.79	—	5.43
1" x 3/4"	P1@.206	Ea	4.29	5.03	—	9.32
1" x 1/2"	P1@.206	Ea	4.88	5.03	—	9.91
1¼" x 1"	P1@.258	Ea	5.10	6.30	—	11.40
1½" x 1¼"	P1@.308	Ea	6.50	7.53	—	14.03
2" x 1½"	P1@.390	Ea	12.00	9.53	—	21.53

Type K copper union with brazed joints

Description	Craft@Hrs	Unit	Material $	Labor $	Equipment $	Total $
1/2"	P1@.146	Ea	3.87	3.57	—	7.44
3/4"	P1@.205	Ea	4.62	5.01	—	9.63
1"	P1@.263	Ea	10.35	6.43	—	16.78
1¼"	P1@.322	Ea	18.82	7.87	—	26.69
1½"	P1@.380	Ea	23.30	9.29	—	32.59
2"	P1@.507	Ea	34.89	12.40	—	47.29

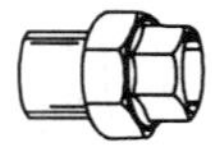

Type K copper dielectric union with brazed joints

Description	Craft@Hrs	Unit	Material $	Labor $	Equipment $	Total $
1/2"	P1@.146	Ea	3.63	3.57	—	7.20
3/4"	P1@.205	Ea	3.63	5.01	—	8.64
1"	P1@.263	Ea	7.53	6.43	—	13.96
1¼"	P1@.322	Ea	12.87	7.87	—	20.74
1½"	P1@.380	Ea	17.50	9.29	—	26.79
2"	P1@.507	Ea	24.31	12.40	—	36.71

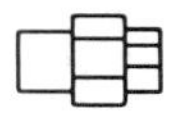

Type K copper pressure cap with brazed joints

Description	Craft@Hrs	Unit	Material $	Labor $	Equipment $	Total $
1/2"	P1@.083	Ea	.31	2.03	—	2.34
3/4"	P1@.116	Ea	.63	2.83	—	3.46
1"	P1@.149	Ea	1.47	3.64	—	5.11
1¼"	P1@.182	Ea	2.09	4.45	—	6.54
1½"	P1@.215	Ea	3.05	5.25	—	8.30
2"	P1@.286	Ea	5.56	6.99	—	12.55
2½"	P1@.352	Ea	14.00	8.60	—	22.60
3"	P1@.418	Ea	21.88	10.20	—	32.08
4"	P1@.550	Ea	33.50	13.40	—	46.90

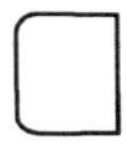

Type K copper coupling with brazed joints

Description	Craft@Hrs	Unit	Material $	Labor $	Equipment $	Total $
1/2"	P1@.129	Ea	.13	3.15	—	3.28
3/4"	P1@.181	Ea	.33	4.42	—	4.75
1"	P1@.232	Ea	1.39	5.67	—	7.06
1¼"	P1@.284	Ea	2.42	6.94	—	9.36
1½"	P1@.335	Ea	3.51	8.19	—	11.70
2"	P1@.447	Ea	5.33	10.90	—	16.23
2½"	P1@.550	Ea	12.42	13.40	—	25.82
3"	P1@.654	Ea	18.98	16.00	—	34.98
4"	P1@.860	Ea	37.47	21.00	—	58.47

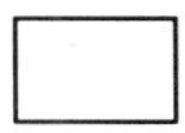

Copper, Type K with Brazed Joints

Class 125 bronze body gate valve, brazed joints

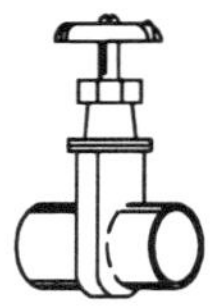

Description	Craft@Hrs	Unit	Material $	Labor $	Equipment $	Total $
1/2"	P1@.240	Ea	6.06	5.86	—	11.92
3/4"	P1@.300	Ea	7.58	7.33	—	14.91
1"	P1@.360	Ea	10.50	8.80	—	19.30
1¼"	P1@.480	Ea	13.40	11.70	—	25.10
1½"	P1@.540	Ea	18.00	13.20	—	31.20
2"	P1@.600	Ea	30.10	14.70	—	44.80
2½"	P1@1.00	Ea	49.90	24.40	—	74.30
3"	P1@1.50	Ea	71.60	36.70	—	108.30

Class 125 iron body gate valve, flanged joints

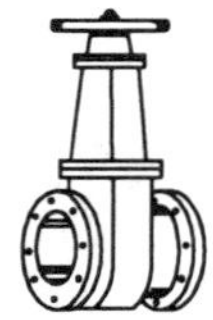

Description	Craft@Hrs	Unit	Material $	Labor $	Equipment $	Total $
2"	P1@.500	Ea	101.00	12.20	—	113.20
2½"	P1@.600	Ea	137.00	14.70	—	151.70
3"	P1@.750	Ea	148.00	18.30	—	166.30
4"	P1@1.35	Ea	218.00	33.00	—	251.00
5"	P1@2.00	Ea	417.00	48.90	—	465.90
6"	P1@2.50	Ea	417.00	61.10	—	478.10

Class 125 bronze body globe valve, with brazed joints

Description	Craft@Hrs	Unit	Material $	Labor $	Equipment $	Total $
1/2"	P1@.240	Ea	15.80	5.86	—	21.66
3/4"	P1@.300	Ea	21.30	7.33	—	28.63
1"	P1@.360	Ea	30.50	8.80	—	39.30
1¼"	P1@.480	Ea	42.90	11.70	—	54.60
1½"	P1@.540	Ea	57.50	13.20	—	70.70
2"	P1@.600	Ea	93.60	14.70	—	108.30

Class 125 iron body globe valve, flanged joints

Description	Craft@Hrs	Unit	Material $	Labor $	Equipment $	Total $
2½"	P1@.650	Ea	220.00	15.90	—	235.90
3"	P1@.750	Ea	264.00	18.30	—	282.30
4"	P1@1.35	Ea	355.00	33.00	—	388.00

200 PSIG iron body butterfly valve, lug type

Description	Craft@Hrs	Unit	Material $	Labor $	Equipment $	Total $
2"	P1@.450	Ea	84.50	11.00	—	95.50
2½"	P1@.450	Ea	86.50	11.00	—	97.50
3"	P1@.550	Ea	91.70	13.40	—	105.10
4"	P1@.550	Ea	114.00	13.40	—	127.40

200 PSIG iron body butterfly valve, wafer type

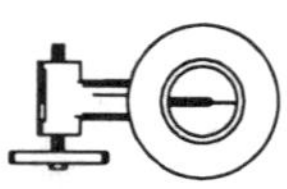

Description	Craft@Hrs	Unit	Material $	Labor $	Equipment $	Total $
2"	P1@.450	Ea	76.70	11.00	—	87.70
2½"	P1@.450	Ea	78.70	11.00	—	89.70
3"	P1@.550	Ea	84.50	13.40	—	97.90
4"	P1@.550	Ea	102.00	13.40	—	115.40

Class 125 bronze body 2-piece ball valve, brazed joints

Description	Craft@Hrs	Unit	Material $	Labor $	Equipment $	Total $
1/2"	P1@.240	Ea	3.56	5.86	—	9.42
3/4"	P1@.300	Ea	4.75	7.33	—	12.08
1"	P1@.360	Ea	8.21	8.80	—	17.01
1¼"	P1@.480	Ea	13.70	11.70	—	25.40
1½"	P1@.540	Ea	18.50	13.20	—	31.70
2"	P1@.600	Ea	23.30	14.70	—	38.00

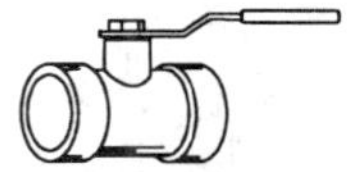

Class 125 bronze body swing check valve, brazed joints

Description	Craft@Hrs	Unit	Material $	Labor $	Equipment $	Total $
1/2"	P1@.240	Ea	8.00	5.86	—	13.86
3/4"	P1@.300	Ea	11.40	7.33	—	18.73
1"	P1@.360	Ea	15.00	8.80	—	23.80
1¼"	P1@.480	Ea	21.40	11.70	—	33.10
1½"	P1@.540	Ea	30.00	13.20	—	43.20
2"	P1@.600	Ea	50.00	14.70	—	64.70
2½"	P1@1.00	Ea	100.00	24.40	—	124.40
3"	P1@1.50	Ea	144.00	36.70	—	180.70

Class 125 iron body swing check valve, flanged joints

Description	Craft@Hrs	Unit	Material $	Labor $	Equipment $	Total $
2"	P1@.500	Ea	115.00	12.20	—	127.20
2½"	P1@.600	Ea	148.00	14.70	—	162.70
3"	P1@.750	Ea	183.00	18.30	—	201.30
4"	P1@1.35	Ea	271.00	33.00	—	304.00

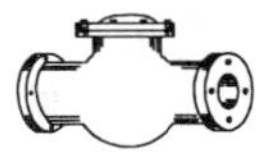

Class 125 iron body silent check valve, wafer type

Description	Craft@Hrs	Unit	Material $	Labor $	Equipment $	Total $
2"	P1@.500	Ea	85.00	12.20	—	97.20
2½"	P1@.600	Ea	95.00	14.70	—	109.70
3"	P1@.750	Ea	108.00	18.30	—	126.30
4"	P1@1.35	Ea	146.00	33.00	—	179.00

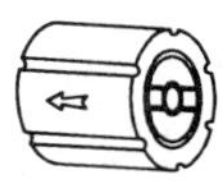

Class 125 bronze body strainer, threaded joints

Description	Craft@Hrs	Unit	Material $	Labor $	Equipment $	Total $
1/2"	P1@.230	Ea	12.40	5.62	—	18.02
3/4"	P1@.260	Ea	16.30	6.35	—	22.65
1"	P1@.330	Ea	20.20	8.06	—	28.26
1¼"	P1@.440	Ea	28.10	10.80	—	38.90
1½"	P1@.495	Ea	42.20	12.10	—	54.30
2"	P1@.550	Ea	73.00	13.40	—	86.40

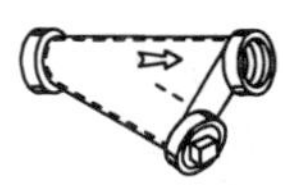

Class 125 iron body strainer, flanged joints

Description	Craft@Hrs	Unit	Material $	Labor $	Equipment $	Total $
2"	P1@.500	Ea	62.00	12.20	—	74.20
2½"	P1@.600	Ea	69.00	14.70	—	83.70
3"	P1@.750	Ea	80.00	18.30	—	98.30
4"	P1@1.35	Ea	137.00	33.00	—	170.00

Copper, Type K with Brazed Joints

Installation of 2-way control valve, threaded joints

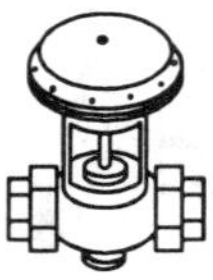

Description	Craft@Hrs	Unit	Material $	Labor $	Equipment $	Total $
1/2"	P1@.210	Ea	—	5.13	—	5.13
3/4"	P1@.275	Ea	—	6.72	—	6.72
1"	P1@.350	Ea	—	8.55	—	8.55
1¼"	P1@.430	Ea	—	10.50	—	10.50
1½"	P1@.505	Ea	—	12.30	—	12.30
2"	P1@.675	Ea	—	16.50	—	16.50
2½"	P1@.830	Ea	—	20.30	—	20.30
3"	P1@.990	Ea	—	24.20	—	24.20

Installation of 3-way control valve, threaded joints

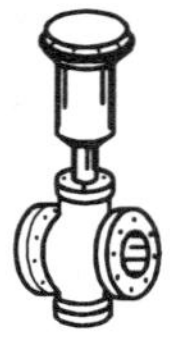

Description	Craft@Hrs	Unit	Material $	Labor $	Equipment $	Total $
1/2"	P1@.260	Ea	—	6.35	—	6.35
3/4"	P1@.365	Ea	—	8.92	—	8.92
1"	P1@.475	Ea	—	11.60	—	11.60
1¼"	P1@.575	Ea	—	14.10	—	14.10
1½"	P1@.680	Ea	—	16.60	—	16.60
2"	P1@.910	Ea	—	22.20	—	22.20
2½"	P1@1.12	Ea	—	27.40	—	27.40
3"	P1@1.33	Ea	—	32.50	—	32.50

Companion flange, 150 pound cast brass

Description	Craft@Hrs	Unit	Material $	Labor $	Equipment $	Total $
2"	P1@.290	Ea	169.00	7.09	—	176.09
2½"	P1@.380	Ea	208.00	9.29	—	217.29
3"	P1@.460	Ea	208.00	11.20	—	219.20
4"	P1@.600	Ea	332.00	14.70	—	346.70

Bolt and gasket sets

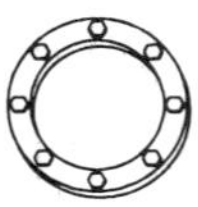

Description	Craft@Hrs	Unit	Material $	Labor $	Equipment $	Total $
2"	P1@.500	Ea	1.50	12.20	—	13.70
2½"	P1@.650	Ea	1.75	15.89	—	17.64
3"	P1@.750	Ea	3.00	18.33	—	21.33
4"	P1@1.00	Ea	5.00	24.44	—	29.44

Thermometer with well

Description	Craft@Hrs	Unit	Material $	Labor $	Equipment $	Total $
7"	P1@.250	Ea	15.00	6.11	—	21.11
9"	P1@.250	Ea	15.50	6.11	—	21.61

Dial type pressure gauge

Description	Craft@Hrs	Unit	Material $	Labor $	Equipment $	Total $
2½"	P1@.200	Ea	15.50	4.89	—	20.39
3½"	P1@.200	Ea	20.50	4.89	—	25.39

Description	Craft@Hrs	Unit	Material $	Labor $	Equipment $	Total $
Pressure/temperature tap						
	P1@.150	Ea	7.30	3.67	—	10.97

Description	Craft@Hrs	Unit	Material $	Labor $	Equipment $	Total $
Hanger with swivel assembly						
1/2"	P1@.450	Ea	1.60	11.00	—	12.60
3/4"	P1@.450	Ea	1.75	11.00	—	12.75
1"	P1@.450	Ea	1.85	11.00	—	12.85
1¼"	P1@.450	Ea	1.90	11.00	—	12.90
1½"	P1@.450	Ea	2.05	11.00	—	13.05
2"	P1@.450	Ea	2.15	11.00	—	13.15
2½"	P1@.450	Ea	2.90	11.00	—	13.90
3"	P1@.450	Ea	3.60	11.00	—	14.60
4"	P1@.450	Ea	3.95	11.00	—	14.95

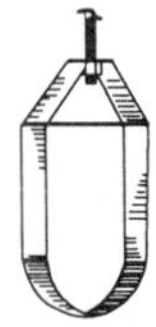

Description	Craft@Hrs	Unit	Material $	Labor $	Equipment $	Total $
Riser clamp						
1/2"	P1@.100	Ea	.95	2.44	—	3.39
3/4"	P1@.100	Ea	1.39	2.44	—	3.83
1"	P1@.100	Ea	1.41	2.44	—	3.85
1¼"	P1@.105	Ea	1.70	2.57	—	4.27
1½"	P1@.110	Ea	1.79	2.69	—	4.48
2"	P1@.115	Ea	1.89	2.81	—	4.70
2½"	P1@.120	Ea	2.00	2.93	—	4.93
3"	P1@.120	Ea	2.18	2.93	—	5.11
4"	P1@.125	Ea	2.75	3.05	—	5.80

Copper, Type K with Soft-Soldered Joints

Type K hard-drawn copper pipe with wrought copper fittings and soft-soldered joints is used in a wide variety of plumbing and HVAC systems such as potable water, heating hot water, chilled water and A.C. condensate.

Soft-soldered joints are those made with solders having melting points in the 350 degree F. to 500 degree F. range. Maximum working pressure/temperature relationships for soft-soldered joints are approximately as follows:

Maximum Working Pressures (PSIG)*				
Soft-solder Type	Water Temperature (degrees F.)	Nominal Pipe Size (inches)		
		Up to 1	1¼ to 2	2½ to 4
50-50 tin-lead**	100	200	175	150
	150	150	125	100
	200	100	90	75
	250	85	75	50
95-5 tin-antimony	100	500	400	300
	150	400	350	275
	200	300	250	200
	250	200	175	150

**For copper pipe and solder-type fittings using soft-solders melting at approximately 350 degrees F. to 500 degrees F.*

***The use of any solder containing lead is not allowed in potable water systems.*

This section has been arranged to save the estimator's time by including all normally-used system components such as pipe, fittings, valves, hanger assemblies, riser clamps and miscellaneous items under one heading. Additional items can be found under "Plumbing and Piping Specialties." The cost estimates in this section are based on the conditions, limitations and wage rates described in the section "How to Use This Book" beginning on page 5.

Type K copper pipe with soft-soldered joints

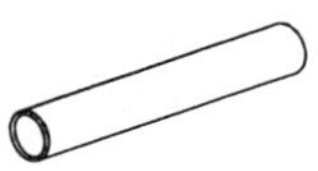

Description	Craft@Hrs	Unit	Material $	Labor $	Equipment $	Total $
1/2"	P1@.032	LF	.95	.78	—	1.73
3/4"	P1@.035	LF	1.75	.86	—	2.61
1"	P1@.038	LF	2.20	.93	—	3.13
1¼"	P1@.042	LF	2.95	1.03	—	3.98
1½"	P1@.046	LF	3.20	1.12	—	4.32
2"	P1@.053	LF	5.75	1.30	—	7.05
2½"	P1@.060	LF	7.45	1.47	—	8.92
3"	P1@.066	LF	9.80	1.61	—	11.41
4"	P1@.080	LF	17.55	1.95	—	19.50

Description	Craft@Hrs	Unit	Material $	Labor $	Equipment $	Total $

Type K copper 45 degree ell C x C with soft-soldered joints

Description	Craft@Hrs	Unit	Material $	Labor $	Equipment $	Total $
1/2"	P1@.107	Ea	.36	2.61	—	2.97
3/4"	P1@.150	Ea	.66	3.67	—	4.33
1"	P1@.193	Ea	2.50	4.72	—	7.22
1¼"	P1@.236	Ea	3.40	5.77	—	9.17
1½"	P1@.278	Ea	6.02	6.79	—	12.81
2"	P1@.371	Ea	9.85	9.07	—	18.92
2½"	P1@.457	Ea	25.44	11.20	—	36.64
3"	P1@.543	Ea	33.45	13.30	—	46.75
4"	P1@.714	Ea	48.75	17.40	—	66.15

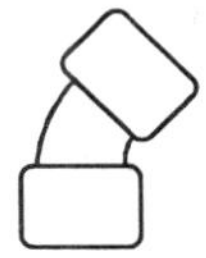

Type K copper 90 degree ell C x C with soft-soldered joints

Description	Craft@Hrs	Unit	Material $	Labor $	Equipment $	Total $
1/2"	P1@.107	Ea	.18	2.61	—	2.79
3/4"	P1@.150	Ea	.26	3.67	—	3.93
1"	P1@.193	Ea	2.28	4.72	—	7.00
1¼"	P1@.236	Ea	3.71	5.77	—	9.48
1½"	P1@.278	Ea	5.62	6.79	—	12.41
2"	P1@.371	Ea	10.05	9.07	—	19.12
2½"	P1@.457	Ea	20.81	11.20	—	32.01
3"	P1@.543	Ea	26.30	13.30	—	39.60
4"	P1@.714	Ea	42.25	17.40	—	59.65

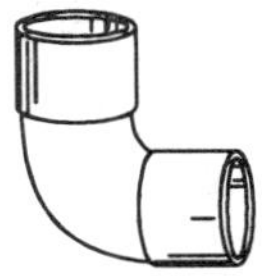

Type K copper 90 degree ell Ftg. x C with soft-soldered joint

Description	Craft@Hrs	Unit	Material $	Labor $	Equipment $	Total $
1/2"	P1@.107	Ea	.49	2.61	—	3.10
3/4"	P1@.150	Ea	1.09	3.67	—	4.76
1"	P1@.193	Ea	3.25	4.72	—	7.97
1¼"	P1@.236	Ea	4.50	5.77	—	10.27
1½"	P1@.278	Ea	6.75	6.79	—	13.54
2"	P1@.371	Ea	13.50	9.07	—	22.57
2½"	P1@.457	Ea	28.50	11.20	—	39.70
3"	P1@.543	Ea	32.00	13.30	—	45.30
4"	P1@.714	Ea	64.90	17.40	—	82.30

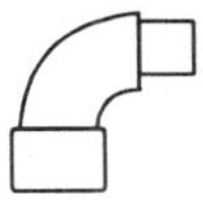

Type K copper tee C x C x C with soft-soldered joints

Description	Craft@Hrs	Unit	Material $	Labor $	Equipment $	Total $
1/2"	P1@.129	Ea	.32	3.15	—	3.47
3/4"	P1@.181	Ea	.79	4.42	—	5.21
1"	P1@.233	Ea	5.40	5.69	—	11.09
1¼"	P1@.285	Ea	8.60	6.96	—	15.56
1½"	P1@.337	Ea	12.48	8.24	—	20.72
2"	P1@.449	Ea	20.00	11.00	—	31.00
2½"	P1@.552	Ea	40.58	13.50	—	54.08
3"	P1@.656	Ea	62.11	16.00	—	78.11
4"	P1@.863	Ea	78.00	21.10	—	99.10

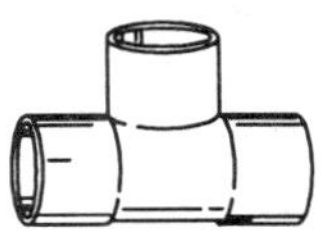

Copper, Type K with Soft-Soldered Joints

Description	Craft@Hrs	Unit	Material $	Labor $	Equipment $	Total $
Type K copper branch reducing tee C x C x C with soft-soldered joints						
1/2" x 3/8"	P1@.121	Ea	2.28	2.96	—	5.24
3/4" x 1/2"	P1@.170	Ea	.79	4.15	—	4.94
1" x 1/2"	P1@.195	Ea	6.35	4.77	—	11.12
1" x 3/4"	P1@.219	Ea	6.35	5.35	—	11.70
1¼" x 1/2"	P1@.225	Ea	9.75	5.50	—	15.25
1¼" x 3/4"	P1@.247	Ea	9.75	6.04	—	15.79
1¼" x 1"	P1@.268	Ea	9.75	6.55	—	16.30
1½" x 1/2"	P1@.272	Ea	11.68	6.65	—	18.33
1½" x 3/4"	P1@.287	Ea	11.68	7.01	—	18.69
1½" x 1"	P1@.302	Ea	11.68	7.38	—	19.06
1½" x 1¼"	P1@.317	Ea	11.68	7.75	—	19.43
2" x 1/2"	P1@.325	Ea	15.77	7.94	—	23.71
2" x 3/4"	P1@.348	Ea	15.77	8.50	—	24.27
2" x 1"	P1@.373	Ea	15.77	9.12	—	24.89
2" x 1¼"	P1@.398	Ea	15.77	9.73	—	25.50
2" x 1½"	P1@.422	Ea	15.77	10.30	—	26.07
2½" x 1/2"	P1@.437	Ea	47.60	10.70	—	58.30
2½" x 3/4"	P1@.453	Ea	47.60	11.10	—	58.70
2½" x 1"	P1@.469	Ea	47.60	11.50	—	59.10
2½" x 1½"	P1@.502	Ea	47.60	12.30	—	59.90
2½" x 2"	P1@.519	Ea	47.60	12.70	—	60.30
3" x 1¼"	P1@.525	Ea	54.26	12.80	—	67.06
3" x 2"	P1@.585	Ea	56.26	14.30	—	70.56
3" x 2½"	P1@.617	Ea	56.26	15.10	—	71.36
4" x 1¼"	P1@.625	Ea	73.00	15.30	—	88.30
4" x 1½"	P1@.671	Ea	73.00	16.40	—	89.40
4" x 2"	P1@.718	Ea	73.00	17.50	—	90.50
4" x 2½"	P1@.765	Ea	73.00	18.70	—	91.70
4" x 3"	P1@.811	Ea	73.00	19.80	—	92.80

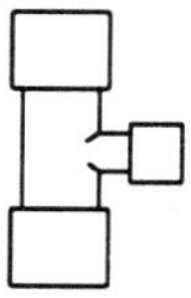

Type K copper reducer with soft-soldered joints

Description	Craft@Hrs	Unit	Material $	Labor $	Equipment $	Total $
1/2" x 3/8"	P1@.124	Ea	.40	3.03	—	3.43
3/4" x 1/2"	P1@.129	Ea	.72	3.15	—	3.87
1" x 3/4"	P1@.172	Ea	1.32	4.20	—	5.52
1¼" x 1/2"	P1@.172	Ea	2.20	4.20	—	6.40
1¼" x 3/4"	P1@.193	Ea	2.20	4.72	—	6.92
1¼" x 1"	P1@.215	Ea	2.20	5.25	—	7.45
1½" x 1/2"	P1@.193	Ea	3.37	4.72	—	8.09
1½" x 3/4"	P1@.214	Ea	3.37	5.23	—	8.60
1½" x 1"	P1@.236	Ea	3.37	5.77	—	9.14
1½" x 1¼"	P1@.257	Ea	3.37	6.28	—	9.65
2" x 1/2"	P1@.239	Ea	5.24	5.84	—	11.08
2" x 3/4"	P1@.261	Ea	5.24	6.38	—	11.62
2" x 1"	P1@.282	Ea	5.24	6.89	—	12.13
2" x 1¼"	P1@.304	Ea	5.24	7.43	—	12.67
2" x 1½"	P1@.325	Ea	5.24	7.94	—	13.18
2½" x 1"	P1@.325	Ea	14.30	7.94	—	22.24
2½" x 1¼"	P1@.347	Ea	14.30	8.48	—	22.78
2½" x 1½"	P1@.368	Ea	14.30	8.99	—	23.29
2½" x 2"	P1@.414	Ea	14.30	10.10	—	24.40
3" x 2"	P1@.457	Ea	16.70	11.20	—	27.90
3" x 2½"	P1@.500	Ea	16.70	12.20	—	28.90
4" x 2"	P1@.543	Ea	32.00	13.30	—	45.30
4" x 2½"	P1@.586	Ea	32.00	14.30	—	46.30
4" x 3"	P1@.629	Ea	32.00	15.40	—	47.40

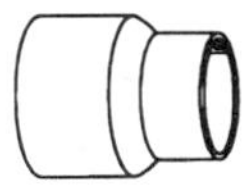

Type K copper adapter C x MPT with soft-soldered joint

Description	Craft@Hrs	Unit	Material $	Labor $	Equipment $	Total $
1/2"	P1@.075	Ea	.97	1.83	—	2.80
3/4"	P1@.105	Ea	1.50	2.57	—	4.07
1"	P1@.134	Ea	3.69	3.27	—	6.96
1¼"	P1@.164	Ea	6.94	4.01	—	10.95
1½"	P1@.194	Ea	7.77	4.74	—	12.51
2"	P1@.259	Ea	10.01	6.33	—	16.34
2½"	P1@.319	Ea	32.60	7.80	—	40.40
3"	P1@.378	Ea	54.00	9.24	—	63.24

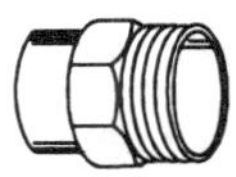

Copper, Type K with Soft-Soldered Joints

Type K copper adapter C x FPT with soft-soldered joint

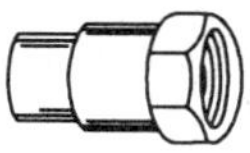

Description	Craft@Hrs	Unit	Material $	Labor $	Equipment $	Total $
1/2"	P1@.075	Ea	1.58	1.83	—	3.41
3/4"	P1@.105	Ea	2.88	2.57	—	5.45
1"	P1@.134	Ea	4.39	3.27	—	7.66
1¼"	P1@.164	Ea	9.14	4.01	—	13.15
1½"	P1@.194	Ea	11.23	4.74	—	15.97
2"	P1@.259	Ea	16.83	6.33	—	23.16
2½"	P1@.319	Ea	42.33	7.80	—	50.13
3"	P1@.378	Ea	59.19	9.24	—	68.43

Type K copper flush bushing with soft-soldered joints

Description	Craft@Hrs	Unit	Material $	Labor $	Equipment $	Total $
1/2" x 3/8"	P1@.124	Ea	.98	3.03	—	4.01
3/4" x 1/2"	P1@.129	Ea	1.64	3.15	—	4.79
1" x 3/4"	P1@.172	Ea	4.29	4.20	—	8.49
1" x 1/2"	P1@.172	Ea	4.88	4.20	—	9.08
1¼" x 1"	P1@.215	Ea	5.10	5.25	—	10.35
1½" x 1¼"	P1@.257	Ea	6.50	6.28	—	12.78
2" x 1½"	P1@.325	Ea	12.00	7.94	—	19.94

Type K copper union with soft-soldered joint

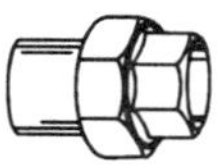

Description	Craft@Hrs	Unit	Material $	Labor $	Equipment $	Total $
1/2"	P1@.121	Ea	3.87	2.96	—	6.83
3/4"	P1@.170	Ea	4.62	4.15	—	8.77
1"	P1@.218	Ea	10.35	5.33	—	15.68
1¼"	P1@.267	Ea	18.82	6.52	—	25.34
1½"	P1@.315	Ea	23.30	7.70	—	31.00
2"	P1@.421	Ea	34.89	10.30	—	45.19

Type K copper dielectric union with soft-soldered joint

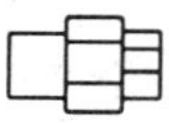

Description	Craft@Hrs	Unit	Material $	Labor $	Equipment $	Total $
1/2"	P1@.121	Ea	3.63	2.96	—	6.59
3/4"	P1@.170	Ea	3.63	4.15	—	7.78
1"	P1@.218	Ea	7.53	5.33	—	12.86
1¼"	P1@.267	Ea	12.87	6.52	—	19.39
1½"	P1@.315	Ea	17.50	7.70	—	25.20
2"	P1@.421	Ea	24.31	10.30	—	34.61

Type K copper cap with soft-soldered joint

Description	Craft@Hrs	Unit	Material $	Labor $	Equipment $	Total $
1/2"	P1@.069	Ea	.31	1.69	—	2.00
3/4"	P1@.096	Ea	.63	2.35	—	2.98
1"	P1@.124	Ea	1.47	3.03	—	4.50
1¼"	P1@.151	Ea	2.09	3.69	—	5.78
1½"	P1@.178	Ea	3.05	4.35	—	7.40
2"	P1@.237	Ea	5.56	5.79	—	11.35
2½"	P1@.292	Ea	14.00	7.14	—	21.14
3"	P1@.347	Ea	21.88	8.48	—	30.36
4"	P1@.457	Ea	33.50	11.20	—	44.70

Description	Craft@Hrs	Unit	Material $	Labor $	Equipment $	Total $
Type K copper coupling with soft-soldered joints						
1/2"	P1@.107	Ea	.13	2.61	—	2.74
3/4"	P1@.150	Ea	.33	3.67	—	4.00
1"	P1@.193	Ea	1.39	4.72	—	6.11
1¼"	P1@.236	Ea	2.42	5.77	—	8.19
1½"	P1@.278	Ea	3.51	6.79	—	10.30
2"	P1@.371	Ea	5.33	9.07	—	14.40
2½"	P1@.457	Ea	12.42	11.20	—	23.62
3"	P1@.543	Ea	18.98	13.30	—	32.28
4"	P1@.714	Ea	37.47	17.40	—	54.87
Class 125 bronze body gate valve, soft-soldered joints						
1/2"	P1@.200	Ea	6.06	4.89	—	10.95
3/4"	P1@.249	Ea	7.58	6.08	—	13.66
1"	P1@.299	Ea	10.50	7.31	—	17.81
1¼"	P1@.398	Ea	13.40	9.73	—	23.13
1½"	P1@.448	Ea	18.00	10.90	—	28.90
2"	P1@.498	Ea	30.10	12.20	—	42.30
2½"	P1@.830	Ea	49.90	20.30	—	70.20
3"	P1@1.24	Ea	71.60	30.30	—	101.90
Class 125 iron body gate valve, flanged joints						
2"	P1@.500	Ea	101.00	12.20	—	113.20
2½"	P1@.600	Ea	137.00	14.70	—	151.70
3"	P1@.750	Ea	148.00	18.30	—	166.30
4"	P1@1.35	Ea	218.00	33.00	—	251.00
5"	P1@2.00	Ea	417.00	48.90	—	465.90
6"	P1@2.50	Ea	417.00	61.10	—	478.10
Class 125 bronze body globe valve, with soft-soldered joints						
1/2"	P1@.200	Ea	15.80	4.89	—	20.69
3/4"	P1@.249	Ea	21.30	6.08	—	27.38
1"	P1@.299	Ea	30.50	7.31	—	37.81
1¼"	P1@.398	Ea	42.90	9.73	—	52.63
1½"	P1@.448	Ea	57.50	10.90	—	68.40
2"	P1@.498	Ea	93.60	12.20	—	105.80
Class 125 iron body globe valve, flanged ends						
2½"	P1@.600	Ea	220.00	14.70	—	234.70
3"	P1@.750	Ea	264.00	18.30	—	282.30
4"	P1@1.35	Ea	352.00	33.00	—	385.00

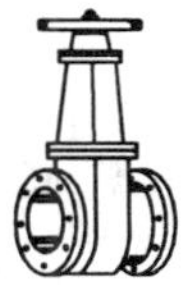

Copper, Type K with Soft-Soldered Joints

200 PSIG iron body butterfly valve, lug type

Description	Craft@Hrs	Unit	Material $	Labor $	Equipment $	Total $
2"	P1@.450	Ea	84.50	11.00	—	95.50
2½"	P1@.450	Ea	86.50	11.00	—	97.50
3"	P1@.550	Ea	91.70	13.40	—	105.10
4"	P1@.550	Ea	114.00	13.40	—	127.40

200 PSIG iron body butterfly valve, wafer type

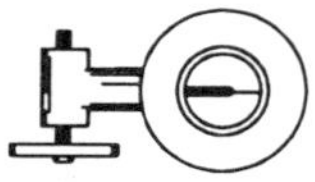

Description	Craft@Hrs	Unit	Material $	Labor $	Equipment $	Total $
2"	P1@.450	Ea	76.70	11.00	—	87.70
2½"	P1@.450	Ea	78.70	11.00	—	89.70
3"	P1@.550	Ea	84.50	13.40	—	97.90
4"	P1@.550	Ea	102.00	13.40	—	115.40

Class 125 bronze body two piece ball valve, soft-soldered joints

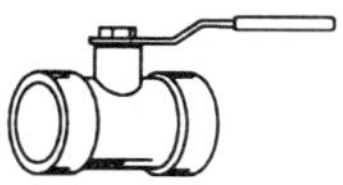

Description	Craft@Hrs	Unit	Material $	Labor $	Equipment $	Total $
1/2"	P1@.200	Ea	3.56	4.89	—	8.45
3/4"	P1@.249	Ea	4.25	6.08	—	10.33
1"	P1@.299	Ea	8.21	7.31	—	15.52
1¼"	P1@.398	Ea	13.70	9.73	—	23.43
1½"	P1@.448	Ea	18.50	10.90	—	29.40
2"	P1@.498	Ea	23.30	12.20	—	35.50

Class 125 bronze body swing check valve, soft-soldered joints

Description	Craft@Hrs	Unit	Material $	Labor $	Equipment $	Total $
1/2"	P1@.200	Ea	8.00	4.89	—	12.89
3/4"	P1@.249	Ea	11.40	6.08	—	17.48
1"	P1@.299	Ea	15.00	7.31	—	22.31
1¼"	P1@.398	Ea	21.40	9.73	—	31.13
1½"	P1@.448	Ea	30.00	10.90	—	40.90
2"	P1@.498	Ea	50.00	12.20	—	62.20
2½"	P1@.830	Ea	100.00	20.30	—	120.30
3"	P1@1.24	Ea	144.00	30.30	—	174.30

Class 125 iron body swing check valve, flanged ends

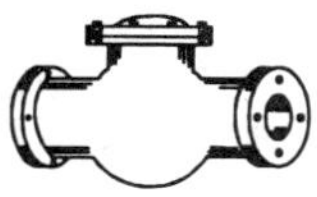

Description	Craft@Hrs	Unit	Material $	Labor $	Equipment $	Total $
2"	P1@.500	Ea	115.00	12.20	—	127.20
2½"	P1@.600	Ea	148.00	14.70	—	162.70
3"	P1@.750	Ea	183.00	18.30	—	201.30
4"	P1@1.35	Ea	271.00	33.00	—	304.00

Class 125 iron body silent check valve, wafer type

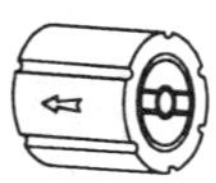

Description	Craft@Hrs	Unit	Material $	Labor $	Equipment $	Total $
2"	P1@.500	Ea	85.00	12.20	—	97.20
2½"	P1@.600	Ea	95.00	14.70	—	109.70
3"	P1@.750	Ea	108.00	18.30	—	126.30
4"	P1@1.35	Ea	146.00	33.00	—	179.00

Class 125 bronze body strainer, threaded ends

Description	Craft@Hrs	Unit	Material $	Labor $	Equipment $	Total $
1/2"	P1@.230	Ea	12.40	5.62	—	18.02
3/4"	P1@.260	Ea	16.30	6.35	—	22.65
1"	P1@.330	Ea	20.20	8.06	—	28.26
1¼"	P1@.440	Ea	28.10	10.80	—	38.90
1½"	P1@.495	Ea	42.20	12.10	—	54.30
2"	P1@.550	Ea	73.00	13.40	—	86.40

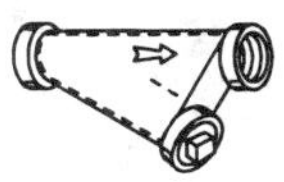

Class 125 iron body strainer, flanged ends

Description	Craft@Hrs	Unit	Material $	Labor $	Equipment $	Total $
2"	P1@.500	Ea	62.00	12.20	—	74.20
2½"	P1@.600	Ea	69.00	14.70	—	83.70
3"	P1@.750	Ea	80.00	18.30	—	98.30
4"	P1@1.35	Ea	137.00	33.00	—	170.00

Installation of copper 2-way control valve, threaded joints

Description	Craft@Hrs	Unit	Material $	Labor $	Equipment $	Total $
1/2"	P1@.210	Ea	—	5.13	—	5.13
3/4"	P1@.275	Ea	—	6.72	—	6.72
1"	P1@.350	Ea	—	8.55	—	8.55
1¼"	P1@.430	Ea	—	10.50	—	10.50
1½"	P1@.505	Ea	—	12.30	—	12.30
2"	P1@.675	Ea	—	16.50	—	16.50
2½"	P1@.830	Ea	—	20.30	—	20.30
3"	P1@.990	Ea	—	24.20	—	24.20

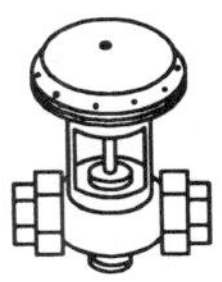

Installation of copper 3-way control valve, threaded joints

Description	Craft@Hrs	Unit	Material $	Labor $	Equipment $	Total $
1/2"	P1@.260	Ea	—	6.35	—	6.35
3/4"	P1@.365	Ea	—	8.92	—	8.92
1"	P1@.475	Ea	—	11.60	—	11.60
1¼"	P1@.575	Ea	—	14.10	—	14.10
1½"	P1@.680	Ea	—	16.60	—	16.60
2"	P1@.910	Ea	—	22.20	—	22.20
2½"	P1@1.12	Ea	—	27.40	—	27.40
3"	P1@1.33	Ea	—	32.50	—	32.50

Companion flange, 150 pound cast brass

Description	Craft@Hrs	Unit	Material $	Labor $	Equipment $	Total $
2"	P1@.290	Ea	169.00	7.09	—	176.09
2½"	P1@.380	Ea	208.00	9.29	—	217.29
3"	P1@.460	Ea	208.00	11.20	—	219.20
4"	P1@.600	Ea	332.00	14.70	—	346.70

Copper, Type K with Soft-Soldered Joints

Description	Craft@Hrs	Unit	Material $	Labor $	Equipment $	Total $
Bolt and gasket sets 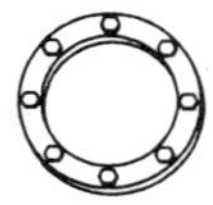						
2"	P1@.500	Ea	1.50	12.20	—	13.70
2½"	P1@.650	Ea	1.75	15.89	—	17.64
3"	P1@.750	Ea	3.00	18.33	—	21.33
4"	P1@1.00	Ea	5.00	24.44	—	29.44
Thermometer with well 						
7"	P1@.250	Ea	15.00	6.11	—	21.11
9"	P1@.250	Ea	15.50	6.11	—	21.61
Dial type pressure gauge 						
2½"	P1@.200	Ea	15.50	4.89	—	20.39
3½"	P1@.200	Ea	20.50	4.89	—	25.39
Pressure/temperature tap 						
	P1@.150	Ea	7.30	3.67	—	10.97
Hanger with swivel assembly 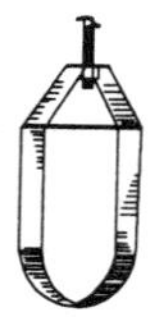						
1/2"	P1@.450	Ea	1.60	11.00	—	12.60
3/4"	P1@.450	Ea	1.75	11.00	—	12.75
1"	P1@.450	Ea	1.85	11.00	—	12.85
1¼"	P1@.450	Ea	1.90	11.00	—	12.90
1½"	P1@.450	Ea	2.05	11.00	—	13.05
2"	P1@.450	Ea	2.15	11.00	—	13.15
2½"	P1@.450	Ea	2.90	11.00	—	13.90
3"	P1@.450	Ea	3.60	11.00	—	14.60
4"	P1@.450	Ea	3.95	11.00	—	14.95
Riser clamp 						
1/2"	P1@.100	Ea	.95	2.44	—	3.39
3/4"	P1@.100	Ea	1.39	2.44	—	3.83
1"	P1@.100	Ea	1.41	2.44	—	3.85
1¼"	P1@.105	Ea	1.70	2.57	—	4.27
1½"	P1@.110	Ea	1.79	2.69	—	4.48
2"	P1@.115	Ea	1.89	2.81	—	4.70
2½"	P1@.120	Ea	2.00	2.93	—	4.93
3"	P1@.120	Ea	2.18	2.93	—	5.11
4"	P1@.125	Ea	2.75	3.05	—	5.80

Type L hard-drawn copper pipe with wrought copper fittings and brazed joints is used in a wide variety of plumbing and HVAC systems such as potable water, heating hot water, chilled water, compressed air, refrigerant and A.C. condensate.

Brazed joints are those made with silver or other alloy filler metals having melting points at, or above, 1,000 degrees F. Maximum working pressure/temperature relationships for brazed joints are approximately as follows:

Maximum Working Pressure (PSIG)*			
Water Temperature (degrees F.)	**Nominal Pipe Size (inches)**		
	Up to 1	**1¼ to 2**	**2½ to 4**
Up to 350	270	190	150

**For copper pipe and solder-type fittings using brazing alloys melting at, or above, 1,000 degrees F.*

This section has been arranged to save the estimator's time by including all normally-used system components such as pipe, fittings, valves, hanger assemblies, riser clamps and miscellaneous items under one heading. Additional items can be found under "Plumbing and Piping Specialties." The cost estimates in this section are based on the conditions, limitations and wage rates described in the section "How to Use This Book" beginning on page 5.

Description	Craft@Hrs	Unit	Material $	Labor $	Equipment $	Total $

Type L copper pipe with brazed joints

Description	Craft@Hrs	Unit	Material $	Labor $	Equipment $	Total $
1/2"	P1@.032	LF	.68	.78	—	1.46
3/4"	P1@.035	LF	1.08	.86	—	1.94
1"	P1@.038	LF	1.50	.93	—	2.43
1¼"	P1@.042	LF	2.13	1.03	—	3.16
1½"	P1@.046	LF	2.73	1.12	—	3.85
2"	P1@.053	LF	4.69	1.30	—	5.99
2½"	P1@.060	LF	6.85	1.47	—	8.32
3"	P1@.066	LF	9.26	1.61	—	10.87
4"	P1@.080	LF	14.64	1.95	—	16.59

Type L copper 45 degree ell C x C with brazed joints

Description	Craft@Hrs	Unit	Material $	Labor $	Equipment $	Total $
1/2"	P1@.129	Ea	.36	3.15	—	3.51
3/4"	P1@.181	Ea	.66	4.42	—	5.08
1"	P1@.232	Ea	2.50	5.67	—	8.17
1¼"	P1@.284	Ea	3.40	6.94	—	10.34
1½"	P1@.335	Ea	6.02	8.19	—	14.21
2"	P1@.447	Ea	9.85	10.90	—	20.75
2½"	P1@.550	Ea	25.44	13.40	—	38.84
3"	P1@.654	Ea	33.45	16.00	—	49.45
4"	P1@.860	Ea	48.75	21.00	—	69.75

Copper, Type L with Brazed Joints

Description	Craft@Hrs	Unit	Material $	Labor $	Equipment $	Total $
Type L copper 90 degree ell C x C with brazed joints						
1/2"	P1@.129	Ea	.18	3.15	—	3.33
3/4"	P1@.181	Ea	.26	4.42	—	4.68
1"	P1@.232	Ea	2.28	5.67	—	7.95
1¼"	P1@.284	Ea	3.71	6.94	—	10.65
1½"	P1@.335	Ea	5.62	8.19	—	13.81
2"	P1@.447	Ea	10.05	10.90	—	20.95
2½"	P1@.550	Ea	20.81	13.40	—	34.21
3"	P1@.654	Ea	26.30	16.00	—	42.30
4"	P1@.860	Ea	42.25	21.00	—	63.25
Type L copper 90 degree ell Ftg. x C with brazed joint						
1/2"	P1@.129	Ea	.49	3.15	—	3.64
3/4"	P1@.181	Ea	1.09	4.42	—	5.51
1"	P1@.232	Ea	3.25	5.67	—	8.92
1¼"	P1@.284	Ea	4.50	6.94	—	11.44
1½"	P1@.335	Ea	6.75	8.19	—	14.94
2"	P1@.447	Ea	13.50	10.90	—	24.40
2½"	P1@.550	Ea	28.50	13.40	—	41.90
3"	P1@.654	Ea	32.00	16.00	—	48.00
4"	P1@.860	Ea	64.90	21.00	—	85.90
Type L copper tee C x C x C with brazed joints						
1/2"	P1@.156	Ea	.32	3.81	—	4.13
3/4"	P1@.204	Ea	.79	4.99	—	5.78
1"	P1@.263	Ea	5.40	6.43	—	11.83
1¼"	P1@.322	Ea	8.60	7.87	—	16.47
1½"	P1@.380	Ea	12.48	9.29	—	21.77
2"	P1@.507	Ea	20.00	12.40	—	32.40
2½"	P1@.624	Ea	40.58	15.20	—	55.78
3"	P1@.741	Ea	62.11	18.10	—	80.21
4"	P1@.975	Ea	78.00	23.80	—	101.80

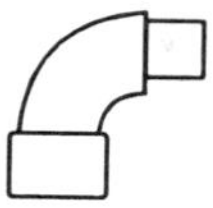

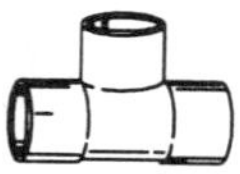

Description	Craft@Hrs	Unit	Material $	Labor $	Equipment $	Total $

Type L copper branch reducing tee C x C x C with brazed joints

Description	Craft@Hrs	Unit	Material $	Labor $	Equipment $	Total $
1/2" x 3/8"	P1@.129	Ea	2.28	3.15	—	5.43
3/4" x 1/2"	P1@.182	Ea	.79	4.45	—	5.24
1" x 1/2"	P1@.209	Ea	6.35	5.11	—	11.46
1" x 3/4"	P1@.234	Ea	6.35	5.72	—	12.07
1¼" x 1/2"	P1@.241	Ea	9.75	5.89	—	15.64
1¼" x 3/4"	P1@.264	Ea	9.75	6.45	—	16.20
1¼" x 1"	P1@.287	Ea	9.75	7.01	—	16.76
1½" x 1/2"	P1@.291	Ea	11.68	7.11	—	18.79
1½" x 3/4"	P1@.307	Ea	11.68	7.50	—	19.18
1½" x 1"	P1@.323	Ea	11.68	7.89	—	19.57
1½" x 1¼"	P1@.339	Ea	11.68	8.28	—	19.96
2" x 1/2"	P1@.348	Ea	15.77	8.50	—	24.27
2" x 3/4"	P1@.373	Ea	15.77	9.12	—	24.89
2" x 1"	P1@.399	Ea	15.77	9.75	—	25.52
2" x 1¼"	P1@.426	Ea	15.77	10.40	—	26.17
2" x 1½"	P1@.452	Ea	15.77	11.00	—	26.77
2½" x 1/2"	P1@.468	Ea	47.60	11.40	—	59.00
2½" x 3/4"	P1@.485	Ea	47.60	11.90	—	59.50
2½" x 1"	P1@.502	Ea	47.60	12.30	—	59.90
2½" x 1½"	P1@.537	Ea	47.60	13.10	—	60.70
2½" x 2"	P1@.555	Ea	47.60	13.60	—	61.20
3" x 1¼"	P1@.562	Ea	54.26	13.70	—	67.96
3" x 2"	P1@.626	Ea	56.26	15.30	—	71.56
3" x 2½"	P1@.660	Ea	56.26	16.10	—	72.36
4" x 1¼"	P1@.669	Ea	73.00	16.30	—	89.30
4" x 1½"	P1@.718	Ea	73.00	17.50	—	90.50
4" x 2"	P1@.768	Ea	73.00	18.80	—	91.80
4" x 2½"	P1@.819	Ea	73.00	20.00	—	93.00
4" x 3"	P1@.868	Ea	73.00	21.20	—	94.20

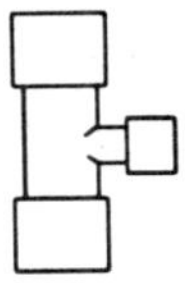

Copper, Type L with Brazed Joints

Type L copper reducer with brazed joints

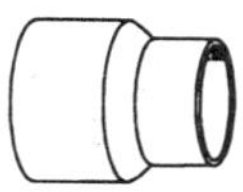

Description	Craft@Hrs	Unit	Material $	Labor $	Equipment $	Total $
1/2" x 3/8"	P1@.148	Ea	.40	3.62	—	4.02
3/4" x 1/2"	P1@.155	Ea	.72	3.79	—	4.51
1" x 3/4"	P1@.206	Ea	1.32	5.03	—	6.35
1¼" x 1/2"	P1@.206	Ea	2.20	5.03	—	7.23
1¼" x 3/4"	P1@.232	Ea	2.20	5.67	—	7.87
1¼" x 1"	P1@.258	Ea	2.20	6.30	—	8.50
1½" x 1/2"	P1@.232	Ea	3.37	5.67	—	9.04
1½" x 3/4"	P1@.257	Ea	3.37	6.28	—	9.65
1½" x 1"	P1@.283	Ea	3.37	6.92	—	10.29
1½" x 1¼"	P1@.308	Ea	3.37	7.53	—	10.90
2" x 1/2"	P1@.287	Ea	5.24	7.01	—	12.25
2" x 3/4"	P1@.277	Ea	5.24	6.77	—	12.01
2" x 1"	P1@.338	Ea	5.24	8.26	—	13.50
2" x 1¼"	P1@.365	Ea	5.24	8.92	—	14.16
2" x 1½"	P1@.390	Ea	5.24	9.53	—	14.77
2½" x 1"	P1@.390	Ea	14.30	9.53	—	23.83
2½" x 1¼"	P1@.416	Ea	14.30	10.20	—	24.50
2½" x 1½"	P1@.442	Ea	14.30	10.80	—	25.10
2½" x 2"	P1@.497	Ea	14.30	12.10	—	26.40
3" x 1¼"	P1@.468	Ea	16.70	11.40	—	28.10
3" x 1½"	P1@.493	Ea	16.70	12.00	—	28.70
3" x 2"	P1@.548	Ea	16.70	13.40	—	30.10
3" x 2½"	P1@.600	Ea	16.70	14.70	—	31.40
4" x 2"	P1@.652	Ea	32.00	15.90	—	47.90
4" x 2½"	P1@.704	Ea	32.00	17.20	—	49.20
4" x 3"	P1@.755	Ea	32.00	18.40	—	50.40

Type L copper adapter C x MPT with brazed joint

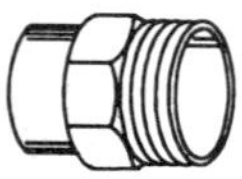

Description	Craft@Hrs	Unit	Material $	Labor $	Equipment $	Total $
1/2"	P1@.090	Ea	.97	2.20	—	3.17
3/4"	P1@.126	Ea	1.50	3.08	—	4.58
1"	P1@.162	Ea	3.69	3.96	—	7.65
1¼"	P1@.198	Ea	6.94	4.84	—	11.78
1½"	P1@.234	Ea	7.77	5.72	—	13.49
2"	P1@.312	Ea	10.01	7.62	—	17.63
2½"	P1@.384	Ea	32.60	9.38	—	41.98
3"	P1@.456	Ea	54.00	11.10	—	65.10

Type L copper adapter C x FPT with brazed joint

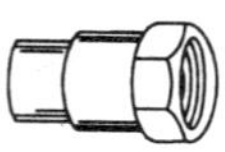

Description	Craft@Hrs	Unit	Material $	Labor $	Equipment $	Total $
1/2"	P1@.090	Ea	1.58	2.20	—	3.78
3/4"	P1@.126	Ea	2.88	3.08	—	5.96
1"	P1@.162	Ea	4.39	3.96	—	8.35
1¼"	P1@.198	Ea	9.14	4.84	—	13.98
1½"	P1@.234	Ea	11.23	5.72	—	16.95
2"	P1@.312	Ea	16.83	7.62	—	24.45
2½"	P1@.384	Ea	42.33	9.38	—	51.71
3"	P1@.456	Ea	59.19	11.10	—	70.29

Description	Craft@Hrs	Unit	Material $	Labor $	Equipment $	Total $

Type L copper flush bushing with brazed joints

Description	Craft@Hrs	Unit	Material $	Labor $	Equipment $	Total $
1/2" x 3/8"	P1@.148	Ea	.98	3.62	—	4.60
3/4" x 1/2"	P1@.155	Ea	1.64	3.79	—	5.43
1" x 3/4"	P1@.206	Ea	4.29	5.03	—	9.32
1" x 1/2"	P1@.206	Ea	4.88	5.03	—	9.91
1¼" x 1"	P1@.258	Ea	5.10	6.30	—	11.40
1½" x 1¼"	P1@.308	Ea	6.50	7.53	—	14.03
2" x 1½"	P1@.390	Ea	12.00	9.53	—	21.53

Type L copper union with brazed joints

Description	Craft@Hrs	Unit	Material $	Labor $	Equipment $	Total $
1/2"	P1@.146	Ea	3.87	3.57	—	7.44
3/4"	P1@.205	Ea	4.62	5.01	—	9.63
1"	P1@.263	Ea	10.35	6.43	—	16.78
1¼"	P1@.322	Ea	18.82	7.87	—	26.69
1½"	P1@.380	Ea	23.30	9.29	—	32.59
2"	P1@.507	Ea	34.89	12.40	—	47.29

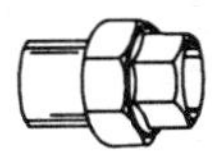

Type L copper dielectric union with brazed joints

Description	Craft@Hrs	Unit	Material $	Labor $	Equipment $	Total $
1/2"	P1@.146	Ea	3.63	3.57	—	7.20
3/4"	P1@.205	Ea	3.63	5.01	—	8.64
1"	P1@.263	Ea	7.53	6.43	—	13.96
1¼"	P1@.322	Ea	12.87	7.87	—	20.74
1½"	P1@.380	Ea	17.50	9.29	—	26.79
2"	P1@.507	Ea	24.31	12.40	—	36.71

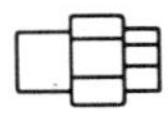

Type L copper cap with brazed joints

Description	Craft@Hrs	Unit	Material $	Labor $	Equipment $	Total $
1/2"	P1@.083	Ea	.31	2.03	—	2.34
3/4"	P1@.116	Ea	.63	2.83	—	3.46
1"	P1@.149	Ea	1.47	3.64	—	5.11
1¼"	P1@.182	Ea	2.09	4.45	—	6.54
1½"	P1@.215	Ea	3.05	5.25	—	8.30
2"	P1@.286	Ea	5.56	6.99	—	12.55
2½"	P1@.352	Ea	14.00	8.60	—	22.60
3"	P1@.418	Ea	21.88	10.20	—	32.08
4"	P1@.550	Ea	33.50	13.40	—	46.90

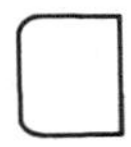

Type L copper coupling with brazed joints

Description	Craft@Hrs	Unit	Material $	Labor $	Equipment $	Total $
1/2"	P1@.129	Ea	.13	3.15	—	3.28
3/4"	P1@.181	Ea	.33	4.42	—	4.75
1"	P1@.232	Ea	1.39	5.67	—	7.06
1¼"	P1@.284	Ea	2.42	6.94	—	9.36
1½"	P1@.335	Ea	3.51	8.19	—	11.70
2"	P1@.447	Ea	5.33	10.90	—	16.23
2½"	P1@.550	Ea	12.42	13.40	—	25.82
3"	P1@.654	Ea	18.98	16.00	—	34.98
4"	P1@.860	Ea	37.47	21.00	—	58.47

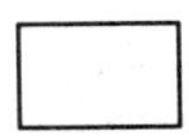

Copper, Type L with Brazed Joints

Description	Craft@Hrs	Unit	Material $	Labor $	Equipment $	Total $

Class 125 bronze body gate valve, brazed joints

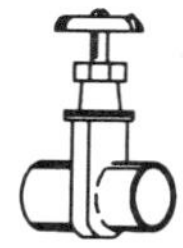

Description	Craft@Hrs	Unit	Material $	Labor $	Equipment $	Total $
1/2"	P1@.240	Ea	6.06	5.86	—	11.92
3/4"	P1@.300	Ea	7.58	7.33	—	14.91
1"	P1@.360	Ea	10.50	8.80	—	19.30
1¼"	P1@.480	Ea	13.40	11.70	—	25.10
1½"	P1@.540	Ea	18.00	13.20	—	31.20
2"	P1@.600	Ea	30.10	14.70	—	44.80
2½"	P1@1.00	Ea	49.90	24.40	—	74.30
3"	P1@1.50	Ea	71.60	36.70	—	108.30

Class 125 iron body gate valve, flanged joints

Description	Craft@Hrs	Unit	Material $	Labor $	Equipment $	Total $
2"	P1@.500	Ea	101.00	12.20	—	113.20
2½"	P1@.600	Ea	137.00	14.70	—	151.70
3"	P1@.750	Ea	148.00	18.30	—	166.30
4"	P1@1.35	Ea	218.00	33.00	—	251.00
5"	P1@2.00	Ea	417.00	48.90	—	465.90
6"	P1@2.50	Ea	417.00	61.10	—	478.10

Class 125 bronze body globe valve, with brazed joints

Description	Craft@Hrs	Unit	Material $	Labor $	Equipment $	Total $
1/2"	P1@.240	Ea	15.80	5.86	—	21.66
3/4"	P1@.300	Ea	21.30	7.33	—	28.63
1"	P1@.360	Ea	30.50	8.80	—	39.30
1¼"	P1@.480	Ea	42.90	11.70	—	54.60
1½"	P1@.540	Ea	57.50	13.20	—	70.70
2"	P1@.600	Ea	93.60	14.70	—	108.30

Class 125 iron body globe valve, flanged joints

Description	Craft@Hrs	Unit	Material $	Labor $	Equipment $	Total $
2½"	P1@.650	Ea	220.00	15.90	—	235.90
3"	P1@.750	Ea	264.00	18.30	—	282.30
4"	P1@1.35	Ea	355.00	33.00	—	388.00

200 PSIG iron body butterfly valve, lug type

Description	Craft@Hrs	Unit	Material $	Labor $	Equipment $	Total $
2"	P1@.450	Ea	84.50	11.00	—	95.50
2½"	P1@.450	Ea	86.50	11.00	—	97.50
3"	P1@.550	Ea	91.70	13.40	—	105.10
4"	P1@.550	Ea	114.00	13.40	—	127.40

200 PSIG iron body butterfly valve, wafer type

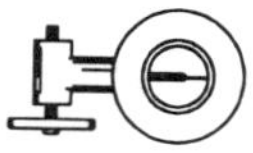

Description	Craft@Hrs	Unit	Material $	Labor $	Equipment $	Total $
2"	P1@.450	Ea	76.70	11.00	—	87.70
2½"	P1@.450	Ea	78.70	11.00	—	89.70
3"	P1@.550	Ea	84.50	13.40	—	97.90
4"	P1@.550	Ea	102.00	13.40	—	115.40

Description	Craft@Hrs	Unit	Material $	Labor $	Equipment $	Total $

Class 125 bronze body 2-piece ball valve, brazed joints

Description	Craft@Hrs	Unit	Material $	Labor $	Equipment $	Total $
1/2"	P1@.240	Ea	3.56	5.86	—	9.42
3/4"	P1@.300	Ea	4.75	7.33	—	12.08
1"	P1@.360	Ea	8.21	8.80	—	17.01
1¼"	P1@.480	Ea	13.70	11.70	—	25.40
1½"	P1@.540	Ea	18.50	13.20	—	31.70
2"	P1@.600	Ea	23.30	14.70	—	38.00

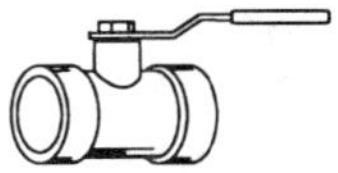

Class 125 bronze body swing check valve, brazed joints

Description	Craft@Hrs	Unit	Material $	Labor $	Equipment $	Total $
1/2"	P1@.240	Ea	8.00	5.86	—	13.86
3/4"	P1@.300	Ea	11.40	7.33	—	18.73
1"	P1@.360	Ea	15.00	8.80	—	23.80
1¼"	P1@.480	Ea	21.40	11.70	—	33.10
1½"	P1@.540	Ea	30.00	13.20	—	43.20
2"	P1@.600	Ea	50.00	14.70	—	64.70
2½"	P1@1.00	Ea	100.00	24.40	—	124.40
3"	P1@1.50	Ea	144.00	36.70	—	180.70

Class 125 iron body swing check valve, flanged joints

Description	Craft@Hrs	Unit	Material $	Labor $	Equipment $	Total $
2"	P1@.500	Ea	115.00	12.20	—	127.20
2½"	P1@.600	Ea	148.00	14.70	—	162.70
3"	P1@.750	Ea	183.00	18.30	—	201.30
4"	P1@1.35	Ea	271.00	33.00	—	304.00

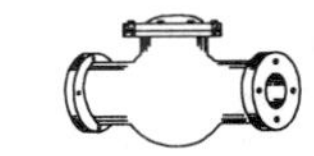

Class 125 iron body silent check valve, wafer type

Description	Craft@Hrs	Unit	Material $	Labor $	Equipment $	Total $
2"	P1@.500	Ea	85.00	12.20	—	97.20
2½"	P1@.600	Ea	95.00	14.70	—	109.70
3"	P1@.750	Ea	108.00	18.30	—	126.30
4"	P1@1.35	Ea	146.00	33.00	—	179.00

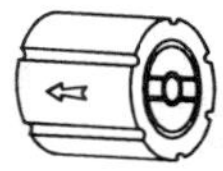

Class 125 bronze body strainer, threaded joints

Description	Craft@Hrs	Unit	Material $	Labor $	Equipment $	Total $
1/2"	P1@.230	Ea	12.40	5.62	—	18.02
3/4"	P1@.260	Ea	16.30	6.35	—	22.65
1"	P1@.330	Ea	20.20	8.06	—	28.26
1¼"	P1@.440	Ea	28.10	10.80	—	38.90
1½"	P1@.495	Ea	42.20	12.10	—	54.30
2"	P1@.550	Ea	73.00	13.40	—	86.40

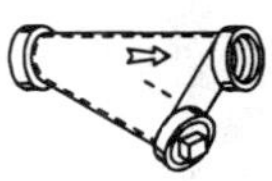

Class 125 iron body strainer, flanged joints

Description	Craft@Hrs	Unit	Material $	Labor $	Equipment $	Total $
2"	P1@.500	Ea	62.00	12.20	—	74.20
2½"	P1@.600	Ea	69.00	14.70	—	83.70
3"	P1@.750	Ea	80.00	18.30	—	98.30
4"	P1@1.35	Ea	137.00	33.00	—	170.00

Copper, Type L with Brazed Joints

Installation of 2-way control valve, threaded joints

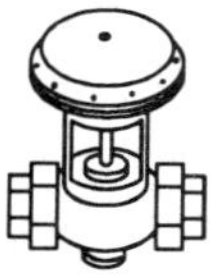

Description	Craft@Hrs	Unit	Material $	Labor $	Equipment $	Total $
1/2"	P1@.210	Ea	—	5.13	—	5.13
3/4"	P1@.275	Ea	—	6.72	—	6.72
1"	P1@.350	Ea	—	8.55	—	8.55
1¼"	P1@.430	Ea	—	10.50	—	10.50
1½"	P1@.505	Ea	—	12.30	—	12.30
2"	P1@.675	Ea	—	16.50	—	16.50
2½"	P1@.830	Ea	—	20.30	—	20.30
3"	P1@.990	Ea	—	24.20	—	24.20

Installation of 3-way control valve, threaded joints

Description	Craft@Hrs	Unit	Material $	Labor $	Equipment $	Total $
1/2"	P1@.260	Ea	—	6.35	—	6.35
3/4"	P1@.365	Ea	—	8.92	—	8.92
1"	P1@.475	Ea	—	11.60	—	11.60
1¼"	P1@.575	Ea	—	14.10	—	14.10
1½"	P1@.680	Ea	—	16.60	—	16.60
2"	P1@.910	Ea	—	22.20	—	22.20
2½"	P1@1.12	Ea	—	27.40	—	27.40
3"	P1@1.33	Ea	—	32.50	—	32.50

Companion flange, 150 pound cast brass

Description	Craft@Hrs	Unit	Material $	Labor $	Equipment $	Total $
2"	P1@.290	Ea	169.00	7.09	—	176.09
2½"	P1@.380	Ea	208.00	9.29	—	217.29
3"	P1@.460	Ea	208.00	11.20	—	219.20
4"	P1@.600	Ea	332.00	14.70	—	346.70

Bolt and gasket sets

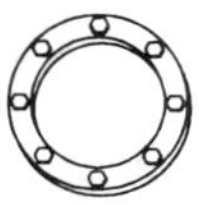

Description	Craft@Hrs	Unit	Material $	Labor $	Equipment $	Total $
2"	P1@.500	Ea	1.50	12.20	—	13.70
2½"	P1@.650	Ea	1.75	15.89	—	17.64
3"	P1@.750	Ea	3.00	18.33	—	21.33
4"	P1@1.00	Ea	5.00	24.44	—	29.44

Thermometer with well

Description	Craft@Hrs	Unit	Material $	Labor $	Equipment $	Total $
7"	P1@.250	Ea	15.00	6.11	—	21.11
9"	P1@.250	Ea	15.50	6.11	—	21.61

Dial type pressure gauge

Description	Craft@Hrs	Unit	Material $	Labor $	Equipment $	Total $
2½"	P1@.200	Ea	15.50	4.89	—	20.39
3½"	P1@.200	Ea	20.50	4.89	—	25.39

Description	Craft@Hrs	Unit	Material $	Labor $	Equipment $	Total $

Pressure/temperature tap

Description	Craft@Hrs	Unit	Material $	Labor $	Equipment $	Total $
	P1@.150	Ea	7.30	3.67	—	10.97

Hanger with swivel assembly

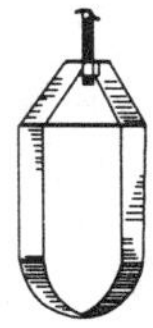

Description	Craft@Hrs	Unit	Material $	Labor $	Equipment $	Total $
1/2"	P1@.450	Ea	1.60	11.00	—	12.60
3/4"	P1@.450	Ea	1.75	11.00	—	12.75
1"	P1@.450	Ea	1.85	11.00	—	12.85
1¼"	P1@.450	Ea	1.90	11.00	—	12.90
1½"	P1@.450	Ea	2.05	11.00	—	13.05
2"	P1@.450	Ea	2.15	11.00	—	13.15
2½"	P1@.450	Ea	2.90	11.00	—	13.90
3"	P1@.450	Ea	3.60	11.00	—	14.60
4"	P1@.450	Ea	3.95	11.00	—	14.95

Riser clamp

Description	Craft@Hrs	Unit	Material $	Labor $	Equipment $	Total $
1/2"	P1@.100	Ea	.95	2.44	—	3.39
3/4"	P1@.100	Ea	1.39	2.44	—	3.83
1"	P1@.100	Ea	1.41	2.44	—	3.85
1¼"	P1@.105	Ea	1.70	2.57	—	4.27
1½"	P1@.110	Ea	1.79	2.69	—	4.48
2"	P1@.115	Ea	1.89	2.81	—	4.70
2½"	P1@.120	Ea	2.00	2.93	—	4.93
3"	P1@.120	Ea	2.18	2.93	—	5.11
4"	P1@.125	Ea	2.75	3.05	—	5.80

Copper, Type L with Soft-Soldered Joints

Type L hard-drawn copper pipe with wrought copper fittings and soft-soldered joints is used in a wide variety of plumbing and HVAC systems such as potable water, heating hot water, chilled water, and A.C. condensate.

Soft-soldered joints are those made with solders having melting points in the 350 degree F. to 500 degree F. range. Maximum working temperature/pressure relationships for soft-soldered joints are approximately as follows:

Maximum Working Pressures (PSIG)*				
Soft-solder Type	Water Temperature (degrees F.)	Nominal Pipe Size (inches)		
		Up to 1	1¼ to 2	2½ x 4
50-50 tin-lead**	100	200	175	150
	150	150	125	100
	200	100	90	75
	250	85	75	50
95-5 tin-antimony	100	500	400	300
	150	400	350	275
	200	300	250	200
	250	200	175	150

**For copper pipe and solder-type fittings using soft-solders melting at approximately 350 degrees F. to 500 degrees F.*

***The use of any solder containing lead is not allowed in potable water systems.*

This section has been arranged to save the estimator's time by including all normally-used system components such as pipe, fittings, valves, hanger assemblies, riser clamps and miscellaneous items under one heading. Additional items can be found under "Plumbing and Piping Specialties." The cost estimates in this section are based on the conditions, limitations and wage rates described in the section "How to Use This Book" beginning on page 5.

Type L copper pipe with soft-soldered joints

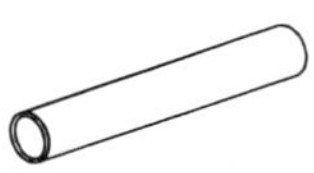

Description	Craft@Hrs	Unit	Material $	Labor $	Equipment $	Total $
1/2"	P1@.032	LF	.68	.78	—	1.46
3/4"	P1@.035	LF	1.08	.86	—	1.94
1"	P1@.038	LF	1.50	.93	—	2.43
1¼"	P1@.042	LF	2.13	1.03	—	3.16
1½"	P1@.046	LF	2.73	1.12	—	3.85
2"	P1@.053	LF	4.69	1.30	—	5.99
2½"	P1@.060	LF	6.85	1.47	—	8.32
3"	P1@.066	LF	9.26	1.61	—	10.87
4"	P1@.080	LF	14.64	1.95	—	16.59

Description	Craft@Hrs	Unit	Material $	Labor $	Equipment $	Total $
Type L copper 45 degree ell C x C with soft-soldered joints						
1/2"	P1@.107	Ea	.36	2.61	—	2.97
3/4"	P1@.150	Ea	.66	3.67	—	4.33
1"	P1@.193	Ea	2.50	4.72	—	7.22
1¼"	P1@.236	Ea	3.40	5.77	—	9.17
1½"	P1@.278	Ea	6.02	6.79	—	12.81
2"	P1@.371	Ea	9.85	9.07	—	18.92
2½"	P1@.457	Ea	25.44	11.20	—	36.64
3"	P1@.543	Ea	33.45	13.30	—	46.75
4"	P1@.714	Ea	48.75	17.40	—	66.15

Description	Craft@Hrs	Unit	Material $	Labor $	Equipment $	Total $
Type L copper 90 degree ell C x C with soft-soldered joints						
1/2"	P1@.107	Ea	.18	2.61	—	2.79
3/4"	P1@.150	Ea	.26	3.67	—	3.93
1"	P1@.193	Ea	2.28	4.72	—	7.00
1¼"	P1@.236	Ea	3.71	5.77	—	9.48
1½"	P1@.278	Ea	5.62	6.79	—	12.41
2"	P1@.371	Ea	10.05	9.07	—	19.12
2½"	P1@.457	Ea	20.81	11.20	—	32.01
3"	P1@.543	Ea	26.30	13.30	—	39.60
4"	P1@.714	Ea	42.25	17.40	—	59.65

Description	Craft@Hrs	Unit	Material $	Labor $	Equipment $	Total $
Type L copper 90 degree ell Ftg. x C with soft-soldered joint						
1/2"	P1@.107	Ea	.49	2.61	—	3.10
3/4"	P1@.150	Ea	1.09	3.67	—	4.76
1"	P1@.193	Ea	3.25	4.72	—	7.97
1¼"	P1@.236	Ea	4.50	5.77	—	10.27
1½"	P1@.278	Ea	6.75	6.79	—	13.54
2"	P1@.371	Ea	13.50	9.07	—	22.57
2½"	P1@.457	Ea	28.50	11.20	—	39.70
3"	P1@.543	Ea	32.00	13.30	—	45.30
4"	P1@.714	Ea	64.90	17.40	—	82.30

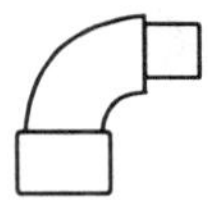

Description	Craft@Hrs	Unit	Material $	Labor $	Equipment $	Total $
Type L copper tee C x C x C with soft-soldered joints						
1/2"	P1@.129	Ea	.32	3.15	—	3.47
3/4"	P1@.181	Ea	.79	4.42	—	5.21
1"	P1@.233	Ea	5.40	5.69	—	11.09
1¼"	P1@.285	Ea	8.60	6.96	—	15.56
1½"	P1@.337	Ea	12.48	8.24	—	20.72
2"	P1@.449	Ea	20.00	11.00	—	31.00
2½"	P1@.552	Ea	40.58	13.50	—	54.08
3"	P1@.656	Ea	62.11	16.00	—	78.11
4"	P1@.863	Ea	78.00	21.10	—	99.10

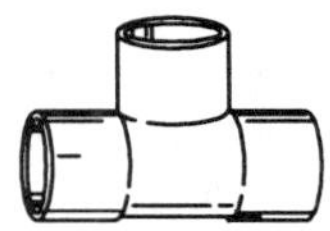

Copper, Type L with Soft-Soldered Joints

Type L copper branch reducing tee C x C x C with soft-soldered joints

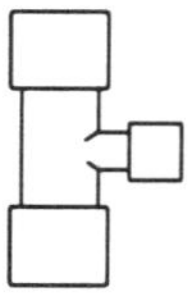

Description	Craft@Hrs	Unit	Material $	Labor $	Equipment $	Total $
1/2" x 3/8"	P1@.121	Ea	2.28	2.96	—	5.24
3/4" x 1/2"	P1@.170	Ea	.79	4.15	—	4.94
1" x 1/2"	P1@.195	Ea	6.35	4.77	—	11.12
1" x 3/4"	P1@.219	Ea	6.35	5.35	—	11.70
1¼" x 1/2"	P1@.225	Ea	9.75	5.50	—	15.25
1¼" x 3/4"	P1@.247	Ea	9.75	6.04	—	15.79
1¼" x 1"	P1@.268	Ea	9.75	6.55	—	16.30
1½" x 1/2"	P1@.272	Ea	11.68	6.65	—	18.33
1½" x 3/4"	P1@.287	Ea	11.68	7.01	—	18.69
1½" x 1"	P1@.302	Ea	11.68	7.38	—	19.06
1½" x 1¼"	P1@.317	Ea	11.68	7.75	—	19.43
2" x 1/2"	P1@.325	Ea	15.77	7.94	—	23.71
2" x 3/4"	P1@.348	Ea	15.77	8.50	—	24.27
2" x 1"	P1@.373	Ea	15.77	9.12	—	24.89
2" x 1¼"	P1@.398	Ea	15.77	9.73	—	25.50
2" x 1½"	P1@.422	Ea	15.77	10.30	—	26.07
2½" x 1/2"	P1@.437	Ea	47.60	10.70	—	58.30
2½" x 3/4"	P1@.453	Ea	47.60	11.10	—	58.70
2½" x 1"	P1@.469	Ea	47.60	11.50	—	59.10
2½" x 1½"	P1@.502	Ea	47.60	12.30	—	59.90
2½" x 2"	P1@.519	Ea	47.60	12.70	—	60.30
3" x 1¼"	P1@.525	Ea	54.26	12.80	—	67.06
3" x 2"	P1@.585	Ea	56.26	14.30	—	70.56
3" x 2½"	P1@.617	Ea	56.26	15.10	—	71.36
4" x 1¼"	P1@.625	Ea	73.00	15.30	—	88.30
4" x 1½"	P1@.671	Ea	73.00	16.40	—	89.40
4" x 2"	P1@.718	Ea	73.00	17.50	—	90.50
4" x 2½"	P1@.765	Ea	73.00	18.70	—	91.70
4" x 3"	P1@.811	Ea	73.00	19.80	—	92.80

Description	Craft@Hrs	Unit	Material $	Labor $	Equipment $	Total $
Type L copper reducer with soft-soldered joints						
1/2" x 3/8"	P1@.124	Ea	.40	3.03	—	3.43
3/4" x 1/2"	P1@.129	Ea	.72	3.15	—	3.87
1" x 3/4"	P1@.172	Ea	1.32	4.20	—	5.52
1¼" x 1/2"	P1@.172	Ea	2.20	4.20	—	6.40
1¼" x 3/4"	P1@.193	Ea	2.20	4.72	—	6.92
1¼" x 1"	P1@.215	Ea	2.20	5.25	—	7.45
1½" x 1/2"	P1@.193	Ea	3.37	4.72	—	8.09
1½" x 3/4"	P1@.214	Ea	3.37	5.23	—	8.60
1½" x 1"	P1@.236	Ea	3.37	5.77	—	9.14
1½" x 1¼"	P1@.257	Ea	3.37	6.28	—	9.65
2" x 1/2"	P1@.239	Ea	5.24	5.84	—	11.08
2" x 3/4"	P1@.261	Ea	5.24	6.38	—	11.62
2" x 1"	P1@.282	Ea	5.24	6.89	—	12.13
2" x 1¼"	P1@.304	Ea	5.24	7.43	—	12.67
2" x 1½"	P1@.325	Ea	5.24	7.94	—	13.18
2½" x 1"	P1@.325	Ea	14.30	7.94	—	22.24
2½" x 1¼"	P1@.347	Ea	14.30	8.48	—	22.78
2½" x 1½"	P1@.368	Ea	14.30	8.99	—	23.29
2½" x 2"	P1@.414	Ea	14.30	10.10	—	24.40
3" x 2"	P1@.457	Ea	16.70	11.20	—	27.90
3" x 2½"	P1@.500	Ea	16.70	12.20	—	28.90
4" x 2"	P1@.543	Ea	32.00	13.30	—	45.30
4" x 2½"	P1@.586	Ea	32.00	14.30	—	46.30
4" x 3"	P1@.629	Ea	32.00	15.40	—	47.40
Type L copper adapter C x MPT with soft-soldered joint						
1/2"	P1@.075	Ea	.97	1.83	—	2.80
3/4"	P1@.105	Ea	1.50	2.57	—	4.07
1"	P1@.134	Ea	3.69	3.27	—	6.96
1¼"	P1@.164	Ea	6.94	4.01	—	10.95
1½"	P1@.194	Ea	7.77	4.74	—	12.51
2"	P1@.259	Ea	10.01	6.33	—	16.34
2½"	P1@.319	Ea	32.60	7.80	—	40.40
3"	P1@.378	Ea	54.00	9.24	—	63.24
Type L copper adapter C x FPT with soft-soldered joint						
1/2"	P1@.075	Ea	1.58	1.83	—	3.41
3/4"	P1@.105	Ea	2.88	2.57	—	5.45
1"	P1@.134	Ea	4.39	3.27	—	7.66
1¼"	P1@.164	Ea	9.14	4.01	—	13.15
1½"	P1@.194	Ea	11.23	4.74	—	15.97
2"	P1@.259	Ea	16.83	6.33	—	23.16
2½"	P1@.319	Ea	42.33	7.80	—	50.13
3"	P1@.378	Ea	59.19	9.24	—	68.43

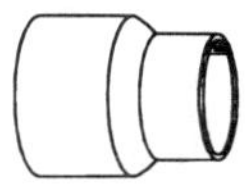

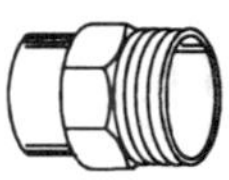

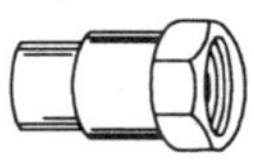

Copper, Type L with Soft-Soldered Joints

Type L copper flush bushing with soft-soldered joints

Description	Craft@Hrs	Unit	Material $	Labor $	Equipment $	Total $
1/2" x 3/8"	P1@.124	Ea	.98	3.03	—	4.01
3/4" x 1/2"	P1@.129	Ea	1.64	3.15	—	4.79
1" x 3/4"	P1@.172	Ea	4.29	4.20	—	8.49
1" x 1/2"	P1@.172	Ea	4.88	4.20	—	9.08
1¼" x 1"	P1@.215	Ea	5.10	5.25	—	10.35
1½" x 1¼"	P1@.257	Ea	6.50	6.28	—	12.78
2" x 1½"	P1@.325	Ea	12.00	7.94	—	19.94

Type L copper union with soft-soldered joint

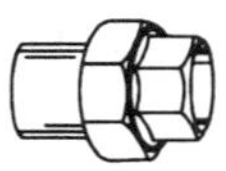

Description	Craft@Hrs	Unit	Material $	Labor $	Equipment $	Total $
1/2"	P1@.121	Ea	3.87	2.96	—	6.83
3/4"	P1@.170	Ea	4.62	4.15	—	8.77
1"	P1@.218	Ea	10.35	5.33	—	15.68
1¼"	P1@.267	Ea	18.82	6.52	—	25.34
1½"	P1@.315	Ea	23.30	7.70	—	31.00
2"	P1@.421	Ea	34.89	10.30	—	45.19

Type L copper dielectric union with soft-soldered joint

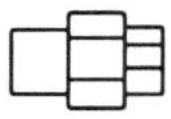

Description	Craft@Hrs	Unit	Material $	Labor $	Equipment $	Total $
1/2"	P1@.121	Ea	3.63	2.96	—	6.59
3/4"	P1@.170	Ea	3.63	4.15	—	7.78
1"	P1@.218	Ea	7.53	5.33	—	12.86
1¼"	P1@.267	Ea	12.87	6.52	—	19.39
1½"	P1@.315	Ea	17.50	7.70	—	25.20
2"	P1@.421	Ea	24.31	10.30	—	34.61

Type L copper cap with soft-soldered joint

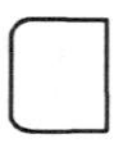

Description	Craft@Hrs	Unit	Material $	Labor $	Equipment $	Total $
1/2"	P1@.069	Ea	.31	1.69	—	2.00
3/4"	P1@.096	Ea	.63	2.35	—	2.98
1"	P1@.124	Ea	1.47	3.03	—	4.50
1¼"	P1@.151	Ea	2.09	3.69	—	5.78
1½"	P1@.178	Ea	3.05	4.35	—	7.40
2"	P1@.237	Ea	5.56	5.79	—	11.35
2½"	P1@.292	Ea	14.00	7.14	—	21.14
3"	P1@.347	Ea	21.88	8.48	—	30.36
4"	P1@.457	Ea	33.50	11.20	—	44.70

Type L copper coupling with soft-soldered joints

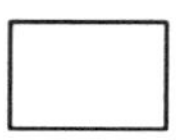

Description	Craft@Hrs	Unit	Material $	Labor $	Equipment $	Total $
1/2"	P1@.107	Ea	.13	2.61	—	2.74
3/4"	P1@.150	Ea	.33	3.67	—	4.00
1"	P1@.193	Ea	1.39	4.72	—	6.11
1¼"	P1@.236	Ea	2.42	5.77	—	8.19
1½"	P1@.278	Ea	3.51	6.79	—	10.30
2"	P1@.371	Ea	5.33	9.07	—	14.40
2½"	P1@.457	Ea	12.42	11.20	—	23.62
3"	P1@.543	Ea	18.98	13.30	—	32.28
4"	P1@.714	Ea	37.47	17.40	—	54.87

Description	Craft@Hrs	Unit	Material $	Labor $	Equipment $	Total $
Class 125 bronze body gate valve, soft-soldered joints						
1/2"	P1@.200	Ea	6.06	4.89	—	10.95
3/4"	P1@.249	Ea	7.58	6.08	—	13.66
1"	P1@.299	Ea	10.50	7.31	—	17.81
1¼"	P1@.398	Ea	13.40	9.73	—	23.13
1½"	P1@.448	Ea	18.00	10.90	—	28.90
2"	P1@.498	Ea	30.10	12.20	—	42.30
2½"	P1@.830	Ea	49.90	20.30	—	70.20
3"	P1@1.24	Ea	71.60	30.30	—	101.90
Class 125 iron body gate valve, flanged ends						
2"	P1@.500	Ea	101.00	12.20	—	113.20
2½"	P1@.600	Ea	137.00	14.70	—	151.70
3"	P1@.750	Ea	148.00	18.30	—	166.30
4"	P1@1.35	Ea	218.00	33.00	—	251.00
5"	P1@2.00	Ea	417.00	48.90	—	465.90
6"	P1@2.50	Ea	417.00	61.10	—	478.10
Class 125 bronze body globe valve, with soft-soldered joints						
1/2"	P1@.200	Ea	15.80	4.89	—	20.69
3/4"	P1@.249	Ea	21.30	6.08	—	27.38
1"	P1@.299	Ea	30.50	7.31	—	37.81
1¼"	P1@.398	Ea	42.90	9.73	—	52.63
1½"	P1@.448	Ea	57.50	10.90	—	68.40
2"	P1@.498	Ea	93.60	12.20	—	105.80
Class 125 iron body globe valve, flanged ends						
2½"	P1@.600	Ea	220.00	14.70	—	234.70
3"	P1@.750	Ea	264.00	18.30	—	282.30
4"	P1@1.35	Ea	352.00	33.00	—	385.00
200 PSIG iron body butterfly valve, lug type						
2"	P1@.450	Ea	84.50	11.00	—	95.50
2½"	P1@.450	Ea	86.50	11.00	—	97.50
3"	P1@.550	Ea	91.70	13.40	—	105.10
4"	P1@.550	Ea	114.00	13.40	—	127.40
200 PSIG iron body butterfly valve, wafer type						
2"	P1@.450	Ea	76.70	11.00	—	87.70
2½"	P1@.450	Ea	78.70	11.00	—	89.70
3"	P1@.550	Ea	84.50	13.40	—	97.90
4"	P1@.550	Ea	102.00	13.40	—	115.40

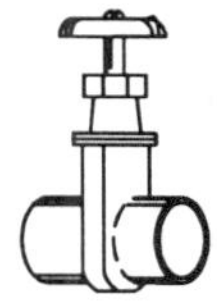
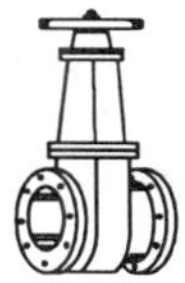

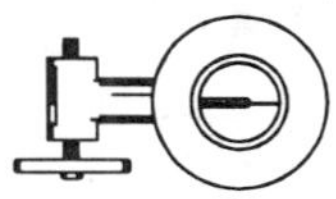

Copper, Type L with Soft-Soldered Joints

Class 125 bronze body two piece ball valve, soft-soldered joints

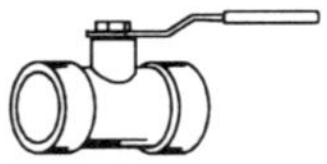

Description	Craft@Hrs	Unit	Material $	Labor $	Equipment $	Total $
1/2"	P1@.200	Ea	3.56	4.89	—	8.45
3/4"	P1@.249	Ea	4.75	6.08	—	10.83
1"	P1@.299	Ea	8.21	7.31	—	15.52
1¼"	P1@.398	Ea	13.70	9.73	—	23.43
1½"	P1@.448	Ea	18.50	10.90	—	29.40
2"	P1@.498	Ea	23.30	12.20	—	35.50

Class 125 bronze body swing check valve, soft-soldered joints

Description	Craft@Hrs	Unit	Material $	Labor $	Equipment $	Total $
1/2"	P1@.200	Ea	8.00	4.89	—	12.89
3/4"	P1@.249	Ea	11.40	6.08	—	17.48
1"	P1@.299	Ea	15.00	7.31	—	22.31
1¼"	P1@.398	Ea	21.40	9.73	—	31.13
1½"	P1@.448	Ea	30.00	10.90	—	40.90
2"	P1@.498	Ea	50.00	12.20	—	62.20
2½"	P1@.830	Ea	100.00	20.30	—	120.30
3"	P1@1.24	Ea	144.00	30.30	—	174.30

Class 125 iron body swing check valve, flanged ends

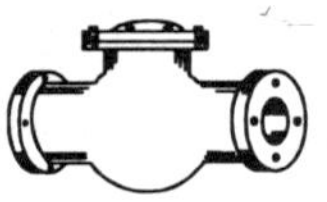

Description	Craft@Hrs	Unit	Material $	Labor $	Equipment $	Total $
2"	P1@.500	Ea	115.00	12.20	—	127.20
2½"	P1@.600	Ea	148.00	14.70	—	162.70
3"	P1@.750	Ea	183.00	18.30	—	201.30
4"	P1@1.35	Ea	271.00	33.00	—	304.00

Class 125 iron body silent check valve, wafer type

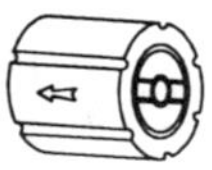

Description	Craft@Hrs	Unit	Material $	Labor $	Equipment $	Total $
2"	P1@.500	Ea	85.00	12.20	—	97.20
2½"	P1@.600	Ea	95.00	14.70	—	109.70
3"	P1@.750	Ea	108.00	18.30	—	126.30
4"	P1@1.35	Ea	146.00	33.00	—	179.00

Class 125 bronze body strainer, threaded ends

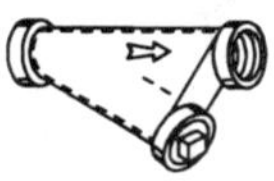

Description	Craft@Hrs	Unit	Material $	Labor $	Equipment $	Total $
1/2"	P1@.230	Ea	12.40	5.62	—	18.02
3/4"	P1@.260	Ea	16.30	6.35	—	22.65
1"	P1@.330	Ea	20.20	8.06	—	28.26
1¼"	P1@.440	Ea	28.10	10.80	—	38.90
1½"	P1@.495	Ea	42.20	12.10	—	54.30
2"	P1@.550	Ea	73.00	13.40	—	86.40

Class 125 iron body strainer, flanged ends

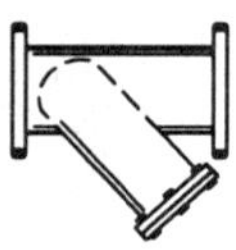

Description	Craft@Hrs	Unit	Material $	Labor $	Equipment $	Total $
2"	P1@.500	Ea	62.00	12.20	—	74.20
2½"	P1@.600	Ea	69.00	14.70	—	83.70
3"	P1@.750	Ea	80.00	18.30	—	98.30
4"	P1@1.35	Ea	137.00	33.00	—	170.00

Description	Craft@Hrs	Unit	Material $	Labor $	Equipment $	Total $
Installation of 2-way control valve, threaded joints						
1/2"	P1@.210	Ea	—	5.13	—	5.13
3/4"	P1@.275	Ea	—	6.72	—	6.72
1"	P1@.350	Ea	—	8.55	—	8.55
1¼"	P1@.430	Ea	—	10.50	—	10.50
1½"	P1@.505	Ea	—	12.30	—	12.30
2"	P1@.675	Ea	—	16.50	—	16.50
2½"	P1@.830	Ea	—	20.30	—	20.30
3"	P1@.990	Ea	—	24.20	—	24.20
Installation of 3-way control valve, threaded joints						
1/2"	P1@.260	Ea	—	6.35	—	6.35
3/4"	P1@.365	Ea	—	8.92	—	8.92
1"	P1@.475	Ea	—	11.60	—	11.60
1¼"	P1@.575	Ea	—	14.10	—	14.10
1½"	P1@.680	Ea	—	16.60	—	16.60
2"	P1@.910	Ea	—	22.20	—	22.20
2½"	P1@1.12	Ea	—	27.40	—	27.40
3"	P1@1.33	Ea	—	32.50	—	32.50
Companion flange, 150 pound cast brass						
2"	P1@.290	Ea	169.00	7.09	—	176.09
2½"	P1@.380	Ea	208.00	9.29	—	217.29
3"	P1@.460	Ea	208.00	11.20	—	219.20
4"	P1@.600	Ea	332.00	14.70	—	346.70
Bolt and gasket sets						
2"	P1@.500	Ea	1.50	12.20	—	13.70
2½"	P1@.650	Ea	1.75	15.89	—	17.64
3"	P1@.750	Ea	3.00	18.33	—	21.33
4"	P1@1.00	Ea	5.00	24.44	—	29.44
Thermometer with well						
7"	P1@.250	Ea	15.00	6.11	—	21.11
9"	P1@.250	Ea	15.50	6.11	—	21.61
Dial type pressure gauge						
2½"	P1@.200	Ea	15.50	4.89	—	20.39
3½"	P1@.200	Ea	20.50	4.89	—	25.39
Pressure/temperature tap						
	P1@.150	Ea	7.30	3.67	—	10.97

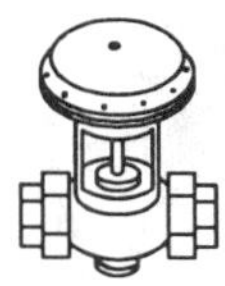

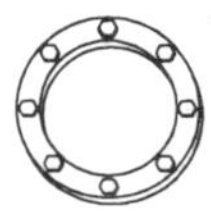

Copper, Type L with Soft-Soldered Joints

Description	Craft@Hrs	Unit	Material $	Labor $	Equipment $	Total $
Hanger with swivel assembly						
1/2"	P1@.450	Ea	1.60	11.00	—	12.60
3/4"	P1@.450	Ea	1.75	11.00	—	12.75
1"	P1@.450	Ea	1.85	11.00	—	12.85
1¼"	P1@.450	Ea	1.90	11.00	—	12.90
1½"	P1@.450	Ea	2.05	11.00	—	13.05
2"	P1@.450	Ea	2.15	11.00	—	13.15
2½"	P1@.450	Ea	2.90	11.00	—	13.90
3"	P1@.450	Ea	3.60	11.00	—	14.60
4"	P1@.450	Ea	3.95	11.00	—	14.95

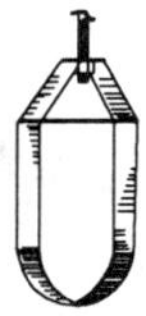

Description	Craft@Hrs	Unit	Material $	Labor $	Equipment $	Total $
Riser clamp						
1/2"	P1@.100	Ea	.95	2.44	—	3.39
3/4"	P1@.100	Ea	1.39	2.44	—	3.83
1"	P1@.100	Ea	1.41	2.44	—	3.85
1¼"	P1@.105	Ea	1.70	2.57	—	4.27
1½"	P1@.110	Ea	1.79	2.69	—	4.48
2"	P1@.115	Ea	1.89	2.81	—	4.70
2½"	P1@.120	Ea	2.00	2.93	—	4.93
3"	P1@.120	Ea	2.18	2.93	—	5.11
4"	P1@.125	Ea	2.75	3.05	—	5.80

Type M hard-drawn copper pipe with wrought copper fittings and brazed joints is used in a wide variety of plumbing and HVAC systems such as potable water, heating hot water and chilled water.

Brazed joints are those made with silver or other alloy filler metals having melting points at, or above, 1,000 degrees F. Maximum working pressure/temperature relationships for brazed joints are approximately as follows:

Maximum Working Pressure (PSIG)*			
Water Temperature	**Nominal Pipe Size (inches)**		
(degrees F.)	**Up to 1**	**1¼ to 2**	**2½ to 4**
Up to 350	270	190	150

**For copper pipe and solder-type fittings using brazing alloys melting at, or above, 1,000 degrees F.*

This section has been arranged to save the estimator's time by including all normally-used system components such as pipe, fittings, valves, hanger assemblies, riser clamps and miscellaneous items under one heading. Additional items can be found under "Plumbing and Piping Specialties." The cost estimates in this section are based on the conditions, limitations and wage rates described in the section "How to Use This Book" beginning on page 5.

Type M copper pipe with brazed joints

Description	Craft@Hrs	Unit	Material $	Labor $	Equipment $	Total $
1/2"	P1@.032	LF	.44	.78	—	1.22
3/4"	P1@.035	LF	.71	.86	—	1.57
1"	P1@.038	LF	1.13	.93	—	2.06
1¼"	P1@.042	LF	1.68	1.03	—	2.71
1½"	P1@.046	LF	2.37	1.12	—	3.49
2"	P1@.053	LF	4.31	1.30	—	5.61
2½"	P1@.060	LF	5.95	1.47	—	7.42
3"	P1@.066	LF	7.56	1.61	—	9.17
4"	P1@.080	LF	12.05	1.95	—	14.00

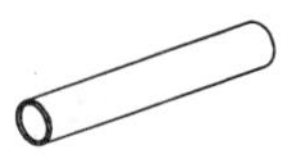

Type M copper 45 degree ell C x C with brazed joints

Description	Craft@Hrs	Unit	Material $	Labor $	Equipment $	Total $
1/2"	P1@.129	Ea	.36	3.15	—	3.51
3/4"	P1@.181	Ea	.66	4.42	—	5.08
1"	P1@.232	Ea	2.50	5.67	—	8.17
1¼"	P1@.284	Ea	3.40	6.94	—	10.34
1½"	P1@.335	Ea	6.02	8.19	—	14.21
2"	P1@.447	Ea	9.85	10.90	—	20.75
2½"	P1@.550	Ea	25.44	13.40	—	38.84
3"	P1@.654	Ea	33.45	16.00	—	49.45
4"	P1@.860	Ea	48.75	21.00	—	69.75

Type M copper 90 degree ell C x C with brazed joints

Description	Craft@Hrs	Unit	Material $	Labor $	Equipment $	Total $
1/2"	P1@.129	Ea	.18	3.15	—	3.33
3/4"	P1@.181	Ea	.26	4.42	—	4.68
1"	P1@.232	Ea	2.28	5.67	—	7.95
1¼"	P1@.284	Ea	3.71	6.94	—	10.65
1½"	P1@.335	Ea	5.62	8.19	—	13.81
2"	P1@.447	Ea	10.05	10.90	—	20.95
2½"	P1@.550	Ea	20.81	13.40	—	34.21
3"	P1@.654	Ea	26.30	16.00	—	42.30
4"	P1@.860	Ea	42.25	21.00	—	63.25

Type M copper 90 degree ell Ftg. x C with brazed joints

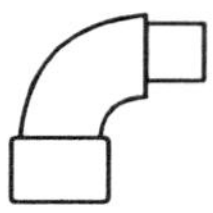

Description	Craft@Hrs	Unit	Material $	Labor $	Equipment $	Total $
1/2"	P1@.129	Ea	.49	3.15	—	3.64
3/4"	P1@.181	Ea	1.09	4.42	—	5.51
1"	P1@.232	Ea	3.25	5.67	—	8.92
1¼"	P1@.284	Ea	4.50	6.94	—	11.44
1½"	P1@.335	Ea	6.75	8.19	—	14.94
2"	P1@.447	Ea	13.50	10.90	—	24.40
2½"	P1@.550	Ea	28.50	13.40	—	41.90
3"	P1@.654	Ea	32.00	16.00	—	48.00
4"	P1@.860	Ea	64.90	21.00	—	85.90

Type M copper tee C x C x C with brazed joints

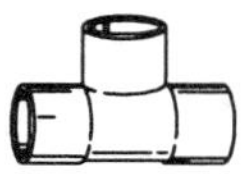

Description	Craft@Hrs	Unit	Material $	Labor $	Equipment $	Total $
1/2"	P1@.156	Ea	.32	3.81	—	4.13
3/4"	P1@.204	Ea	.79	4.99	—	5.78
1"	P1@.263	Ea	5.40	6.43	—	11.83
1¼"	P1@.322	Ea	8.60	7.87	—	16.47
1½"	P1@.380	Ea	12.48	9.29	—	21.77
2"	P1@.507	Ea	20.00	12.40	—	32.40
2½"	P1@.624	Ea	40.58	15.20	—	55.78
3"	P1@.741	Ea	62.11	18.10	—	80.21
4"	P1@.975	Ea	78.00	23.80	—	101.80

Description	Craft@Hrs	Unit	Material $	Labor $	Equipment $	Total $

Type M copper reducing branch tee C x C x C with brazed joints

Description	Craft@Hrs	Unit	Material $	Labor $	Equipment $	Total $
1/2" x 3/8"	P1@.129	Ea	2.28	3.15	—	5.43
3/4" x 1/2"	P1@.182	Ea	.79	4.45	—	5.24
1" x 1/2"	P1@.209	Ea	6.35	5.11	—	11.46
1" x 3/4"	P1@.234	Ea	6.35	5.72	—	12.07
1¼" x 1/2"	P1@.241	Ea	9.75	5.89	—	15.64
1¼" x 3/4"	P1@.264	Ea	9.75	6.45	—	16.20
1¼" x 1"	P1@.287	Ea	9.75	7.01	—	16.76
1½" x 1/2"	P1@.291	Ea	11.68	7.11	—	18.79
1½" x 3/4"	P1@.307	Ea	11.68	7.50	—	19.18
1½" x 1"	P1@.323	Ea	11.68	7.89	—	19.57
1½" x 1¼"	P1@.339	Ea	11.68	8.28	—	19.96
2" x 1/2"	P1@.348	Ea	15.77	8.50	—	24.27
2" x 3/4"	P1@.373	Ea	15.77	9.12	—	24.89
2" x 1"	P1@.399	Ea	15.77	9.75	—	25.52
2" x 1¼"	P1@.426	Ea	15.77	10.40	—	26.17
2" x 1½"	P1@.452	Ea	15.77	11.00	—	26.77
2½" x 1/2"	P1@.468	Ea	47.60	11.40	—	59.00
2½" x 3/4"	P1@.485	Ea	47.60	11.90	—	59.50
2½" x 1"	P1@.502	Ea	47.60	12.30	—	59.90
2½" x 1½"	P1@.537	Ea	47.60	13.10	—	60.70
2½" x 2"	P1@.555	Ea	47.60	13.60	—	61.20
3" x 1¼"	P1@.562	Ea	54.26	13.70	—	67.96
3" x 2"	P1@.626	Ea	56.26	15.30	—	71.56
3" x 2½"	P1@.660	Ea	56.26	16.10	—	72.36
4" x 1¼"	P1@.669	Ea	73.00	16.30	—	89.30
4" x 1½"	P1@.718	Ea	73.00	17.50	—	90.50
4" x 2"	P1@.768	Ea	73.00	18.80	—	91.80
4" x 2½"	P1@.819	Ea	73.00	20.00	—	93.00
4" x 3"	P1@.868	Ea	73.00	21.20	—	94.20

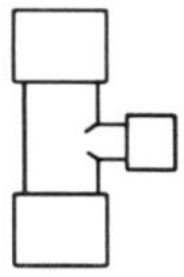

Copper, Type M with Brazed Joints

Type M copper reducer with brazed joints

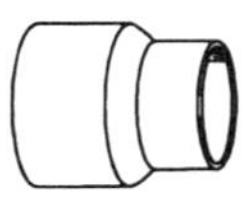

Description	Craft@Hrs	Unit	Material $	Labor $	Equipment $	Total $
1/2" x 3/8"	P1@.148	Ea	.40	3.62	—	4.02
3/4" x 1/2"	P1@.155	Ea	.72	3.79	—	4.51
1" x 3/4"	P1@.206	Ea	1.32	5.03	—	6.35
1¼" x 1/2"	P1@.206	Ea	2.20	5.03	—	7.23
1¼" x 3/4"	P1@.232	Ea	2.20	5.67	—	7.87
1¼" x 1"	P1@.258	Ea	2.20	6.30	—	8.50
1½" x 1/2"	P1@.232	Ea	3.37	5.67	—	9.04
1½" x 3/4"	P1@.257	Ea	3.37	6.28	—	9.65
1½" x 1"	P1@.283	Ea	3.37	6.92	—	10.29
1½" x 1¼"	P1@.308	Ea	3.37	7.53	—	10.90
2" x 1/2"	P1@.287	Ea	5.24	7.01	—	12.25
2" x 3/4"	P1@.277	Ea	5.24	6.77	—	12.01
2" x 1"	P1@.338	Ea	5.24	8.26	—	13.50
2" x 1¼"	P1@.365	Ea	5.24	8.92	—	14.16
2" x 1½"	P1@.390	Ea	5.24	9.53	—	14.77
2½" x 1"	P1@.390	Ea	14.30	9.53	—	23.83
2½" x 1¼"	P1@.416	Ea	14.30	10.20	—	24.50
2½" x 1½"	P1@.442	Ea	14.30	10.80	—	25.10
2½" x 2"	P1@.497	Ea	14.30	12.10	—	26.40
3" x 1¼"	P1@.468	Ea	16.70	11.40	—	28.10
3" x 1½"	P1@.493	Ea	16.70	12.00	—	28.70
3" x 2"	P1@.548	Ea	16.70	13.40	—	30.10
3" x 2½"	P1@.600	Ea	16.70	14.70	—	31.40
4" x 2"	P1@.652	Ea	32.00	15.90	—	47.90
4" x 2½"	P1@.704	Ea	32.00	17.20	—	49.20
4" x 3"	P1@.755	Ea	32.00	18.40	—	50.40

Type M copper adapter C x MPT with brazed joint

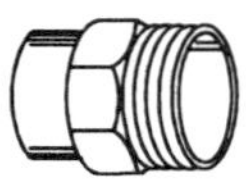

Description	Craft@Hrs	Unit	Material $	Labor $	Equipment $	Total $
1/2"	P1@.090	Ea	.97	2.20	—	3.17
3/4"	P1@.126	Ea	1.50	3.08	—	4.58
1"	P1@.162	Ea	3.69	3.96	—	7.65
1¼"	P1@.198	Ea	6.94	4.84	—	11.78
1½"	P1@.234	Ea	7.77	5.72	—	13.49
2"	P1@.312	Ea	10.01	7.62	—	17.63
2½"	P1@.384	Ea	32.60	9.38	—	41.98
3"	P1@.456	Ea	54.00	11.10	—	65.10

Type M copper adapter C x FPT with brazed joint

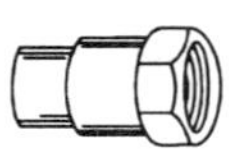

Description	Craft@Hrs	Unit	Material $	Labor $	Equipment $	Total $
1/2"	P1@.090	Ea	1.58	2.20	—	3.78
3/4"	P1@.126	Ea	2.88	3.08	—	5.96
1"	P1@.162	Ea	4.39	3.96	—	8.35
1¼"	P1@.198	Ea	9.14	4.84	—	13.98
1½"	P1@.234	Ea	11.23	5.72	—	16.95
2"	P1@.312	Ea	16.83	7.62	—	24.45
2½"	P1@.384	Ea	42.33	9.38	—	51.71
3"	P1@.456	Ea	59.19	11.10	—	70.29

Description	Craft@Hrs	Unit	Material $	Labor $	Equipment $	Total $
Type M copper flush bushing with brazed joints						
1/2" x 3/8"	P1@.148	Ea	.98	3.62	—	4.60
3/4" x 1/2"	P1@.155	Ea	1.64	3.79	—	5.43
1" x 3/4"	P1@.206	Ea	4.29	5.03	—	9.32
1" x 1/2"	P1@.206	Ea	4.88	5.03	—	9.91
1¼" x 1"	P1@.258	Ea	5.10	6.30	—	11.40
1½" x 1¼"	P1@.308	Ea	6.50	7.53	—	14.03
2" x 1½"	P1@.390	Ea	12.00	9.53	—	21.53
Type M copper union with brazed joints						
1/2"	P1@.146	Ea	3.87	3.57	—	7.44
3/4"	P1@.205	Ea	4.62	5.01	—	9.63
1"	P1@.263	Ea	10.35	6.43	—	16.78
1¼"	P1@.322	Ea	18.82	7.87	—	26.69
1½"	P1@.380	Ea	23.30	9.29	—	32.59
2"	P1@.507	Ea	34.89	12.40	—	47.29
Type M copper dielectric union with brazed joints						
1/2"	P1@.146	Ea	3.63	3.57	—	7.20
3/4"	P1@.205	Ea	3.63	5.01	—	8.64
1"	P1@.263	Ea	7.53	6.43	—	13.96
1¼"	P1@.322	Ea	12.87	7.87	—	20.74
1½"	P1@.380	Ea	17.50	9.29	—	26.79
2"	P1@.507	Ea	24.31	12.40	—	36.71
Type M copper cap with brazed joints						
1/2"	P1@.083	Ea	.31	2.03	—	2.34
3/4"	P1@.116	Ea	.63	2.83	—	3.46
1"	P1@.149	Ea	1.47	3.64	—	5.11
1¼"	P1@.182	Ea	2.09	4.45	—	6.54
1½"	P1@.215	Ea	3.05	5.25	—	8.30
2"	P1@.286	Ea	5.56	6.99	—	12.55
2½"	P1@.352	Ea	14.00	8.60	—	22.60
3"	P1@.418	Ea	21.88	10.20	—	32.08
4"	P1@.550	Ea	33.50	13.40	—	46.90
Type M copper coupling with brazed joints						
1/2"	P1@.129	Ea	.13	3.15	—	3.28
3/4"	P1@.181	Ea	.33	4.42	—	4.75
1"	P1@.232	Ea	1.39	5.67	—	7.06
1¼"	P1@.284	Ea	2.42	6.94	—	9.36
1½"	P1@.335	Ea	3.51	8.19	—	11.70
2"	P1@.447	Ea	5.33	10.90	—	16.23
2½"	P1@.550	Ea	12.42	13.40	—	25.82
3"	P1@.654	Ea	18.98	16.00	—	34.98
4"	P1@.860	Ea	37.47	21.00	—	58.47

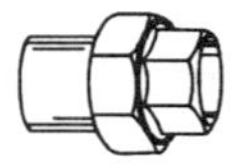
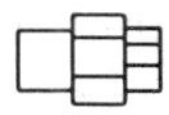

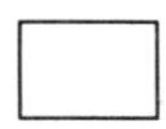

Class 125 bronze body gate valve, brazed joints

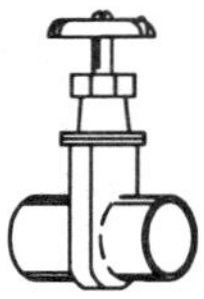

Description	Craft@Hrs	Unit	Material $	Labor $	Equipment $	Total $
1/2"	P1@.240	Ea	6.06	5.86	—	11.92
3/4"	P1@.300	Ea	7.58	7.33	—	14.91
1"	P1@.360	Ea	10.50	8.80	—	19.30
1¼"	P1@.480	Ea	13.40	11.70	—	25.10
1½"	P1@.540	Ea	18.00	13.20	—	31.20
2"	P1@.600	Ea	30.10	14.70	—	44.80
2½"	P1@1.00	Ea	49.90	24.40	—	74.30
3"	P1@1.50	Ea	71.60	36.70	—	108.30

Class 125 iron body gate valve, flanged ends

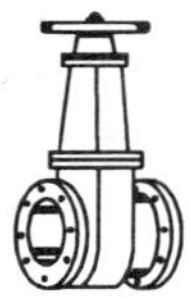

Description	Craft@Hrs	Unit	Material $	Labor $	Equipment $	Total $
2"	P1@.500	Ea	101.00	12.20	—	113.20
2½"	P1@.600	Ea	137.00	14.70	—	151.70
3"	P1@.750	Ea	148.00	18.30	—	166.30
4"	P1@1.35	Ea	218.00	33.00	—	251.00
5"	P1@2.00	Ea	417.00	48.90	—	465.90
6"	P1@2.50	Ea	417.00	61.10	—	478.10

Class 125 bronze body globe valve, brazed joints

Description	Craft@Hrs	Unit	Material $	Labor $	Equipment $	Total $
1/2"	P1@.240	Ea	15.80	5.86	—	21.66
3/4"	P1@.300	Ea	21.30	7.33	—	28.63
1"	P1@.360	Ea	30.50	8.80	—	39.30
1¼"	P1@.480	Ea	42.90	11.70	—	54.60
1½"	P1@.540	Ea	57.50	13.20	—	70.70
2"	P1@.600	Ea	93.60	14.70	—	108.30

Class 125 iron body globe valve, flanged ends

Description	Craft@Hrs	Unit	Material $	Labor $	Equipment $	Total $
2½"	P1@.650	Ea	220.00	15.90	—	235.90
3"	P1@.750	Ea	264.00	18.30	—	282.30
4"	P1@1.35	Ea	355.00	33.00	—	388.00

200 PSIG iron body butterfly valve, lug type

Description	Craft@Hrs	Unit	Material $	Labor $	Equipment $	Total $
2"	P1@.450	Ea	84.50	11.00	—	95.50
2½"	P1@.450	Ea	86.50	11.00	—	97.50
3"	P1@.550	Ea	91.70	13.40	—	105.10
4"	P1@.550	Ea	114.00	13.40	—	127.40

200 PSIG iron body butterfly valve, wafer type

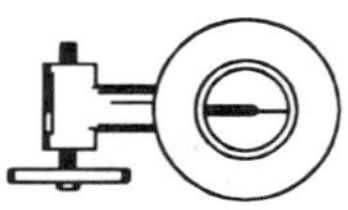

Description	Craft@Hrs	Unit	Material $	Labor $	Equipment $	Total $
2"	P1@.450	Ea	76.70	11.00	—	87.70
2½"	P1@.450	Ea	78.70	11.00	—	89.70
3"	P1@.550	Ea	84.50	13.40	—	97.90
4"	P1@.550	Ea	102.00	13.40	—	115.40

Description	Craft@Hrs	Unit	Material $	Labor $	Equipment $	Total $
Class 125 bronze body 2-piece ball valve, brazed joints						
1/2"	P1@.240	Ea	3.56	5.86	—	9.42
3/4"	P1@.300	Ea	4.75	7.33	—	12.08
1"	P1@.360	Ea	8.21	8.80	—	17.01
1¼"	P1@.480	Ea	13.70	11.70	—	25.40
1½"	P1@.540	Ea	18.50	13.20	—	31.70
2"	P1@.600	Ea	23.30	14.70	—	38.00

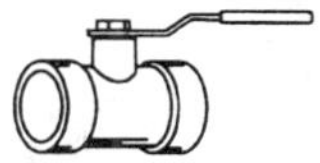

Description	Craft@Hrs	Unit	Material $	Labor $	Equipment $	Total $
Class 125 bronze body swing check valve, brazed joints						
1/2"	P1@.240	Ea	8.00	5.86	—	13.86
3/4"	P1@.300	Ea	11.40	7.33	—	18.73
1"	P1@.360	Ea	15.00	8.80	—	23.80
1¼"	P1@.480	Ea	21.40	11.70	—	33.10
1½"	P1@.540	Ea	30.00	13.20	—	43.20
2"	P1@.600	Ea	50.00	14.70	—	64.70
2½"	P1@1.00	Ea	100.00	24.40	—	124.40
3"	P1@1.50	Ea	144.00	36.70	—	180.70

Description	Craft@Hrs	Unit	Material $	Labor $	Equipment $	Total $
Class 125 iron body swing check valve, flanged ends						
2"	P1@.500	Ea	115.00	12.20	—	127.20
2½"	P1@.600	Ea	148.00	14.70	—	162.70
3"	P1@.750	Ea	183.00	18.30	—	201.30
4"	P1@1.35	Ea	271.00	33.00	—	304.00

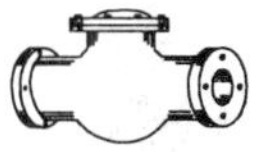

Description	Craft@Hrs	Unit	Material $	Labor $	Equipment $	Total $
Class 125 iron body silent check valve, wafer type						
2"	P1@.500	Ea	85.00	12.20	—	97.20
2½"	P1@.600	Ea	95.00	14.70	—	109.70
3"	P1@.750	Ea	108.00	18.30	—	126.30
4"	P1@1.35	Ea	146.00	33.00	—	179.00

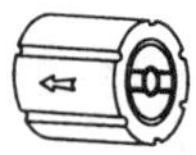

Description	Craft@Hrs	Unit	Material $	Labor $	Equipment $	Total $
Class 125 bronze body strainer, threaded ends						
1/2"	P1@.230	Ea	12.40	5.62	—	18.02
3/4"	P1@.260	Ea	16.30	6.35	—	22.65
1"	P1@.330	Ea	20.20	8.06	—	28.26
1¼"	P1@.440	Ea	28.10	10.80	—	38.90
1½"	P1@.495	Ea	42.20	12.10	—	54.30
2"	P1@.550	Ea	73.00	13.40	—	86.40

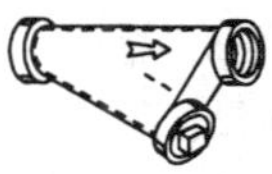

Description	Craft@Hrs	Unit	Material $	Labor $	Equipment $	Total $
Class 125 iron body strainer, flanged ends						
2"	P1@.500	Ea	62.00	12.20	—	74.20
2½"	P1@.600	Ea	69.00	14.70	—	83.70
3"	P1@.750	Ea	80.00	18.30	—	98.30
4"	P1@1.35	Ea	137.00	33.00	—	170.00

Copper, Type M with Brazed Joints

Installation of 2-way control valve, threaded joints

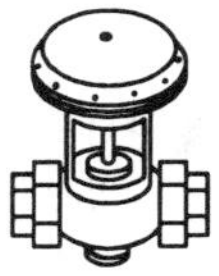

Description	Craft@Hrs	Unit	Material $	Labor $	Equipment $	Total $
1/2"	P1@.210	Ea	—	5.13	—	5.13
3/4"	P1@.275	Ea	—	6.72	—	6.72
1"	P1@.350	Ea	—	8.55	—	8.55
1¼"	P1@.430	Ea	—	10.50	—	10.50
1½"	P1@.505	Ea	—	12.30	—	12.30
2"	P1@.675	Ea	—	16.50	—	16.50
2½"	P1@.830	Ea	—	20.30	—	20.30
3"	P1@.990	Ea	—	24.20	—	24.20

Installation of 3-way control valve, threaded joints

Description	Craft@Hrs	Unit	Material $	Labor $	Equipment $	Total $
1/2"	P1@.260	Ea	—	6.35	—	6.35
3/4"	P1@.365	Ea	—	8.92	—	8.92
1"	P1@.475	Ea	—	11.60	—	11.60
1¼"	P1@.575	Ea	—	14.10	—	14.10
1½"	P1@.680	Ea	—	16.60	—	16.60
2"	P1@.910	Ea	—	22.20	—	22.20
2½"	P1@1.12	Ea	—	27.40	—	27.40
3"	P1@1.33	Ea	—	32.50	—	32.50

Companion flange, 150 pound cast brass

Description	Craft@Hrs	Unit	Material $	Labor $	Equipment $	Total $
2"	P1@.290	Ea	169.00	7.09	—	176.09
2½"	P1@.380	Ea	208.00	9.29	—	217.29
3"	P1@.460	Ea	208.00	11.20	—	219.20
4"	P1@.600	Ea	332.00	14.70	—	346.70

Bolt and gasket sets

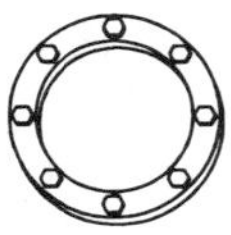

Description	Craft@Hrs	Unit	Material $	Labor $	Equipment $	Total $
2"	P1@.500	Ea	1.50	12.20	—	13.70
2½"	P1@.650	Ea	1.75	15.89	—	17.64
3"	P1@.750	Ea	3.00	18.33	—	21.33
4"	P1@1.00	Ea	5.00	24.44	—	29.44

Thermometer with well

Description	Craft@Hrs	Unit	Material $	Labor $	Equipment $	Total $
7"	P1@.250	Ea	15.00	6.11	—	21.11
9"	P1@.250	Ea	15.50	6.11	—	21.61

Dial type pressure gauge

Description	Craft@Hrs	Unit	Material $	Labor $	Equipment $	Total $
2½"	P1@.200	Ea	15.50	4.89	—	20.39
3½"	P1@.200	Ea	20.50	4.89	—	25.39

Pressure/temperature tap

Description	Craft@Hrs	Unit	Material $	Labor $	Equipment $	Total $
	P1@.150	Ea	7.30	3.67	—	10.97

Description	Craft@Hrs	Unit	Material $	Labor $	Equipment $	Total $

Hanger with swivel assembly

Description	Craft@Hrs	Unit	Material $	Labor $	Equipment $	Total $
1/2"	P1@.450	Ea	1.60	11.00	—	12.60
3/4"	P1@.450	Ea	1.75	11.00	—	12.75
1"	P1@.450	Ea	1.85	11.00	—	12.85
1¼"	P1@.450	Ea	1.90	11.00	—	12.90
1½"	P1@.450	Ea	2.05	11.00	—	13.05
2"	P1@.450	Ea	2.15	11.00	—	13.15
2½"	P1@.450	Ea	2.90	11.00	—	13.90
3"	P1@.450	Ea	3.60	11.00	—	14.60
4"	P1@.450	Ea	3.95	11.00	—	14.95

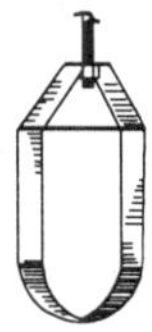

Riser clamp

Description	Craft@Hrs	Unit	Material $	Labor $	Equipment $	Total $
1/2"	P1@.100	Ea	.95	2.44	—	3.39
3/4"	P1@.100	Ea	1.39	2.44	—	3.83
1"	P1@.100	Ea	1.41	2.44	—	3.85
1¼"	P1@.105	Ea	1.70	2.57	—	4.27
1½"	P1@.110	Ea	1.79	2.69	—	4.48
2"	P1@.115	Ea	1.89	2.81	—	4.70
2½"	P1@.120	Ea	2.00	2.93	—	4.93
3"	P1@.120	Ea	2.18	2.93	—	5.11
4"	P1@.125	Ea	2.75	3.05	—	5.80

Copper, Type M with Soft-Soldered Joints

Type M hard-drawn copper pipe with wrought copper fittings and soft-soldered joints is used in a wide variety of plumbing and HVAC systems such as potable water, heating hot water, chilled water, and A.C. condensate.

Soft-soldered joints are those made with solders having melting points in the 350 degree F. to 500 degree F. range. Maximum working pressure/temperature relationships for soft-soldered joints are approximately as follows:

Maximum Working Pressures (PSIG)*				
Soft-solder Type	Water Temperature (degrees F.)	Nominal Pipe Size (inches)		
		Up to 1	1¼ to 2	2½ x 4
50-50 tin-lead**	100	200	175	150
	150	150	125	100
	200	100	90	75
	250	85	75	50
95-5 tin-antimony	100	500	400	300
	150	400	350	275
	200	300	250	200
	250	200	175	150

**For copper pipe and solder-type fittings using soft-solders melting at approximately 350 degrees F. to 500 degrees F.*

***The use of any solder containing lead is not allowed in potable water systems.*

This section has been arranged to save the estimator's time by including all normally-used system components such as pipe, fittings, valves, hanger assemblies, riser clamps and miscellaneous items under one heading. Additional items can be found under "Plumbing and Piping Specialties." The cost estimates in this section are based on the conditions, limitations and wage rates described in the section "How to Use This Book" beginning on page 5.

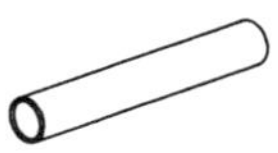

Description	Craft@Hrs	Unit	Material $	Labor $	Equipment $	Total $
Type M copper pipe with soft-soldered joints						
1/2"	P1@.032	LF	.44	.78	—	1.22
3/4"	P1@.035	LF	.71	.86	—	1.57
1"	P1@.038	LF	1.13	.93	—	2.06
1¼"	P1@.042	LF	1.68	1.03	—	2.71
1½"	P1@.046	LF	2.37	1.12	—	3.49
2"	P1@.053	LF	4.31	1.30	—	5.61
2½"	P1@.060	LF	5.95	1.47	—	7.42
3"	P1@.066	LF	7.56	1.61	—	9.17
4"	P1@.080	LF	12.05	1.95	—	14.00

Type M copper 45 degree ell C x C with soft-soldered joints

Description	Craft@Hrs	Unit	Material $	Labor $	Equipment $	Total $
1/2"	P1@.107	Ea	.36	2.61	—	2.97
3/4"	P1@.150	Ea	.66	3.67	—	4.33
1"	P1@.193	Ea	2.50	4.72	—	7.22
1¼"	P1@.236	Ea	3.40	5.77	—	9.17
1½"	P1@.278	Ea	6.02	6.79	—	12.81
2"	P1@.371	Ea	9.85	9.07	—	18.92
2½"	P1@.457	Ea	25.44	11.20	—	36.64
3"	P1@.543	Ea	33.45	13.30	—	46.75
4"	P1@.714	Ea	48.75	17.40	—	66.15

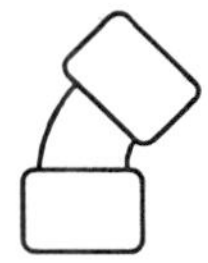

Type M copper 90 degree ell C x C with soft-soldered joints

Description	Craft@Hrs	Unit	Material $	Labor $	Equipment $	Total $
1/2"	P1@.107	Ea	.18	2.61	—	2.79
3/4"	P1@.150	Ea	.26	3.67	—	3.93
1"	P1@.193	Ea	2.28	4.72	—	7.00
1¼"	P1@.236	Ea	3.71	5.77	—	9.48
1½"	P1@.278	Ea	5.62	6.79	—	12.41
2"	P1@.371	Ea	10.05	9.07	—	19.12
2½"	P1@.457	Ea	20.81	11.20	—	32.01
3"	P1@.543	Ea	26.30	13.30	—	39.60
4"	P1@.714	Ea	42.25	17.40	—	59.65

Type M copper 90 degree ell Ftg. x C with soft-soldered joint

Description	Craft@Hrs	Unit	Material $	Labor $	Equipment $	Total $
1/2"	P1@.107	Ea	.49	2.61	—	3.10
3/4"	P1@.150	Ea	1.09	3.67	—	4.76
1"	P1@.193	Ea	3.25	4.72	—	7.97
1¼"	P1@.236	Ea	4.50	5.77	—	10.27
1½"	P1@.278	Ea	6.75	6.79	—	13.54
2"	P1@.371	Ea	13.50	9.07	—	22.57
2½"	P1@.457	Ea	28.50	11.20	—	39.70
3"	P1@.543	Ea	32.00	13.30	—	45.30
4"	P1@.714	Ea	64.90	17.40	—	82.30

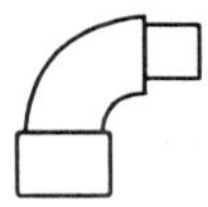

Type M copper tee C x C x C with soft-soldered joints

Description	Craft@Hrs	Unit	Material $	Labor $	Equipment $	Total $
1/2"	P1@.129	Ea	.32	3.15	—	3.47
3/4"	P1@.181	Ea	.79	4.42	—	5.21
1"	P1@.233	Ea	5.40	5.69	—	11.09
1¼"	P1@.285	Ea	8.60	6.96	—	15.56
1½"	P1@.337	Ea	12.48	8.24	—	20.72
2"	P1@.449	Ea	20.00	11.00	—	31.00
2½"	P1@.552	Ea	40.58	13.50	—	54.08
3"	P1@.656	Ea	62.11	16.00	—	78.11
4"	P1@.863	Ea	78.00	21.10	—	99.10

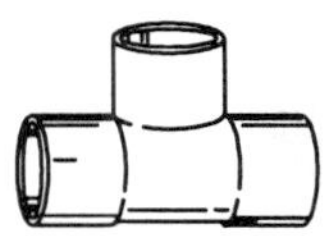

Description	Craft@Hrs	Unit	Material $	Labor $	Equipment $	Total $

Type M copper branch reducing tee C x C x C with soft-soldered joints

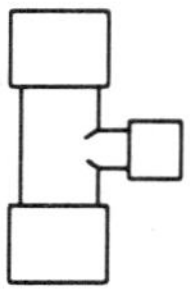

Description	Craft@Hrs	Unit	Material $	Labor $	Equipment $	Total $
1/2" x 3/8"	P1@.121	Ea	2.28	2.96	—	5.24
3/4" x 1/2"	P1@.170	Ea	.79	4.15	—	4.94
1" x 1/2"	P1@.195	Ea	6.35	4.77	—	11.12
1" x 3/4"	P1@.219	Ea	6.35	5.35	—	11.70
1¼" x 1/2"	P1@.225	Ea	9.75	5.50	—	15.25
1¼" x 3/4"	P1@.247	Ea	9.75	6.04	—	15.79
1¼" x 1"	P1@.268	Ea	9.75	6.55	—	16.30
1½" x 1/2"	P1@.272	Ea	11.68	6.65	—	18.33
1½" x 3/4"	P1@.287	Ea	11.68	7.01	—	18.69
1½" x 1"	P1@.302	Ea	11.68	7.38	—	19.06
1½" x 1¼"	P1@.317	Ea	11.68	7.75	—	19.43
2" x 1/2"	P1@.325	Ea	15.77	7.94	—	23.71
2" x 3/4"	P1@.348	Ea	15.77	8.50	—	24.27
2" x 1"	P1@.373	Ea	15.77	9.12	—	24.89
2" x 1¼"	P1@.398	Ea	15.77	9.73	—	25.50
2" x 1½"	P1@.422	Ea	15.77	10.30	—	26.07
2½" x 1/2"	P1@.437	Ea	47.60	10.70	—	58.30
2½" x 3/4"	P1@.453	Ea	47.60	11.10	—	58.70
2½" x 1"	P1@.469	Ea	47.60	11.50	—	59.10
2½" x 1½"	P1@.502	Ea	47.60	12.30	—	59.90
2½" x 2"	P1@.519	Ea	47.60	12.70	—	60.30
3" x 1¼"	P1@.525	Ea	54.26	12.80	—	67.06
3" x 2"	P1@.585	Ea	56.26	14.30	—	70.56
3" x 2½"	P1@.617	Ea	56.26	15.10	—	71.36
4" x 1¼"	P1@.625	Ea	73.00	15.30	—	88.30
4" x 1½"	P1@.671	Ea	73.00	16.40	—	89.40
4" x 2"	P1@.718	Ea	73.00	17.50	—	90.50
4" x 2½"	P1@.765	Ea	73.00	18.70	—	91.70
4" x 3"	P1@.811	Ea	73.00	19.80	—	92.80

Type M copper reducer with soft-soldered joints

Description	Craft@Hrs	Unit	Material $	Labor $	Equipment $	Total $
1/2" x 3/8"	P1@.124	Ea	.40	3.03	—	3.43
3/4" x 1/2"	P1@.129	Ea	.72	3.15	—	3.87
1" x 3/4"	P1@.172	Ea	1.32	4.20	—	5.52
1¼" x 1/2"	P1@.172	Ea	2.20	4.20	—	6.40
1¼" x 3/4"	P1@.193	Ea	2.20	4.72	—	6.92
1¼" x 1"	P1@.215	Ea	2.20	5.25	—	7.45
1½" x 1/2"	P1@.193	Ea	3.37	4.72	—	8.09
1½" x 3/4"	P1@.214	Ea	3.37	5.23	—	8.60
1½" x 1"	P1@.236	Ea	3.37	5.77	—	9.14
1½" x 1¼"	P1@.257	Ea	3.37	6.28	—	9.65
2" x 1/2"	P1@.239	Ea	5.24	5.84	—	11.08
2" x 3/4"	P1@.261	Ea	5.24	6.38	—	11.62
2" x 1"	P1@.282	Ea	5.24	6.89	—	12.13
2" x 1¼"	P1@.304	Ea	5.24	7.43	—	12.67
2" x 1½"	P1@.325	Ea	5.24	7.94	—	13.18
2½" x 1"	P1@.325	Ea	14.30	7.94	—	22.24
2½" x 1¼"	P1@.347	Ea	14.30	8.48	—	22.78
2½" x 1½"	P1@.368	Ea	14.30	8.99	—	23.29
2½" x 2"	P1@.414	Ea	14.30	10.10	—	24.40
3" x 2"	P1@.457	Ea	16.70	11.20	—	27.90
3" x 2½"	P1@.500	Ea	16.70	12.20	—	28.90
4" x 2"	P1@.543	Ea	32.00	13.30	—	45.30
4" x 2½"	P1@.586	Ea	32.00	14.30	—	46.30
4" x 3"	P1@.629	Ea	32.00	15.40	—	47.40

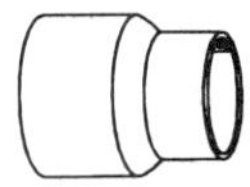

Type M copper adapter C x MPT with soft-soldered joint

Description	Craft@Hrs	Unit	Material $	Labor $	Equipment $	Total $
1/2"	P1@.075	Ea	.97	1.83	—	2.80
3/4"	P1@.105	Ea	1.50	2.57	—	4.07
1"	P1@.134	Ea	3.69	3.27	—	6.96
1¼"	P1@.164	Ea	6.94	4.01	—	10.95
1½"	P1@.194	Ea	7.77	4.74	—	12.51
2"	P1@.259	Ea	10.01	6.33	—	16.34
2½"	P1@.319	Ea	32.60	7.80	—	40.40
3"	P1@.378	Ea	54.00	9.24	—	63.24

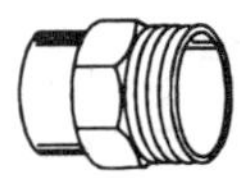

Type M copper adapter C x FPT with soft-soldered joint

Description	Craft@Hrs	Unit	Material $	Labor $	Equipment $	Total $
1/2"	P1@.075	Ea	1.58	1.83	—	3.41
3/4"	P1@.105	Ea	2.88	2.57	—	5.45
1"	P1@.134	Ea	4.39	3.27	—	7.66
1¼"	P1@.164	Ea	9.14	4.01	—	13.15
1½"	P1@.194	Ea	11.23	4.74	—	15.97
2"	P1@.259	Ea	16.83	6.33	—	23.16
2½"	P1@.319	Ea	42.33	7.80	—	50.13
3"	P1@.378	Ea	59.19	9.24	—	68.43

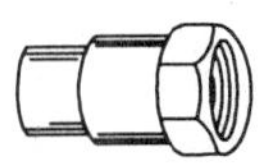

Copper, Type M with Soft-Soldered Joints

Type M copper flush bushing with soft-soldered joints

Description	Craft@Hrs	Unit	Material $	Labor $	Equipment $	Total $
1/2" x 3/8"	P1@.124	Ea	.98	3.03	—	4.01
3/4" x 1/2"	P1@.129	Ea	1.64	3.15	—	4.79
1" x 3/4"	P1@.172	Ea	4.29	4.20	—	8.49
1" x 1/2"	P1@.172	Ea	4.88	4.20	—	9.08
1¼" x 1"	P1@.215	Ea	5.10	5.25	—	10.35
1½" x 1¼"	P1@.257	Ea	6.50	6.28	—	12.78
2" x 1½"	P1@.325	Ea	12.00	7.94	—	19.94

Type M copper union with soft-soldered joint

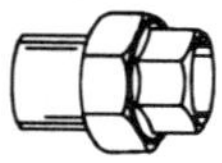

Description	Craft@Hrs	Unit	Material $	Labor $	Equipment $	Total $
1/2"	P1@.121	Ea	3.87	2.96	—	6.83
3/4"	P1@.170	Ea	4.62	4.15	—	8.77
1"	P1@.218	Ea	10.35	5.33	—	15.68
1¼"	P1@.267	Ea	18.82	6.52	—	25.34
1½"	P1@.315	Ea	23.30	7.70	—	31.00
2"	P1@.421	Ea	34.89	10.30	—	45.19

Type M copper dielectric union with soft-soldered joint

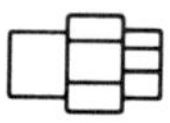

Description	Craft@Hrs	Unit	Material $	Labor $	Equipment $	Total $
1/2"	P1@.121	Ea	3.63	2.96	—	6.59
3/4"	P1@.170	Ea	3.63	4.15	—	7.78
1"	P1@.218	Ea	7.53	5.33	—	12.86
1¼"	P1@.267	Ea	12.87	6.52	—	19.39
1½"	P1@.315	Ea	17.50	7.70	—	25.20
2"	P1@.421	Ea	24.31	10.30	—	34.61

Type M copper cap with soft-soldered joint

Description	Craft@Hrs	Unit	Material $	Labor $	Equipment $	Total $
1/2"	P1@.069	Ea	.31	1.69	—	2.00
3/4"	P1@.096	Ea	.63	2.35	—	2.98
1"	P1@.124	Ea	1.47	3.03	—	4.50
1¼"	P1@.151	Ea	2.09	3.69	—	5.78
1½"	P1@.178	Ea	3.05	4.35	—	7.40
2"	P1@.237	Ea	5.56	5.79	—	11.35
2½"	P1@.292	Ea	14.00	7.14	—	21.14
3"	P1@.347	Ea	21.88	8.48	—	30.36
4"	P1@.457	Ea	33.50	11.20	—	44.70

Type M copper coupling with soft-soldered joints

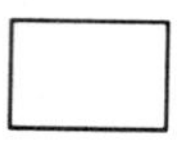

Description	Craft@Hrs	Unit	Material $	Labor $	Equipment $	Total $
1/2"	P1@.107	Ea	.13	2.61	—	2.74
3/4"	P1@.150	Ea	.33	3.67	—	4.00
1"	P1@.193	Ea	1.39	4.72	—	6.11
1¼"	P1@.236	Ea	2.42	5.77	—	8.19
1½"	P1@.278	Ea	3.51	6.79	—	10.30
2"	P1@.371	Ea	5.33	9.07	—	14.40
2½"	P1@.457	Ea	12.42	11.20	—	23.62
3"	P1@.543	Ea	18.98	13.30	—	32.28
4"	P1@.714	Ea	37.47	17.40	—	54.87

Description	Craft@Hrs	Unit	Material $	Labor $	Equipment $	Total $
Class 125 bronze body gate valve with soft-soldered joints						
1/2"	P1@.200	Ea	6.06	4.89	—	10.95
3/4"	P1@.249	Ea	7.58	6.08	—	13.66
1"	P1@.299	Ea	10.50	7.31	—	17.81
1¼"	P1@.398	Ea	13.40	9.73	—	23.13
1½"	P1@.448	Ea	18.00	10.90	—	28.90
2"	P1@.498	Ea	30.10	12.20	—	42.30
2½"	P1@.830	Ea	49.90	20.30	—	70.20
3"	P1@1.24	Ea	71.60	30.30	—	101.90
Class 125 iron body gate valve with flanged ends						
2"	P1@.500	Ea	101.00	12.20	—	113.20
2½"	P1@.600	Ea	137.00	14.70	—	151.70
3"	P1@.750	Ea	148.00	18.30	—	166.30
4"	P1@1.35	Ea	218.00	33.00	—	251.00
5"	P1@2.00	Ea	417.00	48.90	—	465.90
6"	P1@2.50	Ea	417.00	61.10	—	478.10
Class 125 bronze body globe valve with soft-soldered joints						
1/2"	P1@.200	Ea	15.80	4.89	—	20.69
3/4"	P1@.249	Ea	21.30	6.08	—	27.38
1"	P1@.299	Ea	30.50	7.31	—	37.81
1¼"	P1@.398	Ea	42.90	9.73	—	52.63
1½"	P1@.448	Ea	57.50	10.90	—	68.40
2"	P1@.498	Ea	93.60	12.20	—	105.80
Class 125 iron body globe valve with flanged ends						
2½"	P1@.600	Ea	220.00	14.70	—	234.70
3"	P1@.750	Ea	264.00	18.30	—	282.30
4"	P1@1.35	Ea	352.00	33.00	—	385.00
200 PSIG iron body butterfly valve, lug type						
2"	P1@.450	Ea	84.50	11.00	—	95.50
2½"	P1@.450	Ea	86.50	11.00	—	97.50
3"	P1@.550	Ea	91.70	13.40	—	105.10
4"	P1@.550	Ea	114.00	13.40	—	127.40
200 PSIG iron body butterfly valve, wafer type						
2"	P1@.450	Ea	76.70	11.00	—	87.70
2½"	P1@.450	Ea	78.70	11.00	—	89.70
3"	P1@.550	Ea	84.50	13.40	—	97.90
4"	P1@.550	Ea	102.00	13.40	—	115.40

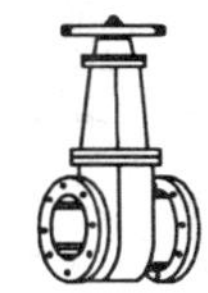
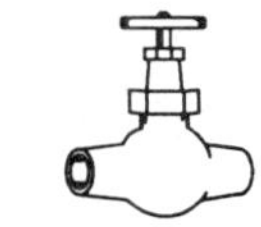

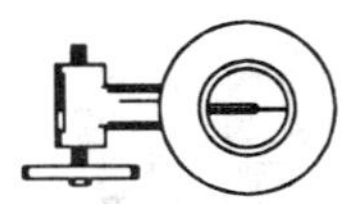

Copper, Type M with Soft-Soldered Joints

Class 125 bronze body 2-piece ball valve with soft-soldered joints

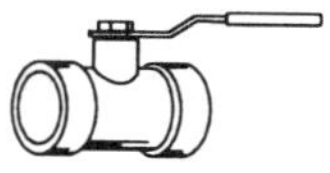

Description	Craft@Hrs	Unit	Material $	Labor $	Equipment $	Total $
1/2"	P1@.200	Ea	3.56	4.89	—	8.45
3/4"	P1@.249	Ea	4.75	6.08	—	10.83
1"	P1@.299	Ea	8.21	7.31	—	15.52
1¼"	P1@.398	Ea	13.70	9.73	—	23.43
1½"	P1@.448	Ea	18.50	10.90	—	29.40
2"	P1@.498	Ea	23.30	12.20	—	35.50

Class 125 bronze body swing check valve with soft-soldered joints

Description	Craft@Hrs	Unit	Material $	Labor $	Equipment $	Total $
1/2"	P1@.200	Ea	8.00	4.89	—	12.89
3/4"	P1@.249	Ea	11.40	6.08	—	17.48
1"	P1@.299	Ea	15.00	7.31	—	22.31
1¼"	P1@.398	Ea	21.40	9.73	—	31.13
1½"	P1@.448	Ea	30.00	10.90	—	40.90
2"	P1@.498	Ea	50.00	12.20	—	62.20
2½"	P1@.830	Ea	100.00	20.30	—	120.30
3"	P1@1.24	Ea	144.00	30.30	—	174.30

Class 125 iron body swing check valve, flanged ends

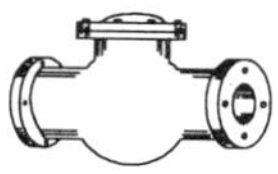

Description	Craft@Hrs	Unit	Material $	Labor $	Equipment $	Total $
2"	P1@.500	Ea	115.00	12.20	—	127.20
2½"	P1@.600	Ea	148.00	14.70	—	162.70
3"	P1@.750	Ea	183.00	18.30	—	201.30
4"	P1@1.35	Ea	271.00	33.00	—	304.00

Class 125 iron body silent check valve, wafer type

Description	Craft@Hrs	Unit	Material $	Labor $	Equipment $	Total $
2"	P1@.500	Ea	85.00	12.20	—	97.20
2½"	P1@.600	Ea	95.00	14.70	—	109.70
3"	P1@.750	Ea	108.00	18.30	—	126.30
4"	P1@1.35	Ea	146.00	33.00	—	179.00

Class 125 bronze body strainer, threaded ends

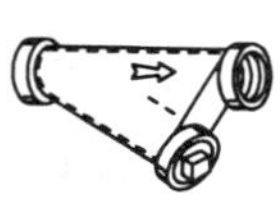

Description	Craft@Hrs	Unit	Material $	Labor $	Equipment $	Total $
1/2"	P1@.230	Ea	12.40	5.62	—	18.02
3/4"	P1@.260	Ea	16.30	6.35	—	22.65
1"	P1@.330	Ea	20.20	8.06	—	28.26
1¼"	P1@.440	Ea	28.10	10.80	—	38.90
1½"	P1@.495	Ea	42.20	12.10	—	54.30
2"	P1@.550	Ea	73.00	13.40	—	86.40

Class 125 iron body strainer, flanged ends

Description	Craft@Hrs	Unit	Material $	Labor $	Equipment $	Total $
2"	P1@.500	Ea	62.00	12.20	—	74.20
2½"	P1@.600	Ea	69.00	14.70	—	83.70
3"	P1@.750	Ea	80.00	18.30	—	98.30
4"	P1@1.35	Ea	137.00	33.00	—	170.00

Installation of 2-way control valve, threaded joints

Description	Craft@Hrs	Unit	Material $	Labor $	Equipment $	Total $
1/2"	P1@.210	Ea	—	5.13	—	5.13
3/4"	P1@.275	Ea	—	6.72	—	6.72
1"	P1@.350	Ea	—	8.55	—	8.55
1¼"	P1@.430	Ea	—	10.50	—	10.50
1½"	P1@.505	Ea	—	12.30	—	12.30
2"	P1@.675	Ea	—	16.50	—	16.50
2½"	P1@.830	Ea	—	20.30	—	20.30
3"	P1@.990	Ea	—	24.20	—	24.20

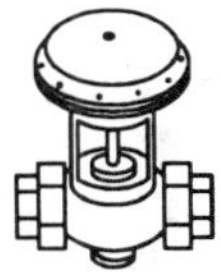

Installation of 3-way control valve, threaded joints

Description	Craft@Hrs	Unit	Material $	Labor $	Equipment $	Total $
1/2"	P1@.260	Ea	—	6.35	—	6.35
3/4"	P1@.365	Ea	—	8.92	—	8.92
1"	P1@.475	Ea	—	11.60	—	11.60
1¼"	P1@.575	Ea	—	14.10	—	14.10
1½"	P1@.680	Ea	—	16.60	—	16.60
2"	P1@.910	Ea	—	22.20	—	22.20
2½"	P1@1.12	Ea	—	27.40	—	27.40
3"	P1@1.33	Ea	—	32.50	—	32.50

Companion flange, 150 pound cast brass

Description	Craft@Hrs	Unit	Material $	Labor $	Equipment $	Total $
2"	P1@.290	Ea	169.00	7.09	—	176.09
2½"	P1@.380	Ea	208.00	9.29	—	217.29
3"	P1@.460	Ea	208.00	11.20	—	219.20
4"	P1@.600	Ea	332.00	14.70	—	346.70

Bolt and gasket sets

Description	Craft@Hrs	Unit	Material $	Labor $	Equipment $	Total $
2"	P1@.500	Ea	1.50	12.20	—	13.70
2½"	P1@.650	Ea	1.75	15.80	—	17.55
3"	P1@.750	Ea	3.00	18.20	—	21.20
4"	P1@1.00	Ea	5.00	24.30	—	29.30

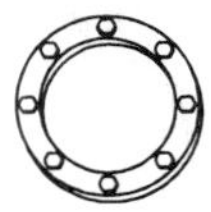

Thermometer with well

Description	Craft@Hrs	Unit	Material $	Labor $	Equipment $	Total $
7"	P1@.250	Ea	15.00	6.11	—	21.11
9"	P1@.250	Ea	15.50	6.11	—	21.61

Dial type pressure gauge

Description	Craft@Hrs	Unit	Material $	Labor $	Equipment $	Total $
2½"	P1@.200	Ea	15.50	4.89	—	20.39
3½"	P1@.200	Ea	20.50	4.89	—	25.39

Pressure/temperature tap

Description	Craft@Hrs	Unit	Material $	Labor $	Equipment $	Total $
	P1@.150	Ea	7.30	3.67	—	10.97

Copper, Type M with Soft-Soldered Joints

Description	Craft@Hrs	Unit	Material $	Labor $	Equipment $	Total $
Hanger with swivel assembly						
1/2"	P1@.450	Ea	1.60	11.00	—	12.60
3/4"	P1@.450	Ea	1.75	11.00	—	12.75
1"	P1@.450	Ea	1.85	11.00	—	12.85
1¼"	P1@.450	Ea	1.90	11.00	—	12.90
1½"	P1@.450	Ea	2.05	11.00	—	13.05
2"	P1@.450	Ea	2.15	11.00	—	13.15
2½"	P1@.450	Ea	2.90	11.00	—	13.90
3"	P1@.450	Ea	3.60	11.00	—	14.60
4"	P1@.450	Ea	3.95	11.00	—	14.95
Riser clamp						
1/2"	P1@.100	Ea	0.95	2.44	—	3.39
3/4"	P1@.100	Ea	1.39	2.44	—	3.83
1"	P1@.100	Ea	1.41	2.44	—	3.85
1¼"	P1@.105	Ea	1.70	2.57	—	4.27
1½"	P1@.110	Ea	1.79	2.69	—	4.48
2"	P1@.115	Ea	1.89	2.81	—	4.70
2½"	P1@.120	Ea	2.00	2.93	—	4.93
3"	P1@.120	Ea	2.18	2.93	—	5.11
4"	P1@.125	Ea	2.75	3.05	—	5.80

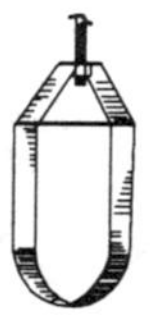

PVC (Polyvinyl Chloride) Schedule 40 pipe with Schedule 40 socket-type fittings and solvent-welded joints is widely used in process, chemical, A.C. condensate, potable water and irrigation piping systems. PVC is available in Type I (normal impact) and Type II (high impact). Type I has a maximum temperature rating of 150 degrees F., and Type II is rated at 140 degrees F.

Consult the manufacturers for maximum pressure ratings and recommended joint solvents for specific applications.

Because of the current unreliability of many plastic valves, most engineers specify standard bronze-body and iron-body valves for use in PVC water piping systems. Plastic valves, however, have to be used in systems conveying liquids that may be injurious to metallic valves.

This section has been arranged to save the estimator's time by including all normally-used system components such as pipe, fittings, valves, hanger assemblies, riser clamps and miscellaneous items under one heading. Additional items can be found under "Plumbing and Piping Specialties." The cost estimates in this section are based on the conditions, limitations and wage rates described in the section "How to Use This Book" beginning on page 5.

Equipment cost, where shown, is $35.00 per hour for a 10 ton hydraulic truck-mounted crane.

Schedule 40 PVC pipe with solvent-weld joints

Description	Craft@Hrs	Unit	Material $	Labor $	Equipment $	Total $
1/2"	P1@.020	LF	.17	.49	—	.66
3/4"	P1@.025	LF	.24	.61	—	.85
1"	P1@.030	LF	.32	.73	—	1.05
1¼"	P1@.035	LF	.39	.86	—	1.25
1½"	P1@.040	LF	.46	.98	—	1.44
2"	P1@.045	LF	.62	1.10	—	1.72
2½"	P1@.050	LF	.98	1.22	—	2.20
3"	P1@.055	LF	1.24	1.34	—	2.58
4"	P1@.070	LF	1.80	1.71	—	3.51
5"	P1@.095	LF	3.10	2.32	—	5.42
6"	P1@.120	LF	3.10	2.93	—	6.03
8"	P1@.160	LF	4.83	3.91	—	8.74

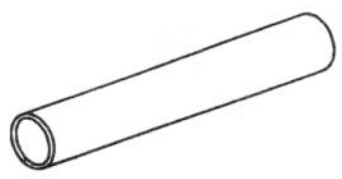

Schedule 40 PVC 45 degree ell S x S with solvent-weld joints

Description	Craft@Hrs	Unit	Material $	Labor $	Equipment $	Total $
1/2"	P1@.100	Ea	.36	2.44	—	2.80
3/4"	P1@.115	Ea	.55	2.81	—	3.36
1"	P1@.120	Ea	.66	2.93	—	3.59
1¼"	P1@.160	Ea	.94	3.91	—	4.85
1½"	P1@.180	Ea	1.17	4.40	—	5.57
2"	P1@.200	Ea	1.58	4.89	—	6.47
2½"	P1@.250	Ea	3.97	6.11	—	10.08
3"	P1@.300	Ea	6.16	7.33	—	13.49
4"	P1@.500	Ea	11.05	12.20	—	23.25
5"	P1@.550	Ea	21.92	13.40	—	35.32
6"	P1@.600	Ea	27.30	14.70	—	42.00
8"	P1@.800	Ea	65.50	19.50	—	85.00

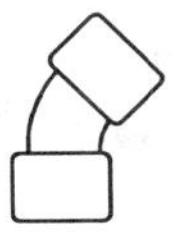

PVC, Schedule 40, with Solvent-Weld Joints

Description	Craft@Hrs	Unit	Material $	Labor $	Equipment $	Total $
Schedule 40 PVC 90 degree ell S x S with solvent-weld joints						
1/2"	P1@.100	Ea	.22	2.44	—	2.66
3/4"	P1@.115	Ea	.24	2.81	—	3.05
1"	P1@.120	Ea	.44	2.93	—	3.37
1¼"	P1@.160	Ea	.77	3.91	—	4.68
1½"	P1@.180	Ea	.83	4.40	—	5.23
2"	P1@.200	Ea	1.30	4.89	—	6.19
2½"	P1@.250	Ea	3.96	6.11	—	10.07
3"	P1@.300	Ea	4.73	7.33	—	12.06
4"	P1@.500	Ea	8.48	12.20	—	20.68
5"	P1@.550	Ea	21.92	13.40	—	35.32
6"	P1@.600	Ea	26.98	14.70	—	41.68
8"	P1@.800	Ea	69.42	19.50	—	88.92

Description	Craft@Hrs	Unit	Material $	Labor $	Equipment $	Total $
Schedule 40 PVC tee S x S x S with solvent-weld joints						
1/2"	P1@.130	Ea	.28	3.18	—	3.46
3/4"	P1@.140	Ea	.31	3.42	—	3.73
1"	P1@.170	Ea	.58	4.15	—	4.73
1¼"	P1@.220	Ea	.91	5.38	—	6.29
1½"	P1@.250	Ea	1.01	6.11	—	7.12
2"	P1@.280	Ea	1.30	6.84	—	8.14
2½"	P1@.350	Ea	5.30	8.55	—	13.85
3"	P1@.420	Ea	6.89	10.30	—	17.19
4"	P1@.560	Ea	12.58	13.70	—	26.28
5"	P1@.710	Ea	30.40	17.40	—	47.80
6"	P1@.840	Ea	44.22	20.50	—	64.72
8"	P1@1.12	Ea	98.28	27.40	—	125.68

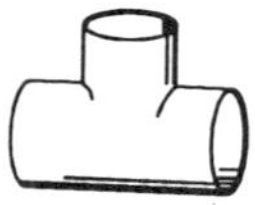

Description	Craft@Hrs	Unit	Material $	Labor $	Equipment $	Total $

Schedule 40 PVC reducing tee S x S x S with solvent-weld joints

Description	Craft@Hrs	Unit	Material $	Labor $	Equipment $	Total $
3/4" x 1/2"	P1@.140	Ea	.36	3.42	—	3.78
1" x 1/2"	P1@.170	Ea	.66	4.15	—	4.81
1" x 3/4"	P1@.170	Ea	1.00	4.15	—	5.15
1¼" x 1/2"	P1@.220	Ea	1.00	5.38	—	6.38
1¼" x 3/4"	P1@.220	Ea	1.00	5.38	—	6.38
1¼" x 1"	P1@.220	Ea	1.00	5.38	—	6.38
1½" x 1/2"	P1@.250	Ea	1.71	6.11	—	7.82
1½" x 3/4"	P1@.250	Ea	1.71	6.11	—	7.82
1½" x 1"	P1@.250	Ea	1.71	6.11	—	7.82
1½" x 1¼"	P1@.250	Ea	1.71	6.11	—	7.82
2" x 1/2"	P1@.280	Ea	2.45	6.84	—	9.29
2" x 3/4"	P1@.280	Ea	2.45	6.84	—	9.29
2" x 1"	P1@.280	Ea	2.45	6.84	—	9.29
2" x 1¼"	P1@.280	Ea	2.45	6.84	—	9.29
2" x 1½"	P1@.280	Ea	2.45	6.84	—	9.29
2½" x 1"	P1@.350	Ea	5.05	8.55	—	13.60
2½" x 1¼"	P1@.350	Ea	5.05	8.55	—	13.60
2½" x 1½"	P1@.350	Ea	5.05	8.55	—	13.60
2½" x 2"	P1@.350	Ea	5.05	8.55	—	13.60
3" x 2"	P1@.420	Ea	7.56	10.30	—	17.86
3" x 2½"	P1@.420	Ea	7.56	10.30	—	17.86
4" x 2"	P1@.560	Ea	12.58	13.70	—	26.28
4" x 2½"	P1@.560	Ea	12.58	13.70	—	26.28
4" x 3"	P1@.560	Ea	12.58	13.70	—	26.28
5" x 3"	P1@.710	Ea	22.65	17.40	—	40.05
5" x 4"	P1@.710	Ea	22.65	17.40	—	40.05
6" x 3"	P1@.840	Ea	36.80	20.50	—	57.30
6" x 4"	P1@.840	Ea	36.80	20.50	—	57.30
6" x 5"	P1@.840	Ea	36.80	20.50	—	57.30
8" x 4"	P1@1.12	Ea	82.00	27.40	—	109.40
8" x 5"	P1@1.12	Ea	82.00	27.40	—	109.40
8" x 6"	P1@1.12	Ea	82.00	27.40	—	109.40

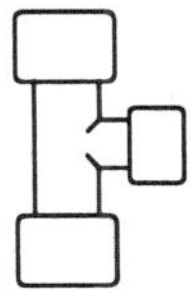

PVC, Schedule 40, with Solvent-Weld Joints

Schedule 40 PVC reducing bushing SPIG x S with solvent-weld joints

Description	Craft@Hrs	Unit	Material $	Labor $	Equipment $	Total $
3/4" x 1/2"	P1@.050	Ea	.23	1.22	—	1.45
1" x 1/2"	P1@.060	Ea	.42	1.47	—	1.89
1" x 3/4"	P1@.060	Ea	.42	1.47	—	1.89
1¼" x 1/2"	P1@.080	Ea	.55	1.95	—	2.50
1¼" x 3/4"	P1@.080	Ea	.55	1.95	—	2.50
1¼" x 1"	P1@.080	Ea	.55	1.95	—	2.50
1½" x 1/2"	P1@.100	Ea	.58	2.44	—	3.02
1½" x 3/4"	P1@.100	Ea	.58	2.44	—	3.02
1½" x 1"	P1@.100	Ea	.58	2.44	—	3.02
1½" x 1¼"	P1@.100	Ea	.58	2.44	—	3.02
2" x 1"	P1@.110	Ea	.97	2.69	—	3.66
2" x 1¼"	P1@.110	Ea	.97	2.69	—	3.66
2" x 1½"	P1@.110	Ea	.97	2.69	—	3.66
2½" x 1¼"	P1@.130	Ea	1.55	3.18	—	4.73
2½" x 1½"	P1@.130	Ea	1.55	3.18	—	4.73
2½" x 2"	P1@.130	Ea	1.55	3.18	—	4.73
3" x 2"	P1@.150	Ea	2.30	3.67	—	5.97
3" x 2½"	P1@.150	Ea	2.30	3.67	—	5.97
4" x 3"	P1@.200	Ea	5.15	4.89	—	10.04
5" x 3"	P1@.250	Ea	10.25	6.11	—	16.36
5" x 4"	P1@.250	Ea	10.25	6.11	—	16.36
6" x 4"	P1@.300	Ea	14.90	7.33	—	22.23
6" x 5"	P1@.300	Ea	14.90	7.33	—	22.23
8" x 4"	P1@.400	Ea	44.45	9.77	—	54.22
8" x 5"	P1@.400	Ea	44.45	9.77	—	54.22

Schedule 40 PVC adapter MPT x S with solvent-weld joints

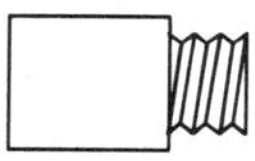

Description	Craft@Hrs	Unit	Material $	Labor $	Equipment $	Total $
1/2"	P1@.060	Ea	.20	1.47	—	1.67
3/4"	P1@.065	Ea	.22	1.59	—	1.81
1"	P1@.070	Ea	.39	1.71	—	2.10
1¼"	P1@.100	Ea	.48	2.44	—	2.92
1½"	P1@.110	Ea	.64	2.69	—	3.33
2"	P1@.120	Ea	2.13	2.93	—	5.06
2½"	P1@.150	Ea	2.47	3.67	—	6.14
3"	P1@.180	Ea	3.61	4.40	—	8.01
4"	P1@.240	Ea	4.60	5.86	—	10.46
5"	P1@.370	Ea	7.80	9.04	—	16.84
6"	P1@.420	Ea	12.00	10.30	—	22.30

Description	Craft@Hrs	Unit	Material $	Labor $	Equipment $	Total $

Schedule 40 PVC adapter FPT x S with solvent-weld joints

Description	Craft@Hrs	Unit	Material $	Labor $	Equipment $	Total $
1/2"	P1@.110	Ea	.24	2.69	—	2.93
3/4"	P1@.120	Ea	.31	2.93	—	3.24
1"	P1@.130	Ea	.36	3.18	—	3.54
1¼"	P1@.170	Ea	.56	4.15	—	4.71
1½"	P1@.210	Ea	.64	5.13	—	5.77
2"	P1@.290	Ea	.86	7.09	—	7.95
2½"	P1@.330	Ea	2.17	8.06	—	10.23
3"	P1@.470	Ea	4.34	11.50	—	15.84
4"	P1@.630	Ea	8.03	15.40	—	23.43
5"	P1@.890	Ea	12.48	21.70	—	34.18
6"	P1@.970	Ea	17.73	23.70	—	41.43

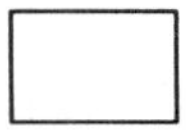

Schedule 40 PVC cap S with solvent-weld joints

Description	Craft@Hrs	Unit	Material $	Labor $	Equipment $	Total $
1/2"	P1@.060	Ea	.20	1.47	—	1.67
3/4"	P1@.065	Ea	.23	1.59	—	1.82
1"	P1@.070	Ea	.36	1.71	—	2.07
1¼"	P1@.100	Ea	.55	2.44	—	2.99
1½"	P1@.110	Ea	.56	2.69	—	3.25
2"	P1@.120	Ea	.67	2.93	—	3.60
2½"	P1@.150	Ea	2.13	3.67	—	5.80
3"	P1@.180	Ea	2.33	4.40	—	6.73
4"	P1@.240	Ea	5.30	5.86	—	11.16
5"	P1@.370	Ea	8.90	9.04	—	17.94
6"	P1@.420	Ea	12.70	10.30	—	23.00

Schedule 40 PVC plug MPT with solvent-weld joints

Description	Craft@Hrs	Unit	Material $	Labor $	Equipment $	Total $
1/2"	P1@.150	Ea	.59	3.67	—	4.26
3/4"	P1@.160	Ea	.63	3.91	—	4.54
1"	P1@.180	Ea	1.03	4.40	—	5.43
1¼"	P1@.210	Ea	1.08	5.13	—	6.21
1½"	P1@.230	Ea	1.16	5.62	—	6.78
2"	P1@.240	Ea	1.50	5.86	—	7.36
2½"	P1@.260	Ea	2.08	6.35	—	8.43
3"	P1@.300	Ea	3.16	7.33	—	10.49
4"	P1@.400	Ea	7.12	9.77	—	16.89

PVC, Schedule 40, with Solvent-Weld Joints

Description	Craft@Hrs	Unit	Material $	Labor $	Equipment $	Total $
Schedule 40 PVC coupling S x S with solvent-weld joints 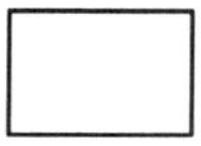						
1/2"	P1@.100	Ea	.14	2.44	—	2.58
3/4"	P1@.115	Ea	.20	2.81	—	3.01
1"	P1@.120	Ea	.35	2.93	—	3.28
1¼"	P1@.160	Ea	.48	3.91	—	4.39
1½"	P1@.180	Ea	.50	4.40	—	4.90
2"	P1@.200	Ea	.78	4.89	—	5.67
2½"	P1@.250	Ea	1.71	6.11	—	7.82
3"	P1@.300	Ea	2.70	7.33	—	10.03
4"	P1@.500	Ea	3.88	12.20	—	16.08
5"	P1@.550	Ea	7.12	13.40	—	20.52
6"	P1@.600	Ea	12.30	14.70	—	27.00
8"	P1@.800	Ea	22.97	19.50	—	42.47
Schedule 40 PVC union S x S with solvent-weld joints 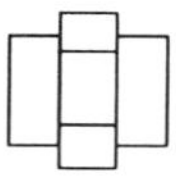						
1/2"	P1@.125	Ea	1.80	3.05	—	4.85
3/4"	P1@.140	Ea	2.25	3.42	—	5.67
1"	P1@.160	Ea	2.33	3.91	—	6.24
1¼"	P1@.200	Ea	4.80	4.89	—	9.69
1½"	P1@.225	Ea	5.48	5.50	—	10.98
2"	P1@.250	Ea	8.50	6.11	—	14.61
2½"	P1@.275	Ea	12.50	6.72	—	19.22
3"	P1@.300	Ea	15.60	7.33	—	22.93
Class 125 bronze body gate valve, threaded ends 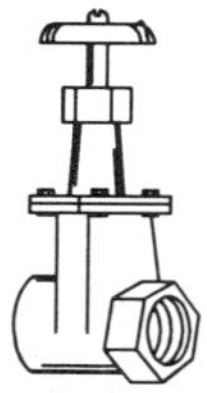						
1/2"	P1@.210	Ea	6.06	5.13	—	11.19
3/4"	P1@.250	Ea	7.58	6.11	—	13.69
1"	P1@.300	Ea	10.50	7.33	—	17.83
1¼"	P1@.400	Ea	13.40	9.77	—	23.17
1½"	P1@.450	Ea	18.00	11.00	—	29.00
2"	P1@.500	Ea	30.10	12.20	—	42.30
2½"	P1@.750	Ea	49.90	18.30	—	68.20
3"	P1@.950	Ea	71.60	23.20	—	94.80
Class 125 iron body gate valve, flanged ends 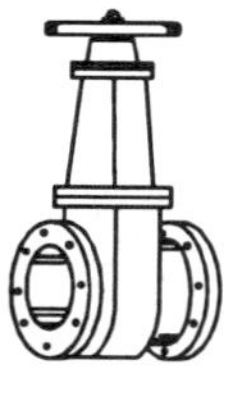						
2"	P1@.500	Ea	101.00	12.20	—	113.20
2½"	P1@.600	Ea	137.00	14.70	—	151.70
3"	P1@.750	Ea	148.00	18.30	—	166.30
4"	P1@1.35	Ea	218.00	33.00	—	251.00
5"	ER@2.00	Ea	417.00	52.20	10.00	479.20
6"	ER@2.50	Ea	417.00	65.20	12.50	494.70
8"	ER@3.00	Ea	680.00	78.20	15.00	773.20

Class 125 bronze body globe valve, threaded ends

Description	Craft@Hrs	Unit	Material $	Labor $	Equipment $	Total $
1/2"	P1@.210	Ea	15.80	5.13	—	20.93
3/4"	P1@.250	Ea	21.30	6.11	—	27.41
1"	P1@.300	Ea	30.50	7.33	—	37.83
1¼"	P1@.400	Ea	42.90	9.77	—	52.67
1½"	P1@.450	Ea	57.50	11.00	—	68.50
2"	P1@.500	Ea	93.60	12.20	—	105.80

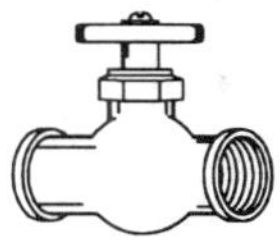

200 PSIG iron body butterfly valve, lug type

Description	Craft@Hrs	Unit	Material $	Labor $	Equipment $	Total $
2"	P1@.450	Ea	84.50	11.00	—	95.50
2½"	P1@.450	Ea	86.50	11.00	—	97.50
3"	P1@.550	Ea	91.70	13.40	—	105.10
4"	P1@.550	Ea	114.00	13.40	—	127.40
5"	ER@.800	Ea	151.00	20.90	4.00	175.90
6"	ER@.800	Ea	185.00	20.90	4.00	209.90
8"	ER@.900	Ea	252.00	23.50	4.50	280.00

200 PSIG iron body butterfly valve, wafer type

Description	Craft@Hrs	Unit	Material $	Labor $	Equipment $	Total $
2"	P1@.450	Ea	76.70	11.00	—	87.70
2½"	P1@.450	Ea	78.70	11.00	—	89.70
3"	P1@.550	Ea	84.50	13.40	—	97.90
4"	P1@.550	Ea	102.00	13.40	—	115.40
5"	ER@.800	Ea	137.00	20.90	4.00	161.90
6"	ER@.800	Ea	170.00	20.90	4.00	194.90
8"	ER@.900	Ea	237.00	23.50	4.50	265.00

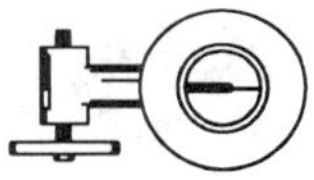

Class 125 bronze body 2-piece ball valve, threaded ends

Description	Craft@Hrs	Unit	Material $	Labor $	Equipment $	Total $
1/2"	P1@.210	Ea	3.56	5.13	—	8.69
3/4"	P1@.250	Ea	4.75	6.11	—	10.86
1"	P1@.300	Ea	8.21	7.33	—	15.54
1¼"	P1@.400	Ea	13.70	9.77	—	23.47
1½"	P1@.450	Ea	18.50	11.00	—	29.50
2"	P1@.500	Ea	23.30	12.20	—	35.50

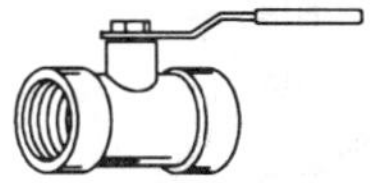

Class 125 bronze body swing check valve, threaded

Description	Craft@Hrs	Unit	Material $	Labor $	Equipment $	Total $
1/2"	P1@.210	Ea	8.00	5.13	—	13.13
3/4"	P1@.250	Ea	11.40	6.11	—	17.51
1"	P1@.300	Ea	15.00	7.33	—	22.33
1¼"	P1@.400	Ea	21.40	9.77	—	31.17
1½"	P1@.450	Ea	30.00	11.00	—	41.00
2"	P1@.500	Ea	50.00	12.20	—	62.20

Class 125 iron body swing check valve, flanged ends

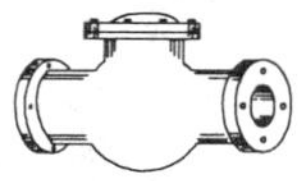

Description	Craft@Hrs	Unit	Material $	Labor $	Equipment $	Total $
2"	P1@.500	Ea	115.00	12.20	—	127.20
2½"	P1@.600	Ea	148.00	14.70	—	162.70
3"	P1@.750	Ea	183.00	18.30	—	201.30
4"	P1@1.35	Ea	271.00	33.00	—	304.00
5"	ER@2.00	Ea	522.00	52.20	10.00	584.20
6"	ER@2.50	Ea	522.00	65.20	12.50	599.70
8"	ER@3.00	Ea	938.00	78.20	15.00	1,031.20

Class 125 iron body silent check valve, wafer type

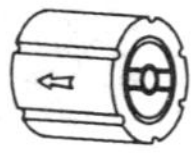

Description	Craft@Hrs	Unit	Material $	Labor $	Equipment $	Total $
2"	P1@.500	Ea	85.00	12.20	—	97.20
2½"	P1@.600	Ea	95.00	14.70	—	109.70
3"	P1@.750	Ea	108.00	18.30	—	126.30
4"	P1@1.35	Ea	146.00	33.00	—	179.00
5"	ER@2.00	Ea	224.00	52.20	10.00	286.20
6"	ER@2.50	Ea	303.00	65.20	12.50	380.70
8"	ER@3.00	Ea	522.00	78.20	15.00	615.20

Class 125 bronze body strainer, threaded ends

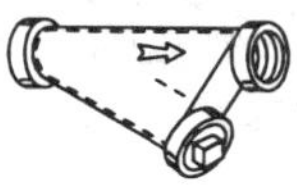

Description	Craft@Hrs	Unit	Material $	Labor $	Equipment $	Total $
1/2"	P1@.210	Ea	12.40	5.13	—	17.53
3/4"	P1@.250	Ea	16.30	6.11	—	22.41
1"	P1@.300	Ea	20.20	7.33	—	27.53
1¼"	P1@.400	Ea	28.10	9.77	—	37.87
1½"	P1@.450	Ea	42.20	11.00	—	53.20
2"	P1@.500	Ea	73.00	12.20	—	85.20

Class 125 iron body strainer, flanged ends

Description	Craft@Hrs	Unit	Material $	Labor $	Equipment $	Total $
2"	P1@.500	Ea	62.00	12.20	—	74.20
2½"	P1@.600	Ea	69.00	14.70	—	83.70
3"	P1@.750	Ea	80.00	18.30	—	98.30
4"	P1@1.35	Ea	137.00	33.00	—	170.00
5"	ER@2.00	Ea	277.00	52.20	10.00	339.20
6"	ER@2.50	Ea	277.00	65.20	12.50	354.70
8"	ER@3.00	Ea	470.00	78.20	15.00	563.20

Description	Craft@Hrs	Unit	Material $	Labor $	Equipment $	Total $
Installation of 2-way control valve, threaded joints						
1/2"	P1@.210	Ea	—	5.13	—	5.13
3/4"	P1@.275	Ea	—	6.72	—	6.72
1"	P1@.350	Ea	—	8.55	—	8.55
1¼"	P1@.430	Ea	—	10.50	—	10.50
1½"	P1@.505	Ea	—	12.30	—	12.30
2"	P1@.675	Ea	—	16.50	—	16.50
2½"	P1@.830	Ea	—	20.30	—	20.30
3"	P1@.990	Ea	—	24.20	—	24.20

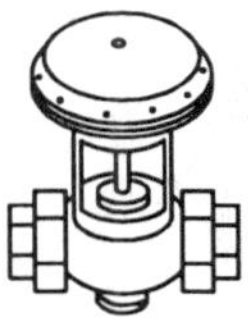

Installation of 3-way control valve, threaded joints

Description	Craft@Hrs	Unit	Material $	Labor $	Equipment $	Total $
1/2"	P1@.260	Ea	—	6.35	—	6.35
3/4"	P1@.365	Ea	—	8.92	—	8.92
1"	P1@.475	Ea	—	11.60	—	11.60
1¼"	P1@.575	Ea	—	14.10	—	14.10
1½"	P1@.680	Ea	—	16.60	—	16.60
2"	P1@.910	Ea	—	22.20	—	22.20
2½"	P1@1.12	Ea	—	27.40	—	27.40
3"	P1@1.33	Ea	—	32.50	—	32.50

PVC companion flanges

Description	Craft@Hrs	Unit	Material $	Labor $	Equipment $	Total $
2"	P1@.140	Ea	2.25	3.42	—	5.67
2½"	P1@.180	Ea	6.75	4.40	—	11.15
3"	P1@.210	Ea	7.00	5.13	—	12.13
4"	P1@.280	Ea	7.25	6.84	—	14.09
5"	P1@.310	Ea	16.10	7.58	—	23.68
6"	P1@.420	Ea	16.10	10.30	—	26.40
8"	P1@.560	Ea	24.00	13.70	—	37.70

Bolt and gasket sets

Description	Craft@Hrs	Unit	Material $	Labor $	Equipment $	Total $
2"	P1@.500	Ea	1.50	12.20	—	13.70
2½"	P1@.650	Ea	1.75	15.90	—	17.65
3"	P1@.750	Ea	3.00	18.30	—	21.30
4"	P1@1.00	Ea	5.00	24.40	—	29.40
5"	P1@1.10	Ea	8.25	26.90	—	35.15
6"	P1@1.20	Ea	8.25	29.30	—	37.55
8"	P1@1.25	Ea	9.35	30.50	—	39.85

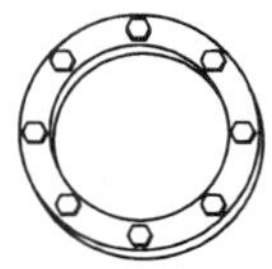

Thermometer with well

Description	Craft@Hrs	Unit	Material $	Labor $	Equipment $	Total $
7"	P1@.250	Ea	15.00	6.11	—	21.11
9"	P1@.250	Ea	15.50	6.11	—	21.61

PVC, Schedule 40, with Solvent-Weld Joints

Description	Craft@Hrs	Unit	Material $	Labor $	Equipment $	Total $

Dial type pressure gauge

Description	Craft@Hrs	Unit	Material $	Labor $	Equipment $	Total $
2½"	P1@.200	Ea	15.50	4.89	—	20.39
3½"	P1@.200	Ea	20.50	4.89	—	25.39

Pressure/temperature tap

Description	Craft@Hrs	Unit	Material $	Labor $	Equipment $	Total $
	P1@.150	Ea	7.30	3.67	—	10.97

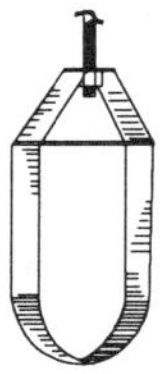

Hanger with swivel assembly

Description	Craft@Hrs	Unit	Material $	Labor $	Equipment $	Total $
1/2"	P1@.450	Ea	1.60	11.00	—	12.60
3/4"	P1@.450	Ea	1.75	11.00	—	12.75
1"	P1@.450	Ea	1.85	11.00	—	12.85
1¼"	P1@.450	Ea	1.90	11.00	—	12.90
1½"	P1@.450	Ea	2.05	11.00	—	13.05
2"	P1@.450	Ea	2.15	11.00	—	13.15
2½"	P1@.450	Ea	2.90	11.00	—	13.90
3"	P1@.450	Ea	3.60	11.00	—	14.60
4"	P1@.450	Ea	3.95	11.00	—	14.95
5"	P1@.500	Ea	5.75	12.20	—	17.95
6"	P1@.500	Ea	6.45	12.20	—	18.65
8"	P1@.550	Ea	8.15	13.40	—	21.55

Riser clamp

Description	Craft@Hrs	Unit	Material $	Labor $	Equipment $	Total $
1/2"	P1@.100	Ea	.95	2.44	—	3.39
3/4"	P1@.100	Ea	1.39	2.44	—	3.83
1"	P1@.100	Ea	1.41	2.44	—	3.85
1¼"	P1@.105	Ea	1.70	2.57	—	4.27
1½"	P1@.110	Ea	1.79	2.69	—	4.48
2"	P1@.115	Ea	1.89	2.81	—	4.70
2½"	P1@.120	Ea	2.00	2.93	—	4.93
3"	P1@.120	Ea	2.18	2.93	—	5.11
4"	P1@.125	Ea	2.75	3.05	—	5.80
5"	P1@.180	Ea	3.99	4.40	—	8.39
6"	P1@.200	Ea	4.79	4.89	—	9.68
8"	P1@.200	Ea	7.77	4.89	—	12.66

PVC (Polyvinyl Chloride) Schedule 80 pipe with Schedule 80 socket-type fittings and solvent-welded joints is widely used in process, chemical, A.C. condensate, potable water and irrigation piping systems. PVC is available in Type I (normal impact) and Type II (high impact). Type I has a maximum temperature rating of 150 degrees F., and Type II is rated at 140 degrees F.

Schedule 80 PVC can be threaded, but it is more economical to use solvent-welded joints.

Manufacturers should be consulted for maximum pressure ratings and recommended joints solvents for specific applications.

Because of the current unreliability of many plastic valves, most engineers specify standard bronze-body and iron-body valves for use in PVC water piping systems. Plastic valves, however, have to be used in systems conveying liquids that may be injurious to metallic valves.

This section has been arranged to save the estimator's time by including all normally-used system components such as pipe, fittings, valves, hanger assemblies, riser clamps and miscellaneous items under one heading. Additional items can be found under "Plumbing and Piping Specialties." The cost estimates in this section are based on the conditions, limitations and wage rates described in the section "How to Use This Book" beginning on page 5.

Equipment cost, where shown, is $35.00 per hour for a 10 ton hydraulic truck-mounted crane.

Schedule 80 PVC pipe with solvent-weld joints

Description	Craft@Hrs	Unit	Material $	Labor $	Equipment $	Total $
1/2"	P1@.021	LF	.21	.51	—	.72
3/4"	P1@.026	LF	.29	.64	—	.93
1"	P1@.032	LF	.42	.78	—	1.20
1¼"	P1@.037	LF	.50	.90	—	1.40
1½"	P1@.042	LF	.59	1.03	—	1.62
2"	P1@.047	LF	.79	1.15	—	1.94
2½"	P1@.052	LF	1.26	1.27	—	2.53
3"	P1@.057	LF	1.69	1.39	—	3.08
4"	P1@.074	LF	2.47	1.81	—	4.28
6"	P1@.160	LF	4.69	3.91	—	8.60
8"	P1@.190	LF	7.14	4.64	—	11.78

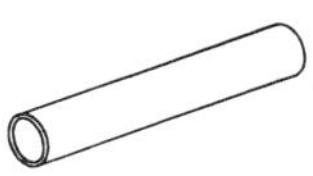

Schedule 80 PVC 45 degree ell S x S with solvent-weld joints

Description	Craft@Hrs	Unit	Material $	Labor $	Equipment $	Total $
1/2"	P1@.105	Ea	1.13	2.57	—	3.70
3/4"	P1@.120	Ea	1.71	2.93	—	4.64
1"	P1@.125	Ea	2.57	3.05	—	5.62
1¼"	P1@.170	Ea	3.28	4.15	—	7.43
1½"	P1@.190	Ea	3.87	4.64	—	8.51
2"	P1@.210	Ea	5.01	5.13	—	10.14
2½"	P1@.262	Ea	10.53	6.40	—	16.93
3"	P1@.315	Ea	12.82	7.70	—	20.52
4"	P1@.520	Ea	23.08	12.70	—	35.78
6"	P1@.630	Ea	29.07	15.40	—	44.47
8"	P1@.840	Ea	62.79	20.50	—	83.29

PVC, Schedule 80, with Solvent-Weld Joints

Description	Craft@Hrs	Unit	Material $	Labor $	Equipment $	Total $

Schedule 80 PVC 90 degree ell S x S with solvent-weld joints

Description	Craft@Hrs	Unit	Material $	Labor $	Equipment $	Total $
1/2"	P1@.105	Ea	.60	2.57	—	3.17
3/4"	P1@.120	Ea	.77	2.93	—	3.70
1"	P1@.125	Ea	1.23	3.05	—	4.28
1¼"	P1@.170	Ea	1.64	4.15	—	5.79
1½"	P1@.190	Ea	1.76	4.64	—	6.40
2"	P1@.210	Ea	2.13	5.13	—	7.26
2½"	P1@.262	Ea	4.97	6.40	—	11.37
3"	P1@.315	Ea	5.60	7.70	—	13.30
4"	P1@.520	Ea	8.50	12.70	—	21.20
6"	P1@.630	Ea	24.18	15.40	—	39.58
8"	P1@.840	Ea	66.30	20.50	—	86.80

Schedule 80 PVC tee S x S x S with solvent-weld joints

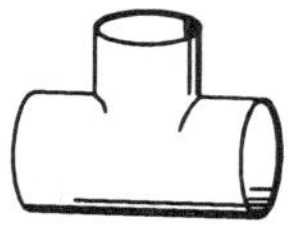

Description	Craft@Hrs	Unit	Material $	Labor $	Equipment $	Total $
1/2"	P1@.135	Ea	1.68	3.30	—	4.98
3/4"	P1@.145	Ea	1.76	3.54	—	5.30
1"	P1@.180	Ea	2.20	4.40	—	6.60
1¼"	P1@.230	Ea	6.05	5.62	—	11.67
1½"	P1@.265	Ea	6.05	6.48	—	12.53
2"	P1@.295	Ea	7.57	7.21	—	14.78
2½"	P1@.368	Ea	8.23	8.99	—	17.22
3"	P1@.440	Ea	10.29	10.80	—	21.09
4"	P1@.585	Ea	11.92	14.30	—	26.22
6"	P1@.880	Ea	40.55	21.50	—	62.05
8"	P1@1.18	Ea	94.00	28.80	—	122.80

Schedule 80 PVC reducing tee S x S x S with solvent-weld joints

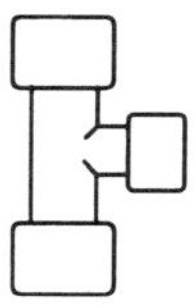

Description	Craft@Hrs	Unit	Material $	Labor $	Equipment $	Total $
3/4" x 1/2"	P1@.145	Ea	1.90	3.54	—	5.44
1" x 1/2"	P1@.175	Ea	2.05	4.28	—	6.33
1" x 3/4"	P1@.175	Ea	2.05	4.28	—	6.33
1¼" x 1/2"	P1@.230	Ea	6.75	5.62	—	12.37
1¼" x 3/4"	P1@.230	Ea	6.75	5.62	—	12.37
1¼" x 1"	P1@.230	Ea	6.75	5.62	—	12.37
1½" x 3/4"	P1@.260	Ea	7.25	6.35	—	13.60
1½" x 1"	P1@.260	Ea	7.25	6.35	—	13.60
1½" x 1¼"	P1@.260	Ea	7.25	6.35	—	13.60
2" x 1/2"	P1@.290	Ea	8.90	7.09	—	15.99
2" x 3/4"	P1@.290	Ea	8.90	7.09	—	15.99
2" x 1"	P1@.290	Ea	8.90	7.09	—	15.99
2" x 1½"	P1@.290	Ea	8.90	7.09	—	15.99
3" x 2"	P1@.430	Ea	12.50	10.50	—	23.00
4" x 2"	P1@.570	Ea	14.30	13.90	—	28.20
4" x 3"	P1@.570	Ea	14.30	13.90	—	28.20
6" x 4"	P1@.850	Ea	48.00	20.80	—	68.80
8" x 4"	P1@1.16	Ea	108.00	28.30	—	136.30
8" x 5"	P1@1.16	Ea	108.00	28.30	—	136.30
8" x 6"	P1@1.16	Ea	108.00	28.30	—	136.30

Schedule 80 PVC reducing bushing SPIG x S with solvent-weld joints

Description	Craft@Hrs	Unit	Material $	Labor $	Equipment $	Total $
3/4" x 1/2"	P1@.053	Ea	.35	1.30	—	1.65
1" x 1/2"	P1@.063	Ea	1.00	1.54	—	2.54
1" x 3/4"	P1@.063	Ea	1.00	1.54	—	2.54
1¼" x 1/2"	P1@.084	Ea	1.60	2.05	—	3.65
1¼" x 3/4"	P1@.084	Ea	1.60	2.05	—	3.65
1¼" x 1"	P1@.084	Ea	1.60	2.05	—	3.65
1½" x 1/2"	P1@.105	Ea	2.13	2.57	—	4.70
1½" x 3/4"	P1@.105	Ea	2.13	2.57	—	4.70
1½" x 1"	P1@.105	Ea	2.13	2.57	—	4.70
1½" x 1¼"	P1@.105	Ea	2.13	2.57	—	4.70
2" x 1"	P1@.116	Ea	3.05	2.83	—	5.88
2" x 1¼"	P1@.116	Ea	3.05	2.83	—	5.88
2" x 1½"	P1@.116	Ea	3.05	2.83	—	5.88
3" x 2"	P1@.158	Ea	8.35	3.86	—	12.21
4" x 2½"	P1@.210	Ea	11.57	5.13	—	16.70
4" x 3"	P1@.210	Ea	11.57	5.13	—	16.70
6" x 3"	P1@.315	Ea	16.07	7.70	—	23.77
6" x 4"	P1@.315	Ea	16.07	7.70	—	23.77
8" x 4"	P1@.420	Ea	39.00	10.30	—	49.30
8" x 6"	P1@.420	Ea	39.00	10.30	—	49.30

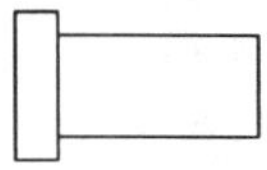

Schedule 80 PVC adapter MPT x S with solvent-weld joints

Description	Craft@Hrs	Unit	Material $	Labor $	Equipment $	Total $
1/2"	P1@.063	Ea	1.26	1.54	—	2.80
3/4"	P1@.068	Ea	1.39	1.66	—	3.05
1"	P1@.074	Ea	2.40	1.81	—	4.21
1¼"	P1@.105	Ea	2.81	2.57	—	5.38
1½"	P1@.116	Ea	4.04	2.83	—	6.87
2"	P1@.126	Ea	5.84	3.08	—	8.92
2½"	P1@.158	Ea	6.67	3.86	—	10.53
3"	P1@.189	Ea	7.38	4.62	—	12.00
4"	P1@.252	Ea	13.11	6.16	—	19.27
5"	P1@.385	Ea	18.25	9.41	—	27.66
6"	P1@.435	Ea	42.00	10.60	—	52.60

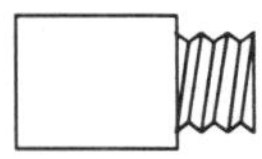

Schedule 80 PVC adapter FPT x S with solvent-weld joints

Description	Craft@Hrs	Unit	Material $	Labor $	Equipment $	Total $
1/2"	P1@.116	Ea	1.01	2.83	—	3.84
3/4"	P1@.126	Ea	1.50	3.08	—	4.58
1"	P1@.137	Ea	2.21	3.35	—	5.56
1¼"	P1@.179	Ea	3.58	4.37	—	7.95
1½"	P1@.221	Ea	4.39	5.40	—	9.79
2"	P1@.305	Ea	7.66	7.45	—	15.11
2½"	P1@.345	Ea	12.09	8.43	—	20.52
3"	P1@.490	Ea	13.61	12.00	—	25.61
4"	P1@.655	Ea	23.40	16.00	—	39.40
5"	P1@.920	Ea	38.00	22.50	—	60.50
6"	P1@1.10	Ea	46.00	26.90	—	72.90

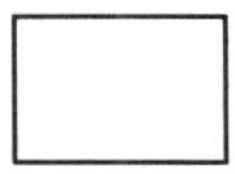

PVC, Schedule 80, with Solvent-Weld Joints

Schedule 80 PVC cap S with solvent-weld joints

Description	Craft@Hrs	Unit	Material $	Labor $	Equipment $	Total $
1/2"	P1@.063	Ea	1.06	1.54	—	2.60
3/4"	P1@.070	Ea	1.11	1.71	—	2.82
1"	P1@.075	Ea	1.97	1.83	—	3.80
1¼"	P1@.105	Ea	2.39	2.57	—	4.96
1½"	P1@.115	Ea	2.39	2.81	—	5.20
2"	P1@.125	Ea	4.70	3.05	—	7.75
2½"	P1@.158	Ea	9.60	3.86	—	13.46
3"	P1@.189	Ea	11.25	4.62	—	15.87
4"	P1@.250	Ea	18.96	6.11	—	25.07
5"	P1@.390	Ea	39.25	9.53	—	48.78
6"	P1@.450	Ea	47.20	11.00	—	58.20

Schedule 80 PVC plug MPT with solvent-weld joints

Description	Craft@Hrs	Unit	Material $	Labor $	Equipment $	Total $
1/2"	P1@.155	Ea	1.02	3.79	—	4.81
3/4"	P1@.165	Ea	1.06	4.03	—	5.09
1"	P1@.190	Ea	1.30	4.64	—	5.94
1¼"	P1@.220	Ea	1.90	5.38	—	7.28
1½"	P1@.240	Ea	2.32	5.86	—	8.18
2"	P1@.255	Ea	2.38	6.23	—	8.61
2½"	P1@.275	Ea	5.95	6.72	—	12.67
3"	P1@.320	Ea	7.75	7.82	—	15.57
4"	P1@.425	Ea	15.08	10.40	—	25.48

Schedule 80 PVC coupling S x S with solvent-weld joints

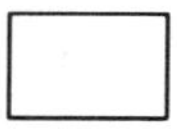

Description	Craft@Hrs	Unit	Material $	Labor $	Equipment $	Total $
1/2"	P1@.105	Ea	1.08	2.57	—	3.65
3/4"	P1@.120	Ea	1.46	2.93	—	4.39
1"	P1@.125	Ea	1.50	3.05	—	4.55
1¼"	P1@.170	Ea	2.29	4.15	—	6.44
1½"	P1@.190	Ea	2.47	4.64	—	7.11
2"	P1@.210	Ea	2.64	5.13	—	7.77
2½"	P1@.262	Ea	6.51	6.40	—	12.91
3"	P1@.315	Ea	7.48	7.70	—	15.18
4"	P1@.520	Ea	9.37	12.70	—	22.07
6"	P1@.630	Ea	20.15	15.40	—	35.55
8"	P1@.840	Ea	27.70	20.50	—	48.20

Schedule 80 PVC union S x S with solvent-weld joints

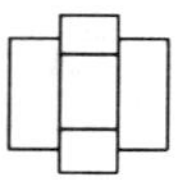

Description	Craft@Hrs	Unit	Material $	Labor $	Equipment $	Total $
1/2"	P1@.125	Ea	2.18	3.05	—	5.23
3/4"	P1@.140	Ea	3.00	3.42	—	6.42
1"	P1@.160	Ea	3.17	3.91	—	7.08
1¼"	P1@.200	Ea	6.30	4.89	—	11.19
1½"	P1@.225	Ea	7.12	5.50	—	12.62
2"	P1@.250	Ea	9.66	6.11	—	15.77
2½"	P1@.275	Ea	15.25	6.72	—	21.97
3"	P1@.300	Ea	17.97	7.33	—	25.30

Class 125 bronze body gate valve, threaded ends

Description	Craft@Hrs	Unit	Material $	Labor $	Equipment $	Total $
1/2"	P1@.210	Ea	6.06	5.13	—	11.19
3/4"	P1@.250	Ea	7.58	6.11	—	13.69
1"	P1@.300	Ea	10.50	7.33	—	17.83
1¼"	P1@.400	Ea	13.40	9.77	—	23.17
1½"	P1@.450	Ea	18.00	11.00	—	29.00
2"	P1@.500	Ea	30.10	12.20	—	42.30
2½"	P1@.750	Ea	49.90	18.30	—	68.20
3"	P1@.950	Ea	71.60	23.20	—	94.80

Class 125 iron body gate valve, flanged ends

Description	Craft@Hrs	Unit	Material $	Labor $	Equipment $	Total $
2"	P1@.500	Ea	101.00	12.20	—	113.20
2½"	P1@.600	Ea	137.00	14.70	—	151.70
3"	P1@.750	Ea	148.00	18.30	—	166.30
4"	P1@1.35	Ea	218.00	33.00	—	251.00
5"	ER@2.00	Ea	417.00	52.20	10.00	479.20
6"	ER@2.50	Ea	417.00	65.20	12.50	494.70
8"	ER@3.00	Ea	680.00	78.20	15.00	773.20

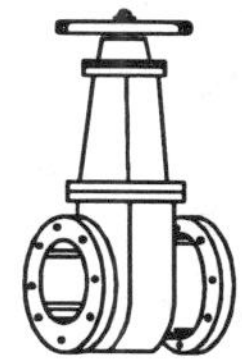

Class 125 bronze body globe valve, threaded ends

Description	Craft@Hrs	Unit	Material $	Labor $	Equipment $	Total $
1/2"	P1@.210	Ea	15.80	5.13	—	20.93
3/4"	P1@.250	Ea	21.30	6.11	—	27.41
1"	P1@.300	Ea	30.50	7.33	—	37.83
1¼"	P1@.400	Ea	42.90	9.77	—	52.67
1½"	P1@.450	Ea	57.50	11.00	—	68.50
2"	P1@.500	Ea	93.60	12.20	—	105.80

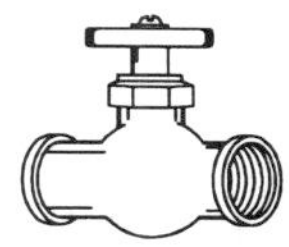

200 PSIG iron body butterfly valve, lug type

Description	Craft@Hrs	Unit	Material $	Labor $	Equipment $	Total $
2"	P1@.450	Ea	84.50	11.00	—	95.50
2½"	P1@.450	Ea	86.50	11.00	—	97.50
3"	P1@.550	Ea	91.70	13.40	—	105.10
4"	P1@.550	Ea	114.00	13.40	—	127.40
5"	ER@.800	Ea	151.00	20.90	4.00	175.90
6"	ER@.800	Ea	185.00	20.90	4.00	209.90
8"	ER@.900	Ea	252.00	23.50	4.50	280.00

200 PSIG iron body butterfly valve, wafer type

Description	Craft@Hrs	Unit	Material $	Labor $	Equipment $	Total $
2"	P1@.450	Ea	76.70	11.00	—	87.70
2½"	P1@.450	Ea	78.70	11.00	—	89.70
3"	P1@.550	Ea	84.50	13.40	—	97.90
4"	P1@.550	Ea	102.00	13.40	—	115.40
5"	ER@.800	Ea	137.00	20.90	4.00	161.90
6"	ER@.800	Ea	170.00	20.90	4.00	194.90
8"	ER@.900	Ea	237.00	23.50	4.50	265.00

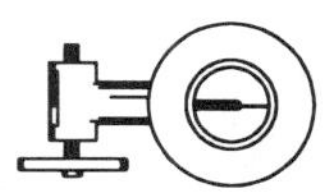

PVC, Schedule 80, with Solvent-Weld Joints

Class 125 bronze body 2-piece ball valve, threaded ends

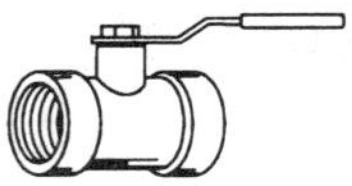

Description	Craft@Hrs	Unit	Material $	Labor $	Equipment $	Total $
1/2"	P1@.210	Ea	3.56	5.13	—	8.69
3/4"	P1@.250	Ea	4.75	6.11	—	10.86
1"	P1@.300	Ea	8.21	7.33	—	15.54
1¼"	P1@.400	Ea	13.70	9.77	—	23.47
1½"	P1@.450	Ea	18.50	11.00	—	29.50
2"	P1@.500	Ea	23.30	12.20	—	35.50

Class 125 bronze body swing check valve, threaded ends

Description	Craft@Hrs	Unit	Material $	Labor $	Equipment $	Total $
1/2"	P1@.210	Ea	8.00	5.13	—	13.13
3/4"	P1@.250	Ea	11.40	6.11	—	17.51
1"	P1@.300	Ea	15.00	7.33	—	22.33
1¼"	P1@.400	Ea	21.40	9.77	—	31.17
1½"	P1@.450	Ea	30.00	11.00	—	41.00
2"	P1@.500	Ea	50.00	12.20	—	62.20

Class 125 iron body swing check valve, flanged ends

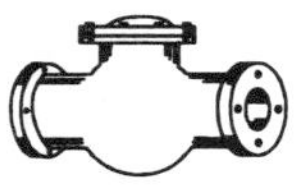

Description	Craft@Hrs	Unit	Material $	Labor $	Equipment $	Total $
2"	P1@.500	Ea	115.00	12.20	—	127.20
2½"	P1@.600	Ea	148.00	14.70	—	162.70
3"	P1@.750	Ea	183.00	18.30	—	201.30
4"	P1@1.35	Ea	271.00	33.00	—	304.00
5"	ER@2.00	Ea	522.00	52.20	10.00	584.20
6"	ER@2.50	Ea	522.00	65.20	12.50	599.70
8"	ER@3.00	Ea	938.00	78.20	15.00	1,031.20

Class 125 iron body silent check valve, wafer type

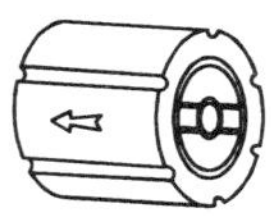

Description	Craft@Hrs	Unit	Material $	Labor $	Equipment $	Total $
2"	P1@.500	Ea	85.00	12.20	—	97.20
2½"	P1@.600	Ea	95.00	14.70	—	109.70
3"	P1@.750	Ea	108.00	18.30	—	126.30
4"	P1@1.35	Ea	146.00	33.00	—	179.00
5"	ER@2.00	Ea	224.00	52.20	10.00	286.20
6"	ER@2.50	Ea	303.00	65.20	12.50	380.70
8"	ER@3.00	Ea	522.00	78.20	15.00	615.20

Class 125 bronze body strainer, threaded ends

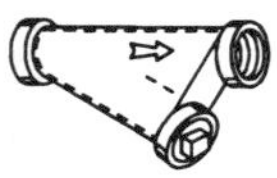

Description	Craft@Hrs	Unit	Material $	Labor $	Equipment $	Total $
1/2"	P1@.210	Ea	12.40	5.13	—	17.53
3/4"	P1@.250	Ea	16.30	6.11	—	22.41
1"	P1@.300	Ea	20.20	7.33	—	27.53
1¼"	P1@.400	Ea	28.10	9.77	—	37.87
1½"	P1@.450	Ea	42.20	11.00	—	53.20
2"	P1@.500	Ea	73.00	12.20	—	85.20

Class 125 iron body strainer, flanged ends

Description	Craft@Hrs	Unit	Material $	Labor $	Equipment $	Total $
2"	P1@.500	Ea	62.00	12.20	—	74.20
2½"	P1@.600	Ea	69.00	14.70	—	83.70
3"	P1@.750	Ea	80.00	18.30	—	98.30
4"	P1@1.35	Ea	137.00	33.00	—	170.00
5"	ER@2.00	Ea	277.00	52.20	10.00	339.20
6"	ER@2.50	Ea	277.00	65.20	12.50	354.70
8"	ER@3.00	Ea	470.00	78.20	15.00	563.20

Installation of 2-way control valve, threaded joints

Description	Craft@Hrs	Unit	Material $	Labor $	Equipment $	Total $
1/2"	P1@.210	Ea	—	5.13	—	5.13
3/4"	P1@.275	Ea	—	6.72	—	6.72
1"	P1@.350	Ea	—	8.55	—	8.55
1¼"	P1@.430	Ea	—	10.50	—	10.50
1½"	P1@.505	Ea	—	12.30	—	12.30
2"	P1@.675	Ea	—	16.50	—	16.50
2½"	P1@.830	Ea	—	20.30	—	20.30
3"	P1@.990	Ea	—	24.20	—	24.20

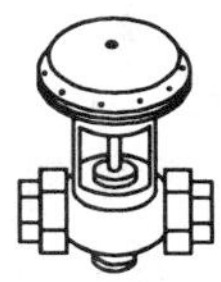

Installation of 3-way control valve, threaded joints

Description	Craft@Hrs	Unit	Material $	Labor $	Equipment $	Total $
1/2"	P1@.260	Ea	—	6.35	—	6.35
3/4"	P1@.365	Ea	—	8.92	—	8.92
1"	P1@.475	Ea	—	11.60	—	11.60
1¼"	P1@.575	Ea	—	14.10	—	14.10
1½"	P1@.680	Ea	—	16.60	—	16.60
2"	P1@.910	Ea	—	22.20	—	22.20
2½"	P1@1.12	Ea	—	27.40	—	27.40
3"	P1@1.33	Ea	—	32.50	—	32.50

PVC companion flange

Description	Craft@Hrs	Unit	Material $	Labor $	Equipment $	Total $
2"	P1@.140	Ea	2.25	3.42	—	5.67
2½"	P1@.180	Ea	6.75	4.40	—	11.15
3"	P1@.210	Ea	7.00	5.13	—	12.13
4"	P1@.280	Ea	7.25	6.84	—	14.09
5"	P1@.310	Ea	16.10	7.58	—	23.68
6"	P1@.420	Ea	16.10	10.30	—	26.40
8"	P1@.560	Ea	24.00	13.70	—	37.70

Bolt and gasket set

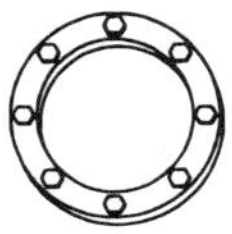

Description	Craft@Hrs	Unit	Material $	Labor $	Equipment $	Total $
2"	P1@.500	Ea	1.50	12.20	—	13.70
2½"	P1@.650	Ea	1.75	15.90	—	17.65
3"	P1@.750	Ea	3.00	18.30	—	21.30
4"	P1@1.00	Ea	5.00	24.40	—	29.40
5"	P1@1.10	Ea	8.25	26.90	—	35.15
6"	P1@1.20	Ea	8.25	29.30	—	37.55
8"	P1@1.25	Ea	9.35	30.50	—	39.85

Thermometer with well

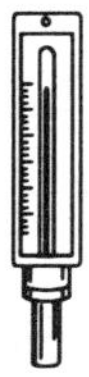

Description	Craft@Hrs	Unit	Material $	Labor $	Equipment $	Total $
7"	P1@.250	Ea	15.00	6.11	—	21.11
9"	P1@.250	Ea	15.50	6.11	—	21.61

Dial type pressure gauge

Description	Craft@Hrs	Unit	Material $	Labor $	Equipment $	Total $
2½"	P1@.200	Ea	15.50	4.89	—	20.39
3½"	P1@.200	Ea	20.50	4.89	—	25.39

Pressure/temperature tap

Description	Craft@Hrs	Unit	Material $	Labor $	Equipment $	Total $
	P1@.140	Ea	7.30	3.42	—	10.72

Hanger with swivel assembly

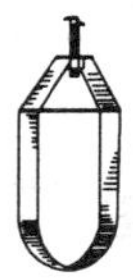

Description	Craft@Hrs	Unit	Material $	Labor $	Equipment $	Total $
1/2"	P1@.450	Ea	1.60	11.00	—	12.60
3/4"	P1@.450	Ea	1.75	11.00	—	12.75
1"	P1@.450	Ea	1.85	11.00	—	12.85
1¼"	P1@.450	Ea	1.90	11.00	—	12.90
1½"	P1@.450	Ea	2.05	11.00	—	13.05
2"	P1@.450	Ea	2.15	11.00	—	13.15
2½"	P1@.450	Ea	2.90	11.00	—	13.90
3"	P1@.450	Ea	3.60	11.00	—	14.60
4"	P1@.450	Ea	3.95	11.00	—	14.95
5"	P1@.500	Ea	5.75	12.20	—	17.95
6"	P1@.500	Ea	6.45	12.20	—	18.65
8"	P1@.550	Ea	8.15	13.40	—	21.55

Riser clamp

Description	Craft@Hrs	Unit	Material $	Labor $	Equipment $	Total $
1/2"	P1@.100	Ea	.95	2.44	—	3.39
3/4"	P1@.100	Ea	1.39	2.44	—	3.83
1"	P1@.100	Ea	1.41	2.44	—	3.85
1¼"	P1@.105	Ea	1.70	2.57	—	4.27
1½"	P1@.110	Ea	1.79	2.69	—	4.48
2"	P1@.115	Ea	1.89	2.81	—	4.70
2½"	P1@.120	Ea	2.00	2.93	—	4.93
3"	P1@.120	Ea	2.18	2.93	—	5.11
4"	P1@.125	Ea	2.75	3.05	—	5.80
5"	P1@.180	Ea	3.99	4.40	—	8.39
6"	P1@.200	Ea	4.79	4.89	—	9.68
8"	P1@.200	Ea	7.77	4.89	—	12.66

Primed steel cam-lock non-fire rated access doors

Description	Craft@Hrs	Unit	Material $	Labor $	Equipment $	Total $
8" x 8"	P1@.400	Ea	29.20	9.77	—	38.97
12" x 12"	P1@.500	Ea	33.70	12.20	—	45.90
16" x 16"	P1@.600	Ea	41.50	14.70	—	56.20
18" x 18"	P1@.800	Ea	47.60	19.50	—	67.10
24" x 24"	P1@1.25	Ea	70.50	30.50	—	101.00
36" x 36"	P1@1.60	Ea	130.00	39.10	—	169.10
Add for Allen key lock	—		4.50	—	—	4.50
Add for cylinder lock	—		11.10	—	—	11.10

Primed steel cam-lock 1½ hour fire rated access doors

Description	Craft@Hrs	Unit	Material $	Labor $	Equipment $	Total $
8" x 8"	P1@.400	Ea	108.00	9.77	—	117.77
12" x 12"	P1@.500	Ea	113.00	12.20	—	125.20
16" x 16"	P1@.600	Ea	134.00	14.70	—	148.70
18" x 18"	P1@.800	Ea	137.00	19.50	—	156.50
24" x 24"	P1@1.25	Ea	191.00	30.50	—	221.50
24" x 36"	P1@1.50	Ea	247.00	36.70	—	283.70
48" x 48"	P1@2.00	Ea	381.00	48.90	—	429.90
Add for cylinder lock			11.40	—	—	11.40

Threaded automatic air vent

Description	Craft@Hrs	Unit	Material $	Labor $	Equipment $	Total $
Float type	P1@.150	Ea	5.65	3.67	—	9.32
35 PSIG	P1@.150	Ea	37.00	3.67	—	40.67
150 PSIG	P1@.150	Ea	60.90	3.67	—	64.57

Threaded manual air vent

Description	Craft@Hrs	Unit	Material $	Labor $	Equipment $	Total $
175 PSIG	P1@.150	Ea	3.50	3.67	—	7.17

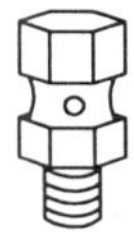

150 pound forged steel slip-on companion flange, welding type

Description	Craft@Hrs	Unit	Material $	Labor $	Equipment $	Total $
2½"	P1@.610	Ea	18.50	14.90	—	33.40
3"	P1@.730	Ea	18.50	17.80	—	36.30
4"	P1@.980	Ea	25.50	23.90	—	49.40
6"	ER@1.47	Ea	33.68	38.30	7.35	79.33
8"	ER@1.77	Ea	62.53	46.20	8.85	117.58
10"	ER@2.20	Ea	103.50	57.40	11.00	171.90
12"	ER@2.64	Ea	154.60	68.90	13.20	236.70

150 pound forged steel companion flange, threaded

Description	Craft@Hrs	Unit	Material $	Labor $	Equipment $	Total $
2"	P1@.290	Ea	19.86	7.09	—	26.95
2½"	P1@.380	Ea	29.89	9.29	—	39.18
3"	P1@.460	Ea	33.32	11.20	—	44.52
4"	P1@.600	Ea	44.28	14.70	—	58.98
6"	ER@.680	Ea	65.50	17.70	3.40	86.60
8"	ER@.760	Ea	91.26	19.80	3.80	114.86

300 pound forged steel slip-on companion flange, welding type

Description	Craft@Hrs	Unit	Material $	Labor $	Equipment $	Total $
2½"	P1@.810	Ea	21.20	19.80	—	41.00
3"	P1@.980	Ea	21.20	23.90	—	45.10
4"	P1@1.30	Ea	31.20	31.80	—	63.00
6"	ER@1.95	Ea	56.50	50.90	9.75	117.15
8"	ER@2.35	Ea	80.00	61.30	11.80	153.10
10"	ER@2.90	Ea	132.00	75.60	14.50	222.10
12"	ER@3.50	Ea	210.00	91.30	17.50	318.80

PVC companion flange

Description	Craft@Hrs	Unit	Material $	Labor $	Equipment $	Total $
2"	P1@.140	Ea	2.25	3.42	—	5.67
2½"	P1@.180	Ea	6.75	4.40	—	11.15
3"	P1@.210	Ea	7.00	5.13	—	12.13
4"	P1@.280	Ea	7.25	6.84	—	14.09
5"	P1@.310	Ea	16.10	7.58	—	23.68
6"	P1@.420	Ea	16.10	10.30	—	26.40
8"	P1@.560	Ea	24.00	13.70	—	37.70

Bolt and gasket set

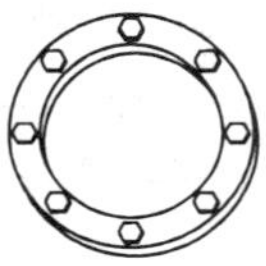

Description	Craft@Hrs	Unit	Material $	Labor $	Equipment $	Total $
2"	P1@.500	Ea	1.50	12.20	—	13.70
2½"	P1@.650	Ea	1.75	15.90	—	17.65
3"	P1@.750	Ea	3.00	18.30	—	21.30
4"	P1@1.00	Ea	5.00	24.40	—	29.40
5"	P1@1.10	Ea	5.75	26.90	—	32.65
6"	P1@1.20	Ea	5.75	29.30	—	35.05
8"	P1@1.25	Ea	7.50	30.50	—	38.00
10"	P1@1.70	Ea	9.00	41.50	—	50.50
12"	P1@2.20	Ea	12.00	53.80	—	65.80

Dielectric union, brazed joints

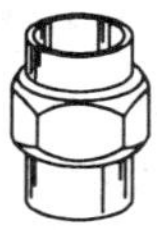

Description	Craft@Hrs	Unit	Material $	Labor $	Equipment $	Total $
1/2"	P1@.146	Ea	3.63	3.57	—	7.20
3/4"	P1@.205	Ea	3.63	5.01	—	8.64
1"	P1@.263	Ea	7.53	6.43	—	13.96
1¼"	P1@.322	Ea	12.87	7.87	—	20.74
1½"	P1@.380	Ea	17.50	9.29	—	26.79
2"	P1@.507	Ea	24.30	12.40	—	36.70

125/150# galvanized ASME expansion tank

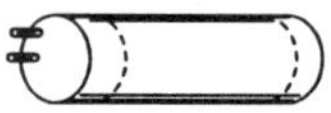

Description	Craft@Hrs	Unit	Material $	Labor $	Equipment $	Total $
15 gallon	P1@1.75	Ea	50.00	42.80	—	92.80
30 gallon	P1@2.00	Ea	90.00	48.90	—	138.90
40 gallon	P1@2.35	Ea	100.00	57.40	—	157.40
60 gallon	P1@2.65	Ea	135.00	64.80	—	199.80
80 gallon	P1@2.95	Ea	250.00	72.10	—	322.10
100 gallon	P1@3.25	Ea	300.00	79.40	—	379.40
120 gallon	P1@3.50	Ea	325.00	85.50	—	410.50

Expansion tank fitting

Description	Craft@Hrs	Unit	Material $	Labor $	Equipment $	Total $
3/4"	P1@.300	Ea	32.40	7.33	—	39.73

Lead roof pipe flashing

Description	Craft@Hrs	Unit	Material $	Labor $	Equipment $	Total $
1½"	P1@.250	Ea	14.00	6.11	—	20.11
2"	P1@.250	Ea	18.40	6.11	—	24.51
3"	P1@.250	Ea	21.30	6.11	—	27.41
4"	P1@.250	Ea	23.50	6.11	—	29.61
6"	P1@.350	Ea	30.00	8.55	—	38.55
8"	P1@.400	Ea	38.10	9.77	—	47.87

Roof flashing, 20" x 20"

Description	Craft@Hrs	Unit	Material $	Labor $	Equipment $	Total $
Pitched roof, neoprene all pipe	P1@.300	Ea	9.00	7.33	—	16.33
Pitched roof, aluminum/neoprene all pipe	P1@.300	Ea	9.50	7.33	—	16.83
Flat roof, aluminum/neoprene all pipe	P1@.300	Ea	18.00	7.33	—	25.33

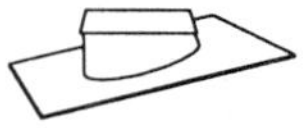

Braided stainless steel pipe connector, solder ends, brazed joints

Description	Craft@Hrs	Unit	Material $	Labor $	Equipment $	Total $
1/2"	P1@.240	Ea	21.50	5.86	—	27.36
3/4"	P1@.300	Ea	28.70	7.33	—	36.03
1"	P1@.360	Ea	37.20	8.80	—	46.00
1¼"	P1@.480	Ea	43.00	11.70	—	54.70
1½"	P1@.540	Ea	49.10	13.20	—	62.30
2"	P1@.600	Ea	65.60	14.70	—	80.30

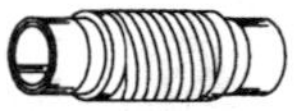

Braided stainless steel pipe connector, threaded joints

Description	Craft@Hrs	Unit	Material $	Labor $	Equipment $	Total $
1/2"	P1@.210	Ea	24.30	5.13	—	29.43
3/4"	P1@.250	Ea	26.30	6.11	—	32.41
1"	P1@.300	Ea	28.40	7.33	—	35.73
1¼"	P1@.400	Ea	29.00	9.77	—	38.77
1½"	P1@.450	Ea	39.20	11.00	—	50.20
2"	P1@.500	Ea	50.60	12.20	—	62.80

Braided stainless steel pipe connector, flanged joints

Description	Craft@Hrs	Unit	Material $	Labor $	Equipment $	Total $
2"	P1@.500	Ea	58.10	12.20	—	70.30
2½"	P1@.630	Ea	71.60	15.40	—	87.00
3"	P1@.750	Ea	92.50	18.30	—	110.80
4"	P1@1.50	Ea	122.00	36.70	—	158.70
6"	P1@2.50	Ea	220.00	61.10	—	281.10
8"	P1@3.00	Ea	491.00	73.30	—	564.30

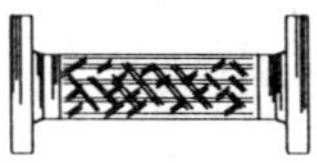

Galvanized steel band pipe hanger

Description	Craft@Hrs	Unit	Material $	Labor $	Equipment $	Total $
1/2"	P1@.450	Ea	.75	11.00	—	11.75
3/4"	P1@.450	Ea	.75	11.00	—	11.75
1"	P1@.450	Ea	.75	11.00	—	11.75
1¼"	P1@.450	Ea	.80	11.00	—	11.80
1½"	P1@.450	Ea	.85	11.00	—	11.85
2"	P1@.450	Ea	.90	11.00	—	11.90
2½"	P1@.450	Ea	1.10	11.00	—	12.10
3"	P1@.450	Ea	1.25	11.00	—	12.25
4"	P1@.450	Ea	1.60	11.00	—	12.60
6"	P1@.500	Ea	2.40	12.20	—	14.60
8"	P1@.550	Ea	3.75	13.40	—	17.15

Trapeze pipe hanger

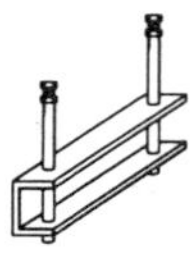

Description	Craft@Hrs	Unit	Material $	Labor $	Equipment $	Total $
24"	P1@1.00	Ea	20.40	24.40	—	44.80
36"	P1@1.25	Ea	26.10	30.50	—	56.60
48"	P1@1.50	Ea	39.60	36.70	—	76.30

Wall bracket support with anchors and bolts

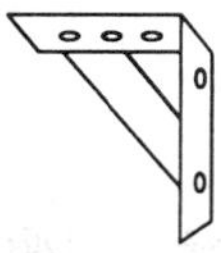

Description	Craft@Hrs	Unit	Material $	Labor $	Equipment $	Total $
6" x 6"	P1@.600	Ea	9.60	14.70	—	24.30
8" x 8"	P1@.650	Ea	11.00	15.90	—	26.90
10" x 10"	P1@.700	Ea	12.30	17.10	—	29.40
12" x 12"	P1@.800	Ea	15.00	19.50	—	34.50
15" x 15"	P1@.950	Ea	17.20	23.20	—	40.40
18" x 18"	P1@1.10	Ea	24.30	26.90	—	51.20
24" x 24"	P1@1.60	Ea	35.30	39.10	—	74.40

Galvanized U-bolts with nuts

Description	Craft@Hrs	Unit	Material $	Labor $	Equipment $	Total $
1/2"	P1@.160	Ea	.78	3.91	—	4.69
3/4"	P1@.165	Ea	.78	4.03	—	4.81
1"	P1@.170	Ea	.86	4.15	—	5.01
1¼"	P1@.175	Ea	.98	4.28	—	5.26
1½"	P1@.180	Ea	1.18	4.40	—	5.58
2"	P1@.190	Ea	1.25	4.64	—	5.89
2½"	P1@.210	Ea	1.95	5.13	—	7.08
3"	P1@.250	Ea	2.25	6.11	—	8.36
4"	P1@.300	Ea	2.55	7.33	—	9.88
6"	P1@.400	Ea	3.98	9.77	—	13.75

Nail-on wire pipe hooks

Description	Craft@Hrs	Unit	Material $	Labor $	Equipment $	Total $
1/2"	P1@.100	Ea	.05	2.44	—	2.49
3/4"	P1@.100	Ea	.06	2.44	—	2.50
1"	P1@.100	Ea	.07	2.44	—	2.51

Galvanized steel pipe sleeves

Description	Craft@Hrs	Unit	Material $	Labor $	Equipment $	Total $
1"	P1@.120	Ea	1.88	2.93	—	4.81
1¼"	P1@.125	Ea	2.10	3.05	—	5.15
1½"	P1@.125	Ea	2.36	3.05	—	5.41
2"	P1@.130	Ea	2.55	3.18	—	5.73
2½"	P1@.150	Ea	2.59	3.67	—	6.26
3"	P1@.180	Ea	2.66	4.40	—	7.06
4"	P1@.220	Ea	3.02	5.38	—	8.40
5"	P1@.250	Ea	3.75	6.11	—	9.86
6"	P1@.270	Ea	4.05	6.60	—	10.65
8"	P1@.270	Ea	4.66	6.60	—	11.26
10"	P1@.290	Ea	5.43	7.09	—	12.52
12"	P1@.310	Ea	7.36	7.58	—	14.94
14"	P1@.330	Ea	13.80	8.06	—	21.86

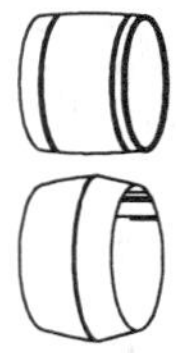

Dial-type pressure gauge

Description	Craft@Hrs	Unit	Material $	Labor $	Equipment $	Total $
2½"	P1@.200	Ea	15.50	4.89	—	20.39
3½"	P1@.200	Ea	20.50	4.89	—	25.39

Riser clamp

Description	Craft@Hrs	Unit	Material $	Labor $	Equipment $	Total $
1/2"	P1@.100	Ea	.95	2.44	—	3.39
3/4"	P1@.100	Ea	1.39	2.44	—	3.83
1"	P1@.100	Ea	1.41	2.44	—	3.85
1¼"	P1@.105	Ea	1.70	2.57	—	4.27
1½"	P1@.110	Ea	1.79	2.69	—	4.48
2"	P1@.115	Ea	1.89	2.81	—	4.70
2½"	P1@.120	Ea	2.00	2.93	—	4.93
3"	P1@.120	Ea	2.18	2.93	—	5.11
4"	P1@.125	Ea	2.75	3.05	—	5.80
5"	P1@.180	Ea	3.99	4.40	—	8.39
6"	P1@.200	Ea	4.79	4.89	—	9.68
8"	P1@.200	Ea	7.77	4.89	—	12.66
10"	P1@.250	Ea	11.60	6.11	—	17.71
12"	P1@.250	Ea	13.75	6.11	—	19.86

15 PSIG float and thermostatic steam trap, threaded

Description	Craft@Hrs	Unit	Material $	Labor $	Equipment $	Total $
3/4"	P1@.250	Ea	102.00	6.11	—	108.11
1"	P1@.300	Ea	102.00	7.33	—	109.33
1¼"	P1@.400	Ea	123.00	9.77	—	132.77
1½"	P1@.450	Ea	230.00	11.00	—	241.00
2"	P1@.500	Ea	297.00	12.20	—	309.20

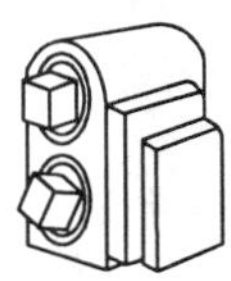

15 PSIG inverted bucket steam trap, threaded joints

Description	Craft@Hrs	Unit	Material $	Labor $	Equipment $	Total $
1/2"	P1@.210	Ea	91.00	5.13	—	96.13
3/4"	P1@.250	Ea	223.00	6.11	—	229.11
1"	P1@.300	Ea	320.00	7.33	—	327.33
1¼"	P1@.400	Ea	530.00	9.77	—	539.77
1½"	P1@.450	Ea	530.00	11.00	—	541.00
2"	P1@.500	Ea	798.00	12.20	—	810.20

Thermometer with well

Description	Craft@Hrs	Unit	Material $	Labor $	Equipment $	Total $
7"	P1@.250	Ea	15.00	6.11	—	21.11
9"	P1@.250	Ea	15.50	6.11	—	21.61

Class 125 bronze body 2-piece ball valve, brazed joints

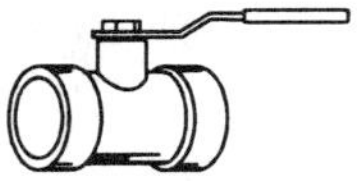

Description	Craft@Hrs	Unit	Material $	Labor $	Equipment $	Total $
1½"	P1@.240	Ea	3.56	5.86	—	9.42
3/4"	P1@.300	Ea	4.75	7.33	—	12.08
1"	P1@.360	Ea	8.21	8.80	—	17.01
1¼"	P1@.480	Ea	13.70	11.70	—	25.40
1½"	P1@.540	Ea	18.50	13.20	—	31.70
2"	P1@.600	Ea	23.30	14.70	—	38.00

Class 125 bronze body 2-piece ball valve, soft-soldered joints

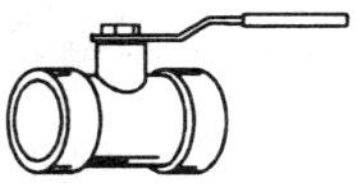

Description	Craft@Hrs	Unit	Material $	Labor $	Equipment $	Total $
1/2"	P1@.200	Ea	3.56	4.89	—	8.45
3/4"	P1@.249	Ea	4.75	6.08	—	10.83
1"	P1@.299	Ea	8.21	7.31	—	15.52
1¼"	P1@.308	Ea	13.70	7.53	—	21.23
1½"	P1@.498	Ea	18.50	12.20	—	30.70
2"	P1@.498	Ea	23.30	12.20	—	35.50

Class 125 bronze body 2-piece ball valve, threaded joints

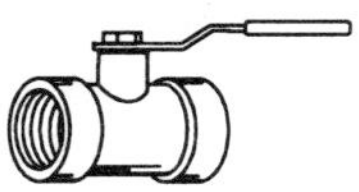

Description	Craft@Hrs	Unit	Material $	Labor $	Equipment $	Total $
1/2"	P1@.210	Ea	3.56	5.13	—	8.69
3/4"	P1@.250	Ea	4.75	6.11	—	10.86
1"	P1@.300	Ea	8.21	7.33	—	15.54
1¼"	P1@.400	Ea	13.70	9.77	—	23.47
1½"	P1@.450	Ea	18.50	11.00	—	29.50
2"	P1@.500	Ea	23.30	12.20	—	35.50

Class 150, 600 pound W.O.G. bronze body 2-piece ball valve, threaded joints

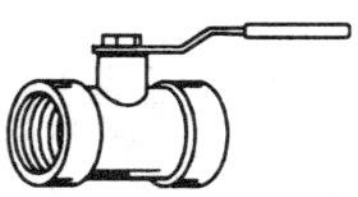

Description	Craft@Hrs	Unit	Material $	Labor $	Equipment $	Total $
1/2"	P1@.210	Ea	5.27	5.13	—	10.40
3/4"	P1@.250	Ea	8.61	6.11	—	14.72
1"	P1@.300	Ea	10.90	7.33	—	18.23
1¼"	P1@.400	Ea	13.70	9.77	—	23.47
1½"	P1@.450	Ea	18.50	11.00	—	29.50
2"	P1@.500	Ea	23.30	12.20	—	35.50

200 PSIG iron body butterfly valve, lug type

Description	Craft@Hrs	Unit	Material $	Labor $	Equipment $	Total $
2"	P1@.450	Ea	84.50	11.00	—	95.50
2½"	P1@.450	Ea	86.50	11.00	—	97.50
3"	P1@.550	Ea	91.70	13.40	—	105.10
4"	P1@.505	Ea	114.00	12.30	—	126.30
6"	ER@.800	Ea	185.00	20.90	4.00	209.90
8"	ER@.800	Ea	252.00	20.90	4.00	276.90
10"	ER@.900	Ea	350.00	23.50	4.50	378.00
12"	ER@1.00	Ea	460.00	26.10	5.00	491.10

200 PSIG iron body butterfly valve, wafer type

Description	Craft@Hrs	Unit	Material $	Labor $	Equipment $	Total $
2"	P1@.450	Ea	76.70	11.00	—	87.70
2½"	P1@.450	Ea	78.70	11.00	—	89.70
3"	P1@.550	Ea	84.50	13.40	—	97.90
4"	P1@.550	Ea	102.00	13.40	—	115.40
6"	ER@.800	Ea	170.00	20.90	4.00	194.90
8"	ER@.800	Ea	237.00	20.90	4.00	261.90
10"	ER@.900	Ea	332.00	23.50	4.50	360.00
12"	ER@1.00	Ea	436.00	26.10	5.00	467.10

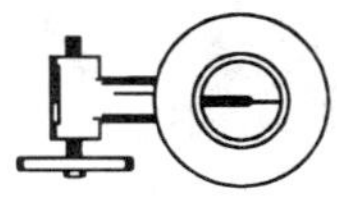

Class 125 bronze body swing check valve, with brazed joints

Description	Craft@Hrs	Unit	Material $	Labor $	Equipment $	Total $
1/2"	P1@.240	Ea	8.00	5.86	—	13.86
3/4"	P1@.300	Ea	11.40	7.33	—	18.73
1"	P1@.360	Ea	15.00	8.80	—	23.80
1¼"	P1@.480	Ea	21.40	11.70	—	33.10
1½"	P1@.540	Ea	30.00	13.20	—	43.20
2"	P1@.600	Ea	50.00	14.70	—	64.70
2½"	P1@1.00	Ea	100.00	24.40	—	124.40
3"	P1@1.50	Ea	144.00	36.70	—	180.70

Class 125 bronze body swing check valve with soft-soldered joints

Description	Craft@Hrs	Unit	Material $	Labor $	Equipment $	Total $
1/2"	P1@.200	Ea	8.00	4.89	—	12.89
3/4"	P1@.249	Ea	11.40	6.08	—	17.48
1"	P1@.299	Ea	15.00	7.31	—	22.31
1¼"	P1@.398	Ea	21.40	9.73	—	31.13
1½"	P1@.448	Ea	30.00	10.90	—	40.90
2"	P1@.498	Ea	50.00	12.20	—	62.20
2½"	P1@.830	Ea	100.00	20.30	—	120.30
3"	P1@1.24	Ea	144.00	30.30	—	174.30

Class 125 bronze body swing check valve, threaded joints

Description	Craft@Hrs	Unit	Material $	Labor $	Equipment $	Total $
1/2"	P1@.210	Ea	8.00	5.13	—	13.13
3/4"	P1@.250	Ea	11.40	6.11	—	17.51
1"	P1@.300	Ea	15.00	7.33	—	22.33
1¼"	P1@.400	Ea	21.40	9.77	—	31.17
1½"	P1@.450	Ea	30.00	11.00	—	41.00
2"	P1@.500	Ea	50.00	12.20	—	62.20

Description	Craft@Hrs	Unit	Material $	Labor $	Equipment $	Total $

Class 300 bronze body swing check valve, threaded joints

Description	Craft@Hrs	Unit	Material $	Labor $	Equipment $	Total $
1/2"	P1@.210	Ea	54.00	5.13	—	59.13
3/4"	P1@.250	Ea	64.00	6.11	—	70.11
1"	P1@.300	Ea	78.00	7.33	—	85.33
1¼"	P1@.400	Ea	113.00	9.77	—	122.77
1½"	P1@.450	Ea	160.00	11.00	—	171.00
2"	P1@.500	Ea	225.00	12.20	—	237.20

Class 125 iron body swing check valve, flanged joints

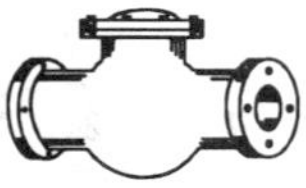

Description	Craft@Hrs	Unit	Material $	Labor $	Equipment $	Total $
2"	P1@.500	Ea	115.00	12.20	—	127.20
2½"	P1@.600	Ea	148.00	14.70	—	162.70
3"	P1@.750	Ea	183.00	18.30	—	201.30
4"	P1@1.35	Ea	271.00	33.00	—	304.00
6"	ER@2.50	Ea	526.00	65.20	12.50	603.70
8"	ER@3.00	Ea	938.00	78.20	15.00	1,031.20
10"	ER@4.00	Ea	1,535.00	104.00	20.00	1,659.00
12"	ER@4.50	Ea	2,100.00	117.00	22.50	2,239.50

Class 250 iron body swing check valve, flanged joints

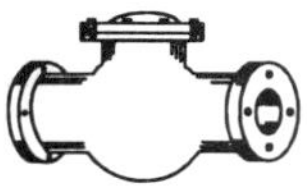

Description	Craft@Hrs	Unit	Material $	Labor $	Equipment $	Total $
2"	P1@.500	Ea	300.00	12.20	—	312.20
2½"	P1@.600	Ea	350.00	14.70	—	364.70
3"	P1@.750	Ea	435.00	18.30	—	453.30
4"	P1@1.35	Ea	535.00	33.00	—	568.00
6"	ER@2.50	Ea	1,020.00	65.20	12.50	1,097.70
8"	ER@3.00	Ea	1,900.00	78.20	15.00	1,993.20

Class 125 iron body silent check valve, wafer type

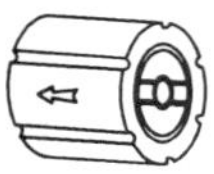

Description	Craft@Hrs	Unit	Material $	Labor $	Equipment $	Total $
2"	P1@.500	Ea	85.00	12.20	—	97.20
2½"	P1@.600	Ea	95.00	14.70	—	109.70
3"	P1@.750	Ea	108.00	18.30	—	126.30
4"	P1@1.35	Ea	146.00	33.00	—	179.00
5"	P1@2.00	Ea	224.00	48.90	—	272.90
6"	P1@2.50	Ea	303.00	61.10	—	364.10
8"	P1@3.00	Ea	522.00	73.30	—	595.30

Class 125 iron body silent check valve, flanged joints

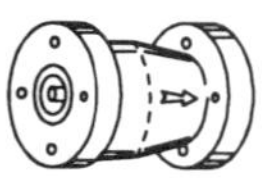

Description	Craft@Hrs	Unit	Material $	Labor $	Equipment $	Total $
2"	P1@.500	Ea	85.00	12.20	—	97.20
2½"	P1@.600	Ea	97.00	14.70	—	111.70
3"	P1@.750	Ea	110.00	18.30	—	128.30
4"	P1@1.35	Ea	145.00	33.00	—	178.00
6"	ER@2.50	Ea	260.00	65.20	12.50	337.70
8"	ER@3.00	Ea	470.00	78.20	15.00	563.20
10"	ER@4.00	Ea	730.00	104.00	20.00	854.00

Description	Craft@Hrs	Unit	Material $	Labor $	Equipment $	Total $

Class 250 iron body silent check valve, flanged joints

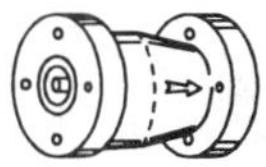

Description	Craft@Hrs	Unit	Material $	Labor $	Equipment $	Total $
2"	P1@.500	Ea	125.00	12.20	—	137.20
2½"	P1@.600	Ea	145.00	14.70	—	159.70
3"	P1@.750	Ea	160.00	18.30	—	178.30
4"	P1@1.35	Ea	220.00	33.00	—	253.00
6"	ER@2.50	Ea	385.00	65.20	12.50	462.70
8"	ER@3.00	Ea	700.00	78.20	15.00	793.20
10"	ER@4.00	Ea	1,100.00	104.00	20.00	1,224.00

Class 125 bronze body gate valve, solder ends

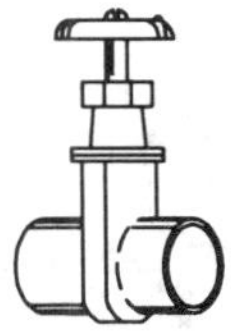

Description	Craft@Hrs	Unit	Material $	Labor $	Equipment $	Total $
1/2"	P1@.240	Ea	6.06	5.86	—	11.92
3/4"	P1@.300	Ea	7.58	7.33	—	14.91
1"	P1@.360	Ea	10.50	8.80	—	19.30
1¼"	P1@.480	Ea	13.40	11.70	—	25.10
1½"	P1@.540	Ea	18.00	13.20	—	31.20
2"	P1@.600	Ea	30.10	14.70	—	44.80
2½"	P1@1.00	Ea	49.90	24.40	—	74.30
3"	P1@1.50	Ea	71.60	36.70	—	108.30

Class 125 bronze body gate valve, threaded joints

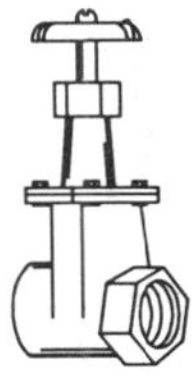

Description	Craft@Hrs	Unit	Material $	Labor $	Equipment $	Total $
1/2"	P1@.210	Ea	6.06	5.13	—	11.19
3/4"	P1@.250	Ea	7.58	6.11	—	13.69
1"	P1@.300	Ea	10.50	7.33	—	17.83
1¼"	P1@.400	Ea	13.40	9.77	—	23.17
1½"	P1@.450	Ea	18.00	11.00	—	29.00
2"	P1@.500	Ea	30.10	12.20	—	42.30
2½"	P1@.750	Ea	49.90	18.30	—	68.20
3"	P1@.950	Ea	71.60	23.20	—	94.80

Class 300 bronze body gate valve, threaded joints

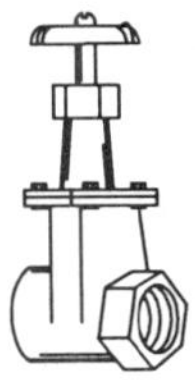

Description	Craft@Hrs	Unit	Material $	Labor $	Equipment $	Total $
1/2"	P1@.210	Ea	14.20	5.13	—	19.33
3/4"	P1@.250	Ea	18.50	6.11	—	24.61
1"	P1@.300	Ea	22.45	7.33	—	29.78
1¼"	P1@.400	Ea	30.67	9.77	—	40.44
1½"	P1@.450	Ea	38.50	11.00	—	49.50
2"	P1@.500	Ea	56.60	12.20	—	68.80
2½"	P1@.750	Ea	115.00	18.30	—	133.30
3"	P1@.950	Ea	164.00	23.20	—	187.20

Class 125 iron body gate valve, flanged joints

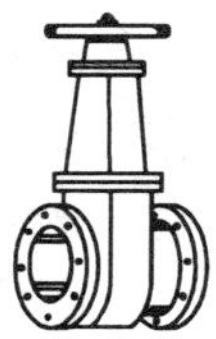

Description	Craft@Hrs	Unit	Material $	Labor $	Equipment $	Total $
2"	P1@.500	Ea	101.00	12.20	—	113.20
2½"	P1@.600	Ea	137.00	14.70	—	151.70
3"	P1@.750	Ea	148.00	18.30	—	166.30
4"	P1@1.35	Ea	218.00	33.00	—	251.00
6"	ER@2.50	Ea	417.00	65.20	12.50	494.70
8"	ER@3.00	Ea	680.00	78.20	15.00	773.20
10"	ER@4.00	Ea	1,120.00	104.00	20.00	1,244.00
12"	ER@4.50	Ea	1,540.00	117.00	22.50	1,679.50

Class 250 iron body gate valve, flanged joints

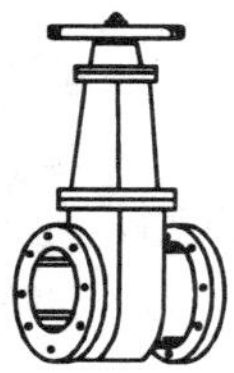

Description	Craft@Hrs	Unit	Material $	Labor $	Equipment $	Total $
2"	P1@.500	Ea	369.00	12.20	—	381.20
2½"	P1@.600	Ea	268.00	14.70	—	282.70
3"	P1@.750	Ea	338.00	18.30	—	356.30
4"	P1@1.35	Ea	570.00	33.00	—	603.00
6"	ER@2.50	Ea	1,061.00	65.20	12.50	1,138.70
8"	ER@3.00	Ea	1,645.00	78.20	15.00	1,738.20
10"	ER@4.00	Ea	2,925.00	104.00	20.00	3,049.00
12"	ER@4.50	Ea	4,775.00	117.00	22.50	4,914.50

Class 125 bronze body globe valve, brazed joints

Description	Craft@Hrs	Unit	Material $	Labor $	Equipment $	Total $
1/2"	P1@.240	Ea	15.80	5.86	—	21.66
3/4"	P1@.300	Ea	21.30	7.33	—	28.63
1"	P1@.360	Ea	30.50	8.80	—	39.30
1¼"	P1@.480	Ea	42.90	11.70	—	54.60
1½"	P1@.540	Ea	57.50	13.20	—	70.70
2"	P1@.600	Ea	93.60	14.70	—	108.30

Class 125 bronze body globe valve, soft-soldered joints

Description	Craft@Hrs	Unit	Material $	Labor $	Equipment $	Total $
1/2"	P1@.200	Ea	15.80	4.89	—	20.69
3/4"	P1@.249	Ea	21.30	6.08	—	27.38
1"	P1@.299	Ea	30.50	7.31	—	37.81
1¼"	P1@.398	Ea	42.90	9.73	—	52.63
1½"	P1@.448	Ea	57.50	10.90	—	68.40
2"	P1@.498	Ea	93.60	12.20	—	105.80

Class 125 bronze body globe valve, threaded ends

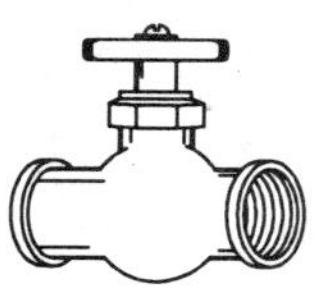

Description	Craft@Hrs	Unit	Material $	Labor $	Equipment $	Total $
1/2"	P1@.210	Ea	15.80	5.13	—	20.93
3/4"	P1@.250	Ea	21.30	6.11	—	27.41
1"	P1@.300	Ea	30.50	7.33	—	37.83
1¼"	P1@.400	Ea	42.90	9.77	—	52.67
1½"	P1@.450	Ea	57.50	11.00	—	68.50
2"	P1@.500	Ea	93.60	12.20	—	105.80

Class 300 bronze body globe valve, threaded joints

Description	Craft@Hrs	Unit	Material $	Labor $	Equipment $	Total $
1/2"	P1@.210	Ea	42.40	5.13	—	47.53
3/4"	P1@.250	Ea	53.65	6.11	—	59.76
1"	P1@.300	Ea	76.50	7.33	—	83.83
1¼"	P1@.400	Ea	110.10	9.77	—	119.87
1½"	P1@.450	Ea	148.00	11.00	—	159.00
2"	P1@.500	Ea	205.00	12.20	—	217.20

Class 125 iron body globe valve, flanged joints

Description	Craft@Hrs	Unit	Material $	Labor $	Equipment $	Total $
2"	P1@.500	Ea	142.00	12.20	—	154.20
2½"	P1@.600	Ea	220.00	14.70	—	234.70
3"	P1@.750	Ea	264.00	18.30	—	282.30
4"	P1@1.35	Ea	352.00	33.00	—	385.00
6"	ER@2.50	Ea	619.00	65.20	12.50	696.70
8"	ER@3.00	Ea	952.00	78.20	15.00	1,045.20
10"	ER@4.00	Ea	1,450.00	104.00	20.00	1,574.00

Class 250 iron body globe valve, flanged joints

Description	Craft@Hrs	Unit	Material $	Labor $	Equipment $	Total $
2"	P1@.500	Ea	310.00	12.20	—	322.20
2½"	P1@.600	Ea	386.00	14.70	—	400.70
3"	P1@.750	Ea	460.00	18.30	—	478.30
4"	P1@1.35	Ea	635.00	33.00	—	668.00
6"	ER@2.50	Ea	1,230.00	65.20	12.50	1,307.70
8"	ER@3.00	Ea	2,780.00	78.20	15.00	2,873.20

Class 125 bronze body strainer, threaded joints

Description	Craft@Hrs	Unit	Material $	Labor $	Equipment $	Total $
1/2"	P1@.230	Ea	12.40	5.62	—	18.02
3/4"	P1@.260	Ea	16.30	6.35	—	22.65
1"	P1@.330	Ea	20.20	8.06	—	28.26
1¼"	P1@.440	Ea	28.10	10.80	—	38.90
1½"	P1@.495	Ea	42.20	12.10	—	54.30
2"	P1@.550	Ea	73.00	13.40	—	86.40

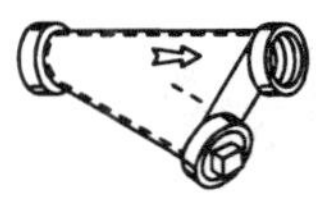

Class 125 iron body strainer, flanged joints

Description	Craft@Hrs	Unit	Material $	Labor $	Equipment $	Total $
2"	P1@.500	Ea	62.00	12.20	—	74.20
2½"	P1@.600	Ea	69.00	14.70	—	83.70
3"	P1@.750	Ea	80.00	18.30	—	98.30
4"	P1@1.35	Ea	137.00	33.00	—	170.00
6"	ER@2.50	Ea	277.00	65.20	12.50	354.70
8"	ER@3.00	Ea	470.00	78.20	15.00	563.20

Class 250 iron body strainer, flanged joints

Description	Craft@Hrs	Unit	Material $	Labor $	Equipment $	Total $
2"	P1@.500	Ea	105.00	12.20	—	117.20
2½"	P1@.600	Ea	115.00	14.70	—	129.70
3"	P1@.750	Ea	136.00	18.30	—	154.30
4"	P1@1.35	Ea	240.00	33.00	—	273.00
6"	ER@2.50	Ea	385.00	65.20	12.50	462.70
8"	ER@3.00	Ea	705.00	78.20	15.00	798.20

Description	Craft@Hrs	Unit	Material $	Labor $	Equipment $	Total $

Installation of 2-way control valve, threaded joints

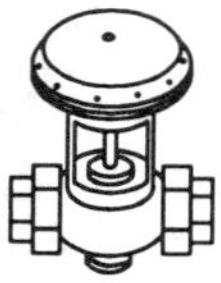

Description	Craft@Hrs	Unit	Material $	Labor $	Equipment $	Total $
1/2"	P1@.210	Ea	—	5.13	—	5.13
3/4"	P1@.250	Ea	—	6.11	—	6.11
1"	P1@.300	Ea	—	7.33	—	7.33
1¼"	P1@.400	Ea	—	9.77	—	9.77
1½"	P1@.450	Ea	—	11.00	—	11.00
2"	P1@.500	Ea	—	12.20	—	12.20
2½"	P1@.830	Ea	—	20.30	—	20.30
3"	P1@.990	Ea	—	24.20	—	24.20

Installation of 2-way control valve, flanged joints

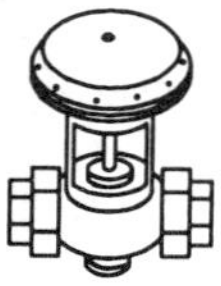

Description	Craft@Hrs	Unit	Material $	Labor $	Equipment $	Total $
2"	P1@.500	Ea	—	12.20	—	12.20
2½"	P1@.600	Ea	—	14.70	—	14.70
3"	P1@.750	Ea	—	18.30	—	18.30
4"	P1@1.35	Ea	—	33.00	—	33.00
6"	P1@2.50	Ea	—	61.10	—	61.10
8"	P1@3.00	Ea	—	73.30	—	73.30
10"	P1@4.00	Ea	—	97.70	—	97.70
12"	P1@4.50	Ea	—	110.00	—	110.00

Installation of 3-way control valve, threaded joints

Description	Craft@Hrs	Unit	Material $	Labor $	Equipment $	Total $
1/2"	P1@.260	Ea	—	6.35	—	6.35
3/4"	P1@.365	Ea	—	8.92	—	8.92
1"	P1@.475	Ea	—	11.60	—	11.60
1¼"	P1@.575	Ea	—	14.10	—	14.10
1½"	P1@.680	Ea	—	16.60	—	16.60
2"	P1@.910	Ea	—	22.20	—	22.20
2½"	P1@1.12	Ea	—	27.40	—	27.40
3"	P1@1.33	Ea	—	32.50	—	32.50

Installation of 3-way control valve, flanged joints

Description	Craft@Hrs	Unit	Material $	Labor $	Equipment $	Total $
2"	P1@.910	Ea	—	22.20	—	22.20
2½"	P1@1.12	Ea	—	27.40	—	27.40
3"	P1@1.33	Ea	—	32.50	—	32.50
4"	P1@2.00	Ea	—	48.90	—	48.90
6"	P1@3.70	Ea	—	90.40	—	90.40
8"	P1@4.40	Ea	—	108.00	—	108.00
10"	P1@5.90	Ea	—	144.00	—	144.00
12"	P1@6.50	Ea	—	159.00	—	159.00

Bronze body circuit balance valve, threaded joints

Description	Craft@Hrs	Unit	Material $	Labor $	Equipment $	Total $
1/2"	P1@.210	Ea	24.00	5.13	—	29.13
3/4"	P1@.250	Ea	25.50	6.11	—	31.61
1"	P1@.300	Ea	30.00	7.33	—	37.33
1¼"	P1@.400	Ea	37.50	9.77	—	47.27
1½"	P1@.400	Ea	48.00	9.77	—	57.77
2"	P1@.500	Ea	65.00	12.20	—	77.20

Description	Craft@Hrs	Unit	Material $	Labor $	Equipment $	Total $
Lever handled gas valve, threaded joints						
1/2"	P1@.210	Ea	15.38	5.13	—	20.51
3/4"	P1@.250	Ea	16.95	6.11	—	23.06
1"	P1@.300	Ea	21.31	7.33	—	28.64
1¼"	P1@.400	Ea	28.25	9.77	—	38.02
1½"	P1@.450	Ea	38.00	11.00	—	49.00
2"	P1@.500	Ea	55.50	12.20	—	67.70
Plug-type gas valve, flanged joints, lubricated						
2"	P1@.500	Ea	141.00	12.20	—	153.20
2½"	P1@.600	Ea	230.00	14.70	—	244.70
3"	P1@.750	Ea	321.00	18.30	—	339.30
4"	P1@1.35	Ea	440.00	33.00	—	473.00
6"	P1@2.50	Ea	785.00	61.10	—	846.10
3/4" hose bibb						
Low quality	P1@.350	Ea	4.68	8.55	—	13.23
Med. quality	P1@.350	Ea	6.80	8.55	—	15.35
High quality	P1@.350	Ea	9.00	8.55	—	17.55
Water pressure reducing valve, threaded joints, 25-75 PSI						
1/2"	P1@.210	Ea	42.00	5.13	—	47.13
3/4"	P1@.250	Ea	54.00	6.11	—	60.11
1"	P1@.300	Ea	90.00	7.33	—	97.33
1¼"	P1@.400	Ea	108.00	9.77	—	117.77
1½"	P1@.450	Ea	155.00	11.00	—	166.00
2"	P1@.500	Ea	245.00	12.20	—	257.20
Reduced pressure backflow preventer, threaded joints						
3/4"	P1@1.00	Ea	210.00	24.40	—	234.40
1"	P1@1.25	Ea	260.00	30.50	—	290.50
1¼"	P1@1.50	Ea	390.00	36.70	—	426.70
1½"	P1@2.00	Ea	410.00	48.90	—	458.90
2"	P1@2.12	Ea	520.00	51.80	—	571.80
Water hammer arrester, threaded joints						
1/2"	P1@.160	Ea	20.10	3.91	—	24.01
3/4"	P1@.190	Ea	23.10	4.64	—	27.74
1"	P1@.230	Ea	59.10	5.62	—	64.72
1¼"	P1@.300	Ea	68.50	7.33	—	75.83
1½"	P1@.340	Ea	98.00	8.31	—	106.31
2"	P1@.380	Ea	149.00	9.29	—	158.29

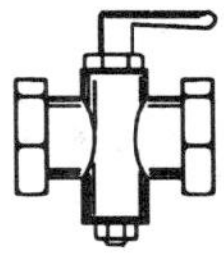

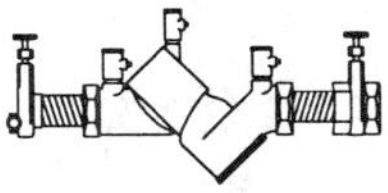

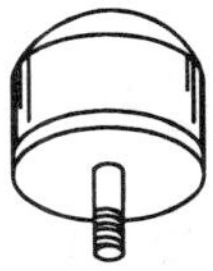

Description	Craft@Hrs	Unit	Material $	Labor $	Equipment $	Total $
Carbon steel weldolet, Schedule 40						
1/2"	P1@.330	Ea	6.68	8.06	—	14.74
3/4"	P1@.500	Ea	6.89	12.20	—	19.09
1"	P1@.670	Ea	7.21	16.40	—	23.61
1¼"	P1@.830	Ea	8.77	20.30	—	29.07
1½"	P1@1.00	Ea	8.77	24.40	—	33.17
2"	P1@1.33	Ea	9.60	32.50	—	42.10
2½"	P1@1.67	Ea	21.00	40.80	—	61.80
3"	P1@2.00	Ea	24.00	48.90	—	72.90
4"	P1@2.67	Ea	31.00	65.20	—	96.20
6"	P1@4.00	Ea	128.00	97.70	—	225.70
8"	P1@4.80	Ea	189.00	117.00	—	306.00
10"	P1@6.00	Ea	260.00	147.00	—	407.00
12"	P1@7.20	Ea	320.00	176.00	—	496.00
Carbon steel weldolet, Schedule 80						
1/2"	P1@.440	Ea	7.34	10.80	—	18.14
3/4"	P1@.670	Ea	7.55	16.40	—	23.95
1"	P1@.890	Ea	8.23	21.70	—	29.93
1¼"	P1@1.11	Ea	10.41	27.10	—	37.51
1½"	P1@1.33	Ea	10.41	32.50	—	42.91
2"	P1@1.77	Ea	11.10	43.30	—	54.40
2½"	P1@2.22	Ea	25.00	54.30	—	79.30
3"	P1@2.66	Ea	25.65	65.00	—	90.65
4"	P1@3.55	Ea	57.00	86.80	—	143.80
6"	P1@5.32	Ea	180.00	130.00	—	310.00
8"	P1@6.38	Ea	240.00	156.00	—	396.00
Carbon steel threadolet, Schedule 40						
3/4"	P1@.330	Ea	4.14	8.06	—	12.20
1"	P1@.440	Ea	4.60	10.80	—	15.40
1¼"	P1@.560	Ea	6.40	13.70	—	20.10
1½"	P1@.670	Ea	7.20	16.40	—	23.60
2"	P1@.890	Ea	8.40	21.70	—	30.10
2½"	P1@1.11	Ea	25.75	27.10	—	52.85
Carbon steel threadolet, Schedule 80						
1/2	P1@.300	Ea	6.20	7.33	—	13.53
3/4"	P1@.440	Ea	6.85	10.80	—	17.65
1"	P1@.590	Ea	8.45	14.40	—	22.85
1¼"	P1@.740	Ea	38.00	18.10	—	56.10
1½"	P1@.890	Ea	38.00	21.70	—	59.70
2"	P1@1.18	Ea	50.00	28.80	—	78.80

Cast Iron, DWV, Service Weight, No-Hub with Coupled Joints

Service weight cast iron soil pipe with hub-less coupled joints is widely used for non-pressurized underground and in-building drain, waste and vent (DWV) systems.

This section has been arranged to save the estimator's time by including all normally-used system components such as pipe, fittings, hanger assemblies and riser clamps under one heading. Additional items can be found under "Plumbing and Piping Specialties." The cost estimates in this section are based on the conditions, limitations and wage rates described in the section "How to Use This Book" beginning on page 5.

Equipment cost, where shown, is $35 per hour for a 10-ton hydraulic truck-mounted crane.

DWV cast iron no-hub pipe

Description	Craft@Hrs	Unit	Material $	Labor $	Equipment $	Total $
2"	P1@.050	LF	3.42	1.22	—	4.64
3"	P1@.060	LF	4.08	1.47	—	5.55
4"	P1@.080	LF	5.33	1.95	—	7.28
6"	ER@.110	LF	9.88	2.87	.55	13.30
8"	ER@.130	LF	18.35	3.39	.65	22.39
10"	ER@.140	LF	32.00	3.65	.70	36.35

DWV cast iron no-hub 1/8 bend

Description	Craft@Hrs	Unit	Material $	Labor $	Equipment $	Total $
2"	P1@.250	Ea	4.08	6.11	—	10.19
3"	P1@.350	Ea	5.33	8.55	—	13.88
4"	P1@.500	Ea	6.73	12.20	—	18.93
6"	ER@.650	Ea	21.69	17.00	3.25	41.94
8"	ER@.710	Ea	57.87	18.50	3.55	79.92
10"	ER@.800	Ea	86.26	20.90	4.00	111.16

DWV cast iron no-hub 1/4 bend

Description	Craft@Hrs	Unit	Material $	Labor $	Equipment $	Total $
2"	P1@.250	Ea	4.99	6.11	—	11.10
3"	P1@.350	Ea	6.20	8.55	—	14.75
4"	P1@.500	Ea	9.43	12.20	—	21.63
6"	ER@.650	Ea	22.95	17.00	3.25	43.20
8"	ER@.710	Ea	88.42	18.50	3.55	110.47
10"	ER@.710	Ea	97.73	18.50	3.55	119.78

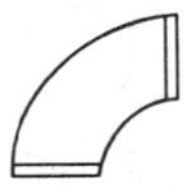

DWV cast iron no-hub long sweep 1/4 bend

Description	Craft@Hrs	Unit	Material $	Labor $	Equipment $	Total $
2"	P1@.250	Ea	8.69	6.11	—	14.80
3"	P1@.350	Ea	10.30	8.55	—	18.85
4"	P1@.500	Ea	16.50	12.20	—	28.70
5"	ER@.580	Ea	30.40	15.10	2.90	48.40
6"	ER@.650	Ea	37.10	17.00	3.25	57.35

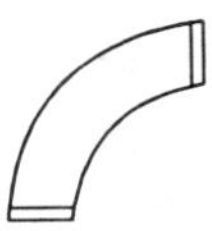

DWV cast iron no-hub low heel outlet 1/4 bend

Description	Craft@Hrs	Unit	Material $	Labor $	Equipment $	Total $
3"	P1@.470	Ea	8.82	11.50	—	20.32
4"	P1@.650	Ea	10.70	15.90	—	26.60

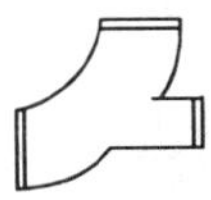

Cast Iron, DWV, Service Weight, No-Hub with Coupled Joints

Description	Craft@Hrs	Unit	Material $	Labor $	Equipment $	Total $
DWV cast iron no-hub closet bend 						
3"	P1@.350	Ea	20.40	8.55	—	28.95
4"	P1@.500	Ea	20.80	12.20	—	33.00
DWV cast iron no-hub closet flange 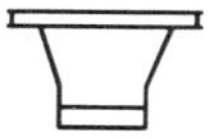						
4" x 3"	P1@.500	Ea	29.75	12.20	—	41.95
DWV cast iron no-hub P-trap 						
2"	P1@.250	Ea	8.06	6.11	—	14.17
3"	P1@.350	Ea	25.11	8.55	—	33.66
4"	P1@.500	Ea	31.12	12.20	—	43.32
6"	ER@.650	Ea	110.30	17.00	3.25	130.55
DWV cast iron no-hub wye 						
2"	P1@.380	Ea	6.36	9.29	—	15.65
3"	P1@.520	Ea	8.91	12.70	—	21.61
4"	P1@.700	Ea	13.89	17.10	—	30.99
6"	ER@.900	Ea	37.67	23.50	4.50	65.67
8"	ER@1.15	Ea	95.28	30.00	5.75	131.03
10"	ER@1.50	Ea	229.86	39.10	7.50	276.46
DWV cast iron no-hub reducing wye 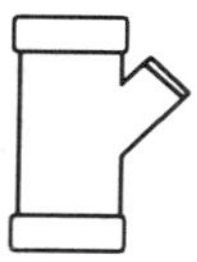						
3" x 2"	P1@.470	Ea	8.58	11.50	—	20.08
4" x 2"	P1@.600	Ea	11.32	14.70	—	26.02
4" x 3"	P1@.650	Ea	12.03	15.90	—	27.93
6" x 2"	ER@.770	Ea	30.57	20.10	3.85	54.52
6" x 3"	ER@.770	Ea	33.17	20.10	3.85	57.12
6" x 4"	ER@.800	Ea	32.57	20.90	4.00	57.47
8" x 2"	ER@1.00	Ea	77.20	26.10	5.00	108.30
8" x 3"	ER@1.00	Ea	78.35	26.10	5.00	109.45
8" x 4"	ER@1.00	Ea	79.72	26.10	5.00	110.82
8" x 6"	ER@1.10	Ea	83.26	28.70	5.50	117.46
DWV cast iron no-hub combination upright wye and 1/8 bend 						
2"	P1@.380	Ea	15.83	9.29	—	25.12
3"	P1@.520	Ea	25.39	12.70	—	38.09
4"	P1@.700	Ea	33.50	17.10	—	50.60

Description	Craft@Hrs	Unit	Material $	Labor $	Equipment $	Total $

DWV cast iron no-hub combination upright reducing wye and 1/8 bend

Description	Craft@Hrs	Unit	Material $	Labor $	Equipment $	Total $
3" x 2"	P1@.470	Ea	27.57	11.50	—	39.07
4" x 2"	P1@.600	Ea	32.77	14.70	—	47.47
4" x 3"	P1@.650	Ea	34.25	15.90	—	50.15

DWV cast iron no-hub double wye

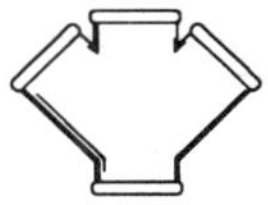

Description	Craft@Hrs	Unit	Material $	Labor $	Equipment $	Total $
2"	P1@.500	Ea	19.65	12.20	—	31.85
3"	P1@.700	Ea	19.80	17.10	—	36.90
4"	P1@.950	Ea	30.85	23.20	—	54.05
6"	ER@1.10	Ea	106.00	28.70	5.50	140.20
8"	ER@1.45	Ea	195.00	37.80	7.25	240.05

DWV cast iron no-hub double reducing wye

Description	Craft@Hrs	Unit	Material $	Labor $	Equipment $	Total $
3" x 2"	P1@.520	Ea	19.65	12.70	—	32.35
4" x 2"	P1@.740	Ea	27.30	18.10	—	45.40
4" x 3"	P1@1.00	Ea	28.25	24.40	—	52.65
6" x 4"	ER@1.75	Ea	104.00	45.60	8.75	158.35

DWV cast iron no-hub combination wye and 1/8 bend

Description	Craft@Hrs	Unit	Material $	Labor $	Equipment $	Total $
2"	P1@.380	Ea	12.22	9.29	—	21.51
3"	P1@.520	Ea	17.40	12.70	—	30.10
4"	P1@.700	Ea	27.85	17.10	—	44.95
6"	ER@.900	Ea	95.00	23.50	4.50	123.00
8"	ER@1.15	Ea	190.00	30.00	5.60	225.60

DWV cast iron no-hub combination reducing wye and 1/8 bend

Description	Craft@Hrs	Unit	Material $	Labor $	Equipment $	Total $
3" x 2"	P1@.470	Ea	14.75	11.50	—	26.25
4" x 2"	P1@.600	Ea	22.66	14.70	—	37.36
4" x 3"	P1@.650	Ea	24.10	15.90	—	40.00
6" x 2"	ER@.770	Ea	107.00	20.10	3.85	130.95
6" x 3"	ER@.770	Ea	112.00	20.10	3.85	135.95
6" x 4"	ER@.800	Ea	115.00	20.90	4.00	139.90
8" x 4"	ER@1.20	Ea	210.00	31.30	6.00	247.30

DWV cast iron no-hub double combination wye and 1/8 bend

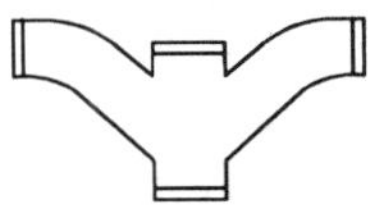

Description	Craft@Hrs	Unit	Material $	Labor $	Equipment $	Total $
2"	P1@.500	Ea	21.10	12.20	—	33.30
3"	P1@.700	Ea	34.00	17.10	—	51.10
4"	P1@.950	Ea	42.00	23.20	—	65.20

Description	Craft@Hrs	Unit	Material $	Labor $	Equipment $	Total $

DWV cast iron no-hub double combination reducing wye and 1/8 bend

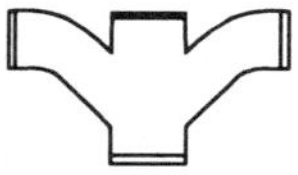

Description	Craft@Hrs	Unit	Material $	Labor $	Equipment $	Total $
3" x 2"	P1@.600	Ea	24.00	14.70	—	38.70
4" x 2"	P1@1.25	Ea	37.95	30.50	—	68.45
4" x 3"	P1@1.35	Ea	39.60	33.00	—	72.60

DWV cast iron no-hub sanitary tee (TY)

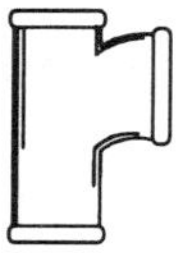

Description	Craft@Hrs	Unit	Material $	Labor $	Equipment $	Total $
2"	P1@.380	Ea	7.37	9.29	—	16.66
3"	P1@.520	Ea	8.91	12.70	—	21.61
4"	P1@.700	Ea	14.55	17.10	—	31.65
6"	ER@.900	Ea	44.50	23.50	4.50	72.50
8"	ER@1.15	Ea	116.80	30.00	5.75	152.55

DWV cast iron no-hub reducing sanitary tee (TY)

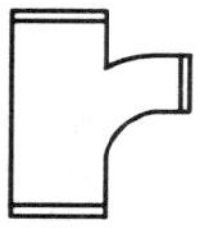

Description	Craft@Hrs	Unit	Material $	Labor $	Equipment $	Total $
3" x 2"	P1@.470	Ea	8.50	11.50	—	20.00
4" x 2"	P1@.600	Ea	11.13	14.70	—	25.83
4" x 3"	P1@.650	Ea	12.96	15.90	—	28.86
6" x 2"	ER@.770	Ea	35.38	20.10	3.85	59.33
6" x 3"	ER@.770	Ea	33.30	20.10	3.85	57.25
6" x 4"	ER@.800	Ea	36.24	20.90	4.00	61.14
8" x 3"	ER@1.00	Ea	92.50	26.10	5.00	123.60
8" x 4"	ER@1.00	Ea	90.00	26.10	5.00	121.10
8" x 6"	ER@1.10	Ea	95.55	28.70	5.50	129.75

DWV cast iron no-hub tapped sanitary tee (Tapped TY)

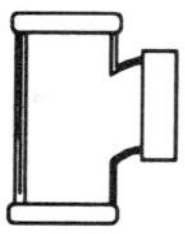

Description	Craft@Hrs	Unit	Material $	Labor $	Equipment $	Total $
2" x 1½"	P1@.200	Ea	20.65	4.89	—	25.54
2" x 2"	P1@.250	Ea	20.65	6.11	—	26.76
3" x 2"	P1@.350	Ea	26.40	8.55	—	34.95
4" x 2"	P1@.425	Ea	31.22	10.40	—	41.62
6" x 2"	ER@.700	Ea	81.00	18.30	3.50	102.80

DWV cast iron no-hub test tee

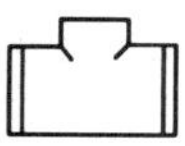

Description	Craft@Hrs	Unit	Material $	Labor $	Equipment $	Total $
2"	P1@.370	Ea	26.23	9.04	—	35.27
3"	P1@.520	Ea	37.12	12.70	—	49.82
4"	P1@.750	Ea	47.23	18.30	—	65.53
6"	ER@1.12	Ea	169.00	29.20	5.60	203.80
8"	ER@1.25	Ea	306.00	32.60	6.25	344.85

DWV cast iron no-hub cross (double TY)

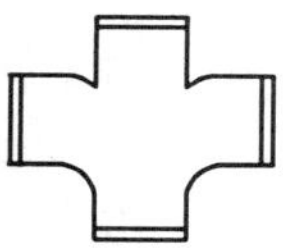

Description	Craft@Hrs	Unit	Material $	Labor $	Equipment $	Total $
2"	P1@.500	Ea	20.75	12.20	—	32.95
3"	P1@.700	Ea	22.25	17.10	—	39.35
4"	P1@.950	Ea	32.00	23.20	—	55.20
6"	ER@1.30	Ea	118.00	33.90	6.50	158.40

DWV cast iron no-hub reducing cross (double TY)

Description	Craft@Hrs	Unit	Material $	Labor $	Equipment $	Total $
3" x 2"	P1@.600	Ea	19.73	14.70	—	34.43
4" x 2"	P1@.900	Ea	27.14	22.00	—	49.14
4" x 3"	P1@.900	Ea	27.85	22.00	—	49.85
6" x 4"	ER@1.20	Ea	109.50	31.30	6.00	146.80
8" x 4"	ER@1.50	Ea	275.00	39.10	7.50	321.60

DWV cast iron no-hub reducer

Description	Craft@Hrs	Unit	Material $	Labor $	Equipment $	Total $
2"	P1@.120	Ea	4.10	2.93	—	7.03
3"	P1@.130	Ea	4.40	3.18	—	7.58
4"	P1@.150	Ea	6.55	3.67	—	10.22
6"	ER@.200	Ea	17.55	5.22	1.00	23.77
8"	ER@.400	Ea	45.00	10.40	2.00	57.40
10"	ER@.650	Ea	69.60	17.00	3.25	89.85

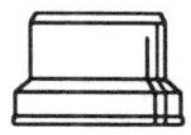

DWV cast iron no-hub cap

Description	Craft@Hrs	Unit	Material $	Labor $	Equipment $	Total $
2"	P1@.125	Ea	2.50	3.05	—	5.55
3"	P1@.175	Ea	4.30	4.28	—	8.58
4"	P1@.250	Ea	5.60	6.11	—	11.71
6"	ER@.375	Ea	10.50	9.78	1.88	22.16
8"	ER@.500	Ea	19.60	13.00	2.50	35.10
10"	ER@.750	Ea	30.20	19.60	3.75	53.55

MJ (mechanical joint) coupling with gasket (labor included with fittings)

Description	Craft@Hrs	Unit	Material $	Labor $	Equipment $	Total $
2"	—	Ea	4.85	9.72	—	14.57
3"	—	Ea	5.30	10.90	—	16.20
4"	—	Ea	5.70	10.90	—	16.60
6"	—	Ea	15.50	12.20	—	27.70
8"	—	Ea	28.75	13.40	—	42.15
10"	—	Ea	40.00	15.80	—	55.80

Hanger with swivel assembly

Description	Craft@Hrs	Unit	Material $	Labor $	Equipment $	Total $
2"	P1@.400	Ea	2.60	9.77	—	12.37
3"	P1@.450	Ea	3.35	11.00	—	14.35
4"	P1@.450	Ea	4.15	11.00	—	15.15
5"	P1@.500	Ea	6.40	12.20	—	18.60
6"	P1@.500	Ea	7.15	12.20	—	19.35
8"	P1@.550	Ea	9.40	13.40	—	22.80
10"	P1@.650	Ea	12.25	15.90	—	28.15

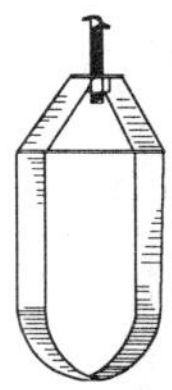

Riser clamp

Description	Craft@Hrs	Unit	Material $	Labor $	Equipment $	Total $
2"	P1@.115	Ea	1.89	2.81	—	4.70
3"	P1@.120	Ea	2.18	2.93	—	5.11
4"	P1@.125	Ea	2.75	3.05	—	5.80
5"	P1@.180	Ea	3.99	4.40	—	8.39
6"	P1@.200	Ea	4.79	4.89	—	9.68
8"	P1@.200	Ea	7.77	4.89	—	12.66
10"	P1@.250	Ea	11.60	6.11	—	17.71

Cast Iron, DWV, Service Weight, Hub & Spigot with Gasketed Joints

Service weight hub and spigot cast iron soil pipe with gasketed joints is widely used for non-pressurized underground and in-building drain, waste and vent (DWV) systems.

This section has been arranged to save the estimator's time by including all normally-used system components such as pipe, fittings, valves, hanger assemblies and riser clamps under one heading. Additional items can be found under "Plumbing and Piping Specialties." The cost estimates in this section are based on the conditions, limitations and wage rates described in the section "How to Use This Book" beginning on page 5.

Equipment cost, where shown, is $35 per hour for a 10-ton hydraulic truck-mounted crane.

Description	Craft@Hrs	Unit	Material $	Labor $	Equipment $	Total $
DWV cast iron pipe, hub and spigot, gasketed joints						
2"	P1@.060	LF	5.52	1.47	—	6.99
3"	P1@.070	LF	6.10	1.71	—	7.81
4"	P1@.090	LF	7.96	2.20	—	10.16
6"	ER@.110	LF	15.21	2.87	.55	18.63
8"	ER@.140	LF	25.57	3.65	.70	29.92
10"	ER@.180	LF	40.91	4.69	.90	46.50
12"	ER@.230	LF	52.80	6.00	1.15	59.95
15"	ER@.280	LF	78.94	7.30	1.40	87.64
DWV cast iron 1/8 bend, hub and spigot, gasketed joints						
2"	P1@.300	Ea	6.38	7.33	—	13.71
3"	P1@.400	Ea	8.83	9.77	—	18.60
4"	P1@.550	Ea	12.30	13.40	—	25.70
6"	ER@.700	Ea	31.00	18.30	3.50	52.80
8"	ER@1.15	Ea	72.80	30.00	5.75	108.55
10"	ER@1.25	Ea	99.45	32.60	6.25	138.30
12"	ER@2.25	Ea	230.00	58.70	11.30	300.00
15"	ER@2.50	Ea	383.00	65.20	12.50	460.70
DWV cast iron 1/4 bend, hub and spigot, gasketed joints						
2"	P1@.300	Ea	7.15	7.33	—	14.48
3"	P1@.400	Ea	11.00	9.77	—	20.77
4"	P1@.550	Ea	19.35	13.40	—	32.75
6"	ER@.900	Ea	36.73	23.50	4.50	64.73
8"	ER@1.15	Ea	98.15	30.00	5.75	133.90
10"	ER@1.25	Ea	135.85	32.60	6.25	174.70
12"	ER@2.25	Ea	285.00	58.70	11.30	355.00
15"	ER@2.50	Ea	576.55	65.20	12.50	654.25
DWV cast iron heel inlet 1/4 bend, hub and spigot, gasketed joints						
3" x 2"	P1@.550	Ea	13.50	13.40	—	26.90
4" x 2"	P1@.700	Ea	16.20	17.10	—	33.30

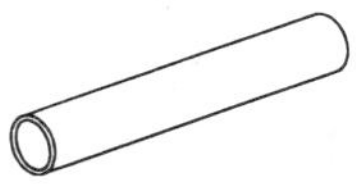

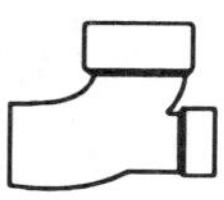

Cast Iron, DWV, Service Weight, Hub & Spigot with Gasketed Joints

Description	Craft@Hrs	Unit	Material $	Labor $	Equipment $	Total $

DWV cast iron 1/16 bend, hub and spigot, gasketed joints

Description	Craft@Hrs	Unit	Material $	Labor $	Equipment $	Total $
2"	P1@.300	Ea	4.07	7.33	—	11.40
3"	P1@.400	Ea	6.77	9.77	—	16.54
4"	P1@.550	Ea	9.43	13.40	—	22.83
6"	ER@.700	Ea	15.90	18.30	3.50	37.70
8"	ER@1.15	Ea	55.70	30.00	5.75	91.45

DWV cast iron closet bend, hub and spigot, gasketed joints

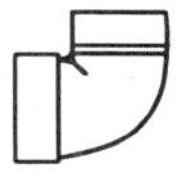

Description	Craft@Hrs	Unit	Material $	Labor $	Equipment $	Total $
4" x 4"	P1@.400	Ea	23.50	9.77	—	33.27

DWV cast iron closet flange, hub and spigot, gasketed joints

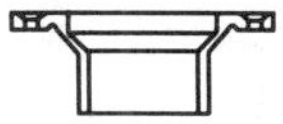

Description	Craft@Hrs	Unit	Material $	Labor $	Equipment $	Total $
4" x 2"	P1@.075	Ea	7.62	1.83	—	9.45
4" x 3"	P1@.100	Ea	10.40	2.44	—	12.84
4" x 4"	P1@.125	Ea	13.60	3.05	—	16.65

DWV cast iron regular P-trap, hub and spigot, gasketed joints

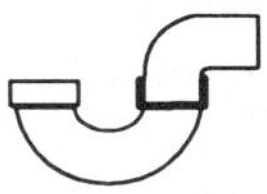

Description	Craft@Hrs	Unit	Material $	Labor $	Equipment $	Total $
2"	P1@.300	Ea	27.30	7.33	—	34.63
3"	P1@.400	Ea	38.00	9.77	—	47.77
4"	P1@.550	Ea	58.50	13.40	—	71.90
6"	ER@.700	Ea	128.30	18.30	3.50	150.10
8"	ER@1.15	Ea	380.25	30.00	5.75	416.00
10"	ER@2.00	Ea	520.00	52.20	10.00	582.20

DWV cast iron sanitary tee, (TY) hub and spigot, gasketed joints

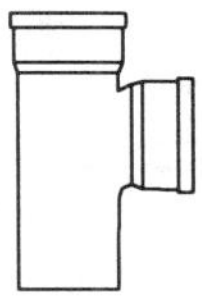

Description	Craft@Hrs	Unit	Material $	Labor $	Equipment $	Total $
2"	P1@.400	Ea	13.10	9.77	—	22.87
3"	P1@.600	Ea	23.00	14.70	—	37.70
4"	P1@.750	Ea	46.00	18.30	—	64.30
6"	ER@.900	Ea	104.00	23.50	4.50	132.00
8"	ER@1.50	Ea	153.00	39.10	7.50	199.60

DWV cast iron sanitary reducing tee, (TY) hub and spigot, gasketed joints

Description	Craft@Hrs	Unit	Material $	Labor $	Equipment $	Total $
3" x 2"	P1@.400	Ea	18.61	9.77	—	28.38
4" x 2"	P1@.650	Ea	48.60	15.90	—	64.50
4" x 3"	P1@.700	Ea	48.60	17.10	—	65.70
6" x 2"	ER@.750	Ea	65.00	19.60	3.75	88.35
6" x 3"	ER@.800	Ea	65.00	20.90	4.00	89.90
6" x 4"	ER@.850	Ea	65.00	22.20	4.25	91.45

Description	Craft@Hrs	Unit	Material $	Labor $	Equipment $	Total $

DWV cast iron sanitary tapped tee, (tapped TY) hub and spigot, gasketed joints

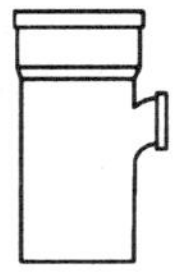

Description	Craft@Hrs	Unit	Material $	Labor $	Equipment $	Total $
2"	P1@.300	Ea	11.10	7.33	—	18.43
3"	P1@.400	Ea	16.40	9.77	—	26.17
4"	P1@.550	Ea	16.20	13.40	—	29.60

DWV cast iron combination wye and 1/8 bend, hub and spigot, gasketed joints

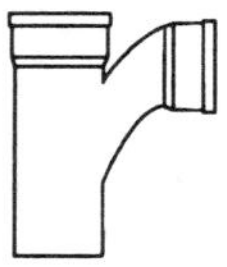

Description	Craft@Hrs	Unit	Material $	Labor $	Equipment $	Total $
2"	P1@.400	Ea	11.50	9.77	—	21.27
3"	P1@.600	Ea	17.50	14.70	—	32.20
4"	P1@.750	Ea	24.20	18.30	—	42.50
5"	ER@.850	Ea	45.80	22.20	4.25	72.25
6"	ER@.900	Ea	58.00	23.50	4.50	86.00

DWV cast iron combination reducing wye and 1/8 bend, hub and spigot, gasketed joints

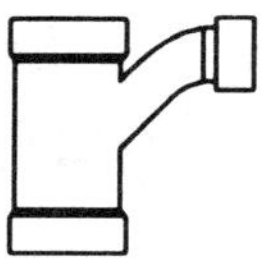

Description	Craft@Hrs	Unit	Material $	Labor $	Equipment $	Total $
3" x 2"	P1@.550	Ea	13.10	13.40	—	26.50
4" x 2"	P1@.650	Ea	17.90	15.90	—	33.80
4" x 3"	P1@.700	Ea	20.90	17.10	—	38.00
5" x 2"	ER@.700	Ea	36.90	18.30	3.50	58.70
5" x 4"	ER@.800	Ea	43.80	20.90	4.00	68.70
6" x 2"	ER@.750	Ea	37.40	19.60	3.75	60.75
6" x 3"	ER@.750	Ea	39.80	19.60	3.75	63.15
6" x 4"	ER@.800	Ea	41.80	20.90	4.00	66.70

DWV cast iron combination double wye and 1/8 bend, hub and spigot, gasketed joints

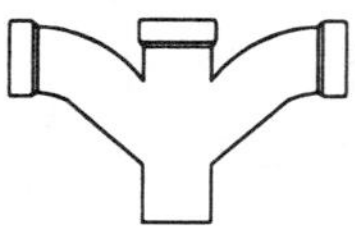

Description	Craft@Hrs	Unit	Material $	Labor $	Equipment $	Total $
2"	P1@.500	Ea	22.30	12.20	—	34.50
3"	P1@.750	Ea	30.30	18.30	—	48.60
4"	P1@.950	Ea	43.20	23.20	—	66.40
6"	ER@1.10	Ea	149.00	28.70	5.50	183.20

DWV cast iron combination double reducing wye and 1/8 bend, hub and spigot, gasketed joints

Description	Craft@Hrs	Unit	Material $	Labor $	Equipment $	Total $
3" x 2"	P1@.550	Ea	24.40	13.40	—	37.80
4" x 2"	P1@.650	Ea	30.60	15.90	—	46.50
4" x 3"	P1@.700	Ea	34.80	17.10	—	51.90
6" x 4"	ER@.800	Ea	97.50	20.90	4.00	122.40

Cast Iron, DWV, Service Weight, Hub & Spigot with Gasketed Joints

DWV cast iron wye, hub and spigot, gasketed joints

Description	Craft@Hrs	Unit	Material $	Labor $	Equipment $	Total $
2"	P1@.400	Ea	10.69	9.77	—	20.46
3"	P1@.600	Ea	18.48	14.70	—	33.18
4"	P1@.750	Ea	24.66	18.30	—	42.96
6"	ER@.900	Ea	59.15	23.50	4.50	87.15
8"	ER@1.50	Ea	131.95	39.10	7.50	178.55
10"	ER@2.00	Ea	223.00	52.20	10.00	285.20
12"	ER@2.50	Ea	500.00	65.20	12.50	577.70
15"	ER@3.00	Ea	995.00	78.20	15.00	1,088.20

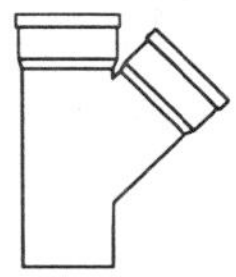

DWV cast iron reducing wye, hub and spigot, gasketed joints

Description	Craft@Hrs	Unit	Material $	Labor $	Equipment $	Total $
4" x 2"	P1@.650	Ea	18.65	15.90	—	34.55
4" x 3"	P1@.700	Ea	22.88	17.10	—	39.98
6" x 2"	ER@.750	Ea	47.45	19.60	3.75	70.80
6" x 3"	ER@.750	Ea	47.45	19.60	3.75	70.80
6" x 4"	ER@.800	Ea	47.90	20.90	4.00	72.80
8" x 4"	ER@1.35	Ea	110.50	35.20	6.75	152.45
8" x 6"	ER@1.45	Ea	117.35	37.80	7.25	162.40
10" x 4"	ER@1.85	Ea	187.80	48.30	9.25	245.35
10" x 6"	ER@1.90	Ea	196.30	49.60	9.50	255.40
10" x 8"	ER@1.95	Ea	208.50	50.90	9.80	269.20
12" x 6"	ER@2.35	Ea	353.30	61.30	11.80	426.40
12" x 8"	ER@2.40	Ea	386.10	62.60	12.00	460.70
12" x 10"	ER@2.45	Ea	417.30	63.90	12.30	493.50
15" x 6"	ER@2.85	Ea	571.00	74.30	14.30	659.60
15" x 8"	ER@2.85	Ea	575.90	74.30	14.30	664.50
15" x 10"	ER@2.90	Ea	648.05	75.60	14.50	738.15
15" x 12"	ER@2.95	Ea	654.00	76.90	14.80	745.70

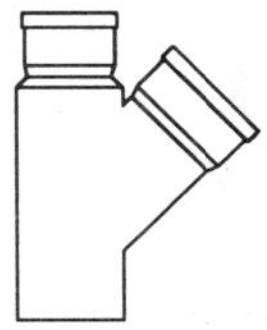

DWV cast iron double wye, hub and spigot, gasketed joints

Description	Craft@Hrs	Unit	Material $	Labor $	Equipment $	Total $
2"	P1@.500	Ea	13.90	12.20	—	26.10
3"	P1@.750	Ea	24.10	18.30	—	42.40
4"	P1@.950	Ea	33.40	23.20	—	56.60
6"	ER@1.10	Ea	78.00	28.70	5.50	112.20
8"	ER@1.85	Ea	191.00	48.30	9.25	248.55
10"	ER@2.50	Ea	268.00	65.20	12.50	345.70
12"	ER@2.75	Ea	387.00	71.70	13.80	472.50
15"	ER@3.50	Ea	1,170.00	91.30	17.50	1,278.80

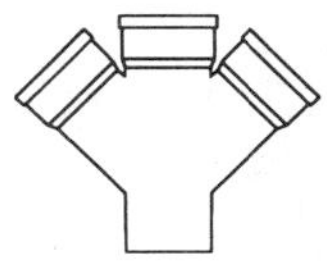

Cast Iron, DWV, Service Weight, Hub & Spigot with Gasketed Joints

DWV cast iron reducing double wye, hub and spigot, gasketed joints

Description	Craft@Hrs	Unit	Material $	Labor $	Equipment $	Total $
3" x 2"	P1@.550	Ea	20.50	13.40	—	33.90
4" x 2"	P1@.650	Ea	24.80	15.90	—	40.70
4" x 3"	P1@.700	Ea	27.60	17.10	—	44.70
6" x 4"	ER@.800	Ea	58.90	20.90	4.00	83.80

DWV cast iron reducer, hub and spigot, gasketed joints

Description	Craft@Hrs	Unit	Material $	Labor $	Equipment $	Total $
3"	P1@.150	Ea	7.15	3.67	—	10.82
4"	P1@.200	Ea	18.66	4.89	—	23.55
6"	ER@.320	Ea	28.92	8.35	1.60	38.87
8"	ER@.350	Ea	61.75	9.13	1.75	72.63
10"	ER@.570	Ea	130.00	14.90	2.85	147.75
12"	ER@.620	Ea	172.00	16.20	3.10	191.30
15"	ER@1.12	Ea	318.00	29.20	5.60	352.80

Gaskets for hub and spigot joints

Description	Craft@Hrs	Unit	Material $	Labor $	Equipment $	Total $
2"	—	Ea	1.79	—	—	1.79
3"	—	Ea	2.37	—	—	2.37
4"	—	Ea	3.02	—	—	3.02
5"	—	Ea	4.65	—	—	4.65
6"	—	Ea	4.88	—	—	4.88
8"	—	Ea	10.70	—	—	10.70
10"	—	Ea	16.50	—	—	16.50
12"	—	Ea	21.00	—	—	21.00
15"	—	Ea	25.00	—	—	25.00

Hanger with swivel assembly

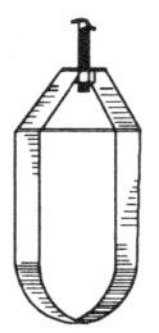

Description	Craft@Hrs	Unit	Material $	Labor $	Equipment $	Total $
2"	P1@.400	Ea	2.60	9.77	—	12.37
3"	P1@.450	Ea	3.35	11.00	—	14.35
4"	P1@.450	Ea	4.15	11.00	—	15.15
5"	P1@.500	Ea	6.40	12.20	—	18.60
6"	P1@.500	Ea	7.15	12.20	—	19.35
8"	P1@.550	Ea	9.40	13.40	—	22.80
10"	P1@.650	Ea	12.25	15.90	—	28.15
12"	P1@.750	Ea	16.95	18.30	—	35.25
15"	P1@.850	Ea	22.50	20.80	—	43.30

Riser clamp

Description	Craft@Hrs	Unit	Material $	Labor $	Equipment $	Total $
2"	P1@.115	Ea	1.89	2.81	—	4.70
3"	P1@.120	Ea	2.18	2.93	—	5.11
4"	P1@.125	Ea	2.75	3.05	—	5.80
5"	P1@.180	Ea	3.99	4.40	—	8.39
6"	P1@.200	Ea	4.79	4.89	—	9.68
8"	P1@.200	Ea	7.77	4.89	—	12.66
10"	P1@.250	Ea	11.60	6.11	—	17.71
12"	P1@.250	Ea	13.75	6.11	—	19.86

Copper drainage tubing is used for non-pressurized above grade, interior building drain, waste and vent (DWV) systems.

Because of its thin walls, it must be handled carefully and protected from damage during building construction.

This section has been arranged to save the estimator's time by including all normally-used system components such as pipe, fittings, hanger assemblies and riser clamps under one heading. Additional items can be found under "Plumbing and Piping Specialties." The cost estimates in this section are based on the conditions, limitations and wage rates described in the section "How to Use This Book" beginning on page 5.

DWV copper pipe, soft-soldered joints

Description	Craft@Hrs	Unit	Material $	Labor $	Equipment $	Total $
1¼"	P1@.040	LF	1.65	.98	—	2.63
1½"	P1@.045	LF	2.07	1.10	—	3.17
2"	P1@.050	LF	2.75	1.22	—	3.97
3"	P1@.060	LF	4.34	1.47	—	5.81
4"	P1@.070	LF	5.51	1.71	—	7.22

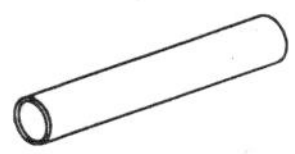

DWV copper 1/8 bend, soft-soldered joints

Description	Craft@Hrs	Unit	Material $	Labor $	Equipment $	Total $
1¼"	P1@.170	Ea	4.50	4.15	—	8.65
1½"	P1@.200	Ea	4.15	4.89	—	9.04
2"	P1@.270	Ea	7.55	6.60	—	14.15
3"	P1@.400	Ea	18.70	9.77	—	28.47
4"	P1@.540	Ea	76.48	13.20	—	89.68

DWV copper 1/4 bend, soft-soldered joints

Description	Craft@Hrs	Unit	Material $	Labor $	Equipment $	Total $
1¼"	P1@.170	Ea	5.05	4.15	—	9.20
1½"	P1@.200	Ea	5.75	4.89	—	10.64
2"	P1@.270	Ea	9.35	6.60	—	15.95
3"	P1@.400	Ea	24.50	9.77	—	34.27
4"	P1@.540	Ea	110.25	13.20	—	123.45

DWV copper closet flange, soft-soldered joints

Description	Craft@Hrs	Unit	Material $	Labor $	Equipment $	Total $
4" x 3"	P1@.680	Ea	24.10	16.60	—	40.70
4" x 4"	P1@.980	Ea	29.60	23.90	—	53.50

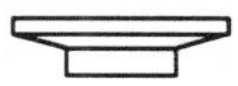

DWV copper P-trap, soft-soldered joints

Description	Craft@Hrs	Unit	Material $	Labor $	Equipment $	Total $
1¼"	P1@.240	Ea	34.10	5.86	—	39.96
1½"	P1@.240	Ea	29.63	5.86	—	35.49
2"	P1@.240	Ea	45.50	5.86	—	51.36
3"	P1@.320	Ea	130.00	7.82	—	137.82

Copper, DWV, with Soft-Soldered Joints

DWV copper sanitary tee, (TY) soft-soldered joints

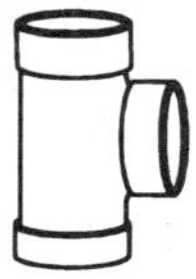

Description	Craft@Hrs	Unit	Material $	Labor $	Equipment $	Total $
1½"	P1@.240	Ea	10.53	5.86	—	16.39
2"	P1@.320	Ea	14.42	7.82	—	22.24
3"	P1@.490	Ea	49.60	12.00	—	61.60
4"	P1@.650	Ea	125.75	15.90	—	141.65

DWV copper sanitary cross, soft-soldered joints

Description	Craft@Hrs	Unit	Material $	Labor $	Equipment $	Total $
1½"	P1@.640	Ea	70.30	15.60	—	85.90
2"	P1@.900	Ea	101.00	22.00	—	123.00
3"	P1@1.36	Ea	285.00	33.20	—	318.20

DWV copper combination wye and 1/8 bend, soft-soldered joints

Description	Craft@Hrs	Unit	Material $	Labor $	Equipment $	Total $
1½"	P1@.240	Ea	22.15	5.86	—	28.01
2"	P1@.320	Ea	37.90	7.82	—	45.72
3"	P1@.490	Ea	126.00	12.00	—	138.00
4"	P1@.650	Ea	280.00	15.90	—	295.90

DWV copper wye, soft-soldered joints

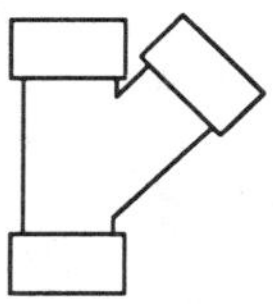

Description	Craft@Hrs	Unit	Material $	Labor $	Equipment $	Total $
1½"	P1@.240	Ea	17.15	5.86	—	23.01
2"	P1@.320	Ea	24.75	7.82	—	32.57
3"	P1@.490	Ea	61.70	12.00	—	73.70
4"	P1@.650	Ea	125.75	15.90	—	141.65

DWV copper cleanout with threaded plug

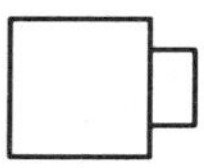

Description	Craft@Hrs	Unit	Material $	Labor $	Equipment $	Total $
1¼"	P1@.100	Ea	8.75	2.44	—	11.19
1½"	P1@.160	Ea	13.00	3.91	—	16.91
2"	P1@.230	Ea	17.00	5.62	—	22.62

DWV copper female adapter

Description	Craft@Hrs	Unit	Material $	Labor $	Equipment $	Total $
1½"	P1@.130	Ea	6.75	3.18	—	9.93
2"	P1@.170	Ea	12.75	4.15	—	16.90
3"	P1@.230	Ea	51.30	5.62	—	56.92
4"	P1@.300	Ea	87.50	7.33	—	94.83

DWV copper male adapter

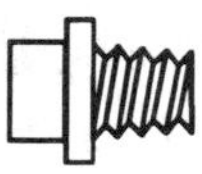

Description	Craft@Hrs	Unit	Material $	Labor $	Equipment $	Total $
1½"	P1@.130	Ea	7.70	3.18	—	10.88
2"	P1@.170	Ea	9.15	4.15	—	13.30
3"	P1@.230	Ea	44.80	5.62	—	50.42
4"	P1@.300	Ea	75.00	7.33	—	82.33

DWV copper soil adapter (copper x MJ)

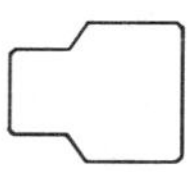

Description	Craft@Hrs	Unit	Material $	Labor $	Equipment $	Total $
2"	P1@.170	Ea	11.10	4.15	—	15.25
3"	P1@.250	Ea	22.40	6.11	—	28.51
4"	P1@.320	Ea	42.00	7.82	—	49.82

DWV copper test cap, soft-soldered joint

Description	Craft@Hrs	Unit	Material $	Labor $	Equipment $	Total $
1½"	P1@.100	Ea	.25	2.44	—	2.69
2"	P1@.130	Ea	.37	3.18	—	3.55
3"	P1@.200	Ea	.46	4.89	—	5.35
4"	P1@.250	Ea	.98	6.11	—	7.09

DWV copper test tee, soft-soldered joints

Description	Craft@Hrs	Unit	Material $	Labor $	Equipment $	Total $
1½"	P1@.200	Ea	25.15	4.89	—	30.04
2"	P1@.270	Ea	32.75	6.60	—	39.35
3"	P1@.400	Ea	88.20	9.77	—	97.97

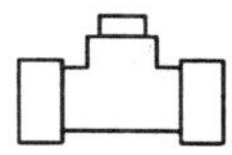

DWV copper fitting reducer, soft-soldered joints

Description	Craft@Hrs	Unit	Material $	Labor $	Equipment $	Total $
1½"	P1@.100	Ea	4.15	2.44	—	6.59
2"	P1@.140	Ea	5.30	3.42	—	8.72
3"	P1@.200	Ea	18.75	4.89	—	23.64
4"	P1@.270	Ea	29.50	6.60	—	36.10

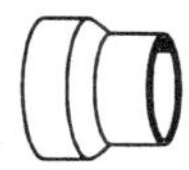

DWV copper coupling, soft-soldered joints

Description	Craft@Hrs	Unit	Material $	Labor $	Equipment $	Total $
1¼"	P1@.170	Ea	2.20	4.15	—	6.35
1½"	P1@.200	Ea	2.55	4.89	—	7.44
2"	P1@.270	Ea	3.65	6.60	—	10.25
3"	P1@.400	Ea	6.75	9.77	—	16.52
4"	P1@.540	Ea	26.40	13.20	—	39.60

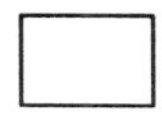

Hanger with swivel assembly

Description	Craft@Hrs	Unit	Material $	Labor $	Equipment $	Total $
1¼"	P1@.450	Ea	1.90	11.00	—	12.90
1½"	P1@.450	Ea	2.05	11.00	—	13.05
2"	P1@.450	Ea	2.15	11.00	—	13.15
3"	P1@.450	Ea	3.60	11.00	—	14.60
4"	P1@.450	Ea	3.95	11.00	—	14.95

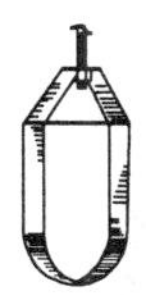

Riser clamp

Description	Craft@Hrs	Unit	Material $	Labor $	Equipment $	Total $
1¼"	P1@.105	Ea	1.70	2.57	—	4.27
1½"	P1@.110	Ea	1.79	2.69	—	4.48
2"	P1@.115	Ea	1.89	2.81	—	4.70
3"	P1@.120	Ea	2.18	2.93	—	5.11
4"	P1@.125	Ea	2.75	3.05	—	5.80

ABS, DWV with Solvent-Weld Joints

ABS (Acrylonitrile Butadiene Styrene) ASTM D 2661-73 is made from various grades of rubber-based resins and is primarily used for non-pressurized drain, waste and vent (DWV) systems.

ABS can also be used for low-temperature (100 degree F. maximum) water distribution systems, but consult the manufacturers for the proper ASTM type, maximum pressure limitations and recommended joint solvent.

The current Uniform Plumbing Code (UPC) states in Sections 401 and 503 that “ABS piping systems shall be limited to those structures where combustible construction is allowed.” This does not preclude its use for underground DWV systems.

This section has been arranged to save the estimator’s time by including all normally used system components such as pipe, fittings, hanger assemblies and riser clamps under one heading. The cost estimates in this section are based on the conditions, limitations and wage rates described in the section “How to Use This Book” beginning on page 5.

Description	Craft@Hrs	Unit	Material $	Labor $	Equipment $	Total $
ABS DWV pipe with solvent-weld joints						
1½"	P1@.020	LF	.30	.49	—	.79
2"	P1@.025	LF	.48	.61	—	1.09
3"	P1@.030	LF	.83	.73	—	1.56
4"	P1@.040	LF	1.36	.98	—	2.34
6"	P1@.055	LF	2.50	1.34	—	3.84
ABS DWV 1/8 bend with solvent-weld joints						
1½"	P1@.120	Ea	.37	2.93	—	3.30
2"	P1@.125	Ea	.67	3.05	—	3.72
3"	P1@.175	Ea	1.62	4.28	—	5.90
4"	P1@.250	Ea	3.06	6.11	—	9.17
6"	P1@.325	Ea	24.37	7.94	—	32.31
ABS DWV 1/8 street bend with solvent-weld joints						
1½"	P1@.060	Ea	.40	1.47	—	1.87
2"	P1@.065	Ea	.78	1.59	—	2.37
3"	P1@.088	Ea	1.72	2.15	—	3.87
4"	P1@.125	Ea	4.73	3.05	—	7.78
6"	P1@.163	Ea	23.66	3.98	—	27.64
ABS DWV 1/4 bend with solvent-weld joints						
1½"	P1@.120	Ea	.45	2.93	—	3.38
2"	P1@.125	Ea	.77	3.05	—	3.82
3"	P1@.175	Ea	1.94	4.28	—	6.22
4"	P1@.250	Ea	4.41	6.11	—	10.52
6"	P1@.325	Ea	25.00	7.94	—	32.94
ABS DWV 1/4 street bend with solvent-weld joints						
1½"	P1@.060	Ea	.51	1.47	—	1.98
2"	P1@.070	Ea	1.19	1.71	—	2.90
3"	P1@.088	Ea	2.31	2.15	—	4.46
4"	P1@.125	Ea	6.77	3.05	—	9.82
6"	P1@.175	Ea	26.00	4.28	—	30.28

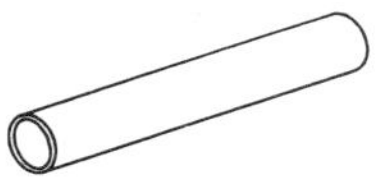

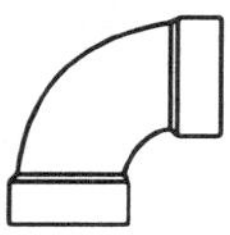
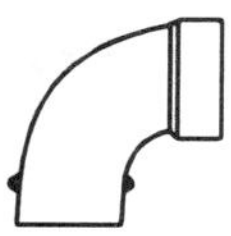

ABS DWV 1/4 bend heel outlet with solvent-weld joints

Description	Craft@Hrs	Unit	Material $	Labor $	Equipment $	Total $
3" x 2"	P1@.235	Ea	8.10	5.74	—	13.84
4" x 2"	P1@.325	Ea	14.70	7.94	—	22.64

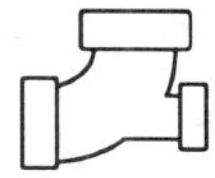

ABS DWV closet bend with solvent-weld joints

Description	Craft@Hrs	Unit	Material $	Labor $	Equipment $	Total $
3" x 4"	P1@.250	Ea	6.60	6.11	—	12.71

ABS DWV closet flange with solvent-weld joints

Description	Craft@Hrs	Unit	Material $	Labor $	Equipment $	Total $
4"	P1@.125	Ea	2.95	3.05	—	6.00

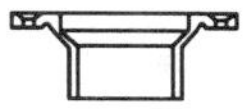

ABS DWV P-trap with solvent-weld joints

Description	Craft@Hrs	Unit	Material $	Labor $	Equipment $	Total $
1½"	P1@.150	Ea	1.35	3.67	—	5.02
2"	P1@.190	Ea	3.35	4.64	—	7.99
3"	P1@.260	Ea	7.90	6.35	—	14.25
4"	P1@.350	Ea	15.50	8.55	—	24.05

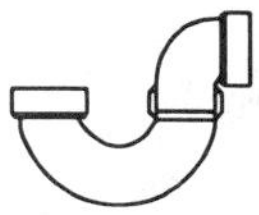

ABS DWV sanitary tee (TY) with solvent-weld joints

Description	Craft@Hrs	Unit	Material $	Labor $	Equipment $	Total $
1½"	P1@.150	Ea	.65	3.67	—	4.32
2"	P1@.190	Ea	1.50	4.64	—	6.14
3"	P1@.260	Ea	2.89	6.35	—	9.24
4"	P1@.350	Ea	6.16	8.55	—	14.71
6"	P1@.450	Ea	27.95	11.00	—	38.95

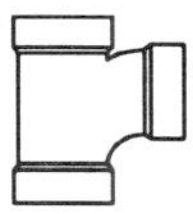

ABS DWV reducing sanitary tee (TY) with solvent-weld joints

Description	Craft@Hrs	Unit	Material $	Labor $	Equipment $	Total $
2" x 1½"	P1@.160	Ea	1.09	3.91	—	5.00
3" x 1½"	P1@.210	Ea	1.87	5.13	—	7.00
3" x 2"	P1@.235	Ea	2.07	5.74	—	7.81
4" x 2"	P1@.300	Ea	5.93	7.33	—	13.26
4" x 3"	P1@.325	Ea	7.10	7.94	—	15.04

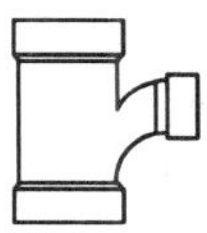

ABS DWV double sanitary tee (TY) with solvent-weld joints

Description	Craft@Hrs	Unit	Material $	Labor $	Equipment $	Total $
1½"	P1@.210	Ea	2.00	5.13	—	7.13
2"	P1@.250	Ea	3.00	6.11	—	9.11
3"	P1@.350	Ea	4.75	8.55	—	13.30
4"	P1@.475	Ea	18.77	11.60	—	30.37

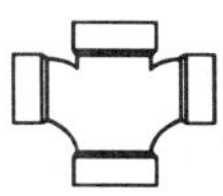

ABS DWV combination wye and 1/8 bend with solvent-weld joints

Description	Craft@Hrs	Unit	Material $	Labor $	Equipment $	Total $
1½"	P1@.150	Ea	3.95	3.67	—	7.62
2"	P1@.190	Ea	5.12	4.64	—	9.76
3"	P1@.260	Ea	6.90	6.35	—	13.25
4"	P1@.350	Ea	12.00	8.55	—	20.55

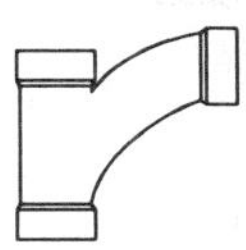

ABS, DWV with Solvent-Weld Joints

ABS DWV combination reducing wye and 1/8 bend with solvent-weld joints

Description	Craft@Hrs	Unit	Material $	Labor $	Equipment $	Total $
2" x 1½"	P1@.160	Ea	5.70	3.91	—	9.61
3" x 1½"	P1@.210	Ea	8.60	5.13	—	13.73
3" x 2"	P1@.235	Ea	6.10	5.74	—	11.84
4" x 2"	P1@.300	Ea	11.00	7.33	—	18.33
4" x 3"	P1@.325	Ea	13.00	7.94	—	20.94

ABS DWV bushing with solvent-weld joints

Description	Craft@Hrs	Unit	Material $	Labor $	Equipment $	Total $
2" x 1½"	P1@.062	Ea	.35	1.52	—	1.87
3" x 1½"	P1@.088	Ea	1.68	2.15	—	3.83
3" x 2"	P1@.088	Ea	1.21	2.15	—	3.36
4" x 2"	P1@.125	Ea	2.44	3.05	—	5.49
4" x 3"	P1@.125	Ea	2.44	3.05	—	5.49
6" x 4"	P1@.163	Ea	12.19	3.98	—	16.17

ABS DWV FIP adapter

Description	Craft@Hrs	Unit	Material $	Labor $	Equipment $	Total $
1½"	P1@.080	Ea	.51	1.95	—	2.46
2"	P1@.085	Ea	.94	2.08	—	3.02
3"	P1@.155	Ea	2.23	3.79	—	6.02
4"	P1@.220	Ea	4.27	5.38	—	9.65
6"	P1@.285	Ea	16.42	6.96	—	23.38

ABS DWV test cap

Description	Craft@Hrs	Unit	Material $	Labor $	Equipment $	Total $
1½"	P1@.030	Ea	.27	.73	—	1.00
2"	P1@.040	Ea	.42	.98	—	1.40
3"	P1@.050	Ea	.66	1.22	—	1.88
4"	P1@.090	Ea	1.03	2.20	—	3.23

ABS DWV cleanout plug

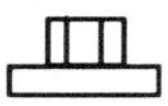

Description	Craft@Hrs	Unit	Material $	Labor $	Equipment $	Total $
1½"	P1@.090	Ea	1.19	2.20	—	3.39
2"	P1@.120	Ea	1.75	2.93	—	4.68
3"	P1@.240	Ea	3.12	5.86	—	8.98
4"	P1@.320	Ea	6.75	7.82	—	14.57
6"	P1@.450	Ea	21.60	11.00	—	32.60

ABS DWV cleanout tee with solvent-weld joints

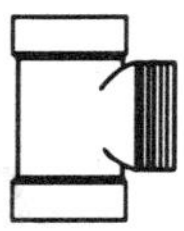

Description	Craft@Hrs	Unit	Material $	Labor $	Equipment $	Total $
1½"	P1@.120	Ea	1.85	2.93	—	4.78
2"	P1@.125	Ea	3.00	3.05	—	6.05
3"	P1@.175	Ea	4.90	4.28	—	9.18
4"	P1@.250	Ea	10.05	6.11	—	16.16

ABS DWV wye with solvent-weld joints

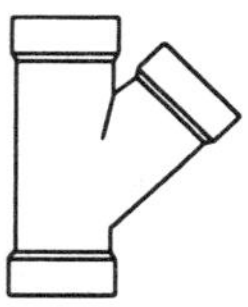

Description	Craft@Hrs	Unit	Material $	Labor $	Equipment $	Total $
1½"	P1@.150	Ea	.86	3.67	—	4.53
2"	P1@.190	Ea	1.62	4.64	—	6.26
3"	P1@.260	Ea	3.00	6.35	—	9.35
4"	P1@.350	Ea	6.95	8.55	—	15.50
6"	P1@.450	Ea	29.35	11.00	—	40.35

ABS DWV reducing wye with solvent-weld joints

Description	Craft@Hrs	Unit	Material $	Labor $	Equipment $	Total $
2" x 1½"	P1@.160	Ea	1.50	3.91	—	5.41
3" x 1½"	P1@.210	Ea	2.11	5.13	—	7.24
3" x 2"	P1@.235	Ea	2.34	5.74	—	8.08
4" x 2"	P1@.300	Ea	5.00	7.33	—	12.33
4" x 3"	P1@.325	Ea	6.28	7.94	—	14.22
6" x 4"	P1@.400	Ea	24.15	9.77	—	33.92

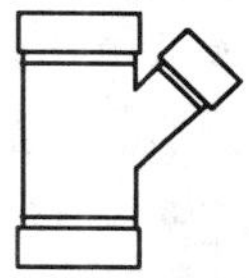

ABS DWV reducer with solvent-weld joints

Description	Craft@Hrs	Unit	Material $	Labor $	Equipment $	Total $
2" x 1½"	P1@.125	Ea	.35	3.05	—	3.40
3" x 1½"	P1@.150	Ea	1.68	3.67	—	5.35
3" x 2"	P1@.150	Ea	1.21	3.67	—	4.88
4" x 2"	P1@.185	Ea	2.44	4.52	—	6.96
4" x 3"	P1@.210	Ea	2.40	5.13	—	7.53

ABS DWV coupling with solvent-weld joints

Description	Craft@Hrs	Unit	Material $	Labor $	Equipment $	Total $
1½"	P1@.120	Ea	.28	2.93	—	3.21
2"	P1@.125	Ea	.43	3.05	—	3.48
3"	P1@.175	Ea	1.00	4.28	—	5.28
4"	P1@.250	Ea	1.80	6.11	—	7.91
6"	P1@.325	Ea	10.62	7.94	—	18.56

Hanger with swivel assembly

Description	Craft@Hrs	Unit	Material $	Labor $	Equipment $	Total $
1½"	P1@.450	Ea	2.05	11.00	—	13.05
2"	P1@.450	Ea	2.15	11.00	—	13.15
3"	P1@.450	Ea	3.60	11.00	—	14.60
4"	P1@.450	Ea	3.95	11.00	—	14.95
6"	P1@.450	Ea	6.45	11.00	—	17.45

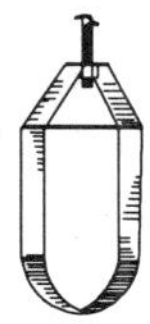

Riser clamp

Description	Craft@Hrs	Unit	Material $	Labor $	Equipment $	Total $
1½"	P1@.110	Ea	1.79	2.69	—	4.48
2"	P1@.115	Ea	1.89	2.81	—	4.70
3"	P1@.120	Ea	2.18	2.93	—	5.11
4"	P1@.125	Ea	2.75	3.05	—	5.80
6"	P1@.130	Ea	4.79	3.18	—	7.97

Galvanized steel pipe sleeves

Description	Craft@Hrs	Unit	Material $	Labor $	Equipment $	Total $
2"	P1@.130	Ea	2.55	3.18	—	5.73
2½"	P1@.150	Ea	2.59	3.67	—	6.26
3"	P1@.180	Ea	2.66	4.40	—	7.06
4"	P1@.220	Ea	3.02	5.38	—	8.40
5"	P1@.250	Ea	3.75	6.11	—	9.86
6"	P1@.270	Ea	4.05	6.60	—	10.65

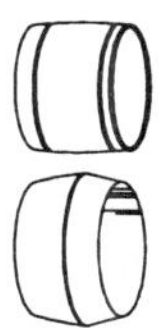

Lead roof pipe flashing

Description	Craft@Hrs	Unit	Material $	Labor $	Equipment $	Total $
1½"	P1@.400	Ea	14.00	9.77	—	23.77
2"	P1@.450	Ea	18.40	11.00	—	29.40
3"	P1@.500	Ea	21.30	12.20	—	33.50
4"	P1@.550	Ea	23.50	13.40	—	36.90

PVC, DWV with Gasketed Bell and Spigot Joints

PVC (Polyvinyl Chloride) is polymerized vinyl chloride produced from acetylene and anhydrous hydrochloric acid.

PVC sewer pipe (ASTM D-3034) 10' lay length, is for underground installation and is used primarily to convey sewage and rain water. It is of the bell & spigot type with joints made water-tight by means of an elastomeric gasket (o-ring) factory-installed in the hub end of the pipe and fittings.

This section includes pipe and all commonly-used fittings. Additional items can be found under "Plumbing and Piping Specialties." The cost estimates in this section are based on the conditions, limitations, and wage rates described in the section "How to Use This Book" beginning on page 5.

PVC SDR35 sewer pipe with bell and spigot gasketed joints

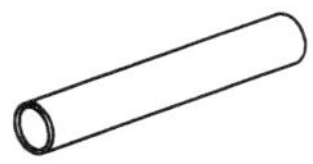

Description	Craft@Hrs	Unit	Material $	Labor $	Equipment $	Total $
4"	P1@.050	LF	.71	1.22	—	1.93
6"	P1@.070	LF	1.56	1.71	—	3.27
8"	P1@.090	LF	2.77	2.20	—	4.97
10"	P1@.110	LF	4.36	2.69	—	7.05
12"	P1@.140	LF	6.24	3.42	—	9.66
15"	P1@.180	LF	9.35	4.40	—	13.75
18"	P1@.230	LF	14.40	5.62	—	20.02

PVC sewer pipe 1/16 bend B x B with bell and spigot gasketed joints

Description	Craft@Hrs	Unit	Material $	Labor $	Equipment $	Total $
4"	P1@.320	Ea	3.17	7.82	—	10.99
6"	P1@.400	Ea	6.23	9.77	—	16.00
8"	P1@.480	Ea	18.20	11.70	—	29.90
10"	P1@.800	Ea	49.70	19.50	—	69.20
12"	P1@1.06	Ea	64.90	25.90	—	90.80
15"	P1@1.34	Ea	172.00	32.70	—	204.70
18"	P1@1.60	Ea	250.00	39.10	—	289.10

PVC sewer pipe 1/16 bend B x S with bell and spigot gasketed joints

Description	Craft@Hrs	Unit	Material $	Labor $	Equipment $	Total $
4"	P1@.320	Ea	3.01	7.82	—	10.83
6"	P1@.400	Ea	5.93	9.77	—	15.70
8"	P1@.480	Ea	19.10	11.70	—	30.80
10"	P1@.800	Ea	48.50	19.50	—	68.00
12"	P1@1.06	Ea	63.50	25.90	—	89.40
15"	P1@1.34	Ea	137.00	32.70	—	169.70
18"	P1@1.60	Ea	210.00	39.10	—	249.10

PVC pipe sewer 1/8 bend B x B with bell and spigot gasketed joints

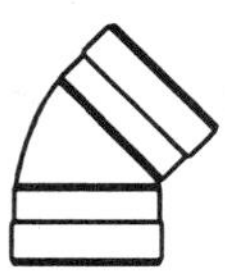

Description	Craft@Hrs	Unit	Material $	Labor $	Equipment $	Total $
4"	P1@.320	Ea	3.19	7.82	—	11.01
6"	P1@.400	Ea	6.44	9.77	—	16.21
8"	P1@.480	Ea	17.20	11.70	—	28.90
10"	P1@.800	Ea	46.70	19.50	—	66.20
12"	P1@1.06	Ea	67.90	25.90	—	93.80
15"	P1@1.34	Ea	152.00	32.70	—	184.70
18"	P1@1.60	Ea	243.00	39.10	—	282.10

Description	Craft@Hrs	Unit	Material $	Labor $	Equipment $	Total $
PVC sewer pipe 1/8 bend B x S with bell and spigot gasketed joints						
4"	P1@.320	Ea	2.85	7.82	—	10.67
6"	P1@.400	Ea	5.71	9.77	—	15.48
8"	P1@.480	Ea	17.20	11.70	—	28.90
10"	P1@.800	Ea	46.00	19.50	—	65.50
12"	P1@1.06	Ea	65.80	25.90	—	91.70
15"	P1@1.34	Ea	119.00	32.70	—	151.70
18"	P1@1.60	Ea	201.00	39.10	—	240.10

Description	Craft@Hrs	Unit	Material $	Labor $	Equipment $	Total $
PVC sewer pipe 1/4 bend B x B with bell and spigot gasketed joints						
4"	P1@.320	Ea	4.56	7.82	—	12.38
6"	P1@.400	Ea	8.38	9.77	—	18.15
8"	P1@.480	Ea	23.20	11.70	—	34.90
10"	P1@.800	Ea	78.00	19.50	—	97.50
12"	P1@1.06	Ea	101.00	25.90	—	126.90
15"	P1@1.34	Ea	214.00	32.70	—	246.70
18"	P1@1.60	Ea	360.00	39.10	—	399.10

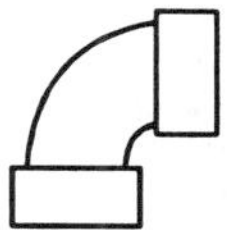

Description	Craft@Hrs	Unit	Material $	Labor $	Equipment $	Total $
PVC sewer pipe 1/4 bend B x S with bell and spigot gasketed joints						
4"	P1@.320	Ea	4.20	7.82	—	12.02
6"	P1@.400	Ea	8.86	9.77	—	18.63
8"	P1@.480	Ea	24.00	11.70	—	35.70
10"	P1@.800	Ea	77.60	19.50	—	97.10
12"	P1@1.06	Ea	98.60	25.90	—	124.50
15"	P1@1.34	Ea	194.00	32.70	—	226.70
18"	P1@1.60	Ea	334.00	39.10	—	373.10

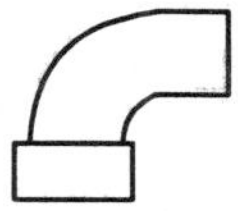

Description	Craft@Hrs	Unit	Material $	Labor $	Equipment $	Total $
PVC sewer pipe coupling B x B with bell and spigot gasketed joints						
4"	P1@.320	Ea	3.90	7.82	—	11.72
6"	P1@.400	Ea	7.82	9.77	—	17.59
8"	P1@.480	Ea	13.30	11.70	—	25.00
10"	P1@.800	Ea	29.60	19.50	—	49.10
12"	P1@1.06	Ea	42.90	25.90	—	68.80
15"	P1@1.34	Ea	88.70	32.70	—	121.40
18"	P1@1.60	Ea	173.00	39.10	—	212.10

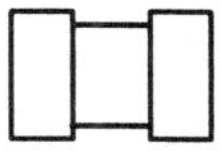

Description	Craft@Hrs	Unit	Material $	Labor $	Equipment $	Total $
PVC sewer pipe wye B x B x B with bell and spigot gasketed joints						
4"	P1@.480	Ea	5.30	11.70	—	17.00
6"	P1@.600	Ea	12.20	14.70	—	26.90
8"	P1@.720	Ea	33.50	17.60	—	51.10
10"	P1@1.20	Ea	98.60	29.30	—	127.90
12"	P1@1.59	Ea	138.00	38.90	—	176.90
15"	P1@2.01	Ea	240.00	49.10	—	289.10
18"	P1@2.40	Ea	428.00	58.60	—	486.60

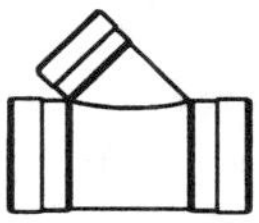

PVC, DWV with Gasketed Bell and Spigot Joints

Description	Craft@Hrs	Unit	Material $	Labor $	Equipment $	Total $
PVC sewer pipe wye B x S x B with bell and spigot gasketed joints						
10"	P1@1.20	Ea	96.90	29.30	—	126.20
12"	P1@1.59	Ea	137.00	38.90	—	175.90
15"	P1@2.01	Ea	223.00	49.10	—	272.10
18"	P1@2.40	Ea	393.00	58.60	—	451.60
PVC sewer pipe reducing wye B x B x B with bell and spigot gasketed joints						
6" x 4"	P1@.560	Ea	10.80	13.70	—	24.50
8" x 4"	P1@.640	Ea	16.10	15.60	—	31.70
8" x 6"	P1@.680	Ea	19.20	16.60	—	35.80
10" x 4"	P1@.960	Ea	48.80	23.50	—	72.30
10" x 6"	P1@1.00	Ea	49.60	24.40	—	74.00
10" x 8"	P1@1.04	Ea	78.50	25.40	—	103.90
12" x 4"	P1@1.22	Ea	70.40	29.80	—	100.20
12" x 6"	P1@1.26	Ea	72.60	30.80	—	103.40
12" x 8"	P1@1.30	Ea	108.00	31.80	—	139.80
12" x 10"	P1@1.46	Ea	132.00	35.70	—	167.70
15" x 4"	P1@1.50	Ea	106.00	36.70	—	142.70
15" x 6"	P1@1.54	Ea	127.00	37.60	—	164.60
15" x 8"	P1@1.58	Ea	142.00	38.60	—	180.60
15" x 10"	P1@1.74	Ea	170.00	42.50	—	212.50
15" x 12"	P1@1.87	Ea	190.00	45.70	—	235.70
18" x 4"	P1@1.76	Ea	243.00	43.00	—	286.00
18" x 6"	P1@1.80	Ea	247.00	44.00	—	291.00
18" x 8"	P1@1.84	Ea	255.00	45.00	—	300.00
18" x 10"	P1@2.00	Ea	287.00	48.90	—	335.90
18" x 12"	P1@2.13	Ea	309.00	52.10	—	361.10
18" x 15"	P1@2.27	Ea	330.00	55.50	—	385.50
PVC sewer pipe reducing wye B x S x B with bell and spigot gasketed joints						
10" x 4"	P1@.960	Ea	47.40	23.50	—	70.90
10" x 6"	P1@1.00	Ea	50.00	24.40	—	74.40
10" x 8"	P1@1.04	Ea	80.80	25.40	—	106.20
12" x 4"	P1@1.22	Ea	71.40	29.80	—	101.20
12" x 6"	P1@1.26	Ea	73.10	30.80	—	103.90
12" x 8"	P1@1.30	Ea	107.00	31.80	—	138.80
12" x 10"	P1@1.46	Ea	131.00	35.70	—	166.70
15" x 4"	P1@1.50	Ea	110.00	36.70	—	146.70
15" x 6"	P1@1.54	Ea	124.00	37.60	—	161.60
15" x 8"	P1@1.58	Ea	144.00	38.60	—	182.60
15" x 10"	P1@1.74	Ea	166.00	42.50	—	208.50
15" x 12"	P1@1.87	Ea	192.00	45.70	—	237.70
18" x 4"	P1@1.76	Ea	228.00	43.00	—	271.00
18" x 6"	P1@1.80	Ea	247.00	44.00	—	291.00
18" x 8"	P1@1.84	Ea	255.00	45.00	—	300.00
18" x 10"	P1@2.00	Ea	287.00	48.90	—	335.90
18" x 12"	P1@2.13	Ea	309.00	52.10	—	361.10
18" x 15"	P1@2.27	Ea	330.00	55.50	—	385.50

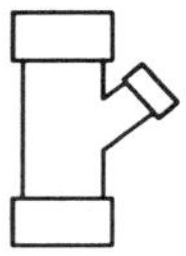

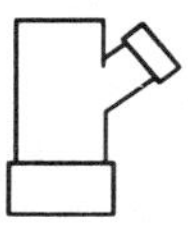

Description	Craft@Hrs	Unit	Material $	Labor $	Equipment $	Total $

PVC sewer pipe tee-wye B x B x B with bell and spigot gasketed joints

Description	Craft@Hrs	Unit	Material $	Labor $	Equipment $	Total $
4"	P1@.480	Ea	5.93	11.70	—	17.63
6"	P1@.600	Ea	14.90	14.70	—	29.60
6" x 4"	P1@.560	Ea	12.80	13.70	—	26.50
8" x 4"	P1@.640	Ea	16.30	15.60	—	31.90
8" x 6"	P1@.680	Ea	18.70	16.60	—	35.30

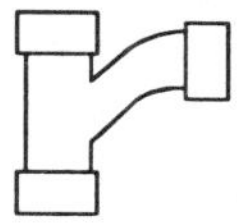

PVC sewer pipe tee B x S x B with bell and spigot gasketed joints

Description	Craft@Hrs	Unit	Material $	Labor $	Equipment $	Total $
10"	P1@1.20	Ea	82.30	29.30	—	111.60
12"	P1@1.59	Ea	114.00	38.90	—	152.90
15"	P1@2.01	Ea	180.00	49.10	—	229.10
18"	P1@2.40	Ea	338.00	58.60	—	396.60

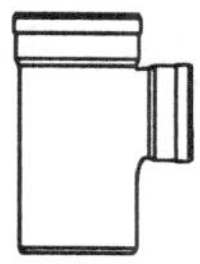

PVC sewer pipe reducing tee B x S x B with bell and spigot gasketed joints

Description	Craft@Hrs	Unit	Material $	Labor $	Equipment $	Total $
10" x 4"	P1@.960	Ea	51.90	23.50	—	75.40
10" x 6"	P1@1.00	Ea	53.90	24.40	—	78.30
10" x 8"	P1@1.04	Ea	84.20	25.40	—	109.60
12" x 4"	P1@1.22	Ea	63.60	29.80	—	93.40
12" x 6"	P1@1.26	Ea	64.80	30.80	—	95.60
12" x 8"	P1@1.30	Ea	89.40	31.80	—	121.20
12" x 10"	P1@1.46	Ea	107.00	35.70	—	142.70
15" x 4"	P1@1.50	Ea	103.00	36.70	—	139.70
15" x 6"	P1@1.54	Ea	109.00	37.60	—	146.60
15" x 8"	P1@1.58	Ea	111.00	38.60	—	149.60
15" x 10"	P1@1.74	Ea	131.00	42.50	—	173.50
15" x 12"	P1@1.87	Ea	142.00	45.70	—	187.70
18" x 4"	P1@1.76	Ea	260.00	43.00	—	303.00
18" x 6"	P1@1.80	Ea	269.00	44.00	—	313.00
18" x 8"	P1@1.84	Ea	282.00	45.00	—	327.00
18" x 10"	P1@2.00	Ea	292.00	48.90	—	340.90
18" x 12"	P1@2.13	Ea	301.00	52.10	—	353.10
18" x 15"	P1@2.27	Ea	320.00	55.50	—	375.50

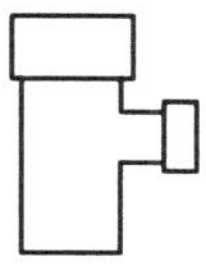

PVC, DWV with Gasketed Bell and Spigot Joints

PVC sewer pipe reducer S x B with bell and spigot gasketed joints

Description	Craft@Hrs	Unit	Material $	Labor $	Equipment $	Total $
6" x 4"	P1@.360	Ea	5.71	8.80	—	14.51
8" x 4"	P1@.400	Ea	15.60	9.77	—	25.37
8" x 6"	P1@.440	Ea	17.50	10.80	—	28.30
10" x 4"	P1@.560	Ea	40.80	13.70	—	54.50
10" x 6"	P1@.600	Ea	42.40	14.70	—	57.10
10" x 8"	P1@.640	Ea	50.40	15.60	—	66.00
12" x 4"	P1@.690	Ea	54.40	16.90	—	71.30
12" x 6"	P1@.730	Ea	55.70	17.80	—	73.50
12" x 8"	P1@.770	Ea	63.90	18.80	—	82.70
12" x 10"	P1@.930	Ea	70.70	22.70	—	93.40
15" x 4"	P1@.830	Ea	90.70	20.30	—	111.00
15" x 6"	P1@.870	Ea	94.20	21.30	—	115.50
15" x 8"	P1@.910	Ea	107.00	22.20	—	129.20
15" x 10"	P1@1.07	Ea	107.00	26.10	—	133.10
15" x 12"	P1@1.20	Ea	123.00	29.30	—	152.30
18" x 8"	P1@1.04	Ea	132.00	25.40	—	157.40
18" x 10"	P1@1.20	Ea	133.00	29.30	—	162.30
18" x 12"	P1@1.33	Ea	135.00	32.50	—	167.50
18" x 15"	P1@1.47	Ea	137.00	35.90	—	172.90

PVC sewer pipe adapter B x S with bell and spigot gasketed joints

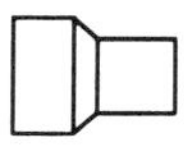

Description	Craft@Hrs	Unit	Material $	Labor $	Equipment $	Total $
4"	P1@.320	Ea	3.34	7.82	—	11.16
6"	P1@.400	Ea	6.20	9.77	—	15.97
8"	P1@.480	Ea	15.80	11.70	—	27.50
10"	P1@.800	Ea	19.80	19.50	—	39.30
12"	P1@1.06	Ea	30.10	25.90	—	56.00
15"	P1@1.34	Ea	45.50	32.70	—	78.20
18"	P1@1.60	Ea	100.00	39.10	—	139.10

PVC sewer pipe cap with bell and spigot gasketed joints

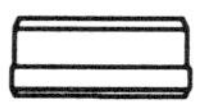

Description	Craft@Hrs	Unit	Material $	Labor $	Equipment $	Total $
4"	P1@.160	Ea	2.03	3.91	—	5.94
6"	P1@.200	Ea	3.83	4.89	—	8.72
8"	P1@.240	Ea	10.30	5.86	—	16.16
10"	P1@.400	Ea	32.30	9.77	—	42.07
12"	P1@.530	Ea	48.70	13.00	—	61.70
15"	P1@.670	Ea	78.80	16.40	—	95.20
18"	P1@.800	Ea	114.00	19.50	—	133.50

PVC sewer pipe test plug with bell and spigot gasketed joints

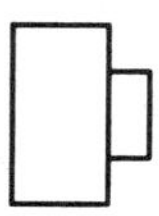

Description	Craft@Hrs	Unit	Material $	Labor $	Equipment $	Total $
4"	P1@.160	Ea	1.54	3.91	—	5.45
6"	P1@.200	Ea	2.36	4.89	—	7.25
8"	P1@.240	Ea	8.40	5.86	—	14.26
10"	P1@.400	Ea	28.00	9.77	—	37.77
12"	P1@.530	Ea	33.40	13.00	—	46.40
15"	P1@.670	Ea	65.80	16.40	—	82.20
18"	P1@.800	Ea	92.80	19.50	—	112.30

Polypropylene is used almost exclusively for acid and laboratory drain, waste and vent (DWV) systems.

Joints are made by applying low-voltage current to electrical resistance coils imbedded in propylene collars, which are slipped over the ends of the pipe, then inserted into the hubs of the fittings. Compression clamps are placed around the assemblies and tightened to compress the joints prior to applying electrical current from the power unit to fuse the joint.

This section has been arranged to save the estimator's time by including all normally-used system components such as pipe, fittings, hanger assemblies, and riser clamps under one heading. The cost estimates in this section are based on the conditions, limitations and wage rates described in the section "How to Use This Book" beginning on page 5.

Schedule 40 DWV polypropylene pipe with heat-fusioned joints

Description	Craft@Hrs	Unit	Material $	Labor $	Equipment $	Total $
1½"	P1@.050	LF	2.21	1.22	—	3.43
2"	P1@.055	LF	3.00	1.34	—	4.34
3"	P1@.060	LF	5.40	1.47	—	6.87
4"	P1@.080	LF	7.66	1.95	—	9.61
6"	P1@.110	LF	13.80	2.69	—	16.49

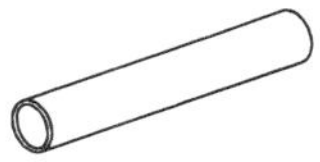

Schedule 40 DWV polypropylene 1/8 bend with heat-fusioned joints

Description	Craft@Hrs	Unit	Material $	Labor $	Equipment $	Total $
1½"	P1@.350	Ea	9.73	8.55	—	18.28
2"	P1@.400	Ea	11.90	9.77	—	21.67
3"	P1@.450	Ea	21.80	11.00	—	32.80
4"	P1@.500	Ea	24.70	12.20	—	36.90
6"	P1@.750	Ea	68.40	18.30	—	86.70

Schedule 40 DWV polypropylene 1/4 bend with heat-fusioned joints

Description	Craft@Hrs	Unit	Material $	Labor $	Equipment $	Total $
1½"	P1@.350	Ea	9.90	8.55	—	18.45
2"	P1@.400	Ea	12.30	9.77	—	22.07
3"	P1@.450	Ea	20.90	11.00	—	31.90
4"	P1@.500	Ea	33.20	12.20	—	45.40
6"	P1@.750	Ea	81.70	18.30	—	100.00

Schedule 40 DWV polypropylene long sweep 1/4 bend with heat-fusioned joints

Description	Craft@Hrs	Unit	Material $	Labor $	Equipment $	Total $
1½"	P1@.350	Ea	10.20	8.55	—	18.75
2"	P1@.400	Ea	14.00	9.77	—	23.77
3"	P1@.450	Ea	23.90	11.00	—	34.90
4"	P1@.500	Ea	34.30	12.20	—	46.50

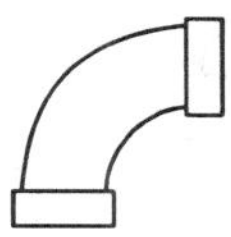

Schedule 40 DWV polypropylene P-trap with heat-fusioned joints

Description	Craft@Hrs	Unit	Material $	Labor $	Equipment $	Total $
1½"	P1@.500	Ea	19.30	12.20	—	31.50
2"	P1@.560	Ea	27.00	13.70	—	40.70
3"	P1@.630	Ea	48.90	15.40	—	64.30
4"	P1@.700	Ea	87.00	17.10	—	104.10

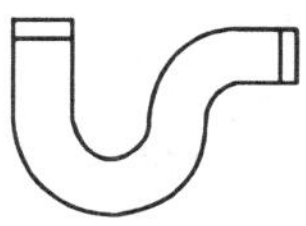

Description	Craft@Hrs	Unit	Material $	Labor $	Equipment $	Total $

Schedule 40 DWV polypropylene sanitary tee (TY) with heat-fusioned joints

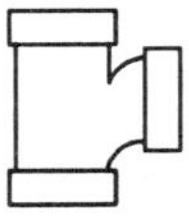

Description	Craft@Hrs	Unit	Material $	Labor $	Equipment $	Total $
1½"	P1@.500	Ea	12.30	12.20	—	24.50
2"	P1@.560	Ea	14.70	13.70	—	28.40
3"	P1@.630	Ea	24.50	15.40	—	39.90
4"	P1@.700	Ea	44.20	17.10	—	61.30

Schedule 40 DWV polypropylene reducing sanitary tee (TY) with heat-fusioned joints

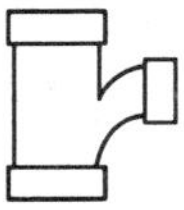

Description	Craft@Hrs	Unit	Material $	Labor $	Equipment $	Total $
2" x 1½"	P1@.510	Ea	14.70	12.50	—	27.20
3" x 1½"	P1@.530	Ea	25.60	13.00	—	38.60
3" x 2"	P1@.580	Ea	27.50	14.20	—	41.70
4" x 2"	P1@.650	Ea	39.50	15.90	—	55.40
4" x 3"	P1@.700	Ea	41.20	17.10	—	58.30
6" x 4"	P1@.850	Ea	74.80	20.80	—	95.60

Schedule 40 DWV polypropylene combination wye and 1/8 bend with heat-fusioned joints

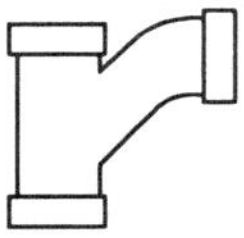

Description	Craft@Hrs	Unit	Material $	Labor $	Equipment $	Total $
1½"	P1@.500	Ea	16.40	12.20	—	28.60
2"	P1@.560	Ea	21.90	13.70	—	35.60
3"	P1@.630	Ea	34.90	15.40	—	50.30
4"	P1@.700	Ea	49.60	17.10	—	66.70

Schedule 40 DWV polypropylene combination reducing wye and 1/8 bend with heat-fusioned joints

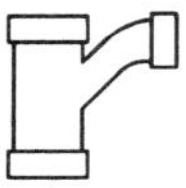

Description	Craft@Hrs	Unit	Material $	Labor $	Equipment $	Total $
2" x 1½"	P1@.500	Ea	19.70	12.20	—	31.90
3" x 1½"	P1@.520	Ea	29.00	12.70	—	41.70
3" x 2"	P1@.570	Ea	31.40	13.90	—	45.30
4" x 2"	P1@.580	Ea	43.50	14.20	—	57.70
4" x 3"	P1@.630	Ea	45.70	15.40	—	61.10

Schedule 40 DWV polypropylene wye with heat-fusioned joints

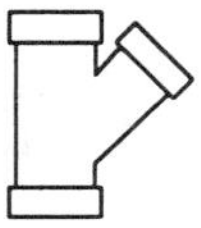

Description	Craft@Hrs	Unit	Material $	Labor $	Equipment $	Total $
1½"	P1@.500	Ea	12.50	12.20	—	24.70
2"	P1@.560	Ea	17.90	13.70	—	31.60
3"	P1@.630	Ea	31.40	15.40	—	46.80
4"	P1@.700	Ea	45.70	17.10	—	62.80
6"	P1@1.05	Ea	116.00	25.70	—	141.70

Schedule 40 DWV polypropylene reducing wye with heat-fusioned joints

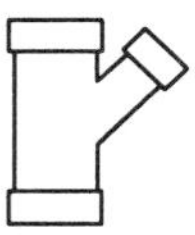

Description	Craft@Hrs	Unit	Material $	Labor $	Equipment $	Total $
2" x 1½"	P1@.510	Ea	17.60	12.50	—	30.10
3" x 2"	P1@.580	Ea	29.20	14.20	—	43.40
4" x 2"	P1@.600	Ea	42.10	14.70	—	56.80
4" x 3"	P1@.650	Ea	43.90	15.90	—	59.80
6" x 2"	P1@.900	Ea	73.90	22.00	—	95.90
6" x 3"	P1@.900	Ea	74.50	22.00	—	96.50
6" x 4"	P1@.950	Ea	75.40	23.20	—	98.60

Schedule 40 DWV polypropylene female adapter with heat-fusioned joints

Description	Craft@Hrs	Unit	Material $	Labor $	Equipment $	Total $
1½"	P1@.280	Ea	6.09	6.84	—	12.93
2"	P1@.330	Ea	8.29	8.06	—	16.35
3"	P1@.350	Ea	13.70	8.55	—	22.25
4"	P1@.400	Ea	26.30	9.77	—	36.07

Schedule 40 DWV polypropylene male adapter with heat-fusioned joints

Description	Craft@Hrs	Unit	Material $	Labor $	Equipment $	Total $
1½"	P1@.280	Ea	6.02	6.84	—	12.86
2"	P1@.330	Ea	7.29	8.06	—	15.35
3"	P1@.350	Ea	12.00	8.55	—	20.55
4"	P1@.400	Ea	25.00	9.77	—	34.77

Schedule 40 DWV polypropylene cleanout adapter with heat-fusioned joints

Description	Craft@Hrs	Unit	Material $	Labor $	Equipment $	Total $
1½"	P1@.210	Ea	6.41	5.13	—	11.54
2"	P1@.280	Ea	7.55	6.84	—	14.39
3"	P1@.430	Ea	13.90	10.50	—	24.40
4"	P1@.640	Ea	24.00	15.60	—	39.60
6"	P1@.850	Ea	41.50	20.80	—	62.30

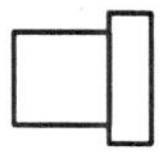

Schedule 40 DWV polypropylene plug with heat-fusioned joints

Description	Craft@Hrs	Unit	Material $	Labor $	Equipment $	Total $
1½"	P1@.060	Ea	2.22	1.47	—	3.69
2"	P1@.065	Ea	2.42	1.59	—	4.01
3"	P1@.070	Ea	4.23	1.71	—	5.94
4"	P1@.090	Ea	4.75	2.20	—	6.95

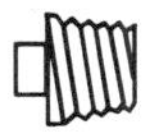

Schedule 40 DWV polypropylene coupling with heat-fusioned joints

Description	Craft@Hrs	Unit	Material $	Labor $	Equipment $	Total $
1½"	P1@.350	Ea	7.74	8.55	—	16.29
2"	P1@.400	Ea	9.66	9.77	—	19.43
3"	P1@.450	Ea	12.40	11.00	—	23.40
4"	P1@.500	Ea	17.60	12.20	—	29.80
6"	P1@.750	Ea	27.90	18.30	—	46.20

Schedule 40 DWV polypropylene reducer with heat-fusioned joints

Description	Craft@Hrs	Unit	Material $	Labor $	Equipment $	Total $
2" x 1½"	P1@.360	Ea	8.25	8.80	—	17.05
3" x 2"	P1@.410	Ea	9.51	10.00	—	19.51
4" x 2"	P1@.430	Ea	21.10	10.50	—	31.60
4" x 3"	P1@.450	Ea	21.60	11.00	—	32.60

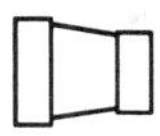

Polypropylene, Schedule 40, with Heat-Fusioned Joints

Schedule 40 DWV polypropylene flange with heat-fusioned joints

Description	Craft@Hrs	Unit	Material $	Labor $	Equipment $	Total $
2"	P1@.350	Ea	27.10	8.55	—	35.65
3"	P1@.380	Ea	55.80	9.29	—	65.09
4"	P1@.430	Ea	72.40	10.50	—	82.90
6"	P1@.630	Ea	96.00	15.40	—	111.40

Schedule 40 DWV polypropylene bolt and gasket set

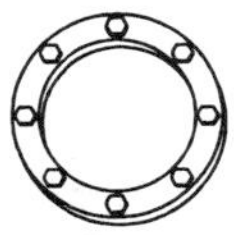

Description	Craft@Hrs	Unit	Material $	Labor $	Equipment $	Total $
2"	P1@.500	Ea	3.52	12.20	—	15.72
2½"	P1@.650	Ea	3.72	15.90	—	19.62
3"	P1@.750	Ea	3.90	18.30	—	22.20
4"	P1@1.00	Ea	7.07	24.40	—	31.47

Hanger with swivel assembly

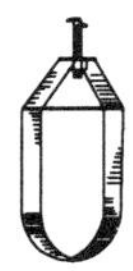

Description	Craft@Hrs	Unit	Material $	Labor $	Equipment $	Total $
1½"	P1@.450	Ea	2.05	11.00	—	13.05
2"	P1@.450	Ea	2.15	11.00	—	13.15
3"	P1@.450	Ea	3.60	11.00	—	14.60
4"	P1@.450	Ea	3.95	11.00	—	14.95
6"	P1@.500	Ea	6.45	12.20	—	18.65

Riser clamp

Description	Craft@Hrs	Unit	Material $	Labor $	Equipment $	Total $
1½"	P1@.110	Ea	1.79	2.69	—	4.48
2"	P1@.115	Ea	1.89	2.81	—	4.70
3"	P1@.120	Ea	2.18	2.93	—	5.11
4"	P1@.125	Ea	2.75	3.05	—	5.80
6"	P1@.200	Ea	4.79	4.89	—	9.68

Pipe sleeve

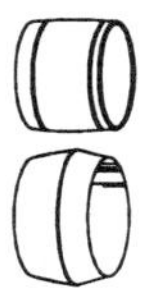

Description	Craft@Hrs	Unit	Material $	Labor $	Equipment $	Total $
2"	P1@.130	Ea	2.55	3.18	—	5.73
2½"	P1@.150	Ea	2.59	3.67	—	6.26
3"	P1@.180	Ea	2.66	4.40	—	7.06
4"	P1@.220	Ea	3.02	5.38	—	8.40
5"	P1@.250	Ea	3.75	6.11	—	9.86
6"	P1@.270	Ea	4.05	6.60	—	10.65
8"	P1@.270	Ea	4.66	6.60	—	11.26

Lead roof pipe flashing

Description	Craft@Hrs	Unit	Material $	Labor $	Equipment $	Total $
1½"	P1@.250	Ea	14.00	6.11	—	20.11
2"	P1@.250	Ea	18.40	6.11	—	24.51
3"	P1@.250	Ea	21.30	6.11	—	27.41
4"	P1@.250	Ea	23.50	6.11	—	29.61

Floor drain, cast iron or plastic body, nickel-bronze grate. Add for rough-in (P trap, fittings, hangers etc).

Description	Craft@Hrs	Unit	Material $	Labor $	Equipment $	Total $
3" x 2"	P1@.500	Ea	28.00	12.20	—	40.20
5" x 3"	P1@.500	Ea	35.00	12.20	—	47.20
6" x 4"	P1@.500	Ea	45.00	12.20	—	57.20
8" x 6"	P1@.500	Ea	60.00	12.20	—	72.20

Area drain, cast iron or plastic body, cast iron grate. Add for rough-in (P trap, fittings, hangers etc).

Description	Craft@Hrs	Unit	Material $	Labor $	Equipment $	Total $
10" x 3"	P1@.650	Ea	60.00	15.90	—	75.90
12" x 4"	P1@.650	Ea	65.00	15.90	—	80.90
14" x 6"	P1@.700	Ea	70.00	17.10	—	87.10

Roof and overflow drain, cast iron or plastic body, with plastic dome, flow control weir and deck clamp. Add for rough-in (pipe, fittings, hangers etc).

Description	Craft@Hrs	Unit	Material $	Labor $	Equipment $	Total $
8" x 3"	P1@1.50	Ea	80.00	36.70	—	116.70
10" x 4"	P1@1.50	Ea	85.00	36.70	—	121.70
12" x 6"	P1@1.65	Ea	85.00	40.30	—	125.30

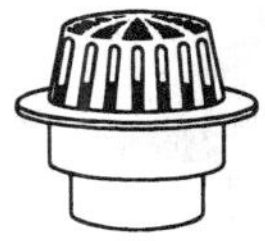

Planter drain, cast iron. Add for rough-in (pipe, fittings, hangers etc).

Description	Craft@Hrs	Unit	Material $	Labor $	Equipment $	Total $
2"	P1@1.00	Ea	15.40	24.40	—	39.80
3"	P1@1.10	Ea	17.70	26.90	—	44.60
4"	P1@1.15	Ea	21.80	28.10	—	49.90
6"	P1@1.25	Ea	31.20	30.50	—	61.70

Planter drain, plastic. Add for rough-in (pipe, fittings, hangers etc).

Description	Craft@Hrs	Unit	Material $	Labor $	Equipment $	Total $
4"	P1@.800	Ea	4.75	19.50	—	24.25
5"	P1@1.00	Ea	5.05	24.40	—	29.45
6"	P1@1.05	Ea	8.31	25.70	—	34.01
8"	P1@1.10	Ea	12.60	26.90	—	39.50

Cleanouts

Cleanout, inline, (Barrett) cast iron, (MJ) brass plug. Add for MJ couplings.

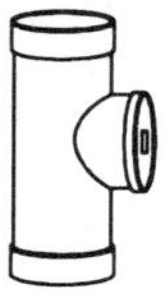

Description	Craft@Hrs	Unit	Material $	Labor $	Equipment $	Total $
2"	P1@.350	Ea	25.00	8.55	—	33.55
3"	P1@.350	Ea	32.00	8.55	—	40.55
4"	P1@.400	Ea	45.00	9.77	—	54.77
6"	P1@.400	Ea	160.00	9.77	—	169.77

Cleanout, in-line, (Barrett) ABS/PVC

Description	Craft@Hrs	Unit	Material $	Labor $	Equipment $	Total $
2"	P1@.300	Ea	6.60	7.33	—	13.93
3"	P1@.300	Ea	10.60	7.33	—	17.93
4"	P1@.350	Ea	19.20	8.55	—	27.75
6"	P1@.350	Ea	75.00	8.55	—	83.55
8"	P1@.400	Ea	146.00	9.77	—	155.77
10"	P1@.550	Ea	178.00	13.40	—	191.40

Cleanout, end of line (Malcolm), cast iron body, brass top. Add for MJ coupling.

Description	Craft@Hrs	Unit	Material $	Labor $	Equipment $	Total $
2" raised	P1@.250	Ea	9.00	6.11	—	15.11
2" recessed	P1@.250	Ea	22.50	6.11	—	28.61
3" raised	P1@.250	Ea	10.50	6.11	—	16.61
3" recessed	P1@.300	Ea	33.00	7.33	—	40.33
4" raised	P1@.300	Ea	14.00	7.33	—	21.33
4" recessed	P1@.350	Ea	43.00	8.55	—	51.55
6" raised	P1@.350	Ea	44.00	8.55	—	52.55
6" recessed	P1@.400	Ea	82.00	9.77	—	91.77
8" raised	P1@.450	Ea	96.00	11.00	—	107.00

Cleanout, end of line (Malcolm), ABS/PVC

Description	Craft@Hrs	Unit	Material $	Labor $	Equipment $	Total $
2"	P1@.250	Ea	3.50	6.11	—	9.61
3"	P1@.250	Ea	5.85	6.11	—	11.96
4"	P1@.300	Ea	10.60	7.33	—	17.93
6"	P1@.300	Ea	44.50	7.33	—	51.83
8"	P1@.400	Ea	80.00	9.77	—	89.77
10"	P1@.450	Ea	93.00	11.00	—	104.00
12"	P1@.550	Ea	108.00	13.40	—	121.40

Description	Craft@Hrs	Unit	Material $	Labor $	Equipment $	Total $

Complete wet sprinkler system (square foot costs). System includes subcontractors' overhead and profit. Cost includes design drawings, zone and supervision valves, sprinkler heads, pipe, and connection to water supply service. Make additional allowances if a fire pump is required. Costs are based on ordinary hazard coverage (110 sf per head). Coverage density can vary depending on room sizes and/or hazard rating. Use these costs for preliminary estimates.

Description	Craft@Hrs	Unit	Material $	Labor $	Equipment $	Total $
Exposed piping to 5000 SF		SF	—	—	—	1.79
Exposed piping 5000-15000 SF		SF	—	—	—	1.52
Exposed piping over 15000 SF		SF	—	—	—	1.34
Concealed piping to 5000 SF		SF	—	—	—	2.00
Concealed piping 5000-15000		SF	—	—	—	1.61
Concealed piping over 15000		SF	—	—	—	1.52
Add for dry system		%	—	—	—	10.00

Complete wet sprinkler system (per head costs). Cost includes sprinkler heads, sprinkler mains, branch piping, and supports. Make additional allowances for zone and supervision valves, alarms, connection to water service and fire pump when required. Costs are based on ordinary hazard 155 degree heads. Use these costs for preliminary estimates.

Description	Craft@Hrs	Unit	Material $	Labor $	Equipment $	Total $
Upright heads to 5000 SF	SL@2.05	Ea	42.00	49.80	—	91.80
Upright heads 5000 to 15000 SF	SL@1.95	Ea	40.00	47.40	—	87.40
Upright heads over 15000 SF	SL@1.85	Ea	38.00	44.90	—	82.90
Pendent heads to 5000 SF	SL@2.45	Ea	48.00	59.50	—	107.50
Pendent heads 5000 to 15000 SF	SL@2.35	Ea	46.00	57.10	—	103.10
Pendent heads over 15000 SF	SL@2.25	Ea	44.00	54.70	—	98.70

Supervision (zone) valves, flanged. Add for alarm trim if required.

Description	Craft@Hrs	Unit	Material $	Labor $	Equipment $	Total $
2½" OS&Y gate valve	SL@1.70	Ea	155.00	41.30	—	196.30
3" OS&Y gate vlv	SL@1.80	Ea	180.00	43.70	—	223.70
4" OS&Y gate vlv	SL@2.20	Ea	210.00	53.50	—	263.50
6" OS&Y gate vlv	SL@2.80	Ea	440.00	68.00	—	508.00
8" OS&Y gate vlv	SL@3.50	Ea	725.00	85.00	—	810.00
2½" butterfly valve (lug)	SL@.600	Ea	85.00	14.60	—	99.60
3" butterfly valve	SL@.700	Ea	95.00	17.00	—	112.00
4" butterfly valve	SL@.850	Ea	125.00	20.70	—	145.70
6" butterfly valve	SL@1.15	Ea	170.00	27.90	—	197.90
8" butterfly valve	SL@1.85	Ea	220.00	44.90	—	264.90

Description	Craft@Hrs	Unit	Material $	Labor $	Equipment $	Total $
Supervision (zone) valves, grooved. Add for alarm trim if required						
4" butterfly valve	SL@.550	Ea	205.00	13.40	—	218.40
6" butterfly valve	SL@.850	Ea	260.00	20.70	—	280.70
8" butterfly valve	SL@1.25	Ea	525.90	30.40	—	556.30
Alarm valves (flanged or grooved)						
4" alarm valve	SL@3.50	Ea	335.00	85.00	—	420.00
6" alarm valve	SL@4.00	Ea	435.00	97.20	—	532.20
8" alarm valve	SL@4.75	Ea	680.00	115.00	—	795.00
4" alarm vlv trim	SL@1.25	Ea	130.00	30.40	—	160.40
6" alarm vlv trim	SL@1.25	Ea	130.00	30.40	—	160.40
8" alarm vlv trim	SL@1.25	Ea	130.00	30.40	—	160.40
4" alarm vlv pkg	SL@4.75	Ea	965.00	115.00	—	1,080.00
6" alarm vlv pkg	SL@5.50	Ea	1,060.00	134.00	—	1,194.00
8" alarm vlv pkg	SL@6.00	Ea	1,390.00	146.00	—	1,536.00
Dry valves and trim (deluge/pre-action system)						
2" dry valve	SL@2.25	Ea	315.00	54.70	—	369.70
3" dry valve	SL@2.75	Ea	415.00	66.80	—	481.80
4" dry valve	SL@3.05	Ea	510.00	74.10	—	584.10
6" dry valve	SL@3.65	Ea	680.00	88.70	—	768.70
2" dry valve trim	SL@1.25	Ea	170.00	30.40	—	200.40
3" dry valve trim	SL@1.25	Ea	170.00	30.40	—	200.40
4" dry valve trim	SL@1.25	Ea	170.00	30.40	—	200.40
6" dry valve trim	SL@1.25	Ea	170.00	30.40	—	200.40
Check valves, grooved						
2½"	SL@.450	Ea	80.00	10.90	—	90.90
3"	SL@.500	Ea	90.00	12.10	—	102.10
4"	SL@.550	Ea	95.00	13.40	—	108.40
6"	SL@.850	Ea	175.00	20.70	—	195.70
Check valves, wafer type, flanged						
2½"	SL@1.65	Ea	60.00	40.10	—	100.10
3"	SL@1.70	Ea	75.00	41.30	—	116.30
4"	SL@2.10	Ea	90.00	51.00	—	141.00
6"	SL@2.60	Ea	130.00	63.20	—	193.20
8"	SL@3.20	Ea	170.00	77.80	—	247.80
10"	SL@3.95	Ea	330.00	96.00	—	426.00
Double check detector valve assembly (flanged)						
2½"	SL@4.00	Ea	745.00	97.20	—	842.20
3"	SL@4.35	Ea	770.00	106.00	—	876.00
4"	SL@4.95	Ea	1,050.00	120.00	—	1,170.00
6"	SL@6.25	Ea	1,585.00	152.00	—	1,737.00
8"	SL@6.95	Ea	2,950.00	169.00	—	3,119.00

Description	Craft@Hrs	Unit	Material $	Labor $	Equipment $	Total $
Switches, flow, supervisory, monitor, pressure						
Flow switch	SL@1.00	Ea	65.00	24.30	—	89.30
Pressure switch	SL@1.00	Ea	80.00	24.30	—	104.30
Monitor switch	SL@1.00	Ea	100.00	24.30	—	124.30
Supervisory sw.	SL@1.00	Ea	45.00	24.30	—	69.30
Sprinkler heads, 155 degree						
Pendent, brass	SL@.350	Ea	2.90	8.50	—	11.40
Pendent, chrome	SL@.350	Ea	3.30	8.50	—	11.80
Upright, brass	SL@.350	Ea	2.80	8.50	—	11.30
Upright, chrome	SL@.350	Ea	3.30	8.50	—	11.80
Sidewall, brass	SL@.350	Ea	4.70	8.50	—	13.20
Sidewall, chrome	SL@.350	Ea	4.30	8.50	—	12.80
Sprinkler heads, dry, 165 degree						
Dry pendent	SL@.450	Ea	26.00	10.90	—	36.90
Dry sidewall	SL@.450	Ea	28.00	10.90	—	38.90
Sprinkler heads, 200 degree						
Pendent, brass	SL@.350	Ea	2.80	8.50	—	11.30
Pendent, chrome	SL@.350	Ea	3.30	8.50	—	11.80
Upright, brass	SL@.350	Ea	2.80	8.50	—	11.30
Upright, chrome	SL@.350	Ea	3.30	8.50	—	11.80
Sidewall, brass	SL@.350	Ea	4.70	8.50	—	13.20
Sidewall, chrome	SL@.350	Ea	5.40	8.50	—	13.90
Sprinkler heads, 286 degree						
Pendent, brass	SL@.350	Ea	2.95	8.50	—	11.45
Pendent, chrome	SL@.350	Ea	3.45	8.50	—	11.95
Upright, brass	SL@.350	Ea	2.95	8.50	—	11.45
Upright, chrome	SL@.350	Ea	3.30	8.50	—	11.80
Sidewall, brass	SL@.350	Ea	4.80	8.50	—	13.30
Sidewall, chrome	SL@.350	Ea	5.50	8.50	—	14.00
Sprinkler heads, 360 degree						
Pendent, brass	SL@.350	Ea	3.30	8.50	—	11.80
Upright, brass	SL@.350	Ea	3.30	8.50	—	11.80
Sprinkler heads, 400 degree						
Pendent, brass	SL@.350	Ea	8.30	8.50	—	16.80
Upright, brass	SL@.350	Ea	8.30	8.50	—	16.80

Description	Craft@Hrs	Unit	Material $	Labor $	Equipment $	Total $

Fire hose cabinet complete assembly. Includes recessed steel cabinet, 2½" and 1½" angle valves, 100' fire hose, hose rack and fog nozzle.

Description	Craft@Hrs	Unit	Material $	Labor $	Equipment $	Total $
FHC, chrome trim	SL@1.95	Ea	345.00	47.40	—	392.40
FHC, brass trim	SL@1.95	Ea	335.00	47.40	—	382.40
FHC, plastic trim	SL@1.95	Ea	300.00	47.40	—	347.40

Fire hose cabinet components

Description	Craft@Hrs	Unit	Material $	Labor $	Equipment $	Total $
Recessed cabinet	SL@1.00	Ea	90.00	24.30	—	114.30
Surface cabinet	SL@.850	Ea	95.00	20.70	—	115.70
Cabinet glass	SL@.350	Ea	11.00	8.50	—	19.50
75' fire hose	SL@.250	Ea	70.00	6.07	—	76.07
100' fire hose	SL@.250	Ea	85.00	6.07	—	91.07
75' hose rack	SL@.350	Ea	30.00	8.50	—	38.50
100' hose rack	SL@.350	Ea	40.00	8.50	—	48.50
1½" brass fire hose angle valve	SL@.350	Ea	36.00	8.50	—	44.50
1½" chrome fire hose angle valve	SL@.350	Ea	42.00	8.50	—	50.50
2½" brass fire hose angle valve	SL@.450	Ea	70.00	10.90	—	80.90
2½" chrome fire hose angle valve	SL@.450	Ea	85.00	10.90	—	95.90
Brass fog nozzle	SL@.150	Ea	25.00	3.64	—	28.64
Plastic fog nozzle	SL@.150	Ea	9.00	3.64	—	12.64
Chrome fog nozz	SL@.150	Ea	18.00	3.64	—	21.64

Description	Craft@Hrs	Unit	Material $	Labor $	Equipment $	Total $
Electric fire pump, skid mounted with controller and jockey pump						
15 KW, 500 gpm	SK@24.0	Ea	24,500.00	622.00	1.00	25,123.00
22 KW, 500 gpm	SK@24.0	Ea	25,500.00	622.00	1.00	26,123.00
30 KW, 750 gpm	SK@26.0	Ea	25,800.00	674.00	1.00	26,475.00
38 KW, 1000 gpm	SK@28.0	Ea	26,300.00	726.00	1.00	27,027.00
56 KW, 1000 gpm	SK@28.0	Ea	27,900.00	726.00	1.00	28,627.00
Diesel fire pump, skid mounted with controller and jockey pump						
500 gpm	SK@32.0	Ea	16,500.00	829.00	1.00	17,330.00
750 gpm	SK@32.0	Ea	22,000.00	829.00	1.00	22,830.00
1000 gpm	SK@34.0	Ea	25,300.00	881.00	1.00	26,182.00
1500 gpm	SK@36.0	Ea	27,300.00	933.00	1.00	28,234.00
2500 gpm	SK@38.0	Ea	34,400.00	985.00	1.00	35,386.00
3000 gpm	SK@44.0	Ea	38,900.00	1,140.00	1.00	40,041.00
4000 gpm	SK@48.0	Ea	54,300.00	1,240.00	1.00	55,541.00
5000 gpm	SK@54.0	Ea	75,100.00	1,400.00	1.00	76,501.00
Excess pressure pump						
1/4 hp	SL@1.35	Ea	125.00	32.80	—	157.80
1/3 hp	SL@1.35	Ea	160.00	32.80	—	192.80
1/2 hp	SL@1.50	Ea	185.00	36.40	—	221.40
Air compressor						
1/2 hp	SL@3.00	Ea	275.00	72.90	—	347.90
3/4 hp	SL@3.00	Ea	310.00	72.90	—	382.90
1 hp	SL@3.50	Ea	360.00	85.00	—	445.00
1½ hp	SL@3.50	Ea	400.00	85.00	—	485.00
2 hp	SL@3.50	Ea	425.00	85.00	—	510.00

Description	Craft@Hrs	Unit	Material $	Labor $	Equipment $	Total $
Siamese connection, flush, fire dept. connection						
4" brass	SL@4.00	Ea	230.00	97.20	—	327.20
4" chrome	SL@4.00	Ea	260.00	97.20	—	357.20
6" brass	SL@4.50	Ea	260.00	109.00	—	369.00
6" chrome	SL@4.50	Ea	290.00	109.00	—	399.00
Siamese connection, surface, fire dept. connection						
4" brass	SL@2.85	Ea	70.00	69.20	—	139.20
6" brass	SL@3.25	Ea	110.00	79.00	—	189.00
Siamese connection, sidewalk, fire dept. connection						
4" brass	SL@3.25	Ea	65.00	79.00	—	144.00
6" brass	SL@3.75	Ea	95.00	91.10	—	186.10
Water motor gong						
	SL@1.50	Ea	125.00	36.40	—	161.40
Multi-purpose fire extinguisher						
2lb	SL@.500	Ea	20.00	12.10	—	32.10
5lb	SL@.500	Ea	28.00	12.10	—	40.10
10lb	SL@.500	Ea	40.00	12.10	—	52.10
20lb	SL@.500	Ea	85.00	12.10	—	97.10
Fire hydrant with spool						
6"	P1@5.00	Ea	795.00	122.00	—	917.00
Indicator post						
6"	P1@6.50	Ea	480.00	159.00	—	639.00

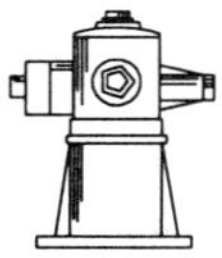

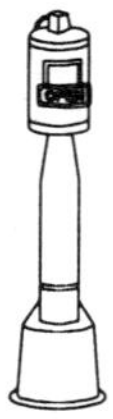

Description	Craft@Hrs	Unit	Material $	Labor $	Equipment $	Total $

Black steel pipe, (A53) installed horizontal, roll grooved, schedule 40. Standard weight, including hangers every 10' and a fitting every 33'. Use these figures for preliminary estimates.

Description	Craft@Hrs	Unit	Material $	Labor $	Equipment $	Total $
2" (50mm)	SL@.160	LF	4.01	3.89	—	7.90
2½" (65mm)	SL@.180	LF	5.65	4.37	—	10.02
3" (75mm)	SL@.210	LF	7.19	5.10	—	12.29
4" (10cm)	SL@.300	LF	9.76	7.29	—	17.05
6" (15cm)	SK@.450	LF	19.50	11.70	0.23	31.43
8" (20cm)	SK@.630	LF	29.00	16.30	0.33	45.63
10" (25cm)	SK@.750	LF	56.10	19.40	0.39	75.89
12" (30cm)	SK@.800	LF	58.30	20.70	0.42	79.42

Black steel pipe, (A53) installed riser, roll grooved, schedule 40. Standard weight, including a riser clamp every other floor and a tee at every floor. Make additional allowances for sleeving or coring as required. Use these figures for preliminary estimates.

Description	Craft@Hrs	Unit	Material $	Labor $	Equipment $	Total $
2" (50mm)	SL@.110	LF	3.85	2.67	—	6.52
2½" (65mm)	SL@.120	LF	5.16	2.92	—	8.08
3" (75mm)	SL@.150	LF	6.16	3.64	—	9.80
4" (10cm)	SL@.190	LF	8.77	4.62	—	13.39
6" (15cm)	SK@.300	LF	18.10	7.78	0.16	26.04
8" (20cm)	SK@.400	LF	28.40	10.40	0.21	39.01
10" (25cm)	SK@.510	LF	44.10	13.20	0.27	57.57
12" (30cm)	SK@.560	LF	53.40	14.50	0.29	68.19

Black steel pipe, (A53) installed horizontal, roll grooved, thin wall, schedule 10. Light weight, including hangers every 10' and a fitting every 33'. Use these figures for preliminary estimates.

Description	Craft@Hrs	Unit	Material $	Labor $	Equipment $	Total $
2" (50mm)	SL@.160	LF	3.33	3.89	—	7.22
2½" (65mm)	SL@.180	LF	4.22	4.37	—	8.59
3" (75mm)	SL@.210	LF	4.98	5.10	—	10.08
4" (10cm)	SL@.300	LF	7.08	7.29	—	14.37
6" (15cm)	SK@.450	LF	14.50	11.70	0.22	26.42
8" (20cm)	SK@.630	LF	26.70	16.30	0.30	43.30

Black steel pipe, (A53) installed riser, roll grooved, thin wall, schedule 10. Light weight, including a riser clamp every other floor and a tee at every floor. Make additional allowances for sleeving or coring as required. Use these figures for preliminary estimates

Description	Craft@Hrs	Unit	Material $	Labor $	Equipment $	Total $
2" (50mm)	SL@.110	LF	3.47	2.67	—	6.14
2½ " (65mm)	SL@.120	LF	4.19	2.92	—	7.11
3" (75mm)	SL@.150	LF	4.88	3.64	—	8.52
4" (10cm)	SL@.190	LF	6.83	4.62	—	11.45
6" (15cm)	SK@.300	LF	14.40	7.78	0.14	22.32
8" (20cm)	SK@.400	LF	28.00	10.40	0.19	38.59

Fire Protection Sprinkler Pipe (Roll Grooved)

Description	Craft@Hrs	Unit	Material $	Labor $	Equipment $	Total $
Black steel pipe only, (A53) roll grooved, schedule 40. Standard weight, no hangers or fittings.						
2" (50mm)	SL@.070	LF	1.98	1.70	—	3.68
2½" (65mm)	SL@.070	LF	3.50	1.70	—	5.20
3" (75mm)	SL@.090	LF	3.89	2.19	—	6.08
4" (10cm)	SL@.130	LF	5.55	3.16	—	8.71
6" (15cm)	SK@.160	LF	8.83	4.15	0.08	13.06
8" (20cm)	SK@.210	LF	13.40	5.44	0.11	18.95
10" (25cm)	SK@.280	LF	19.40	7.26	0.15	26.81
12" (30cm)	SK@.340	LF	27.30	8.81	0.18	36.29
14" (36cm)	SK@.400	LF	31.40	10.40	1.81	43.61
16" (41cm)	SK@.550	LF	36.60	14.30	2.49	53.39
Black steel pipe only, (A53) roll grooved, thin wall, schedule 10. Light weight, no hangers or fittings.						
2" (50mm)	SL@.070	LF	1.49	1.70	—	3.19
2½" (65mm)	SL@.070	LF	2.62	1.70	—	4.32
3" (75mm)	SL@.090	LF	2.92	2.19	—	5.11
4" (10cm)	SL@.130	LF	4.16	3.16	—	7.32
6" (15cm)	SK@.160	LF	6.62	4.15	0.08	10.85
8" (20cm)	SK@.210	LF	11.30	5.44	0.10	16.84
10" (25cm)	SK@.280	LF	16.50	7.26	0.13	23.89
12" (30cm)	SK@.340	LF	23.20	8.81	0.16	32.17
14" (36cm)	SK@.400	LF	26.70	10.40	1.81	38.91
16" (41cm)	SK@.550	LF	31.10	14.30	2.49	47.89
Roll Grooved Fittings (Victaulic)						
90 degree elbow, roll grooved, style #10, (Victaulic)						
2"	SL@.410	Ea	10.00	9.96	—	19.96
2½"	SL@.430	Ea	10.00	10.40	—	20.40
3"	SL@.500	Ea	13.30	12.10	—	25.40
4"	SL@.690	Ea	20.60	16.80	—	37.40
6"	SL@1.19	Ea	51.30	28.90	—	80.20
8"	SL@1.69	Ea	87.80	41.10	—	128.90
10"	SL@2.23	Ea	168.00	54.20	—	222.20
12"	SL@2.48	Ea	267.00	60.30	—	327.30
45 degree elbow, roll grooved, style #11, (Victaulic)						
2"	SL@.410	Ea	10.00	9.96	—	19.96
2½"	SL@.430	Ea	10.00	10.40	—	20.40
3"	SL@.500	Ea	13.30	12.10	—	25.40
4"	SL@.690	Ea	20.60	16.80	—	37.40
6"	SL@1.19	Ea	51.30	28.90	—	80.20
8"	SL@1.69	Ea	87.80	41.10	—	128.90
10"	SL@2.23	Ea	168.00	54.20	—	222.20
12"	SL@2.48	Ea	267.00	60.30	—	327.30

Description	Craft@Hrs	Unit	Material $	Labor $	Equipment $	Total $
22½ degree elbow, roll grooved, style #12, (Victaulic)						
2"	SL@.410	Ea	11.80	9.96	—	21.76
2½"	SL@.430	Ea	11.80	10.40	—	22.20
3"	SL@.500	Ea	12.50	12.10	—	24.60
4"	SL@.690	Ea	23.50	16.80	—	40.30
6"	SL@1.19	Ea	58.70	28.90	—	87.60
8"	SL@1.69	Ea	120.00	41.10	—	161.10
10"	SL@2.23	Ea	169.00	54.20	—	223.20
12"	SL@2.48	Ea	181.00	60.30	—	241.30
Tee, roll grooved, style #20, (Victaulic)						
2"	SL@.510	Ea	19.40	12.40	—	31.80
2½"	SL@.540	Ea	19.40	13.10	—	32.50
3"	SL@.630	Ea	23.40	15.30	—	38.70
4"	SL@.860	Ea	32.70	20.90	—	53.60
6"	SL@1.49	Ea	86.80	36.20	—	123.00
8"	SL@2.11	Ea	146.00	51.30	—	197.30
10"	SL@2.79	Ea	258.00	67.80	—	325.80
12"	SL@3.10	Ea	302.00	75.30	—	377.30
Reducing tee, roll grooved, style #25, (Victaulic)						
2"	SL@.510	Ea	19.40	12.40	—	31.80
2½"	SL@.540	Ea	25.40	13.10	—	38.50
3"	SL@.630	Ea	23.10	15.30	—	38.40
4"	SL@.860	Ea	39.50	20.90	—	60.40
6"	SL@1.49	Ea	83.40	36.20	—	119.60
8"	SL@2.11	Ea	152.00	51.30	—	203.30
10"	SL@2.79	Ea	200.00	67.80	—	267.80
12"	SL@3.10	Ea	221.00	75.30	—	296.30
Reducer, roll grooved, style #50, (Victaulic)						
2"	SL@.410	Ea	6.37	9.96	—	16.33
2½"	SL@.430	Ea	6.37	10.40	—	16.77
3"	SL@.500	Ea	8.99	12.10	—	21.09
4"	SL@.690	Ea	14.90	16.80	—	31.70
6"	SL@1.19	Ea	24.90	28.90	—	53.80
8"	SL@1.69	Ea	44.20	41.10	—	85.30
10"	SL@2.23	Ea	92.40	54.20	—	146.60
Cap, roll grooved, style #60, (Victaulic)						
2"	SL@.270	Ea	5.65	6.56	—	12.21
2½"	SL@.280	Ea	5.65	6.80	—	12.45
3"	SL@.330	Ea	7.09	8.02	—	15.11
4"	SL@.450	Ea	8.22	10.90	—	19.12
6"	SL@.770	Ea	18.40	18.70	—	37.10
8"	SL@1.10	Ea	29.80	26.70	—	56.50
10"	SL@1.61	Ea	127.00	39.10	—	166.10

Description	Craft@Hrs	Unit	Material $	Labor $	Equipment $	Total $
Mechanical tee (saddle tee), roll grooved, style #920, (Victaulic)						
3"	SL@.750	Ea	19.50	18.20	—	37.70
4"	SL@1.00	Ea	22.10	24.30	—	46.40
6"	SL@1.30	Ea	30.30	31.60	—	61.90
8"	SL@1.50	Ea	48.10	36.40	—	84.50
Coupling, roll grooved, 800 PSI, style #77, (Victaulic). Labor included with fittings						
3"	SL@.000	Ea	9.76	0.00	—	9.76
4"	SL@.000	Ea	16.40	0.00	—	16.40
6"	SL@.000	Ea	28.30	0.00	—	28.30
8"	SL@.000	Ea	42.10	0.00	—	42.10
10"	SL@.000	Ea	56.50	0.00	—	56.50
12"	SL@.000	Ea	75.00	0.00	—	75.00
Coupling, roll grooved, light weight, style #75, (Victaulic). Labor included with fittings						
3"	SL@.000	Ea	8.01	0.00	—	8.01
4"	SL@.000	Ea	8.63	0.00	—	8.63
6"	SL@.000	Ea	17.70	0.00	—	17.70
8"	SL@.000	Ea	27.00	0.00	—	27.00
Coupling, roll grooved, zeroflex, style #07, (Victaulic). Labor included with fittings						
3"	SL@.000	Ea	9.24	0.00	—	9.24
4"	SL@.000	Ea	12.00	0.00	—	12.00
6"	SL@.000	Ea	20.50	0.00	—	20.50
8"	SL@.000	Ea	31.10	0.00	—	31.10
10"	SL@.000	Ea	45.70	0.00	—	45.70
12"	SL@.000	Ea	62.90	0.00	—	62.90
Flange, roll grooved, style #741, (Victaulic)						
3"	SL@.500	Ea	30.00	12.10	—	42.10
4"	SL@.690	Ea	42.10	16.80	—	58.90
6"	SL@1.19	Ea	59.30	28.90	—	88.20
8"	SL@1.69	Ea	68.10	41.10	—	109.20
10"	SL@2.23	Ea	101.00	54.20	—	155.20
12"	SL@2.48	Ea	139.00	60.30	—	199.30

Description	Craft@Hrs	Unit	Material $	Labor $	Equipment $	Total $
Schedule 40 carbon steel pipe, threaded and coupled, pipe only. Add for hangers, fittings etc.						
1"	P1@.090	LF	1.16	2.20	—	3.36
1¼"	P1@.100	LF	1.51	2.44	—	3.95
1½"	P1@.110	LF	1.80	2.69	—	4.49
2"	P1@.120	LF	2.43	2.93	—	5.36
125# cast iron 90 degree ell, threaded joints						
1" x 3/4"	P1@.130	Ea	2.40	3.18	—	5.58
1"	P1@.180	Ea	2.20	4.40	—	6.60
1¼"	P1@.240	Ea	3.10	5.86	—	8.96
1½"	P1@.300	Ea	4.35	7.33	—	11.68
2"	P1@.380	Ea	6.60	9.29	—	15.89
125# cast iron 45 degree ell, threaded joints						
1"	P1@.180	Ea	2.80	4.40	—	7.20
1¼"	P1@.240	Ea	4.00	5.86	—	9.86
1½"	P1@.300	Ea	5.15	7.33	—	12.48
2"	P1@.380	Ea	7.00	9.29	—	16.29
125# cast iron tee, threaded joints						
1"	P1@.230	Ea	2.95	5.62	—	8.57
1¼"	P1@.310	Ea	5.35	7.58	—	12.93
1½"	P1@.390	Ea	6.85	9.53	—	16.38
2"	P1@.490	Ea	9.50	12.00	—	21.50
125# cast iron reducing tee, threaded joints						
1" x 3/4"	P1@.220	Ea	3.76	5.38	—	9.14
1¼" x 3/4"	P1@.290	Ea	4.80	7.09	—	11.89
1¼" x 1"	P1@.290	Ea	5.25	7.09	—	12.34
1½" x 3/4"	P1@.370	Ea	6.60	9.04	—	15.64
1½" x 1"	P1@.370	Ea	7.25	9.04	—	16.29
1½" x 1¼"	P1@.370	Ea	8.05	9.04	—	17.09
2" x 1"	P1@.460	Ea	10.50	11.20	—	21.70
2" x 1¼"	P1@.460	Ea	11.10	11.20	—	22.30
2" x 1½"	P1@.460	Ea	11.90	11.20	—	23.10
125# cast iron reducer, threaded joints						
1" x 3/4"	P1@.200	Ea	2.95	4.89	—	7.84
1¼" x 1"	P1@.210	Ea	3.55	5.13	—	8.68
1½" x 1"	P1@.270	Ea	4.35	6.60	—	10.95
1½" x 1¼"	P1@.270	Ea	4.60	6.60	—	11.20
2" x 1¼"	P1@.340	Ea	6.85	8.31	—	15.16
2" x 1½"	P1@.340	Ea	7.20	8.31	—	15.51

Description	Craft@Hrs	Unit	Material $	Labor $	Equipment $	Total $
125# cast iron cross, threaded joints						
1"	P1@.360	Ea	6.90	8.80	—	15.70
1¼"	P1@.480	Ea	11.20	11.70	—	22.90
1½"	P1@.600	Ea	13.80	14.70	—	28.50
2"	P1@.700	Ea	22.80	17.10	—	39.90
125# cast iron cap, threaded joint						
3/4"	P1@.100	Ea	1.45	2.44	—	3.89
1"	P1@.140	Ea	1.75	3.42	—	5.17
1¼"	P1@.180	Ea	2.25	4.40	—	6.65
1½"	P1@.230	Ea	2.85	5.62	—	8.47
2"	P1@.290	Ea	4.45	7.09	—	11.54
125# cast iron plug, threaded joint						
3/4"	P1@.100	Ea	.95	2.44	—	3.39
1"	P1@.140	Ea	1.25	3.42	—	4.67
1¼"	P1@.180	Ea	1.95	4.40	—	6.35
1½"	P1@.230	Ea	2.45	5.62	—	8.07
2"	P1@.290	Ea	3.05	7.09	—	10.14
125# cast iron coupling, threaded joints						
1"	P1@.180	Ea	2.55	4.40	—	6.95
1¼"	P1@.240	Ea	3.15	5.86	—	9.01
1½"	P1@.300	Ea	3.95	7.33	—	11.28
2"	P1@.380	Ea	5.35	9.29	—	14.64

Description	Craft@Hrs	Unit	Material $	Labor $	Equipment $	Total $

Steel pipe nipples — 3" long, standard right hand thread

Description	Craft@Hrs	Unit	Material $	Labor $	Equipment $	Total $
1"	P1@.025	Ea	.58	.61	—	1.19
1¼"	P1@.025	Ea	.75	.61	—	1.36
1½"	P1@.025	Ea	.90	.61	—	1.51
2"	P1@.030	Ea	1.12	.73	—	1.85

Steel pipe nipples — 4" long, standard right hand thread

Description	Craft@Hrs	Unit	Material $	Labor $	Equipment $	Total $
1"	P1@.025	Ea	.69	.61	—	1.30
1¼"	P1@.030	Ea	.87	.73	—	1.60
1½"	P1@.030	Ea	1.09	.73	—	1.82
2"	P1@.030	Ea	1.39	.73	—	2.12

Steel pipe nipples — 6" long, standard right hand thread

Description	Craft@Hrs	Unit	Material $	Labor $	Equipment $	Total $
1"	P1@.030	Ea	.90	.73	—	1.63
1¼"	P1@.030	Ea	1.13	.73	—	1.86
1½"	P1@.030	Ea	1.37	.73	—	2.10
2"	P1@.035	Ea	1.83	.86	—	2.69

Description	Craft@Hrs	Unit	Material $	Labor $	Equipment $	Total $
Chlorinated Poly Vinyl Chloride Pipe (CPVC)						
CPVC pipe only, solvent weld. No hangers or fittings						
3/4"	SL@.055	LF	.50	1.34	—	1.84
1"	SL@.055	LF	.72	1.34	—	2.06
1¼"	SL@.065	LF	1.14	1.58	—	2.72
1½"	SL@.075	LF	1.58	1.82	—	3.40
2"	SL@.085	LF	2.36	2.07	—	4.43
2½"	SL@.110	LF	3.66	2.67	—	6.33
3"	SL@.125	LF	5.61	3.04	—	8.65
CPVC 90 degree elbow, solvent weld						
3/4"	SL@.100	Ea	1.09	2.43	—	3.52
1"	SL@.120	Ea	1.64	2.92	—	4.56
1¼"	SL@.170	Ea	2.10	4.13	—	6.23
1½"	SL@.180	Ea	3.15	4.37	—	7.52
2"	SL@.200	Ea	3.66	4.86	—	8.52
CPVC 45 degree elbow, solvent weld						
3/4"	SL@.100	Ea	1.13	2.43	—	3.56
1"	SL@.120	Ea	1.47	2.92	—	4.39
1¼"	SL@.170	Ea	1.93	4.13	—	6.06
1½"	SL@.180	Ea	2.61	4.37	—	6.98
2"	SL@.200	Ea	3.23	4.86	—	8.09
CPVC tee, solvent weld						
3/4"	SL@.150	Ea	1.20	3.64	—	4.84
1"	SL@.170	Ea	2.04	4.13	—	6.17
1¼"	SL@.230	Ea	3.21	5.59	—	8.80
1½"	SL@.250	Ea	4.25	6.07	—	10.32
2"	SL@.280	Ea	6.62	6.80	—	13.42
CPVC reducing tee, solvent weld						
1"	SL@.170	Ea	1.75	4.13	—	5.88
1¼"	SL@.230	Ea	3.19	5.59	—	8.78
1½"	SL@.250	Ea	4.27	6.07	—	10.34
2"	SL@.280	Ea	6.74	6.80	—	13.54
3"	SL@.350	Ea	10.71	8.50	—	19.21
CPVC coupling, solvent weld						
3/4"	SL@.100	Ea	.79	2.43	—	3.22
1"	SL@.120	Ea	1.06	2.92	—	3.98
1¼"	SL@.170	Ea	1.68	4.13	—	5.81
1½"	SL@.180	Ea	2.13	4.37	—	6.50
2"	SL@.200	Ea	2.68	4.86	—	7.54
3"	SL@.280	Ea	5.47	6.80	—	12.27

Description	Craft@Hrs	Unit	Material $	Labor $	Equipment $	Total $
CPVC head adapter, solvent weld						
3/4" x 1/2"	SL@.150	Ea	1.79	3.64	—	5.43
1" x 1/2"	SL@.165	Ea	1.90	4.01	—	5.91
CPVC cap, solvent weld						
3/4"	SL@.060	Ea	.59	1.46	—	2.05
1"	SL@.065	Ea	.82	1.58	—	2.40
1¼"	SL@.075	Ea	1.39	1.82	—	3.21
1½"	SL@.085	Ea	1.80	2.07	—	3.87
2"	SL@.095	Ea	2.09	2.31	—	4.40
CPVC flange, solvent weld						
1"	SL@.165	Ea	2.09	4.01	—	6.10
1¼"	SL@.175	Ea	2.68	4.25	—	6.93
1½"	SL@.185	Ea	4.15	4.49	—	8.64
2"	SL@.200	Ea	6.30	4.86	—	11.16
3"	SL@.225	Ea	10.33	5.47	—	15.80
CPVC female adapter, solvent weld						
1"	SL@.065	Ea	5.15	1.58	—	6.73
1¼"	SL@.075	Ea	8.38	1.82	—	10.20
1½"	SL@.085	Ea	11.08	2.07	—	13.15
2"	SL@.095	Ea	14.23	2.31	—	16.54

Commercial Boilers

Boilers. Shipped Factory Assembled, Skid Mounted. Costs shown include the boiler shell and structure, I-beam skids, combustion chamber, firetubes, crown sheet, tube sheets (for firetube), water tubes, mud drum(s), steam drum (for watertube boiler), finned watertubes and check valves (for hot water generators), refractory brick lining, explosion doors, combustion air inlets and stack outlet.
Add the cost of the fuel train piping, electrical wiring, refractory, stack, feedwater system, combustion controls, water treatment, expansion tanks (if hot water) or condensate return (if steam) and system start-up from the sections following the sections on the packaged boilers that follow. Costs shown are based on boiler horsepower rating (BHP). Note: Boiler horsepower (BHP) multiplied times 33.5 gives the approximate required 1000 Btu per hour input (MBH input).

Equipment cost, where shown, is $35 per hour for a 10-ton hydraulic truck-mounted crane.

Hot water or steam cast iron gas boiler 30 PSIG. Costs shown are based on boiler horsepower rating (BHP). Set in place only. Make additional allowances for associated appurtenances and pipe connections.

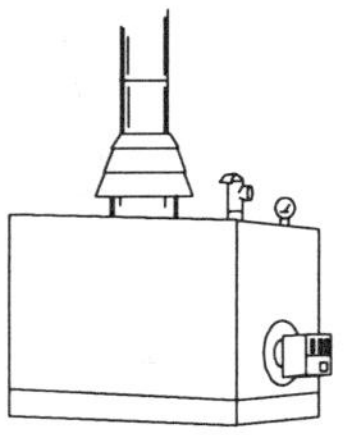

Description	Craft@Hrs	Unit	Material $	Labor $	Equipment $	Total $
4.5 BHP	SN@4.50	Ea	2,180.00	116.00	22.50	2,318.50
5.9 BHP	SN@5.00	Ea	2,740.00	129.00	25.00	2,894.00
9.0 BHP	SN@5.50	Ea	4,080.00	142.00	27.50	4,249.50
15.0 BHP	SN@6.00	Ea	6,340.00	155.00	30.00	6,525.00
20.0 BHP	SN@6.50	Ea	8,390.00	167.00	32.50	8,589.50
30.0 BHP	SN@7.50	Ea	10,900.00	193.00	37.50	11,130.50
45.0 BHP	SN@8.50	Ea	15,800.00	219.00	42.50	16,061.50
60.0 BHP	SN@9.00	Ea	20,600.00	232.00	45.00	20,877.00

Hot water Scotch marine firetube gas boiler 30 PSIG. Costs shown are based on boiler horsepower rating (BHP) Set in place only. Make additional allowances for associated appurtenances and pipe connections.

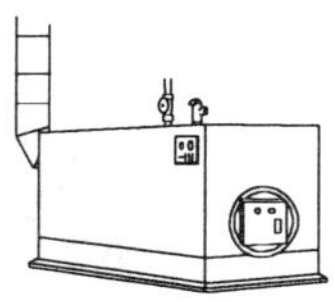

Description	Craft@Hrs	Unit	Material $	Labor $	Equipment $	Total $
30 BHP	SN@10.0	Ea	15,700.00	258.00	37.50	15,995.50
40 BHP	SN@10.0	Ea	16,600.00	258.00	40.00	16,898.00
50 BHP	SN@10.0	Ea	17,300.00	258.00	42.50	17,600.50
60 BHP	SN@12.0	Ea	19,300.00	309.00	45.00	19,654.00
70 BHP	SN@12.0	Ea	20,800.00	309.00	47.50	21,156.50
80 BHP	SN@14.0	Ea	22,000.00	361.00	50.00	22,411.00
100 BHP	SN@14.0	Ea	26,400.00	361.00	60.00	26,821.00
125 BHP	SN@16.0	Ea	28,200.00	412.00	80.00	28,692.00
150 BHP	SN@18.0	Ea	30,200.00	464.00	80.00	30,744.00
200 BHP	SN@18.0	Ea	35,800.00	464.00	80.00	36,344.00
250 BHP	SN@26.0	Ea	39,400.00	670.00	160.00	40,230.00
300 BHP	SN@26.0	Ea	42,300.00	670.00	160.00	43,130.00
350 BHP	SN@26.0	Ea	47,900.00	670.00	160.00	48,730.00
400 BHP	SN@32.0	Ea	49,300.00	824.00	160.00	50,284.00
500 BHP	SN@32.0	Ea	58,800.00	824.00	160.00	59,784.00
600 BHP	SN@32.0	Ea	65,700.00	824.00	160.00	66,684.00
750 BHP	SN@40.0	Ea	69,300.00	1,030.00	320.00	70,650.00

Description	Craft@Hrs	Unit	Material $	Labor $	Equipment $	Total $
Packaged Scotch marine firetube oil or gas fired boiler 150 PSIG. Costs shown are based on boiler horsepower rating (BHP). Set in place only. Make additional allowances for associated appurtenances and pipe connections.						
60 BHP	SN@12.0	Ea	22,000.00	309.00	45.00	22,354.00
70 BHP	SN@12.0	Ea	23,800.00	309.00	47.50	24,156.50
80 BHP	SN@14.0	Ea	25,500.00	361.00	50.00	25,911.00
100 BHP	SN@14.0	Ea	30,300.00	361.00	60.00	30,721.00
125 BHP	SN@16.0	Ea	32,300.00	412.00	80.00	32,792.00
150 BHP	SN@18.0	Ea	34,700.00	464.00	80.00	35,244.00
200 BHP	SN@18.0	Ea	45,500.00	464.00	160.00	46,124.00
250 BHP	SN@26.0	Ea	48,700.00	670.00	160.00	49,530.00
300 BHP	SN@26.0	Ea	55,000.00	670.00	160.00	55,830.00
350 BHP	SN@26.0	Ea	56,700.00	670.00	160.00	57,530.00
400 BHP	SN@32.0	Ea	67,700.00	824.00	160.00	68,684.00
500 BHP	SN@32.0	Ea	75,400.00	824.00	160.00	76,384.00
600 BHP	SN@32.0	Ea	76,800.00	824.00	160.00	77,784.00
700 BHP	SN@40.0	Ea	79,800.00	1,030.00	320.00	81,150.00
Packaged hot water firebox oil or gas fired boiler. Costs shown are based on boiler horsepower rating (BHP). Set in place only. Make additional allowances for associated appurtenances and pipe connections.						
13 BHP	SN@8.00	Ea	8,320.00	206.00	40.00	8,566.00
16 BHP	SN@8.00	Ea	8,460.00	206.00	40.00	8,706.00
19 BHP	SN@8.00	Ea	8,760.00	206.00	40.00	9,006.00
22 BHP	SN@8.00	Ea	8,990.00	206.00	40.00	9,236.00
28 BHP	SN@8.00	Ea	9,340.00	206.00	40.00	9,586.00
34 BHP	SN@8.00	Ea	10,800.00	206.00	40.00	11,046.00
40 BHP	SN@8.00	Ea	11,700.00	206.00	40.00	11,946.00
46 BHP	SN@8.00	Ea	13,500.00	206.00	40.00	13,746.00
52 BHP	SN@8.00	Ea	15,900.00	206.00	40.00	16,146.00
61 BHP	SN@8.00	Ea	17,400.00	206.00	40.00	17,646.00
70 BHP	SN@8.00	Ea	18,600.00	206.00	40.00	18,846.00
80 BHP	SN@8.00	Ea	19,300.00	206.00	40.00	19,546.00
100 BHP	SN@8.00	Ea	21,600.00	206.00	40.00	21,846.00
125 BHP	SN@16.0	Ea	25,700.00	412.00	80.00	26,192.00
150 BHP	SN@16.0	Ea	27,800.00	412.00	80.00	28,292.00
Packaged watertube industrial high pressure boilers. Costs shown are based on boiler horsepower rating (BHP). Set in place only. Make additional allowances for associated appurtenances and pipe connections.						
480 BHP	SN@32.0	Ea	157,000.00	824.00	160.00	157,984.00
526 BHP	SN@32.0	Ea	162,000.00	824.00	160.00	162,984.00
572 BHP	SN@32.0	Ea	168,000.00	824.00	160.00	168,984.00
618 BHP	SN@32.0	Ea	172,000.00	824.00	160.00	172,984.00
664 BHP	SN@32.0	Ea	180,000.00	824.00	160.00	180,984.00
710 BHP	SN@64.0	Ea	183,000.00	1,650.00	320.00	184,970.00
747 BHP	SN@64.0	Ea	185,000.00	1,650.00	320.00	186,970.00
792 BHP	SN@64.0	Ea	191,000.00	1,650.00	320.00	192,970.00
838 BHP	SN@64.0	Ea	197,000.00	1,650.00	320.00	198,970.00
884 BHP	SN@64.0	Ea	203,000.00	1,650.00	320.00	204,970.00
930 BHP	SN@64.0	Ea	207,000.00	1,650.00	320.00	208,970.00
1,000 BHP	SN@64.0	Ea	209,000.00	1,650.00	320.00	210,970.00

Description	Craft@Hrs	Unit	Material $	Labor $	Equipment $	Total $

Commercial Hot Water Boiler Connection. Includes pipe, fittings, water feed and pressure relief (drains), pipe insulation, valves, gauges, vents and thermometers to connect boiler to distribution system.

Description	Craft@Hrs	Unit	Material $	Labor $	Equipment $	Total $
2½" supply	SN@18.0	Ea	1,070.00	464.00	21.78	1,555.78
3" supply	SN@20.5	Ea	1,250.00	528.00	24.80	1,802.80
4" supply	SN@26.0	Ea	1,820.00	670.00	31.46	2,521.46
6" supply	SN@32.0	Ea	2,660.00	824.00	38.72	3,522.72
8" supply	SN@39.5	Ea	3,560.00	1,020.00	47.80	4,627.80
10" supply	SN@48.0	Ea	5,090.00	1,240.00	58.08	6,388.08
12" supply	SN@54.0	Ea	6,270.00	1,390.00	65.34	7,725.34

Commercial Steam Boiler Connection. Includes pipe, fittings, water feed and pressure relief (drains), pipe insulation, valves, gauges, vents and thermometers to connect boiler to distribution system.

Description	Craft@Hrs	Unit	Material $	Labor $	Equipment $	Total $
2½" supply	SN@34.0	Ea	1,920.00	876.00	41.14	2,837.14
3" supply	SN@38.0	Ea	2,030.00	979.00	45.98	3,054.98
4" supply	SN@43.0	Ea	2,340.00	1,110.00	52.03	3,502.03
6" supply	SN@53.0	Ea	3,210.00	1,360.00	64.13	4,634.13
8" supply	SN@62.0	Ea	4,140.00	1,600.00	75.02	5,815.02
10" supply	SN@70.0	Ea	5,640.00	1,800.00	84.70	7,524.70
12" supply	SN@82.0	Ea	7,130.00	2,110.00	99.22	9,339.22

Costs for Combustion Trains for Boilers. Costs shown include the burner, burner head, burner mount flange, associated refractory and gasketing, main fuel valve, draft damper and fan (for a power burner), or burner face plate (for an atmospheric or induced-draft burner). Note: Boiler horsepower (BHP) multiplied times 33.5 gives the approximate required 1000 Btu per hour input (MBH input).

Description	Craft@Hrs	Unit	Material $	Labor $	Equipment $	Total $

Combustion train for Scotch marine firetube boilers. Costs shown are based on boiler horsepower rating (BHP).

Description	Craft@Hrs	Unit	Material $	Labor $	Equipment $	Total $
20 BHP	SN@8.00	Ea	1,280.00	206.00	40.00	1,526.00
30 BHP	SN@8.00	Ea	1,750.00	206.00	40.00	1,996.00
40 BHP	SN@8.00	Ea	2,390.00	206.00	40.00	2,636.00
50 BHP	SN@8.00	Ea	2,990.00	206.00	40.00	3,236.00
60 BHP	SN@8.00	Ea	3,310.00	206.00	40.00	3,556.00
70 BHP	SN@8.00	Ea	3,610.00	206.00	40.00	3,856.00
80 BHP	SN@8.00	Ea	3,680.00	206.00	40.00	3,926.00
100 BHP	SN@8.00	Ea	4,020.00	206.00	40.00	4,266.00
125 BHP	SN@8.00	Ea	4,180.00	206.00	40.00	4,426.00
150 BHP	SN@8.00	Ea	4,630.00	206.00	40.00	4,876.00
200 BHP	SN@8.00	Ea	4,820.00	206.00	40.00	5,066.00
250 BHP	SN@8.00	Ea	5,300.00	206.00	40.00	5,546.00
300 BHP	SN@8.00	Ea	5,940.00	206.00	40.00	6,186.00
350 BHP	SN@8.00	Ea	6,080.00	206.00	40.00	6,326.00
400 BHP	SN@8.00	Ea	6,270.00	206.00	40.00	6,516.00
500 BHP	SN@16.0	Ea	6,810.00	412.00	80.00	7,302.00
600 BHP	SN@16.0	Ea	7,280.00	412.00	80.00	7,772.00
750 BHP	SN@32.0	Ea	7,730.00	824.00	110.00	8,664.00
800 BHP	SN@32.0	Ea	8,070.00	824.00	110.00	9,004.00

Combustion train for package firebox boilers. Costs shown are based on boiler horsepower rating (BHP).

Description	Craft@Hrs	Unit	Material $	Labor $	Equipment $	Total $
13.4 BHP	SN@8.00	Ea	672.00	206.00	40.00	918.00
16.4 BHP	SN@8.00	Ea	886.00	206.00	40.00	1,132.00
17.4 BHP	SN@8.00	Ea	1,050.00	206.00	40.00	1,296.00
22.4 BHP	SN@8.00	Ea	1,240.00	206.00	40.00	1,486.00
28.4 BHP	SN@8.00	Ea	1,340.00	206.00	40.00	1,586.00
34.4 BHP	SN@8.00	Ea	1,470.00	206.00	40.00	1,716.00
40.3 BHP	SN@8.00	Ea	1,600.00	206.00	40.00	1,846.00
46.3 BHP	SN@8.00	Ea	1,760.00	206.00	40.00	2,006.00
52.3 BHP	SN@8.00	Ea	1,960.00	206.00	40.00	2,206.00
61.1 BHP	SN@8.00	Ea	2,120.00	206.00	40.00	2,366.00
70.0 BHP	SN@8.00	Ea	2,170.00	206.00	40.00	2,416.00
80.0 BHP	SN@8.00	Ea	2,270.00	206.00	40.00	2,516.00
100.0 BHP	SN@8.00	Ea	2,390.00	206.00	40.00	2,636.00
125.0 BHP	SN@8.00	Ea	2,550.00	206.00	40.00	2,796.00
150.0 BHP	SN@8.00	Ea	2,790.00	206.00	40.00	3,036.00

Description	Craft@Hrs	Unit	Material $	Labor $	Equipment $	Total $
Natural gas fuel train piping for firetube and watertube boilers. Meets requirements of Underwriters' Laboratories. Shipped pre-assembled. Add the cost of venting and piping to a gas meter. Costs shown are based on boiler horsepower rating (BHP).						
20 BHP	SN@8.00	Ea	340.00	206.00	40.00	586.00
30 BHP	SN@8.00	Ea	467.00	206.00	40.00	713.00
40 BHP	SN@8.00	Ea	714.00	206.00	40.00	960.00
50 BHP	SN@8.00	Ea	851.00	206.00	40.00	1,097.00
60 BHP	SN@8.00	Ea	932.00	206.00	40.00	1,178.00
70 BHP	SN@8.00	Ea	989.00	206.00	40.00	1,235.00
80 BHP	SN@8.00	Ea	1,070.00	206.00	40.00	1,316.00
100 BHP	SN@8.00	Ea	1,240.00	206.00	40.00	1,486.00
125 BHP	SN@8.00	Ea	1,470.00	206.00	40.00	1,716.00
150 BHP	SN@8.00	Ea	1,750.00	206.00	40.00	1,996.00
200 BHP	SN@8.00	Ea	2,090.00	206.00	40.00	2,336.00
250 BHP	SN@8.00	Ea	2,580.00	206.00	40.00	2,826.00
300 BHP	SN@8.00	Ea	2,720.00	206.00	40.00	2,966.00
350 BHP	SN@8.00	Ea	2,920.00	206.00	40.00	3,166.00
400 BHP	SN@8.00	Ea	3,130.00	206.00	40.00	3,376.00
500 BHP	SN@16.0	Ea	3,520.00	412.00	80.00	4,012.00
600 BHP	SN@16.0	Ea	4,070.00	412.00	80.00	4,562.00
750 BHP	SN@16.0	Ea	4,390.00	412.00	80.00	4,882.00
800 BHP	SN@16.0	Ea	4,600.00	412.00	80.00	5,092.00
Natural gas fuel train piping for package firebox boilers. Costs shown are based on boiler horsepower rating (BHP).						
13 BHP	SN@8.00	Ea	456.00	206.00	40.00	702.00
16 BHP	SN@8.00	Ea	521.00	206.00	40.00	767.00
19 BHP	SN@8.00	Ea	560.00	206.00	40.00	806.00
22 BHP	SN@8.00	Ea	615.00	206.00	40.00	861.00
28 BHP	SN@8.00	Ea	648.00	206.00	40.00	894.00
34 BHP	SN@8.00	Ea	714.00	206.00	40.00	960.00
40 BHP	SN@8.00	Ea	851.00	206.00	40.00	1,097.00
45 BHP	SN@8.00	Ea	895.00	206.00	40.00	1,141.00
50 BHP	SN@8.00	Ea	1,070.00	206.00	40.00	1,316.00
60 BHP	SN@8.00	Ea	1,200.00	206.00	40.00	1,446.00
70 BHP	SN@8.00	Ea	1,370.00	206.00	40.00	1,616.00
80 BHP	SN@8.00	Ea	1,430.00	206.00	40.00	1,676.00
100 BHP	SN@8.00	Ea	1,710.00	206.00	40.00	1,956.00
125 BHP	SN@8.00	Ea	1,770.00	206.00	40.00	2,016.00
150 BHP	SN@8.00	Ea	1,930.00	206.00	40.00	2,176.00
Add for code approval:						
Factory Mutual	—	%	7.0	—	—	—
Industrial Risk	—	%	10.0	—	—	—

Oil Fuel Train Piping. Costs shown include recirculating pump, electric oil preheater, regulating valve, check valves, isolating pump stop cocks, oil line pressure relief valve and pressure gauges. Add the cost of concrete pump pad, electrical connections, relief line piping back to the oil tank and tank connecting piping. Add for number 4-6 fuel oil as shown.

Description	Craft@Hrs	Unit	Material $	Labor $	Equipment $	Total $
Oil fuel train piping for boilers. Costs shown are based on boiler horsepower rating (BHP) using number 2 fuel oil.						
20 to 30 BHP	SN@4.00	Ea	730.00	103.00	20.00	853.00
40 to 60 BHP	SN@4.00	Ea	810.00	103.00	20.00	933.00
70 to 80 BHP	SN@4.00	Ea	886.00	103.00	20.00	1,009.00
100 BHP	SN@8.00	Ea	1,140.00	206.00	40.00	1,386.00
125 to 150 BHP	SN@8.00	Ea	1,470.00	206.00	40.00	1,716.00
200 BHP	SN@8.00	Ea	1,950.00	206.00	40.00	2,196.00
250 BHP	SN@8.00	Ea	2,310.00	206.00	40.00	2,556.00
300 BHP	SN@16.0	Ea	2,470.00	412.00	80.00	2,962.00
350 BHP	SN@16.0	Ea	2,760.00	412.00	80.00	3,252.00
400 BHP	SN@16.0	Ea	3,300.00	412.00	80.00	3,792.00
500 BHP	SN@32.0	Ea	3,360.00	824.00	160.00	4,344.00
600 BHP	SN@32.0	Ea	3,510.00	824.00	160.00	4,494.00
750 to 800 BHP	SN@32.0	Ea	3,840.00	824.00	160.00	4,824.00
900 to 1000 BHP	SN@32.0	Ea	4,520.00	824.00	160.00	5,504.00
Add for #4-6 oil	—	%	10.0	—	—	—
Add for tie-in with: dual-fuel burner	—	%	5.0	—	—	—

Costs for Electrical Service for Boilers, Subcontract. Costs shown include typical main power tie-in and fusing, circuit breaker panel, and main bus. Assumes controls, fan, blower switchgear and control console are pre-wired electrical service. These costs include the electrical subcontractor's overhead and profit.

Description	Craft@Hrs	Unit	Material $	Labor $	Equipment $	Total $
Electrical service for firetube and watertube boilers. Costs shown are based on boiler horsepower rating (BHP).						
20 to 100 BHP	—	Ea	—	—	—	565.83
125 to 200 BHP	—	Ea	—	—	—	680.41
250 to 350 BHP	—	Ea	—	—	—	787.93
400 to 500 BHP	—	Ea	—	—	—	894.02
500 to 600 BHP	—	Ea	—	—	—	1,305.32
750 to 800 BHP	—	Ea	—	—	—	2,343.36
Electrical service for package firebox boilers. Costs shown are based on boiler horsepower rating (BHP).						
13 to 70 BHP	—	Ea	—	—	—	542.49
80 to 125 BHP	—	Ea	—	—	—	557.51
150 BHP	—	Ea	—	—	—	680.41

Commercial Boiler Components and Accessories

Costs for Refractory for Boilers, Subcontract. For package Scotch marine firetube, firebox and watertube boilers. Most new boilers are shipped from the factory with the refactory brick installed. If refractory brick is installed on the job site to repair shipping damage or to create firedoor seals, use the figures below. These costs include the refractory subcontractor's overhead and profit. Costs shown are per 100 pounds of refractory used.

Description	Craft@Hrs	Unit	Material $	Labor $	Equipment $	Total $
Repairs and seals	—	100lb	—	—	—	147.54

Costs for Boiler Feedwater Pumps. Costs shown are for the pump only. Add the cost of piping, pump actuating valves and concrete pump base.

Description	Craft@Hrs	Unit	Material $	Labor $	Equipment $	Total $
Feedwater pumps for firetube and watertube boilers. Costs shown are based on boiler horsepower rating (BHP).						
20 to 40 BHP	SN@8.00	Ea	4,430.00	206.00	40.00	4,676.00
50 to 80 BHP	SN@8.00	Ea	5,350.00	206.00	40.00	5,596.00
100 BHP	SN@8.00	Ea	5,920.00	206.00	40.00	6,166.00
125 to 200 BHP	SN@16.0	Ea	6,120.00	412.00	80.00	6,612.00
250 to 400 BHP	SN@32.0	Ea	6,730.00	824.00	160.00	7,714.00
500 BHP	SN@32.0	Ea	7,100.00	824.00	160.00	8,084.00
600 BHP	SN@32.0	Ea	7,380.00	824.00	160.00	8,364.00
750 BHP	SN@32.0	Ea	7,580.00	824.00	160.00	8,564.00
800 BHP	SN@32.0	Ea	10,300.00	824.00	160.00	11,284.00
Feedwater pumps for package firebox boilers. Costs shown are based on boiler horsepower rating (BHP).						
13 to 35 BHP	SN@8.00	Ea	2,370.00	206.00	40.00	2,616.00
40 to 45 BHP	SN@8.00	Ea	2,790.00	206.00	40.00	3,036.00
50 to 60 BHP	SN@8.00	Ea	3,500.00	206.00	40.00	3,746.00
70 to 80 BHP	SN@8.00	Ea	4,170.00	206.00	40.00	4,416.00
100 BHP	SN@8.00	Ea	5,350.00	206.00	40.00	5,596.00
125 BHP	SN@16.0	Ea	5,350.00	412.00	80.00	5,842.00
150 BHP	SN@16.0	Ea	5,920.00	412.00	80.00	6,412.00

Costs for Natural Gas Combustion Controls for Boilers. Costs shown include automatic recycling or continuous pilot, spark igniter and coil (for electric ignition), timer to control the purge cycle, pilot, main burner (hi-lo-off) or main burner (proportional control), flame safeguard and sensor and combustion control panel.

Natural gas combustion controls for firetube and watertube boilers.
Costs shown are based on boiler horsepower rating (BHP).

Description	Craft@Hrs	Unit	Material $	Labor $	Equipment $	Total $
20 to 80 BHP	SN@4.00	Ea	1,820.00	103.00	20.00	1,943.00
100 BHP	SN@4.00	Ea	2,040.00	103.00	20.00	2,163.00
125 to 200 BHP	SN@8.00	Ea	2,040.00	206.00	40.00	2,286.00
250 BHP	SN@16.0	Ea	2,040.00	412.00	80.00	2,532.00
300 BHP	SN@16.0	Ea	2,450.00	412.00	80.00	2,942.00
350 or 400 BHP	SN@16.0	Ea	2,870.00	412.00	80.00	3,362.00
500 BHP	SN@32.0	Ea	2,870.00	824.00	160.00	3,854.00
600 BHP	SN@32.0	Ea	3,430.00	824.00	160.00	4,414.00
750 BHP	SN@64.0	Ea	3,990.00	1,650.00	320.00	5,960.00
800 BHP	SN@64.0	Ea	4,640.00	1,650.00	320.00	6,610.00

Natural gas combustion controls for firebox boilers.
Costs shown are based on boiler horsepower rating (BHP).

Description	Craft@Hrs	Unit	Material $	Labor $	Equipment $	Total $
13.4 to 80.0 BHP	SN@4.00	Ea	1,760.00	103.00	20.00	1,883.00
100.0 BHP	SN@4.00	Ea	2,040.00	103.00	20.00	2,163.00
126.9 BHP	SN@8.00	Ea	2,040.00	206.00	40.00	2,286.00
150.0 BHP	SN@8.00	Ea	3,790.00	206.00	40.00	4,036.00
Add if burner controls are for:						
#2 oil-fired	—	%	—	—	—	+10.0
Heavy oil-fired	—	%	—	—	—	+20.0
Dual fuel burners	—	%	—	—	—	+70.0

Costs for Water Softening Systems for Use with Boilers. Costs shown include the metering valve, tank, pump, and zeolite media. Add the cost of piping to connect the water softening system to the main feedwater tank and pump.

Water softening system for firetube and watertube boilers.
Costs shown are based on boiler horsepower rating (BHP).

Description	Craft@Hrs	Unit	Material $	Labor $	Equipment $	Total $
20 to 80 BHP	SN@4.00	Ea	3,020.00	103.00	20.00	3,143.00
100 BHP	SN@4.00	Ea	3,550.00	103.00	20.00	3,673.00
125 BHP	SN@8.00	Ea	4,110.00	206.00	40.00	4,356.00
150 or 200 BHP	SN@8.00	Ea	5,200.00	206.00	40.00	5,446.00
250 BHP	SN@8.00	Ea	6,140.00	206.00	40.00	6,386.00
300 or 350 BHP	SN@16.0	Ea	7,850.00	412.00	80.00	8,342.00
400 or 500 BHP	SN@16.0	Ea	9,800.00	412.00	80.00	10,292.00
600 BHP	SN@32.0	Ea	11,800.00	824.00	160.00	12,784.00
750 or 800 BHP	SN@64.0	Ea	14,000.00	1,650.00	320.00	15,970.00

Water softening system for package firebox boilers. Costs shown are based on boiler horsepower rating (BHP).

Description	Craft@Hrs	Unit	Material $	Labor $	Equipment $	Total $
13.4 to 80 BHP	SN@4.00	Ea	3,560.00	103.00	20.00	3,683.00
100.0 BHP	SN@4.00	Ea	3,560.00	103.00	20.00	3,683.00
126.9 BHP	SN@8.00	Ea	4,170.00	206.00	40.00	4,416.00
150.0 BHP	SN@8.00	Ea	4,840.00	206.00	40.00	5,086.00

Costs for Start-up, Shakedown, and Calibration of Boilers. Costs shown include minor changes and adjustments to meet requirements of the boiler inspector, adjustment of combustion efficiency at the burner by a control technician, pump and metering calibration testing and steam or hot water metering on start-up.

Start-up, shakedown & calibration of firetube and watertube boilers. Costs shown are based on boiler horsepower rating (BHP.) If a factory technician performs the start-up, add this cost at a minimum of $500 (8 hours at $62.50 per hour).

Description	Craft@Hrs	Unit	Material $	Labor $	Equipment $	Total $
13.4 to 100 BHP	S2@4.00	Ea	—	96.60	—	96.60
125 to 200 BHP	S2@8.00	Ea	—	193.00	—	193.00
250 to 400 BHP	S2@16.0	Ea	—	386.00	—	386.00
500 to 600 BHP	S2@24.0	Ea	—	580.00	—	580.00
750 to 800 BHP	S2@30.0	Ea	—	724.00	—	724.00

Costs for Stack Economizer Units for Boilers. Costs shown include piping connection to existing boiler, economizer fin tube coils, feedwater pump, check valves, thermostatic regulating valve, stack dampers, pressure relief valve, stack mount flanges and insulating enclosure, but no architectural modifications.

Stack waste heat recovery system for boilers. Costs shown are based on boiler horsepower rating (BHP).

Description	Craft@Hrs	Unit	Material $	Labor $	Equipment $	Total $
500 BHP	SN@40.0	Ea	17,100.00	1,030.00	200.00	18,330.00
750 BHP	SN@48.0	Ea	21,300.00	1,240.00	240.00	22,780.00
800 BHP	SN@48.0	Ea	22,900.00	1,240.00	240.00	24,380.00
1,000 BHP	SN@56.0	Ea	28,600.00	1,440.00	280.00	30,320.00

Costs for Condensate Receiver and Pumping Units for Boilers. Costs shown include a stainless steel factory assembled tank and pump with motor, float switch, condensate inlet, feedwater tank, connection, level gauge, thermometer, solenoid valve for direct boiler feed. No concrete mounting pad required.

Description	Craft@Hrs	Unit	Material $	Labor $	Equipment $	Total $
Condensate receiver and pumping unit for boilers. Unit sizing and costs are based on storage for one minute at the designed steam return rate. Set in place only. Make additional allowances for pipe and electrical connections.						
750 lbs per hour	SN@4.00	Ea	2,230.00	103.00	20.00	2,353.00
1,500 lbs per hour	SN@4.00	Ea	2,230.00	103.00	20.00	2,353.00
3,000 lbs per hour	SN@4.00	Ea	2,840.00	103.00	20.00	2,963.00
5,000 lbs per hour	SN@6.00	Ea	3,120.00	155.00	30.00	3,305.00
6,250 lbs per hour	SN@6.00	Ea	3,490.00	155.00	30.00	3,675.00

Costs for Deaerator/Condenser Units for Boilers. Pressurized jet spray type. Costs shown include an ASME storage receiver, stand, jet spray deaerator head, makeup control valve, makeup controller, pressure control valve, level controls, pressure and temperature controls, pumped condensate inlet, pumps and motor, steam inlet and system accessories. Add for a concrete mounting pad and piping to the condensate return line (inlet) and piping to the boiler feedwater system (outlet).

Description	Craft@Hrs	Unit	Material $	Labor $	Equipment $	Total $
Deaerator/condenser unit for boilers. Costs shown are based on boiler horsepower rating (BHP).						
100 BHP	SN@8.00	Ea	23,400.00	206.00	40.00	23,646.00
200 BHP	SN@12.0	Ea	24,300.00	309.00	60.00	24,669.00
300 BHP	SN@14.0	Ea	25,800.00	361.00	70.00	26,231.00
400 BHP	SN@16.0	Ea	29,200.00	412.00	80.00	29,692.00
600 BHP	SN@24.0	Ea	31,200.00	618.00	120.00	31,938.00
800 BHP	SN@36.0	Ea	34,400.00	927.00	180.00	35,507.00
1,000 BHP	SN@48.0	Ea	40,100.00	1,240.00	240.00	41,580.00
Rolairtrol type air separators. Priced by capacity in gallons per minute (GPM) and by pipe connection size. Includes hanging separator but no connection, (2" is a screwed connection, 2-1/2" and up are flanged connections) using small tools.						
56 GPM, 2"	S2@2.50	Ea	421.00	60.40	1.20	482.60
90 GPM, 2½"	S2@2.50	Ea	457.00	60.40	1.20	518.60
170 GPM, 3"	S2@3.00	Ea	600.00	72.40	1.44	673.84
300 GPM, 4"	S2@3.50	Ea	893.00	84.50	1.68	979.18
500 GPM, 6"	S2@4.75	Ea	1,170.00	115.00	2.37	1,287.37
700 GPM, 6"	S2@4.75	Ea	1,520.00	115.00	2.37	1,637.37
1300 GPM, 8	SN@6.00	Ea	2,050.00	155.00	3.00	2,208.00
2000 GPM, 10"	SN@6.00	Ea	2,860.00	155.00	3.00	3,018.00
2750 GPM, 12"	SN@12.0	Ea	3,700.00	309.00	6.00	4,015.00

Commercial Boiler Components and Accessories

Costs for Chemical Feed Duplex Pump, Packaged Units for Boilers. Costs shown include the chemical tanks, electronic metering, valve and stand. Add the cost of piping to the feedwater tank.

Description	Craft@Hrs	Unit	Material $	Labor $	Equipment $	Total $
Chemical feed duplex pump for boilers. Costs shown are based on boiler horsepower rating (BHP).						
20 to 100 BHP	SN@3.00	Ea	863.00	77.30	15.00	955.30
125 to 200 BHP	SN@3.00	Ea	1,120.00	77.30	15.00	1,212.30
200 to 400 BHP	SN@6.00	Ea	1,120.00	155.00	30.00	1,305.00
500 to 600 BHP	SN@6.00	Ea	2,540.00	155.00	30.00	2,725.00
700 to 800 BHP	SN@8.00	Ea	2,840.00	206.00	40.00	3,086.00
1,000 BHP	SN@10.0	Ea	3,850.00	258.00	50.00	4,158.00
Chemical feed simple in-line system for hot water heating system. Costs shown include 3 gallon pot feeder, 2, 3/4" isolation valves, 10' of 3/4" pipe and 2, 3/4" thread-o-lets. Chemicals are drawn in through suction created by hot water system circulation pumps.						
50 to 20 BHP	SN@3.00	Ea	200.00	77.30	15.00	292.30

Costs for Packaged Boiler Feedwater Systems for Boilers. Costs shown include duplex pumps with heater assembly, thermal lining, pressure gauges, makeup feeder valve, ASME feedwater tank and stand, single phase electric motor and level gauge. Add the costs for a concrete mounting pad and piping to the boilers.

Description	Craft@Hrs	Unit	Material $	Labor $	Equipment $	Total $
Packaged boiler feedwater systems. Costs shown are based on boiler horsepower rating (BHP).						
15 to 50 BHP	SN@5.00	Ea	5,920.00	129.00	25.00	6,074.00
60 to 80 BHP	SN@5.00	Ea	6,510.00	129.00	25.00	6,664.00
100 BHP	SN@6.00	Ea	6,620.00	155.00	30.00	6,805.00
125 to 150 BHP	SN@6.00	Ea	7,100.00	155.00	30.00	7,285.00
200 BHP	SN@8.00	Ea	7,320.00	206.00	40.00	7,566.00
250 BHP	SN@8.00	Ea	7,640.00	206.00	40.00	7,886.00
300 BHP	SN@8.00	Ea	8,070.00	206.00	40.00	8,316.00
400 BHP	SN@8.00	Ea	8,340.00	206.00	40.00	8,586.00
500 BHP	SN@8.00	Ea	10,100.00	206.00	40.00	10,346.00
600 BHP	SN@12.0	Ea	10,600.00	309.00	60.00	10,969.00
750 BHP	SN@12.0	Ea	10,500.00	309.00	60.00	10,869.00

Costs for Continuous Blowdown Heat Recovery Systems. Costs shown include a flash receiver (50 PSI ASME rated tank), plate and frame heat exchanger, pneumatic level controller and valve, air filter regulator, water gauge, pressure gauge, backflush piping, pressure relief valve connected to mud drum and drain. Add for concrete pad mount, piping and electrical wiring.

Description	Craft@Hrs	Unit	Material $	Labor $	Equipment $	Total $
Continuous blowdown heat recovery systems. Costs shown by pounds per hour (lbs per hr) of rated makeup capacity.						
5,000 lbs per hr	SN@8.00	Ea	19,900.00	206.00	40.00	20,146.00
10,000 lbs per hr	SN@8.00	Ea	39,300.00	206.00	40.00	39,546.00
25,000 lbs per hr	SN@8.00	Ea	49,700.00	206.00	40.00	49,946.00
50,000 lbs per hr	SN@8.00	Ea	79,100.00	206.00	40.00	79,346.00

Costs for Boiler Stacks. Costs shown assume a 20-foot stack height and include rigging, breeching, connection of stack sections, guy wires where required, vendor-supplied stack supports and insulation. Add the cost of a concrete base.

Description	Craft@Hrs	Unit	Material $	Labor $	Equipment $	Total $
Boiler stack. Costs shown assume a 20-foot stack height.						
10" diameter	SN@12.0	Ea	1,420.00	309.00	60.00	1,789.00
12" diameter	SN@12.0	Ea	1,750.00	309.00	60.00	2,119.00
16" diameter	SN@16.0	Ea	2,120.00	412.00	80.00	2,612.00
20" diameter	SN@20.0	Ea	2,650.00	515.00	100.00	3,265.00
24" diameter	SN@24.0	Ea	3,490.00	618.00	120.00	4,228.00
30" diameter	SN@32.0	Ea	4,140.00	824.00	160.00	5,124.00
56" diameter	SN@40.0	Ea	8,170.00	1,030.00	200.00	9,400.00
Add for each 10' section (or portion thereof) beyond 20' high						
per 10' sect.	—	%	—	—	—	15.0

Commercial Boiler Components and Accessories

Costs for Tankless Indirect Water Heater. Provides domestic hot water from boiler coil insert. Costs shown include heat exchanger, coil, circulating pump, thermostatically controlled valve, bypass piping and air bleeder valve. Add the cost of an equipment base, electrical wiring, piping and an insulated hot water storage tank.

Description	Craft@Hrs	Unit	Material $	Labor $	Equipment $	Total $
Tankless indirect water heater. Costs by gallons per hour (GPH).						
240 GPH	SN@8.00	Ea	8,260.00	206.00	40.00	8,506.00
360 GPH	SN@8.00	Ea	10,400.00	206.00	40.00	10,646.00
480 GPH	SN@10.0	Ea	10,600.00	258.00	50.00	10,908.00
840 GPH	SN@12.0	Ea	24,200.00	309.00	60.00	24,569.00
920 GPH	SN@16.0	Ea	27,100.00	412.00	80.00	27,592.00
1,240 GPH	SN@20.0	Ea	30,200.00	515.00	100.00	30,815.00

Costs for Boiler Trim. Costs shown include the pressure relief valve, level gauge (for steam systems), level controller & low water cutoff, pressure and temperature gauge, feedwater valve & actuator, main valve, blowdown valve, cleanout manholes and gaskets.

Description	Craft@Hrs	Unit	Material $	Labor $	Equipment $	Total $
Boiler trim. Costs shown are per boiler horsepower rating (BHP).						
20 to 100 BHP	SN@4.00	Ea	1,290.00	103.00	20.00	1,413.00
100 to 200 BHP	SN@6.00	Ea	1,460.00	155.00	30.00	1,645.00
200 to 400 BHP	SN@8.00	Ea	1,790.00	206.00	40.00	2,036.00
400 to 600 BHP	SN@8.00	Ea	2,740.00	206.00	40.00	2,986.00
600 to 800 BHP	SN@8.00	Ea	3,230.00	206.00	40.00	3,476.00
800 to 1000 BHP	SN@10.0	Ea	3,240.00	258.00	50.00	3,548.00

Base mounted, bronze fitted centrifugal pump, set in place only. Make additional allowances for pipe connection assembly.

Description	Craft@Hrs	Unit	Material $	Labor $	Equipment $	Total $
1/2 HP	P1@1.50	Ea	456.00	36.70	—	492.70
3/4 HP	P1@1.75	Ea	578.00	42.80	—	620.80
1 HP	P1@2.00	Ea	705.00	48.90	—	753.90
1½ HP	P1@2.40	Ea	852.00	58.60	—	910.60
2 HP	P1@2.70	Ea	985.00	66.00	—	1,051.00
3 HP	P1@3.00	Ea	1,250.00	73.30	—	1,323.30
5 HP	P1@3.50	Ea	1,430.00	85.50	—	1,515.50
7½ HP	P1@4.00	Ea	1,590.00	97.70	—	1,687.70
10 HP	P1@4.50	Ea	2,320.00	110.00	—	2,430.00
15 HP	ER@5.00	Ea	2,680.00	130.00	25.00	2,835.00
20 HP	ER@6.00	Ea	3,110.00	156.00	30.00	3,296.00
25 HP	ER@7.00	Ea	3,240.00	183.00	35.00	3,458.00
30 HP	ER@7.40	Ea	3,840.00	193.00	37.00	4,070.00
40 HP	ER@7.70	Ea	3,810.00	201.00	38.50	4,049.50
50 HP	ER@8.00	Ea	4,630.00	209.00	40.00	4,879.00

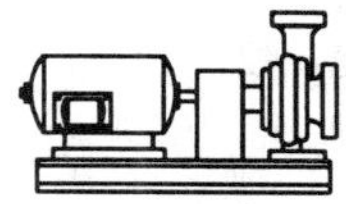

Base mounted centrifugal pump connection assembly. Includes pipe, fittings, pipe insulation, valves, strainers, suction guides and gauges.

Description	Craft@Hrs	Unit	Material $	Labor $	Equipment $	Total $
2½" supply	P1@12.0	Ea	1,080.00	293.00	—	1,373.00
3" supply	P1@13.5	Ea	1,230.00	330.00	—	1,560.00
4" supply	P1@16.0	Ea	1,700.00	391.00	—	2,091.00
6" supply	P1@22.0	Ea	2,820.00	538.00	—	3,358.00
8" supply	P1@28.0	Ea	4,110.00	684.00	—	4,794.00
10" supply	P1@35.0	Ea	6,750.00	855.00	—	7,605.00
12" supply	P1@42.0	Ea	8,940.00	1,030.00	—	9,970.00

Cast iron line-mounted centrifugal pump, including connection and supports.

Description	Craft@Hrs	Unit	Material $	Labor $	Equipment $	Total $
1/6 HP	P1@1.25	Ea	322.00	30.50	—	352.50
1/4 HP	P1@1.40	Ea	413.00	34.20	—	447.20
1/3 HP	P1@1.50	Ea	511.00	36.70	—	547.70
1/2 HP	P1@1.70	Ea	650.00	41.50	—	691.50
3/4 HP	P1@1.80	Ea	876.00	44.00	—	920.00
1 HP	P1@2.00	Ea	1,040.00	48.90	—	1,088.90

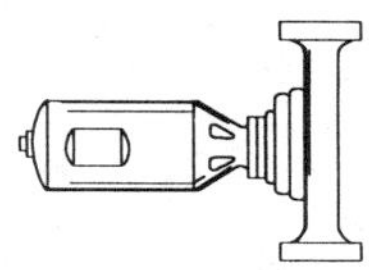

Description	Craft@Hrs	Unit	Material $	Labor $	Equipment $	Total $

2-pass shell and tube type heat exchanger, set in place only. Make additional allowances for pipe connection assembly.

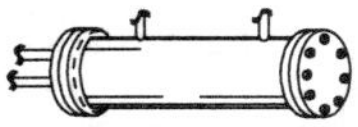

Description	Craft@Hrs	Unit	Material $	Labor $	Equipment $	Total $
4" dia. x 30"	SN@2.00	Ea	743.00	51.50	10.00	804.50
4" dia. x 60"	SN@2.50	Ea	873.00	64.40	12.50	949.90
6" dia. x 30"	SN@3.25	Ea	1,070.00	83.70	16.25	1,169.95
6" dia. x 60"	SN@3.50	Ea	1,330.00	90.10	17.50	1,437.60
8" dia. x 30"	SN@4.25	Ea	1,650.00	109.00	21.25	1,780.25
8" dia. x 60"	SN@4.50	Ea	2,140.00	116.00	22.50	2,278.50
10" dia. x 30"	SN@5.50	Ea	2,430.00	142.00	27.50	2,599.50
10" dia. x 60"	SN@5.75	Ea	3,230.00	148.00	28.75	3,406.75
10" dia. x 90"	SN@6.00	Ea	4,030.00	155.00	30.00	4,215.00
12" dia. x 60"	SN@6.50	Ea	4,230.00	167.00	32.50	4,429.50
12" dia. x 90"	SN@7.00	Ea	5,410.00	180.00	35.00	5,625.00
12" dia. x 120"	SN@7.25	Ea	6,560.00	187.00	36.25	6,783.25
14" dia. x 60"	SN@7.50	Ea	6,260.00	193.00	37.50	6,490.50
14" dia. x 90"	SN@8.00	Ea	7,770.00	206.00	40.00	8,016.00
14" dia. x 120"	SN@8.25	Ea	9,400.00	212.00	41.25	9,653.25
16" dia. x 60"	SN@8.50	Ea	7,680.00	219.00	42.50	7,941.50
16" dia. x 90"	SN@9.00	Ea	9,670.00	232.00	45.00	9,947.00
16" dia. x 120"	SN@9.25	Ea	11,000.00	238.00	46.25	11,284.25

Description	Craft@Hrs	Unit	Material $	Labor $	Equipment $	Total $

Heat exchanger connection assembly, hot water to hot water. Includes pipe, fittings, pipe insulation, valves, thermometers gauges and vents.

Description	Craft@Hrs	Unit	Material $	Labor $	Equipment $	Total $
1½" supply	P1@32.0	Ea	1,700.00	782.00	60.50	2,542.50
2" supply	P1@38.0	Ea	2,100.00	929.00	60.50	3,089.50
2½" supply	P1@50.0	Ea	2,710.00	1,220.00	60.50	3,990.50
3" supply	P1@55.0	Ea	2,870.00	1,340.00	66.55	4,276.55
4" supply	P1@62.0	Ea	3,480.00	1,520.00	75.02	5,075.02
6" supply	P1@85.0	Ea	5,670.00	2,080.00	102.85	7,852.85
8" supply	P1@102.	Ea	7,990.00	2,490.00	123.42	10,603.42
10" supply	P1@130.	Ea	10,400.00	3,180.00	157.30	13,737.30

Heat exchanger connection assembly, steam to hot water. Includes pipe, fittings, pipe insulation, valves, thermometers gauges and vents.

Description	Craft@Hrs	Unit	Material $	Labor $	Equipment $	Total $
1½" supply	P1@22.0	Ea	1,950.00	538.00	60.50	2,548.50
2" supply	P1@26.0	Ea	2,275.00	635.00	60.50	2,970.50
2½" supply	P1@41.0	Ea	2,820.00	1,000.00	49.61	3,869.61
3" supply	P1@44.0	Ea	3,270.00	1,080.00	53.24	4,403.24
4" supply	P1@54.5	Ea	3,700.00	1,330.00	65.94	5,095.94
6" supply	P1@67.0	Ea	5,940.00	1,640.00	81.07	7,661.07
8" supply	P1@80.0	Ea	8,340.00	1,950.00	96.80	10,386.80

Fan Coil Units and Connections

Description	Craft@Hrs	Unit	Material $	Labor $	Equipment $	Total $

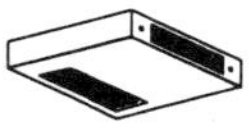

Ceiling-hung horizontal, 2 pipe hydronic fan coil unit with cabinet, set in place only. Make additional allowances for pipe connection assembly.

Description	Craft@Hrs	Unit	Material $	Labor $	Equipment $	Total $
200 CFM	P1@2.00	Ea	384.00	48.90	—	432.90
400 CFM	P1@2.50	Ea	455.00	61.10	—	516.10
600 CFM	P1@3.00	Ea	536.00	73.30	—	609.30
800 CFM	P1@3.25	Ea	676.00	79.40	—	755.40
1,000 CFM	P1@3.50	Ea	849.00	85.50	—	934.50

Less cabinet: Deduct 20% from material cost.
Floor-mounted (vertical): Add 10% to material cost.

Vertical, 4 pipe hydronic fan coil unit, set in place only. Make additional allowances for pipe connection assembly.

Description	Craft@Hrs	Unit	Material $	Labor $	Equipment $	Total $
200 CFM	P1@2.15	Ea	435.00	52.50	—	487.50
400 CFM	P1@2.75	Ea	505.00	67.20	—	572.20
600 CFM	P1@3.25	Ea	595.00	79.40	—	674.40
800 CFM	P1@3.45	Ea	735.00	84.30	—	819.30
1,000 CFM	P1@3.60	Ea	930.00	88.00	—	1,018.00

2 pipe fan coil unit connection. Coil connection includes type L copper pipe and wrought fittings, pipe insulation, condensate drain, isolation valves and 2 way control valve.

Description	Craft@Hrs	Unit	Material $	Labor $	Equipment $	Total $
1/2" supply	P1@2.50	Ea	146.00	61.10	—	207.10
3/4" supply	P1@2.80	Ea	176.00	68.40	—	244.40
1" supply	P1@3.40	Ea	267.00	83.10	—	350.10

4 pipe fan coil unit connection. Coil connection includes type L copper pipe and wrought fittings, pipe insulation, condensate drain, isolation valves and 3 way control valve.

Description	Craft@Hrs	Unit	Material $	Labor $	Equipment $	Total $
1/2" supply	P1@2.75	Ea	176.00	67.20	—	243.20
3/4" supply	P1@2.95	Ea	206.00	72.10	—	278.10
1" supply	P1@3.60	Ea	297.00	88.00	—	385.00

Description	Craft@Hrs	Unit	Material $	Labor $	Equipment $	Total $

Duct-mounted flanged hot water reheat coil, set in place only. Make additional allowances for pipe connection assembly.

Description	Craft@Hrs	Unit	Material $	Labor $	Equipment $	Total $
12" x 6"	SN@1.00	Ea	126.00	25.80	7.50	159.30
12" x 8"	SN@1.00	Ea	133.00	25.80	8.00	166.80
12" x 10"	SN@1.00	Ea	140.00	25.80	8.50	174.30
12" x 12"	SN@1.00	Ea	150.00	25.80	9.00	184.80
18" x 6"	SN@1.25	Ea	160.00	32.20	9.50	201.70
18" x 12"	SN@1.25	Ea	165.00	32.20	10.00	207.20
18" x 18"	SN@1.25	Ea	175.00	32.20	10.50	217.70
24" x 12"	SN@1.50	Ea	190.00	38.60	11.50	240.10
24" x 18"	SN@1.50	Ea	200.00	38.60	12.50	251.10
24" x 24"	SN@1.75	Ea	225.00	45.10	13.75	283.85

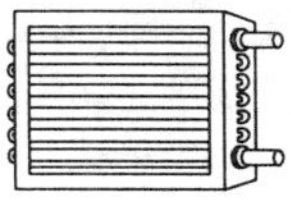

Duct mounted hot water reheat coil connection assembly. Includes pipe, fittings, pipe insulation, valves and strainers.

Description	Craft@Hrs	Unit	Material $	Labor $	Equipment $	Total $
1/2" steel	P1@2.25	Ea	88.50	55.00	—	143.50
3/4" steel	P1@2.35	Ea	104.00	57.40	—	161.40
1" steel	P1@2.65	Ea	136.00	64.80	—	200.80
1¼" steel	P1@3.00	Ea	165.00	73.30	—	238.30
1½" steel	P1@3.25	Ea	175.00	79.40	—	254.40
2" steel	P1@3.75	Ea	195.00	91.60	—	286.60

Duct-mounted slip-in electric reheat coil, set in place only. Make additional allowances for electrical connection.

Description	Craft@Hrs	Unit	Material $	Labor $	Equipment $	Total $
12" x 6"	SN@1.10	Ea	302.00	28.30	5.50	335.80
12" x 8"	SN@1.20	Ea	343.00	30.90	6.00	379.90
12" x 10"	SN@1.30	Ea	419.00	33.50	6.50	459.00
12" x 12"	SN@1.40	Ea	455.00	36.10	7.00	498.10
18" x 6"	SN@1.50	Ea	504.00	38.60	7.50	550.10
18" x 12"	SN@1.60	Ea	547.00	41.20	8.00	596.20
18" x 18"	SN@1.70	Ea	581.00	43.80	8.50	633.30
24" x 12"	SN@1.90	Ea	617.00	48.90	9.50	675.40
24" x 18"	SN@2.10	Ea	652.00	54.10	10.50	716.60
24" x 24"	SN@2.40	Ea	804.00	61.80	12.00	877.80

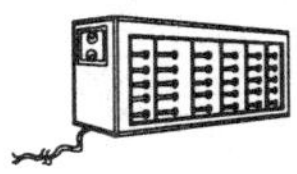

Unit Heaters and Connections

Description	Craft@Hrs	Unit	Material $	Labor $	Equipment $	Total $
Gas-fired unit heater, hang in place only. Make additional allowances for gas connection assembly.						
60mbh, 200cfm	P1@2.00	Ea	523.00	48.90	—	571.90
100mbh, 400cfm	P1@2.50	Ea	612.00	61.10	—	673.10
150mbh, 600cfm	P1@3.00	Ea	739.00	73.30	—	812.30
175mbh, 800cfm	P1@3.25	Ea	804.00	79.40	—	883.40
200mbh, 1000cfm	P1@3.50	Ea	866.00	85.50	—	951.50
250mbh, 1500cfm	P1@3.75	Ea	1,010.00	91.60	—	1,101.60
300mbh, 2000cfm	P1@4.00	Ea	1,130.00	97.70	—	1,227.70
Hot water unit heater, hang in place only. Make additional allowances for pipe connection assembly.						
8 mbh, 400cfm	P1@2.00	Ea	217.00	48.90	0.90	266.80
12 mbh, 400cfm	P1@2.00	Ea	280.00	48.90	0.90	329.80
16 mbh, 400 cfm	P1@2.00	Ea	296.00	48.90	0.90	345.80
30mbh, 700cfm	P1@3.00	Ea	395.00	73.30	1.35	469.65
43mbh, 1040cfm	P1@3.50	Ea	486.00	85.50	1.57	573.07
57mbh, 1370cfm	P1@4.00	Ea	523.00	97.70	1.80	622.50
68mbh, 1715cfm	P1@4.00	Ea	820.00	97.70	1.80	919.50
105mbh, 2360cfm	P1@5.00	Ea	972.00	122.00	2.25	1,096.25
123mbh, 2900cfm	P1@6.00	Ea	1,250.00	147.00	2.70	1,399.70
140mbh, 3530cfm	P1@6.00	Ea	1,450.00	147.00	2.70	1,599.70
156mbh, 4250cfm	P1@6.00	Ea	1,520.00	147.00	2.70	1,669.70
210mbh, 4450cfm	P1@8.00	Ea	2,710.00	195.00	3.60	2,908.60
123mbh, 5200cfm	P1@6.00	Ea	2,240.00	147.00	2.70	2,389.70
257mbh 5350cfm	P1@8.00	Ea	4,570.00	195.00	3.60	4,768.60
Hot water unit heater connection. Includes schedule 40 threaded and coupled supply and return piping with malleable iron fittings, pipe insulation, valves and vents Make additional allowances for electrical connection.						
3/4" supply	P1@2.50	Ea	117.00	61.10	1.12	179.22
1" supply	P1@3.00	Ea	137.00	73.30	1.35	211.65
1¼" supply	P1@3.30	Ea	163.00	80.60	1.49	245.09
1½" supply	P1@3.75	Ea	188.00	91.60	1.69	281.29
2" supply	P1@4.10	Ea	244.00	100.00	1.84	345.84
Gas fired unit heater flue connection						
2" b vent	SN@.090	LF	2.10	2.32	—	4.42
3" b vent	SN@.100	LF	2.59	2.58	—	5.17
4" b vent	SN@.110	LF	3.46	2.83	—	6.29
6" b vent	SN@.130	LF	3.95	3.35	—	7.30
Gas fired unit heater fuel connection. Assembly includes 8' of black steel pipe, malleable fittings, plug valve and regulator.						
3/4" gas conn.	P1@1.30	Ea	44.00	31.80	—	75.80
1" gas conn.	P1@1.60	Ea	47.80	39.10	—	86.90
1¼" gas conn.	P1@2.10	Ea	85.60	51.30	—	136.90
1½" gas conn.	P1@2.50	Ea	98.10	61.10	—	159.20

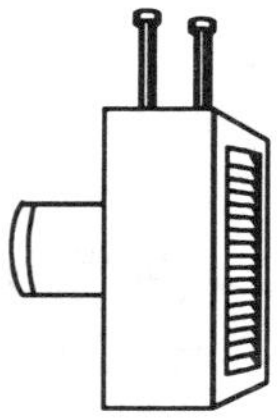

Centrifugal water-cooled chiller, set in place only. Unit comes factory assembled complete. Make additional allowances for cooling tower and pipe connections.

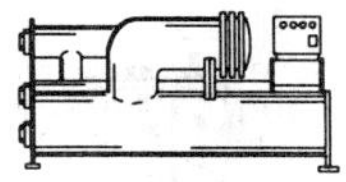

Description	Craft@Hrs	Unit	Material $	Labor $	Equipment $	Total $
100 tons	SN@20.0	Ea	49,400.00	515.00	60.00	49,975.00
150 tons	SN@20.0	Ea	73,700.00	515.00	62.50	74,277.50
200 tons	SN@20.0	Ea	79,700.00	515.00	65.00	80,280.00
250 tons	SN@20.0	Ea	94,800.00	515.00	70.00	95,385.00
300 tons	SN@25.0	Ea	101,000.00	644.00	75.00	101,719.00
350 tons	SN@25.0	Ea	104,000.00	644.00	80.00	104,724.00
400 tons	SN@25.0	Ea	122,000.00	644.00	85.00	122,729.00
500 tons	SN@25.0	Ea	155,000.00	644.00	95.00	155,739.00

Reciprocating water-cooled chiller, set in place only. Unit comes factory assembled complete. Make additional allowances for cooling tower and pipe connections.

Description	Craft@Hrs	Unit	Material $	Labor $	Equipment $	Total $
20 tons	SN@12.0	Ea	15,100.00	309.00	50.00	15,459.00
25 tons	SN@12.0	Ea	17,100.00	309.00	52.50	17,461.50
30 tons	SN@12.0	Ea	18,800.00	309.00	60.00	19,169.00
40 tons	SN@12.0	Ea	22,600.00	309.00	65.00	22,974.00
50 tons	SN@16.0	Ea	31,900.00	412.00	70.00	32,382.00
60 tons	SN@16.0	Ea	41,200.00	412.00	75.00	41,687.00
75 tons	SN@16.0	Ea	44,000.00	412.00	80.00	44,492.00
100 tons	SN@20.0	Ea	47,000.00	515.00	82.50	47,597.50
125 tons	SN@20.0	Ea	56,300.00	515.00	85.00	56,900.00
150 tons	SN@20.0	Ea	65,500.00	515.00	87.50	66,102.50

Water cooled chiller connection assembly. Includes condenser and chilled water pipe, fittings, pipe insulation, valves, thermometers and gauges.

Description	Craft@Hrs	Unit	Material $	Labor $	Equipment $	Total $
2½" chiller conn.	SN@36.0	Ea	1,530.00	927.00	43.56	2,500.56
3" chiller conn.	SN@38.0	Ea	1,680.00	979.00	45.98	2,704.98
4" chiller conn.	SN@51.0	Ea	2,250.00	1,310.00	61.71	3,621.71
6" chiller conn.	SN@65.0	Ea	3,750.00	1,670.00	78.65	5,498.65
8" chiller conn.	SN@77.0	Ea	5,700.00	1,980.00	93.17	7,773.17
10" chiller conn.	SN@100.	Ea	10,500.00	2,580.00	121.00	13,201.00
12" chiller conn.	SN@125.	Ea	12,100.00	3,220.00	151.25	15,471.25

Air cooled condensing unit, set in place only. Unit comes factory assembled complete. Make additional allowances for pipe and electrical connections.

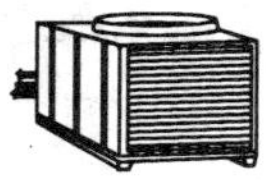

Description	Craft@Hrs	Unit	Material $	Labor $	Equipment $	Total $
3 tons	SN@2.00	Ea	2,200.00	51.50	10.00	2,261.50
5 tons	SN@3.00	Ea	3,780.00	77.30	15.00	3,872.30
7½ tons	SN@4.00	Ea	4,900.00	103.00	20.00	5,023.00
10 tons	SN@4.50	Ea	6,530.00	116.00	22.50	6,668.50
12½ tons	SN@5.00	Ea	7,690.00	129.00	25.00	7,844.00
15 tons	SN@5.50	Ea	8,820.00	142.00	27.50	8,989.50
20 tons	SN@6.00	Ea	11,700.00	155.00	30.00	11,885.00
25 tons	SN@7.00	Ea	14,700.00	180.00	35.00	14,915.00
30 tons	SN@8.00	Ea	17,500.00	206.00	40.00	17,746.00
40 tons	SN@9.00	Ea	23,100.00	232.00	45.00	23,377.00
50 tons	SN@10.0	Ea	28,900.00	258.00	50.00	29,208.00

Galvanized steel cooling tower. Assemble and set in place only. Add for pipe connections.

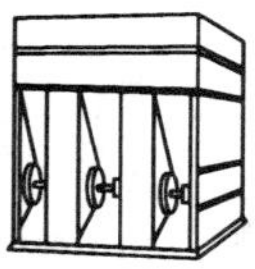

Description	Craft@Hrs	Unit	Material $	Labor $	Equipment $	Total $
10 tons	SN@16.0	Ea	2,250.00	412.00	30.00	2,692.00
15 tons	SN@16.0	Ea	1,670.00	412.00	32.50	2,114.50
20 tons	SN@18.0	Ea	4,020.00	464.00	35.00	4,519.00
25 tons	SN@20.0	Ea	4,960.00	515.00	37.50	5,512.50
30 tons	SN@25.0	Ea	5,650.00	644.00	40.00	6,334.00
40 tons	SN@30.0	Ea	7,210.00	773.00	42.50	8,025.50
50 tons	SN@35.0	Ea	7,900.00	901.00	45.00	8,846.00
60 tons	SN@40.0	Ea	8,310.00	1,030.00	47.50	9,387.50
80 tons	SN@45.0	Ea	9,170.00	1,160.00	50.00	10,380.00
100 tons	SN@50.0	Ea	10,800.00	1,290.00	52.50	12,142.50
125 tons	SN@60.0	Ea	13,000.00	1,550.00	55.00	14,605.00
150 tons	SN@65.0	Ea	15,100.00	1,670.00	57.50	16,827.50
175 tons	SN@70.0	Ea	17,100.00	1,800.00	60.00	18,960.00
200 tons	SN@70.0	Ea	18,700.00	1,800.00	65.00	20,565.00
300 tons	SN@75.0	Ea	27,000.00	1,930.00	80.00	29,010.00
400 tons	SN@80.0	Ea	34,400.00	2,060.00	90.00	36,550.00
500 tons	SN@80.0	Ea	41,300.00	2,060.00	100.00	43,460.00

Redwood or fir induced-draft cooling tower. Assemble and set in place only. Add for pipe connections.

Description	Craft@Hrs	Unit	Material $	Labor $	Equipment $	Total $
100 tons	SN@75.0	Ea	6,400.00	1,930.00	130.00	8,460.00
200 tons	SN@80.0	Ea	10,800.00	2,060.00	160.00	13,020.00
300 tons	SN@85.0	Ea	15,400.00	2,190.00	200.00	17,790.00
400 tons	SN@90.0	Ea	17,600.00	2,320.00	225.00	20,145.00
500 tons	SN@90.0	Ea	20,400.00	2,320.00	250.00	22,970.00

Description	Craft@Hrs	Unit	Material $	Labor $	Equipment $	Total $

Redwood or fir forced draft cooling tower. Assemble and set in place only. Add for pipe connections.

Description	Craft@Hrs	Unit	Material $	Labor $	Equipment $	Total $
100 tons	SN@75.0	Ea	8,760.00	1,930.00	130.00	10,820.00
200 tons	SN@80.0	Ea	16,700.00	2,060.00	160.00	18,920.00
300 tons	SN@85.0	Ea	19,500.00	2,190.00	200.00	21,890.00
400 tons	SN@90.0	Ea	25,800.00	2,320.00	225.00	28,345.00
500 tons	SN@90.0	Ea	31,400.00	2,320.00	250.00	33,970.00

Cooling Tower connection assembly. Includes condenser water pipe, fittings, valves, thermometers and gauges.

Description	Craft@Hrs	Unit	Material $	Labor $	Equipment $	Total $
2½" connection	SN@42.0	Ea	2,800.00	1,080.00	43.56	3,923.56
3" connection	SN@47.0	Ea	3,200.00	1,210.00	45.98	4,455.98
4" connection	SN@62.0	Ea	4,600.00	1,600.00	61.71	6,261.71
6" connection	SN@96.0	Ea	7,300.00	2,470.00	78.65	9,848.65
8" connection	SN@118	Ea	10,600.00	3,040.00	93.17	13,733.17
10" connection	SN@150	Ea	14,300.00	3,860.00	121.00	18,281.00
12" connection	SN@215	Ea	17,500.00	5,540.00	151.25	23,191.25
14" connection	SN@260	Ea	20,200.00	6,700.00	151.25	27,051.25

Carbon Steel, Schedule 40 with 150# Fittings & Butt-Welded Joints

Schedule 40 carbon steel (ASTM A-53 and A-120) pipe with 150 pound carbon steel welding fittings is commonly used for heating hot water, chilled water, steam, and steam condensate systems where operating pressures do not exceed 250 PSIG.

Schedule 40 (ASTM A-53B SML) steel pipe with carbon steel welding fittings can also be used for refrigerant piping systems.

This section has been arranged to save the estimator's time by including all normally-used system components such as pipe, fittings, valves, hanger assemblies, riser clamps and miscellaneous items under one heading. Additional items can be found under "Plumbing and Piping Specialties." The cost estimates in this section are based on the conditions, limitations and wage rates described in the section "How to Use This Book" beginning on page 5.

Equipment cost, where shown, is $35 per hour for a 10-ton hydraulic truck-mounted crane.

Schedule 40 carbon steel pipe, plain end, butt-welded joints

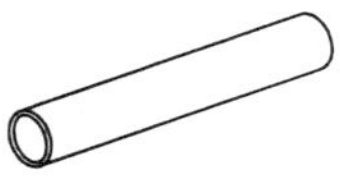

Description	Craft@Hrs	Unit	Material $	Labor $	Equipment $	Total $
1/2"	P1@.070	LF	.44	1.71	—	2.15
3/4"	P1@.080	LF	.50	1.95	—	2.45
1"	P1@.100	LF	.75	2.44	—	3.19
1¼"	P1@.110	LF	.98	2.69	—	3.67
1½"	P1@.120	LF	1.17	2.93	—	4.10
2"	P1@.140	LF	1.58	3.42	—	5.00
2½"	P1@.170	LF	2.45	4.15	—	6.60
3"	P1@.190	LF	3.23	4.64	—	7.87
4"	P1@.300	LF	4.59	7.33	—	11.92
6"	ER@.420	LF	7.90	11.00	2.10	21.00
8"	ER@.510	LF	11.80	13.30	2.55	27.65
10"	ER@.600	LF	16.70	15.60	3.00	35.30
12"	ER@.760	LF	20.30	19.80	3.80	43.90

Schedule 40 carbon steel 45 degree ell, butt-welded joints

Description	Craft@Hrs	Unit	Material $	Labor $	Equipment $	Total $
1/2"	P1@.220	Ea	2.75	5.38	—	8.13
3/4"	P1@.330	Ea	2.75	8.06	—	10.81
1"	P1@.440	Ea	2.75	10.80	—	13.55
1¼"	P1@.560	Ea	2.90	13.70	—	16.60
1½"	P1@.710	Ea	2.90	17.40	—	20.30
2"	P1@.890	Ea	3.00	21.70	—	24.70
2½"	P1@1.11	Ea	3.50	27.10	—	30.60
3"	P1@1.33	Ea	4.25	32.50	—	36.75
4"	P1@1.78	Ea	6.25	43.50	—	49.75
6"	ER@2.67	Ea	16.00	69.60	13.40	99.00
8"	ER@3.20	Ea	30.00	83.50	16.00	129.50
10"	ER@4.00	Ea	46.00	104.00	20.00	170.00
12"	ER@4.80	Ea	67.00	125.00	24.00	216.00

Carbon Steel, Schedule 40 with 150# Fittings & Butt-Welded Joints

Description	Craft@Hrs	Unit	Material $	Labor $	Equipment $	Total $

Schedule 40 carbon steel 90 degree ell, butt-welded joints

Description	Craft@Hrs	Unit	Material $	Labor $	Equipment $	Total $
1/2"	P1@.220	Ea	2.75	5.38	—	8.13
3/4"	P1@.330	Ea	2.75	8.06	—	10.81
1"	P1@.440	Ea	2.75	10.80	—	13.55
1¼"	P1@.560	Ea	2.90	13.70	—	16.60
1½"	P1@.670	Ea	2.90	16.40	—	19.30
2"	P1@.890	Ea	3.20	21.70	—	24.90
2½"	P1@1.11	Ea	4.10	27.10	—	31.20
3"	P1@1.33	Ea	5.50	32.50	—	38.00
4"	P1@1.78	Ea	8.90	43.50	—	52.40
6"	ER@2.67	Ea	25.00	69.60	13.40	108.00
8"	ER@3.20	Ea	45.00	83.50	16.00	144.50
10"	ER@4.00	Ea	81.00	104.00	20.00	205.00
12"	ER@4.80	Ea	120.00	125.00	24.00	269.00

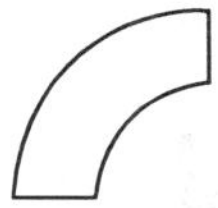

Schedule 40 carbon steel tee, butt-welded joints

Description	Craft@Hrs	Unit	Material $	Labor $	Equipment $	Total $
1/2"	P1@.330	Ea	8.65	8.06	—	16.71
3/4"	P1@.500	Ea	8.65	12.20	—	20.85
1"	P1@.670	Ea	8.65	16.40	—	25.05
1¼"	P1@.830	Ea	10.50	20.30	—	30.80
1½"	P1@1.00	Ea	13.00	24.40	—	37.40
2"	P1@1.33	Ea	10.40	32.50	—	42.90
2½"	P1@1.67	Ea	12.75	40.80	—	53.55
3"	P1@2.00	Ea	14.00	48.90	—	62.90
4"	P1@2.67	Ea	19.00	65.20	—	84.20
6"	ER@4.00	Ea	35.00	104.00	20.00	159.00
8"	ER@4.80	Ea	64.00	125.00	24.00	213.00
10"	ER@6.00	Ea	110.00	156.00	30.00	296.00
12"	ER@7.20	Ea	165.00	188.00	36.00	389.00

Carbon Steel, Schedule 40 with 150# Fittings & Butt-Welded Joints

Schedule 40 carbon steel reducing tee, butt-welded joints

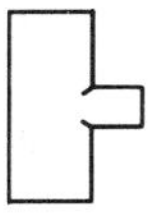

Description	Craft@Hrs	Unit	Material $	Labor $	Equipment $	Total $
1½" x 1/2"	P1@.940	Ea	13.30	23.00	—	36.30
1½" x 3/4"	P1@.940	Ea	13.30	23.00	—	36.30
1½" x 1"	P1@.940	Ea	13.30	23.00	—	36.30
1½" x 1¼"	P1@.940	Ea	13.30	23.00	—	36.30
2" x 3/4"	P1@1.22	Ea	17.70	29.80	—	47.50
2" x 1"	P1@1.22	Ea	17.70	29.80	—	47.50
2" x 1¼"	P1@1.22	Ea	17.70	29.80	—	47.50
2" x 1½"	P1@1.22	Ea	17.70	29.80	—	47.50
2½" x 1"	P1@1.56	Ea	17.90	38.10	—	56.00
2½" x 1¼"	P1@1.56	Ea	17.90	38.10	—	56.00
2½" x 1½"	P1@1.56	Ea	17.90	38.10	—	56.00
2½" x 2"	P1@1.56	Ea	17.90	38.10	—	56.00
3" x 1¼"	P1@1.89	Ea	18.20	46.20	—	64.40
3" x 2"	P1@1.89	Ea	18.20	46.20	—	64.40
3" x 2½"	P1@1.89	Ea	18.20	46.20	—	64.40
4" x 1½"	P1@2.44	Ea	19.60	59.60	—	79.20
4" x 2"	P1@2.44	Ea	19.60	59.60	—	79.20
4" x 2½"	P1@2.44	Ea	19.60	59.60	—	79.20
4" x 3"	P1@2.44	Ea	19.60	59.60	—	79.20
6" x 2½"	ER@3.78	Ea	44.00	98.60	18.90	161.50
6" x 3"	ER@3.78	Ea	44.00	98.60	18.90	161.50
6" x 4"	ER@3.78	Ea	44.00	98.60	18.90	161.50
8" x 3"	ER@4.40	Ea	85.00	115.00	22.00	222.00
8" x 4"	ER@4.40	Ea	85.00	115.00	22.00	222.00
8" x 6"	ER@4.40	Ea	85.00	115.00	22.00	222.00
10" x 4"	ER@5.60	Ea	145.00	146.00	28.00	319.00
10" x 6"	ER@5.60	Ea	145.00	146.00	28.00	319.00
10" x 8"	ER@5.60	Ea	145.00	146.00	28.00	319.00
12" x 6"	ER@6.80	Ea	215.00	177.00	34.00	426.00
12" x 8"	ER@6.80	Ea	215.00	177.00	34.00	426.00
12" x 10"	ER@6.80	Ea	215.00	177.00	34.00	426.00

Schedule 40 carbon steel concentric reducer, butt-welded joints

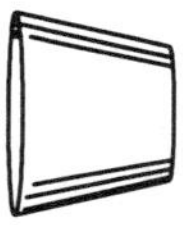

Description	Craft@Hrs	Unit	Material $	Labor $	Equipment $	Total $
3/4"	P1@.295	Ea	5.50	7.21	—	12.71
1"	P1@.395	Ea	6.15	9.65	—	15.80
1¼"	P1@.500	Ea	5.20	12.20	—	17.40
1½"	P1@.600	Ea	5.80	14.70	—	20.50
2"	P1@.800	Ea	5.40	19.50	—	24.90
2½"	P1@1.00	Ea	6.90	24.40	—	31.30
3"	P1@1.20	Ea	5.80	29.30	—	35.10
4"	P1@1.60	Ea	8.20	39.10	—	47.30
6"	ER@2.40	Ea	15.20	62.60	12.00	89.80
8"	ER@2.88	Ea	23.00	75.10	14.40	112.50
10"	ER@3.60	Ea	35.00	93.90	18.00	146.90
12"	ER@4.30	Ea	54.00	112.00	21.50	187.50

Carbon Steel, Schedule 40 with 150# Fittings & Butt-Welded Joints

Description	Craft@Hrs	Unit	Material $	Labor $	Equipment $	Total $

Schedule 40 carbon steel eccentric reducer, butt-welded joints

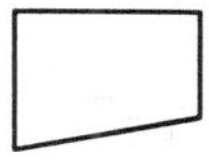

Description	Craft@Hrs	Unit	Material $	Labor $	Equipment $	Total $
3/4"	P1@.295	Ea	6.75	7.21	—	13.96
1"	P1@.395	Ea	8.50	9.65	—	18.15
1¼"	P1@.500	Ea	7.90	12.20	—	20.10
1½"	P1@.600	Ea	8.40	14.70	—	23.10
2"	P1@.800	Ea	6.45	19.50	—	25.95
2½"	P1@1.00	Ea	8.35	24.40	—	32.75
3"	P1@1.20	Ea	7.80	29.30	—	37.10
4"	P1@1.60	Ea	11.00	39.10	—	50.10
6"	ER@2.40	Ea	22.00	62.60	12.00	96.60
8"	ER@2.88	Ea	32.00	75.10	14.40	121.50
10"	ER@3.60	Ea	55.00	93.90	18.00	166.90
12"	ER@4.30	Ea	83.00	112.00	21.50	216.50

Schedule 40 carbon steel cap, butt-welded joints

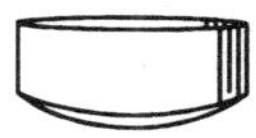

Description	Craft@Hrs	Unit	Material $	Labor $	Equipment $	Total $
1/2"	P1@.150	Ea	6.65	3.67	—	10.32
3/4"	P1@.230	Ea	6.65	5.62	—	12.27
1"	P1@.300	Ea	6.65	7.33	—	13.98
1¼"	P1@.390	Ea	6.65	9.53	—	16.18
1½"	P1@.470	Ea	6.65	11.50	—	18.15
2"	P1@.620	Ea	6.65	15.20	—	21.85
2½"	P1@.770	Ea	6.65	18.80	—	25.45
3"	P1@.930	Ea	6.65	22.70	—	29.35
4"	P1@1.25	Ea	8.65	30.50	—	39.15
6"	ER@1.87	Ea	10.00	48.80	9.35	68.15
8"	ER@2.24	Ea	14.30	58.40	11.20	83.90
10"	ER@2.80	Ea	24.00	73.00	14.00	111.00
12"	ER@3.36	Ea	30.00	87.60	16.80	134.40

Schedule 40 carbon steel union, butt welded joints

Description	Craft@Hrs	Unit	Material $	Labor $	Equipment $	Total $
1/2"	P1@.220	Ea	4.80	5.38	—	10.18
3/4"	P1@.330	Ea	5.80	8.06	—	13.86
1"	P1@.440	Ea	7.50	10.80	—	18.30
1¼"	P1@.560	Ea	12.80	13.70	—	26.50
1½"	P1@.670	Ea	13.90	16.40	—	30.30
2"	P1@.890	Ea	18.60	21.70	—	40.30
2½"	P1@1.11	Ea	40.00	27.10	—	67.10
3"	P1@1.33	Ea	50.00	32.50	—	82.50

Carbon steel weldolet, Schedule 40

Description	Craft@Hrs	Unit	Material $	Labor $	Equipment $	Total $
1/2"	P1@.330	Ea	6.68	8.06	—	14.74
3/4"	P1@.500	Ea	6.89	12.20	—	19.09
1"	P1@.670	Ea	7.21	16.40	—	23.61
1¼"	P1@.830	Ea	8.77	20.30	—	29.07
1½"	P1@1.00	Ea	8.77	24.40	—	33.17
2"	P1@1.33	Ea	9.60	32.50	—	42.10
2½"	P1@1.67	Ea	21.00	40.80	—	61.80
3"	P1@2.00	Ea	24.00	48.90	—	72.90
4"	P1@2.67	Ea	31.00	65.20	—	96.20
6"	P1@4.00	Ea	128.00	97.70	—	225.70
8"	P1@4.80	Ea	189.00	117.00	—	306.00
10"	P1@6.00	Ea	260.00	147.00	—	407.00
12"	P1@7.20	Ea	320.00	176.00	—	496.00

Carbon steel threadolet, Schedule 40

Description	Craft@Hrs	Unit	Material $	Labor $	Equipment $	Total $
3/4"	P1@.330	Ea	4.14	8.06	—	12.20
1"	P1@.440	Ea	4.60	10.80	—	15.40
1¼"	P1@.560	Ea	6.40	13.70	—	20.10
1½"	P1@.670	Ea	7.20	16.40	—	23.60
2"	P1@.890	Ea	8.40	21.70	—	30.10
2½"	P1@1.11	Ea	25.75	27.10	—	52.85
3"	P1@1.33	Ea	29.90	32.50	—	62.40
4"	P1@1.78	Ea	58.76	43.50	—	102.26

150# forged steel companion flange, weld neck

Description	Craft@Hrs	Unit	Material $	Labor $	Equipment $	Total $
2½"	P1@.610	Ea	14.95	14.90	—	29.85
3"	P1@.730	Ea	14.65	17.80	—	32.45
4"	P1@.980	Ea	20.15	23.90	—	44.05
6"	ER@1.47	Ea	32.30	38.30	7.35	77.95
8"	ER@1.77	Ea	51.50	46.20	8.85	106.55
10"	ER@2.20	Ea	85.00	57.40	11.00	153.40
12"	ER@2.64	Ea	125.00	68.90	13.20	207.10

Class 125 bronze body gate valve, threaded ends

Description	Craft@Hrs	Unit	Material $	Labor $	Equipment $	Total $
1/2"	P1@.210	Ea	6.06	5.13	—	11.19
3/4"	P1@.250	Ea	7.58	6.11	—	13.69
1"	P1@.300	Ea	10.50	7.33	—	17.83
1¼"	P1@.400	Ea	13.40	9.77	—	23.17
1½"	P1@.450	Ea	18.00	11.00	—	29.00
2"	P1@.500	Ea	30.10	12.20	—	42.30
2½"	P1@.750	Ea	49.90	18.30	—	68.20
3"	P1@.950	Ea	71.60	23.20	—	94.80

Carbon Steel, Schedule 40 with 150# Fittings & Butt-Welded Joints

Description	Craft@Hrs	Unit	Material $	Labor $	Equipment $	Total $

Class 125 iron body gate valve, flanged ends

Description	Craft@Hrs	Unit	Material $	Labor $	Equipment $	Total $
2"	P1@.500	Ea	101.00	12.20	—	113.20
2½"	P1@.600	Ea	137.00	14.70	—	151.70
3"	P1@.750	Ea	148.00	18.30	—	166.30
4"	P1@1.35	Ea	218.00	33.00	—	251.00
6"	ER@2.50	Ea	417.00	65.20	12.50	494.70
8"	ER@3.00	Ea	680.00	78.20	15.00	773.20
10"	ER@4.00	Ea	1,120.00	104.00	20.00	1,244.00
12"	ER@4.50	Ea	1,540.00	117.00	22.50	1,679.50

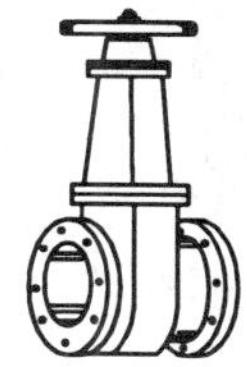

Class 125 bronze body globe valve, threaded ends

Description	Craft@Hrs	Unit	Material $	Labor $	Equipment $	Total $
1/2"	P1@.210	Ea	15.80	5.13	—	20.93
3/4"	P1@.250	Ea	21.30	6.11	—	27.41
1"	P1@.300	Ea	30.50	7.33	—	37.83
1¼"	P1@.400	Ea	42.90	9.77	—	52.67
1½"	P1@.450	Ea	57.50	11.00	—	68.50
2"	P1@.500	Ea	93.60	12.20	—	105.80

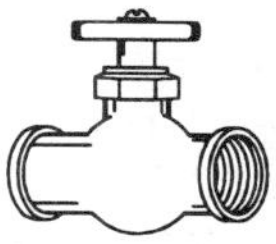

Class 125 iron body globe valve, flanged ends

Description	Craft@Hrs	Unit	Material $	Labor $	Equipment $	Total $
2½"	P1@.600	Ea	220.00	14.70	—	234.70
3"	P1@.750	Ea	264.00	18.30	—	282.30
4"	P1@1.35	Ea	352.00	33.00	—	385.00
6"	ER@2.50	Ea	619.00	65.20	—	684.20
8"	ER@3.00	Ea	952.00	78.20	—	1,030.20
10"	ER@4.00	Ea	1,450.00	104.00	—	1,554.00

200 PSIG iron body butterfly valve, lug type

Description	Craft@Hrs	Unit	Material $	Labor $	Equipment $	Total $
2"	P1@.450	Ea	84.50	11.00	—	95.50
2½"	P1@.450	Ea	86.50	11.00	—	97.50
3"	P1@.550	Ea	91.70	13.40	—	105.10
4"	P1@.550	Ea	114.00	13.40	—	127.40
6"	ER@.800	Ea	185.00	20.90	4.00	209.90
8"	ER@.800	Ea	252.00	20.90	4.00	276.90
10"	ER@.900	Ea	350.00	23.50	4.50	378.00
12"	ER@1.00	Ea	460.00	26.10	5.00	491.10

200 PSIG iron body butterfly valve, wafer type

Description	Craft@Hrs	Unit	Material $	Labor $	Equipment $	Total $
2"	P1@.450	Ea	76.70	11.00	—	87.70
2½"	P1@.450	Ea	78.70	11.00	—	89.70
3"	P1@.550	Ea	84.50	13.40	—	97.90
4"	P1@.550	Ea	102.00	13.40	—	115.40
6"	ER@.800	Ea	170.00	20.90	4.00	194.90
8"	ER@.800	Ea	237.00	20.90	4.00	261.90
10"	ER@.900	Ea	332.00	23.50	4.50	360.00
12"	ER@1.00	Ea	436.00	26.10	5.00	467.10

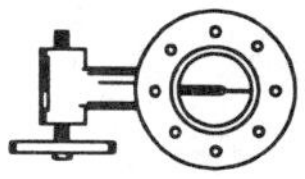

Carbon Steel, Schedule 40 with 150# Fittings & Butt-Welded Joints

Class 125 bronze body swing check valve, threaded

Description	Craft@Hrs	Unit	Material $	Labor $	Equipment $	Total $
1/2"	P1@.210	Ea	8.00	5.13	—	13.13
3/4"	P1@.250	Ea	11.40	6.11	—	17.51
1"	P1@.300	Ea	15.00	7.33	—	22.33
1¼"	P1@.400	Ea	21.40	9.77	—	31.17
1½"	P1@.450	Ea	30.00	11.00	—	41.00
2"	P1@.500	Ea	50.00	12.20	—	62.20

Class 125 iron body swing check valve, flanged joints

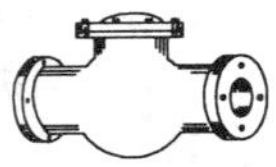

Description	Craft@Hrs	Unit	Material $	Labor $	Equipment $	Total $
2"	P1@.500	Ea	115.00	12.20	—	127.20
2½"	P1@.600	Ea	148.00	14.70	—	162.70
3"	P1@.750	Ea	183.00	18.30	—	201.30
4"	P1@1.35	Ea	271.00	33.00	—	304.00
6"	ER@2.50	Ea	526.00	65.20	12.50	603.70
8"	ER@3.00	Ea	938.00	78.20	15.00	1,031.20
10"	ER@4.00	Ea	1,535.00	104.00	20.00	1,659.00
12"	ER@4.50	Ea	2,100.00	117.00	22.50	2,239.50

Class 125 iron body silent check valve, flanged joints

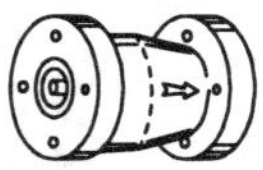

Description	Craft@Hrs	Unit	Material $	Labor $	Equipment $	Total $
2"	P1@.500	Ea	85.00	12.20	—	97.20
2½"	P1@.600	Ea	97.00	14.70	—	111.70
3"	P1@.750	Ea	110.00	18.30	—	128.30
4"	P1@1.35	Ea	145.00	33.00	—	178.00
6"	ER@2.50	Ea	260.00	65.20	12.50	337.70
8"	ER@3.00	Ea	470.00	78.20	15.00	563.20
10"	ER@4.00	Ea	730.00	104.00	20.00	854.00

Class 125 bronze body strainer, threaded ends

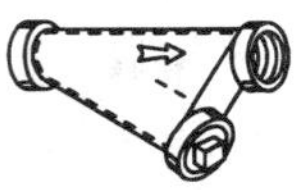

Description	Craft@Hrs	Unit	Material $	Labor $	Equipment $	Total $
1/2"	P1@.210	Ea	12.40	5.13	—	17.53
3/4"	P1@.250	Ea	16.30	6.11	—	22.41
1"	P1@.300	Ea	20.20	7.33	—	27.53
1¼"	P1@.400	Ea	28.10	9.77	—	37.87
1½"	P1@.450	Ea	42.20	11.00	—	53.20
2"	P1@.500	Ea	73.00	12.20	—	85.20

Class 125 iron body strainer, flanged

Description	Craft@Hrs	Unit	Material $	Labor $	Equipment $	Total $
2"	P1@.500	Ea	62.00	12.20	—	74.20
2½"	P1@.600	Ea	69.00	14.70	—	83.70
3"	P1@.750	Ea	80.00	18.30	—	98.30
4"	P1@1.35	Ea	137.00	33.00	—	170.00
6"	ER@2.50	Ea	277.00	65.20	12.50	354.70
8"	ER@3.00	Ea	470.00	78.20	15.00	563.20

Installation of 2-way control valve, threaded joints

Description	Craft@Hrs	Unit	Material $	Labor $	Equipment $	Total $
1/2"	P1@.210	Ea	—	5.13	—	5.13
3/4"	P1@.250	Ea	—	6.11	—	6.11
1"	P1@.300	Ea	—	7.33	—	7.33
1¼"	P1@.400	Ea	—	9.77	—	9.77
1½"	P1@.450	Ea	—	11.00	—	11.00
2"	P1@.500	Ea	—	12.20	—	12.20
2½"	P1@.830	Ea	—	20.30	—	20.30
3"	P1@.990	Ea	—	24.20	—	24.20

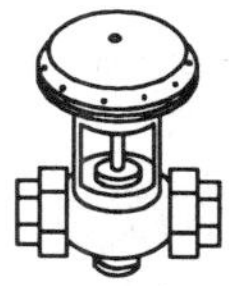

Installation of 2-way control valve, flanged joints

Description	Craft@Hrs	Unit	Material $	Labor $	Equipment $	Total $
2"	P1@.500	Ea	—	12.20	—	12.20
2½"	P1@.600	Ea	—	14.70	—	14.70
3"	P1@.750	Ea	—	18.30	—	18.30
4"	P1@1.35	Ea	—	33.00	—	33.00
6"	P1@2.50	Ea	—	61.10	—	61.10
8"	P1@3.00	Ea	—	73.30	—	73.30
10"	P1@4.00	Ea	—	97.70	—	97.70
12"	P1@4.50	Ea	—	110.00	—	110.00

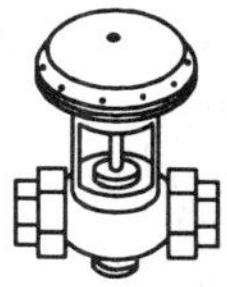

Installation of 3-way control valve, threaded joints

Description	Craft@Hrs	Unit	Material $	Labor $	Equipment $	Total $
1/2"	P1@.260	Ea	—	6.35	—	6.35
3/4"	P1@.365	Ea	—	8.92	—	8.92
1"	P1@.475	Ea	—	11.60	—	11.60
1¼"	P1@.575	Ea	—	14.10	—	14.10
1½"	P1@.680	Ea	—	16.60	—	16.60
2"	P1@.910	Ea	—	22.20	—	22.20
2½"	P1@1.12	Ea	—	27.40	—	27.40
3"	P1@1.33	Ea	—	32.50	—	32.50

Installation of 3-way control valve, flanged joints

Description	Craft@Hrs	Unit	Material $	Labor $	Equipment $	Total $
2"	P1@.910	Ea	—	22.20	—	22.20
2½"	P1@1.12	Ea	—	27.40	—	27.40
3"	P1@1.33	Ea	—	32.50	—	32.50
4"	P1@2.00	Ea	—	48.90	—	48.90
6"	P1@3.70	Ea	—	90.40	—	90.40
8"	P1@4.40	Ea	—	108.00	—	108.00
10"	P1@5.90	Ea	—	144.00	—	144.00
12"	P1@6.50	Ea	—	159.00	—	159.00

Carbon Steel, Schedule 40 with 150# Fittings & Butt-Welded Joints

Bolt and gasket sets

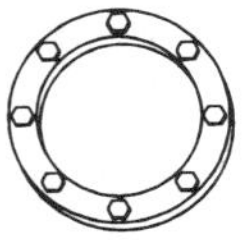

Description	Craft@Hrs	Unit	Material $	Labor $	Equipment $	Total $
2"	P1@.500	Ea	1.50	12.20	—	13.70
2½"	P1@.650	Ea	1.75	15.90	—	17.65
3"	P1@.750	Ea	3.00	18.30	—	21.30
4"	P1@1.00	Ea	5.00	24.40	—	29.40
6"	P1@1.20	Ea	8.25	29.30	—	37.55
8"	P1@1.25	Ea	9.35	30.50	—	39.85
10"	P1@1.70	Ea	16.00	41.50	—	57.50
12"	P1@2.20	Ea	18.00	53.80	—	71.80

Thermometer with well

Description	Craft@Hrs	Unit	Material $	Labor $	Equipment $	Total $
7"	P1@.250	Ea	15.50	6.11	—	21.61
9"	P1@.250	Ea	20.50	6.11	—	26.61

Dial type pressure gauge

Description	Craft@Hrs	Unit	Material $	Labor $	Equipment $	Total $
2½"	P1@.200	Ea	15.50	4.89	—	20.39
3½"	P1@.200	Ea	20.50	4.89	—	25.39

Pressure/temperature tap

Description	Craft@Hrs	Unit	Material $	Labor $	Equipment $	Total $
	P1@.150	Ea	7.30	3.67	—	10.97

Hanger with swivel assembly

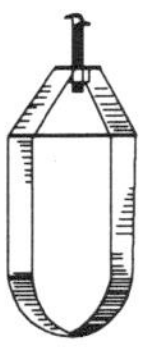

Description	Craft@Hrs	Unit	Material $	Labor $	Equipment $	Total $
2"	P1@.450	Ea	2.60	11.00	—	13.60
2½"	P1@.450	Ea	3.35	11.00	—	14.35
3"	P1@.450	Ea	4.15	11.00	—	15.15
4"	P1@.450	Ea	6.40	11.00	—	17.40
5"	P1@.500	Ea	7.15	12.20	—	19.35
6"	P1@.500	Ea	9.40	12.20	—	21.60
8"	P1@.550	Ea	12.25	13.40	—	25.65
10"	P1@.650	Ea	16.95	15.90	—	32.85
12"	P1@.750	Ea	22.50	18.30	—	40.80

Riser clamp

Description	Craft@Hrs	Unit	Material $	Labor $	Equipment $	Total $
2"	P1@.115	Ea	1.89	2.81	—	4.70
2½"	P1@.120	Ea	2.00	2.93	—	4.93
3"	P1@.120	Ea	2.18	2.93	—	5.11
4"	P1@.125	Ea	2.75	3.05	—	5.80
5"	P1@.180	Ea	3.99	4.40	—	8.39
6"	P1@.200	Ea	4.79	4.89	—	9.68
8"	P1@.200	Ea	7.77	4.89	—	12.66
10"	P1@.250	Ea	11.60	6.11	—	17.71
12"	P1@.250	Ea	13.75	6.11	—	19.86

Carbon Steel, Schedule 40 with 150# M.I. Fittings & Threaded Joints

Schedule 40 carbon steel (ASTM A-53 and A-120) pipe with 150 pound malleable iron threaded fittings is commonly used for heating hot water and chilled water systems where operating pressures do not exceed 125 PSIG and temperatures are below 250 degrees F., for pipe sizes 2 inches and smaller. For pipe sizes 2½ inches and larger, wrought steel fittings with either welded or grooved joints should be used.

This section has been arranged to save the estimator's time by including all normally-used system components such as pipe, fittings, valves, hanger assemblies, riser clamps and miscellaneous items under one heading. Other accessories can be found under "Plumbing and Piping Specialties." The cost estimates in this section are based on the conditions, limitations and wage rates described in the section "How to Use This Book" beginning on page 5.

Description	Craft@Hrs	Unit	Material $	Labor $	Equipment $	Total $

Schedule 40 carbon steel pipe, threaded and coupled

Description	Craft@Hrs	Unit	Material $	Labor $	Equipment $	Total $
1/2"	P1@.060	LF	.68	1.47	—	2.15
3/4"	P1@.070	LF	.77	1.71	—	2.48
1"	P1@.090	LF	1.16	2.20	—	3.36
1¼"	P1@.100	LF	1.51	2.44	—	3.95
1½"	P1@.110	LF	1.80	2.69	—	4.49
2"	P1@.120	LF	2.43	2.93	—	5.36
2½"	P1@.130	LF	3.75	3.18	—	6.93
3"	P1@.150	LF	4.96	3.67	—	8.63
4"	P1@.260	LF	7.06	6.35	—	13.41

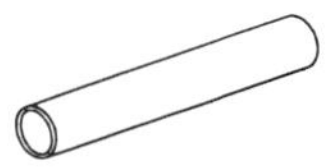

150# malleable iron 45 degree ell, threaded joints

Description	Craft@Hrs	Unit	Material $	Labor $	Equipment $	Total $
1/2"	P1@.120	Ea	1.53	2.93	—	4.46
3/4"	P1@.130	Ea	1.87	3.18	—	5.05
1"	P1@.180	Ea	2.39	4.40	—	6.79
1¼"	P1@.240	Ea	4.21	5.86	—	10.07
1½"	P1@.300	Ea	5.19	7.33	—	12.52
2"	P1@.380	Ea	7.78	9.29	—	17.07
2½"	P1@.510	Ea	22.50	12.50	—	35.00
3"	P1@.640	Ea	29.20	15.60	—	44.80
4"	P1@.850	Ea	56.80	20.80	—	77.60

150# malleable iron 90 degree ell, threaded joints

Description	Craft@Hrs	Unit	Material $	Labor $	Equipment $	Total $
1/2"	P1@.120	Ea	.95	2.93	—	3.88
3/4"	P1@.130	Ea	1.15	3.18	—	4.33
1"	P1@.180	Ea	1.99	4.40	—	6.39
1¼"	P1@.240	Ea	3.26	5.86	—	9.12
1½"	P1@.300	Ea	4.30	7.33	—	11.63
2"	P1@.380	Ea	7.33	9.29	—	16.62
2½"	P1@.510	Ea	16.00	12.50	—	28.50
3"	P1@.640	Ea	23.90	15.60	—	39.50
4"	P1@.850	Ea	55.60	20.80	—	76.40

Carbon Steel, Schedule 40 with 150# M.I. Fittings & Threaded Joints

150# malleable iron tee, threaded joints

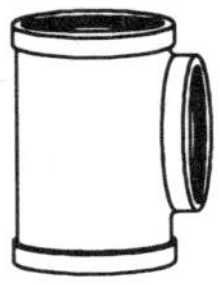

Description	Craft@Hrs	Unit	Material $	Labor $	Equipment $	Total $
1/2"	P1@.180	Ea	1.27	4.40	—	5.67
3/4"	P1@.190	Ea	1.82	4.64	—	6.46
1"	P1@.230	Ea	3.09	5.62	—	8.71
1¼"	P1@.310	Ea	5.08	7.58	—	12.66
1½"	P1@.390	Ea	6.24	9.53	—	15.77
2"	P1@.490	Ea	10.50	12.00	—	22.50
2½"	P1@.660	Ea	22.20	16.10	—	38.30
3"	P1@.830	Ea	32.70	20.30	—	53.00
4"	P1@1.10	Ea	78.40	26.90	—	105.30

150# malleable iron reducing tee, threaded joints

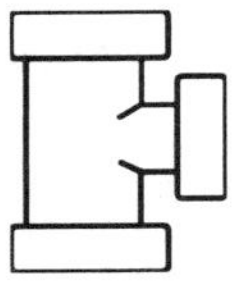

Description	Craft@Hrs	Unit	Material $	Labor $	Equipment $	Total $
3/4" x 1/2"	P1@.180	Ea	2.69	4.40	—	7.09
1" x 1/2"	P1@.220	Ea	3.43	5.38	—	8.81
1" x 3/4"	P1@.220	Ea	3.43	5.38	—	8.81
1¼" x 1/2"	P1@.290	Ea	6.24	7.09	—	13.33
1¼" x 3/4"	P1@.290	Ea	5.48	7.09	—	12.57
1¼" x 1"	P1@.290	Ea	5.93	7.09	—	13.02
1½" x 1/2"	P1@.370	Ea	7.32	9.04	—	16.36
1½" x 3/4"	P1@.370	Ea	6.91	9.04	—	15.95
1½" x 1"	P1@.370	Ea	6.91	9.04	—	15.95
1½" x 1¼"	P1@.370	Ea	8.54	9.04	—	17.58
2" x 1"	P1@.460	Ea	10.80	11.20	—	22.00
2" x 1¼"	P1@.460	Ea	11.70	11.20	—	22.90
2" x 1½"	P1@.460	Ea	11.70	11.20	—	22.90
4" x 1½"	P1@1.05	Ea	81.20	25.70	—	106.90
4" x 2"	P1@1.05	Ea	81.20	25.70	—	106.90
4" x 3"	P1@1.05	Ea	81.20	25.70	—	106.90

150# malleable iron reducer, threaded joints

Description	Craft@Hrs	Unit	Material $	Labor $	Equipment $	Total $
3/4" x 1/2"	P1@.130	Ea	1.71	3.18	—	4.89
1" x 1/2"	P1@.160	Ea	2.66	3.91	—	6.57
1" x 3/4"	P1@.160	Ea	2.66	3.91	—	6.57
1¼" x 3/4"	P1@.210	Ea	2.66	5.13	—	7.79
1¼" x 1"	P1@.210	Ea	3.33	5.13	—	8.46
1½" x 1"	P1@.270	Ea	4.87	6.60	—	11.47
1½" x 1¼"	P1@.270	Ea	4.21	6.60	—	10.81
2" x 1¼"	P1@.340	Ea	6.97	8.31	—	15.28
2" x 1½"	P1@.340	Ea	6.10	8.31	—	14.41
3" x 2"	P1@.530	Ea	18.10	13.00	—	31.10
3" x 2½"	P1@.530	Ea	21.10	13.00	—	34.10
4" x 3"	P1@.750	Ea	21.10	18.30	—	39.40

Carbon Steel, Schedule 40 with 150# M.I. Fittings & Threaded Joints

150# malleable iron cross, threaded joints

Description	Craft@Hrs	Unit	Material $	Labor $	Equipment $	Total $
1/2"	P1@.280	Ea	4.59	6.84	—	11.43
3/4"	P1@.320	Ea	5.61	7.82	—	13.43
1"	P1@.360	Ea	6.90	8.80	—	15.70
1¼"	P1@.480	Ea	11.20	11.70	—	22.90
1½"	P1@.600	Ea	13.80	14.70	—	28.50
2"	P1@.700	Ea	22.80	17.10	—	39.90
2½"	P1@1.02	Ea	42.00	24.90	—	66.90
3"	P1@1.28	Ea	58.30	31.30	—	89.60

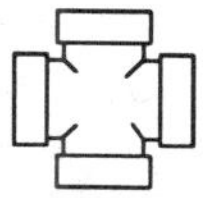

150# malleable iron cap, threaded joint

Description	Craft@Hrs	Unit	Material $	Labor $	Equipment $	Total $
1/2"	P1@.090	Ea	.96	2.20	—	3.16
3/4"	P1@.100	Ea	1.28	2.44	—	3.72
1"	P1@.140	Ea	1.58	3.42	—	5.00
1¼"	P1@.180	Ea	2.03	4.40	—	6.43
1½"	P1@.230	Ea	2.77	5.62	—	8.39
2"	P1@.290	Ea	4.09	7.09	—	11.18
2½"	P1@.380	Ea	9.11	9.29	—	18.40
3"	P1@.480	Ea	13.50	11.70	—	25.20
4"	P1@.640	Ea	23.00	15.60	—	38.60

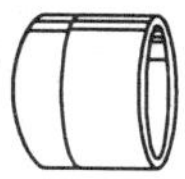

150# malleable iron plug, threaded joint

Description	Craft@Hrs	Unit	Material $	Labor $	Equipment $	Total $
1/2"	P1@.090	Ea	.76	2.20	—	2.96
3/4"	P1@.100	Ea	.81	2.44	—	3.25
1"	P1@.140	Ea	.81	3.42	—	4.23
1¼"	P1@.180	Ea	1.31	4.40	—	5.71
1½"	P1@.230	Ea	1.85	5.62	—	7.47
2"	P1@.290	Ea	2.15	7.09	—	9.24
2½"	P1@.380	Ea	5.23	9.29	—	14.52
3"	P1@.480	Ea	6.07	11.70	—	17.77
4"	P1@.640	Ea	11.20	15.60	—	26.80

150# malleable iron union, threaded joints

Description	Craft@Hrs	Unit	Material $	Labor $	Equipment $	Total $
1/2"	P1@.140	Ea	4.17	3.42	—	7.59
3/4"	P1@.150	Ea	4.79	3.67	—	8.46
1"	P1@.210	Ea	6.25	5.13	—	11.38
1¼"	P1@.280	Ea	8.95	6.84	—	15.79
1½"	P1@.360	Ea	11.20	8.80	—	20.00
2"	P1@.450	Ea	12.90	11.00	—	23.90
2½"	P1@.610	Ea	38.30	14.90	—	53.20
3"	P1@.760	Ea	46.10	18.60	—	64.70

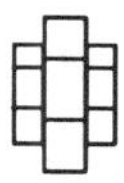

Carbon Steel, Schedule 40 with 150# M.I. Fittings & Threaded Joints

Description	Craft@Hrs	Unit	Material $	Labor $	Equipment $	Total $

150# malleable iron coupling, threaded joints

Description	Craft@Hrs	Unit	Material $	Labor $	Equipment $	Total $
1/2"	P1@.120	Ea	1.28	2.93	—	4.21
3/4"	P1@.130	Ea	1.51	3.18	—	4.69
1"	P1@.180	Ea	2.29	4.40	—	6.69
1¼"	P1@.240	Ea	2.93	5.86	—	8.79
1½"	P1@.300	Ea	3.93	7.33	—	11.26
2"	P1@.380	Ea	5.72	9.29	—	15.01

Steel pipe nipples — Close nipple, standard right hand thread

Description	Craft@Hrs	Unit	Material $	Labor $	Equipment $	Total $
1/2"	P1@.025	Ea	.27	.61	—	.88
3/4"	P1@.025	Ea	.32	.61	—	.93
1"	P1@.025	Ea	.48	.61	—	1.09
1¼"	P1@.025	Ea	.60	.61	—	1.21
1½"	P1@.025	Ea	.72	.61	—	1.33
2"	P1@.025	Ea	.93	.61	—	1.54
2½"	P1@.030	Ea	3.61	.73	—	4.34
3"	P1@.030	Ea	3.93	.73	—	4.66
4"	P1@.040	Ea	6.19	.98	—	7.17
6"	P1@.050	Ea	31.58	1.22	—	32.80
8"	P1@.065	Ea	55.46	1.59	—	57.05

Steel pipe nipples — 1½" long, standard right hand thread

Description	Craft@Hrs	Unit	Material $	Labor $	Equipment $	Total $
1/2"	P1@.025	Ea	.28	.61	—	.89
3/4"	P1@.025	Ea	.33	.61	—	.94

Steel pipe nipples — 2" long, standard right hand thread

Description	Craft@Hrs	Unit	Material $	Labor $	Equipment $	Total $
1/2"	P1@.025	Ea	.28	.61	—	.89
3/4"	P1@.025	Ea	.33	.61	—	.94
1"	P1@.025	Ea	.48	.61	—	1.09
1¼"	P1@.025	Ea	.60	.61	—	1.21
1½"	P1@.025	Ea	.72	.61	—	1.33

Steel pipe nipples — 2½" long, standard right hand thread

Description	Craft@Hrs	Unit	Material $	Labor $	Equipment $	Total $
1/2"	P1@.025	Ea	.31	.61	—	.92
3/4"	P1@.025	Ea	.36	.61	—	.97
1"	P1@.025	Ea	.56	.61	—	1.17
1¼"	P1@.025	Ea	.70	.61	—	1.31
1½"	P1@.025	Ea	.89	.61	—	1.50
2"	P1@.025	Ea	1.03	.61	—	1.64

Description	Craft@Hrs	Unit	Material $	Labor $	Equipment $	Total $

Steel pipe nipples — 3" long, standard right hand thread

Description	Craft@Hrs	Unit	Material $	Labor $	Equipment $	Total $
1/2"	P1@.025	Ea	.35	.61	—	.96
3/4"	P1@.025	Ea	.41	.61	—	1.02
1"	P1@.025	Ea	.58	.61	—	1.19
1¼"	P1@.025	Ea	.75	.61	—	1.36
1½"	P1@.025	Ea	.90	.61	—	1.51
2"	P1@.030	Ea	1.12	.73	—	1.85
2½"	P1@.030	Ea	3.77	.73	—	4.50
3"	P1@.035	Ea	4.24	.86	—	5.10

Steel pipe nipples — 3½" long, standard right hand thread

Description	Craft@Hrs	Unit	Material $	Labor $	Equipment $	Total $
1/2"	P1@.025	Ea	.39	.61	—	1.00
3/4"	P1@.025	Ea	.47	.61	—	1.08
1"	P1@.025	Ea	.65	.61	—	1.26
1¼"	P1@.025	Ea	.86	.61	—	1.47
1½"	P1@.030	Ea	1.07	.73	—	1.80
2"	P1@.030	Ea	1.32	.73	—	2.05
2½"	P1@.035	Ea	4.97	.86	—	5.83
3"	P1@.035	Ea	5.70	.86	—	6.56
4"	P1@.040	Ea	7.78	.98	—	8.76

Steel pipe nipples — 4" long, standard right hand thread

Description	Craft@Hrs	Unit	Material $	Labor $	Equipment $	Total $
1/2"	P1@.025	Ea	.42	.61	—	1.03
3/4"	P1@.025	Ea	.48	.61	—	1.09
1"	P1@.025	Ea	.69	.61	—	1.30
1¼"	P1@.030	Ea	.87	.73	—	1.60
1½"	P1@.030	Ea	1.09	.73	—	1.82
2"	P1@.030	Ea	1.39	.73	—	2.12
2½"	P1@.035	Ea	4.33	.86	—	5.19
3"	P1@.040	Ea	4.85	.98	—	5.83
4"	P1@.040	Ea	7.02	.98	—	8.00

Steel pipe nipples — 4½" long, standard right hand thread

Description	Craft@Hrs	Unit	Material $	Labor $	Equipment $	Total $
1/2"	P1@.025	Ea	.48	.61	—	1.09
3/4"	P1@.025	Ea	.55	.61	—	1.16
1"	P1@.025	Ea	.76	.61	—	1.37
1¼"	P1@.030	Ea	1.18	.73	—	1.91
1½"	P1@.030	Ea	1.44	.73	—	2.17
2"	P1@.030	Ea	1.53	.73	—	2.26
2½"	P1@.035	Ea	5.52	.86	—	6.38
3"	P1@.040	Ea	6.94	.98	—	7.92
4"	P1@.050	Ea	8.82	1.22	—	10.04
6"	P1@.070	Ea	37.97	1.71	—	39.68

Carbon Steel, Schedule 40 with 150# M.I. Fittings & Threaded Joints

Description	Craft@Hrs	Unit	Material $	Labor $	Equipment $	Total $
Steel pipe nipples — 5" long, standard right hand thread						
1/2"	P1@.025	Ea	.49	.61	—	1.10
3/4"	P1@.025	Ea	.55	.61	—	1.16
1"	P1@.030	Ea	.81	.73	—	1.54
1¼"	P1@.030	Ea	1.06	.73	—	1.79
1½"	P1@.030	Ea	1.28	.73	—	2.01
2"	P1@.035	Ea	1.59	.86	—	2.45
2½"	P1@.040	Ea	5.85	.98	—	6.83
3"	P1@.045	Ea	6.05	1.10	—	7.15
4"	P1@.050	Ea	9.56	1.22	—	10.78
6"	P1@.075	Ea	38.56	1.83	—	40.39
8"	P1@.100	Ea	64.25	2.44	—	66.69
Steel pipe nipples — 6" long, standard right hand thread						
1/2"	P1@.025	Ea	.55	.61	—	1.16
3/4"	P1@.025	Ea	.66	.61	—	1.27
1"	P1@.030	Ea	.90	.73	—	1.63
1¼"	P1@.030	Ea	1.13	.73	—	1.86
1½"	P1@.030	Ea	1.37	.73	—	2.10
2"	P1@.035	Ea	1.83	.86	—	2.69
2½"	P1@.040	Ea	5.25	.98	—	6.23
3"	P1@.045	Ea	5.86	1.10	—	6.96
4"	P1@.060	Ea	8.85	1.47	—	10.32
6"	P1@.085	Ea	43.11	2.08	—	45.19
8"	P1@.120	Ea	68.69	2.93	—	71.62
Steel pipe nipples — 8" long, standard right hand thread						
2½"	P1@.050	Ea	6.88	1.22	—	8.10
3"	P1@.055	Ea	8.41	1.34	—	9.75
4"	P1@.070	Ea	13.58	1.71	—	15.29
6"	P1@.100	Ea	45.24	2.44	—	47.68
8"	P1@.150	Ea	77.70	3.67	—	81.37
Steel pipe nipples — 10" long, standard right hand thread						
2½"	P1@.055	Ea	8.24	1.34	—	9.58
3"	P1@.060	Ea	9.99	1.47	—	11.46
4"	P1@.080	Ea	16.27	1.95	—	18.22
6"	P1@.115	Ea	50.23	2.81	—	53.04
8"	P1@.165	Ea	86.66	4.03	—	90.69
Steel pipe nipples — 12" long, standard right hand thread						
2½"	P1@.060	Ea	8.08	1.47	—	9.55
3"	P1@.065	Ea	9.84	1.59	—	11.43
4"	P1@.080	Ea	15.95	1.95	—	17.90
6"	P1@.130	Ea	57.38	3.18	—	60.56
8"	P1@.190	Ea	95.63	4.64	—	100.27

Carbon Steel, Schedule 40 with 150# M.I. Fittings & Threaded Joints

Class 125 bronze body gate valve, threaded ends

Description	Craft@Hrs	Unit	Material $	Labor $	Equipment $	Total $
1/2"	P1@.210	Ea	6.06	5.13	—	11.19
3/4"	P1@.250	Ea	7.58	6.11	—	13.69
1"	P1@.300	Ea	10.50	7.33	—	17.83
1¼"	P1@.400	Ea	13.40	9.77	—	23.17
1½"	P1@.450	Ea	18.00	11.00	—	29.00
2"	P1@.500	Ea	30.10	12.20	—	42.30
2½"	P1@.750	Ea	49.90	18.30	—	68.20
3"	P1@.950	Ea	71.60	23.20	—	94.80

Class 125 iron body gate valve, flanged ends

Description	Craft@Hrs	Unit	Material $	Labor $	Equipment $	Total $
2"	P1@.500	Ea	101.00	12.20	—	113.20
2½"	P1@.600	Ea	137.00	14.70	—	151.70
3"	P1@.750	Ea	148.00	18.30	—	166.30
4"	P1@1.35	Ea	218.00	33.00	—	251.00

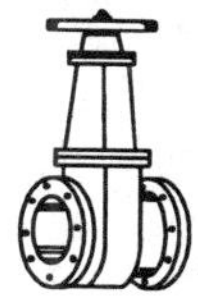

Class 125 bronze body globe valve, threaded ends

Description	Craft@Hrs	Unit	Material $	Labor $	Equipment $	Total $
1/2"	P1@.210	Ea	15.80	5.13	—	20.93
3/4"	P1@.250	Ea	21.30	6.11	—	27.41
1"	P1@.300	Ea	30.50	7.33	—	37.83
1¼"	P1@.400	Ea	42.90	9.77	—	52.67
1½"	P1@.450	Ea	57.50	11.00	—	68.50
2"	P1@.500	Ea	93.60	12.20	—	105.80

Class 125 iron body globe valve, flanged ends

Description	Craft@Hrs	Unit	Material $	Labor $	Equipment $	Total $
2"	P1@.500	Ea	142.00	12.20	—	154.20
2½"	P1@.600	Ea	220.00	14.70	—	234.70
3"	P1@.750	Ea	264.00	18.30	—	282.30
4"	P1@1.35	Ea	352.00	33.00	—	385.00

200 PSIG iron body butterfly valve, lug type

Description	Craft@Hrs	Unit	Material $	Labor $	Equipment $	Total $
2"	P1@.450	Ea	84.50	11.00	—	95.50
2½"	P1@.450	Ea	86.50	11.00	—	97.50
3"	P1@.550	Ea	91.70	13.40	—	105.10
4"	P1@.550	Ea	114.00	13.40	—	127.40

Description	Craft@Hrs	Unit	Material $	Labor $	Equipment $	Total $
200 PSIG iron body butterfly valve, wafer type 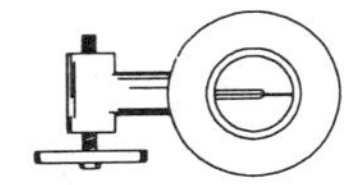						
2"	P1@.450	Ea	76.70	11.00	—	87.70
2½"	P1@.450	Ea	78.70	11.00	—	89.70
3"	P1@.550	Ea	84.50	13.40	—	97.90
4"	P1@.550	Ea	102.00	13.40	—	115.40
Class 125 bronze body 2-piece ball valve, threaded ends 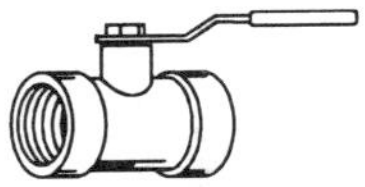						
1/2"	P1@.210	Ea	3.56	5.13	—	8.69
3/4"	P1@.250	Ea	4.75	6.11	—	10.86
1"	P1@.300	Ea	8.21	7.33	—	15.54
1¼"	P1@.400	Ea	13.70	9.77	—	23.47
1½"	P1@.450	Ea	18.50	11.00	—	29.50
2"	P1@.500	Ea	23.30	12.20	—	35.50
Class 125 bronze body swing check valve, threaded 						
1/2"	P1@.210	Ea	8.00	5.13	—	13.13
3/4"	P1@.250	Ea	11.40	6.11	—	17.51
1"	P1@.300	Ea	15.00	7.33	—	22.33
1¼"	P1@.400	Ea	21.40	9.77	—	31.17
1½"	P1@.450	Ea	30.00	11.00	—	41.00
2"	P1@.500	Ea	50.00	12.20	—	62.20
Class 125 iron body swing check valve, flanged ends 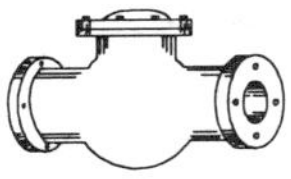						
2"	P1@.500	Ea	115.00	12.20	—	127.20
2½"	P1@.600	Ea	148.00	14.70	—	162.70
3"	P1@.750	Ea	183.00	18.30	—	201.30
4"	P1@1.35	Ea	271.00	33.00	—	304.00
Class 125 iron body silent check valve, wafer type 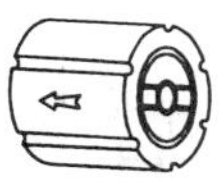						
2"	P1@.500	Ea	85.00	12.20	—	97.20
2½"	P1@.600	Ea	95.00	14.70	—	109.70
3"	P1@.750	Ea	108.00	18.30	—	126.30
4"	P1@1.35	Ea	146.00	33.00	—	179.00
Class 125 bronze body strainer, threaded ends 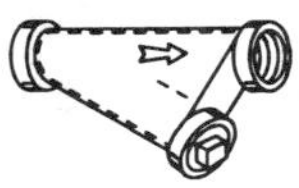						
1/2"	P1@.210	Ea	12.40	5.13	—	17.53
3/4"	P1@.250	Ea	16.30	6.11	—	22.41
1"	P1@.300	Ea	20.20	7.33	—	27.53
1¼"	P1@.400	Ea	28.10	9.77	—	37.87
1½"	P1@.450	Ea	42.20	11.00	—	53.20
2"	P1@.500	Ea	73.00	12.20	—	85.20

Class 125 iron body strainer, flanged ends

Description	Craft@Hrs	Unit	Material $	Labor $	Equipment $	Total $
2"	P1@.500	Ea	62.00	12.20	—	74.20
2½"	P1@.600	Ea	69.00	14.70	—	83.70
3"	P1@.750	Ea	80.00	18.30	—	98.30
4"	P1@1.35	Ea	240.00	33.00	—	273.00

Installation of 2-way control valve, threaded joints

Description	Craft@Hrs	Unit	Material $	Labor $	Equipment $	Total $
1/2"	P1@.210	Ea	—	5.13	—	5.13
3/4"	P1@.275	Ea	—	6.72	—	6.72
1"	P1@.350	Ea	—	8.55	—	8.55
1¼"	P1@.430	Ea	—	10.50	—	10.50
1½"	P1@.505	Ea	—	12.30	—	12.30
2"	P1@.675	Ea	—	16.50	—	16.50
2½"	P1@.830	Ea	—	20.30	—	20.30
3"	P1@.990	Ea	—	24.20	—	24.20

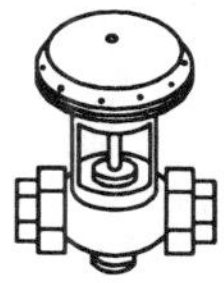

Installation of 3-way control valve, threaded joints

Description	Craft@Hrs	Unit	Material $	Labor $	Equipment $	Total $
1/2"	P1@.260	Ea	—	6.35	—	6.35
3/4"	P1@.365	Ea	—	8.92	—	8.92
1"	P1@.475	Ea	—	11.60	—	11.60
1¼"	P1@.575	Ea	—	14.10	—	14.10
1½"	P1@.680	Ea	—	16.60	—	16.60
2"	P1@.910	Ea	—	22.20	—	22.20
2½"	P1@1.12	Ea	—	27.40	—	27.40
3"	P1@1.33	Ea	—	32.50	—	32.50

Companion flange, threaded

Description	Craft@Hrs	Unit	Material $	Labor $	Equipment $	Total $
2"	P1@.290	Ea	19.36	7.09	—	26.45
2½"	P1@.380	Ea	29.89	9.29	—	39.18
3"	P1@.460	Ea	32.32	11.20	—	43.52
4"	P1@.600	Ea	44.28	14.70	—	58.98

Bolt and gasket sets

Description	Craft@Hrs	Unit	Material $	Labor $	Equipment $	Total $
2"	P1@.500	Ea	1.50	12.20	—	13.70
2½"	P1@.650	Ea	1.75	15.90	—	17.65
3"	P1@.750	Ea	3.00	18.30	—	21.30
4"	P1@1.00	Ea	5.00	24.40	—	29.40

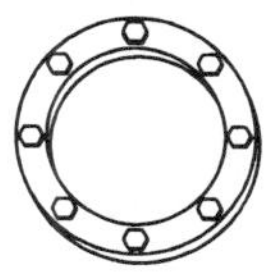

Description	Craft@Hrs	Unit	Material $	Labor $	Equipment $	Total $
Thermometer with well						
7"	P1@.250	Ea	15.00	6.11	—	21.11
9"	P1@.250	Ea	15.50	6.11	—	21.61
Dial type pressure gauge						
2½"	P1@.200	Ea	15.50	4.89	—	20.39
3½"	P1@.200	Ea	20.50	4.89	—	25.39
Pressure/temperature tap						
	P1@.150	Ea	7.30	3.67	—	10.97
Hanger with swivel assembly						
1/2"	P1@.450	Ea	1.75	11.00	—	12.75
3/4"	P1@.450	Ea	1.85	11.00	—	12.85
1"	P1@.450	Ea	1.95	11.00	—	12.95
1¼"	P1@.450	Ea	2.05	11.00	—	13.05
1½"	P1@.450	Ea	2.35	11.00	—	13.35
2"	P1@.450	Ea	2.60	11.00	—	13.60
2½"	P1@.450	Ea	3.35	11.00	—	14.35
3"	P1@.450	Ea	4.15	11.00	—	15.15
4"	P1@.450	Ea	6.40	11.00	—	17.40
Riser clamp						
1/2"	P1@.100	Ea	.95	2.44	—	3.39
3/4"	P1@.100	Ea	1.39	2.44	—	3.83
1"	P1@.100	Ea	1.41	2.44	—	3.85
1¼"	P1@.105	Ea	1.70	2.57	—	4.27
1½"	P1@.110	Ea	1.79	2.69	—	4.48
2"	P1@.115	Ea	1.89	2.81	—	4.70
2½"	P1@.120	Ea	2.00	2.93	—	4.93
3"	P1@.120	Ea	2.18	2.93	—	5.11
4"	P1@.125	Ea	2.75	3.05	—	5.80

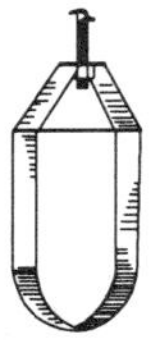

Carbon Steel, Schedule 80 with 300# Fittings & Butt-Welded Joints

Schedule 80 carbon steel (ASTM A-53 and A-120) pipe with 300 pound carbon steel butt-welded fittings are commonly used for steam, steam condensate and re-circulating water systems where operating pressures do not exceed 700 PSIG.

This section has been arranged to save the estimator's time by including all normally-used system components such as pipe, fittings, valves, hanger assemblies, riser clamps and miscellaneous items under one heading. Additional items can be found under "Plumbing and Piping Specialties." The cost estimates in this section are based on the conditions, limitations and wage rates described in the section "How to Use This Book" beginning on page 5.

Equipment cost, where shown, is $35 per hour for a 10-ton hydraulic truck-mounted crane.

Schedule 80 carbon steel pipe, plain end, butt-welded

Description	Craft@Hrs	Unit	Material $	Labor $	Equipment $	Total $
1/2"	P1@.080	LF	.69	1.95	—	2.64
3/4"	P1@.090	LF	.87	2.20	—	3.07
1"	P1@.110	LF	1.18	2.69	—	3.87
1¼"	P1@.120	LF	1.58	2.93	—	4.51
1½"	P1@.130	LF	1.90	3.18	—	5.08
2"	P1@.170	LF	2.64	4.15	—	6.79
2½"	P1@.220	LF	4.05	5.38	—	9.43
3"	P1@.280	LF	5.43	6.84	—	12.27
4"	P1@.340	LF	7.90	8.31	—	16.21
6"	ER@.460	LF	15.00	12.00	2.30	29.30
8"	ER@.560	LF	22.80	14.60	2.80	40.20
10"	ER@.660	LF	39.60	17.20	3.30	60.10
12"	ER@.840	LF	56.30	21.90	4.20	82.40

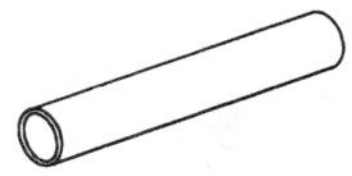

Schedule 80 carbon steel 45 degree ell, butt-welded

Description	Craft@Hrs	Unit	Material $	Labor $	Equipment $	Total $
1/2"	P1@.300	Ea	3.38	7.33	—	10.71
3/4"	P1@.440	Ea	3.38	10.80	—	14.18
1"	P1@.590	Ea	3.38	14.40	—	17.78
1¼"	P1@.740	Ea	3.70	18.10	—	21.80
1½"	P1@.890	Ea	3.86	21.70	—	25.56
2"	P1@1.18	Ea	3.62	28.80	—	32.42
2½"	P1@1.49	Ea	4.95	36.40	—	41.35
3"	P1@1.77	Ea	6.25	43.30	—	49.55
4"	P1@2.37	Ea	9.40	57.90	—	67.30
6"	ER@3.55	Ea	26.40	92.60	17.80	136.80
8"	ER@4.26	Ea	43.00	111.00	21.30	175.30
10"	ER@5.32	Ea	69.00	139.00	26.60	234.60
12"	ER@6.38	Ea	103.00	166.00	31.90	300.90

Schedule 80 carbon steel 90 degree ell, butt-welded

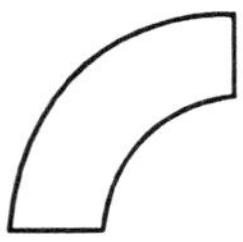

Description	Craft@Hrs	Unit	Material $	Labor $	Equipment $	Total $
1/2"	P1@.300	Ea	3.40	7.33	—	10.73
3/4"	P1@.440	Ea	3.40	10.80	—	14.20
1"	P1@.590	Ea	3.40	14.40	—	17.80
1¼"	P1@.740	Ea	3.90	18.10	—	22.00
1½"	P1@.890	Ea	4.20	21.70	—	25.90
2"	P1@1.18	Ea	6.30	28.80	—	35.10
2½"	P1@1.49	Ea	6.30	36.40	—	42.70
3"	P1@1.77	Ea	8.15	43.30	—	51.45
4"	P1@2.37	Ea	19.60	57.90	—	77.50
6"	ER@3.55	Ea	37.20	92.60	17.80	147.60
8"	ER@4.26	Ea	68.00	111.00	21.30	200.30
10"	ER@5.32	Ea	123.00	139.00	26.60	288.60
12"	ER@6.38	Ea	174.00	166.00	31.90	371.90

Schedule 80 carbon steel tee, butt-welded

Description	Craft@Hrs	Unit	Material $	Labor $	Equipment $	Total $
1/2"	P1@.440	Ea	11.40	10.80	—	22.20
3/4"	P1@.670	Ea	11.40	16.40	—	27.80
1"	P1@.890	Ea	11.40	21.70	—	33.10
1¼"	P1@1.11	Ea	11.40	27.10	—	38.50
1½"	P1@1.33	Ea	13.10	32.50	—	45.60
2"	P1@1.77	Ea	11.20	43.30	—	54.50
2½"	P1@2.22	Ea	26.50	54.30	—	80.80
3"	P1@2.66	Ea	18.70	65.00	—	83.70
4"	P1@3.55	Ea	30.70	86.80	—	117.50
6"	ER@5.32	Ea	49.50	139.00	26.60	215.10
8"	ER@6.38	Ea	99.00	166.00	31.90	296.90
10"	ER@7.98	Ea	160.00	208.00	39.90	407.90
12"	ER@9.58	Ea	233.00	250.00	47.90	530.90

Schedule 80 carbon steel reducing tee, butt-welded

Description	Craft@Hrs	Unit	Material $	Labor $	Equipment $	Total $
1" x 3/4"	P1@.810	Ea	13.85	19.80	—	33.65
1¼" x 1/2"	P1@1.04	Ea	13.85	25.40	—	39.25
1¼" x 3/4"	P1@1.04	Ea	13.85	25.40	—	39.25
1¼" x 1"	P1@1.04	Ea	13.85	25.40	—	39.25
1½" x 1/2"	P1@1.26	Ea	13.85	30.80	—	44.65
1½" x 3/4"	P1@1.26	Ea	13.85	30.80	—	44.65
1½" x 1"	P1@1.26	Ea	13.85	30.80	—	44.65
1½" x 1¼"	P1@1.26	Ea	13.85	30.80	—	44.65
2" x 3/4"	P1@1.63	Ea	15.00	39.80	—	54.80
2" x 1"	P1@1.63	Ea	15.00	39.80	—	54.80
2" x 1¼"	P1@1.63	Ea	15.00	39.80	—	54.80
2" x 1½"	P1@1.63	Ea	15.00	39.80	—	54.80
2½" x 1"	P1@2.07	Ea	21.10	50.60	—	71.70
2½" x 1¼"	P1@2.07	Ea	21.10	50.60	—	71.70
2½" x 1½"	P1@2.07	Ea	21.10	50.60	—	71.70
2½" x 2"	P1@2.07	Ea	21.10	50.60	—	71.70
3" x 1¼"	P1@2.51	Ea	21.10	61.30	—	82.40
3" x 2"	P1@2.51	Ea	21.10	61.30	—	82.40
3" x 2½"	P1@2.51	Ea	21.10	61.30	—	82.40
4" x 1½"	P1@3.25	Ea	33.00	79.40	—	112.40
4" x 2"	P1@3.25	Ea	33.00	79.40	—	112.40
4" x 2½"	P1@3.25	Ea	33.00	79.40	—	112.40
4" x 3"	P1@3.25	Ea	33.00	79.40	—	112.40
6" x 2½"	ER@5.03	Ea	61.00	131.00	25.20	217.20
6" x 3"	ER@5.03	Ea	61.00	131.00	25.20	217.20
6" x 4"	ER@5.03	Ea	61.00	131.00	25.20	217.20
8" x 3"	ER@5.85	Ea	108.00	153.00	29.30	290.30
8" x 4"	ER@5.85	Ea	108.00	153.00	29.30	290.30
8" x 6"	ER@5.85	Ea	108.00	153.00	29.30	290.30
10" x 4"	ER@7.45	Ea	170.00	194.00	37.30	401.30
10" x 6"	ER@7.45	Ea	170.00	194.00	37.30	401.30
10" x 8"	ER@7.45	Ea	170.00	194.00	37.30	401.30
12" x 6"	ER@9.04	Ea	315.00	236.00	45.20	596.20
12" x 8"	ER@9.04	Ea	315.00	236.00	45.20	596.20
12" x 10"	ER@9.04	Ea	315.00	236.00	45.20	596.20

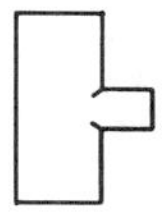

Schedule 80 carbon steel concentric reducer, butt-welded

Description	Craft@Hrs	Unit	Material $	Labor $	Equipment $	Total $
3/4"	P1@.360	Ea	6.75	8.80	—	15.55
1"	P1@.500	Ea	7.20	12.20	—	19.40
1¼"	P1@.665	Ea	6.30	16.30	—	22.60
1½"	P1@.800	Ea	6.70	19.50	—	26.20
2"	P1@1.00	Ea	5.80	24.40	—	30.20
2½"	P1@1.33	Ea	7.60	32.50	—	40.10
3"	P1@1.63	Ea	7.20	39.80	—	47.00
4"	P1@2.00	Ea	9.60	48.90	—	58.50
6"	ER@2.96	Ea	21.00	77.20	14.80	113.00
8"	ER@3.90	Ea	31.00	102.00	19.50	152.50
10"	ER@4.79	Ea	43.00	125.00	24.00	192.00
12"	ER@5.85	Ea	63.00	153.00	29.30	245.30

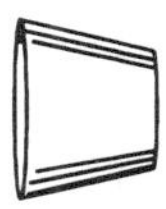

Carbon Steel, Schedule 80 with 300# Fittings & Butt-Welded Joints

Description	Craft@Hrs	Unit	Material $	Labor $	Equipment $	Total $
Schedule 80 carbon steel eccentric reducer, butt-welded						
3/4"	P1@.360	Ea	8.35	8.80	—	17.15
1"	P1@.500	Ea	10.80	12.20	—	23.00
1¼"	P1@.665	Ea	8.70	16.30	—	25.00
1½"	P1@.800	Ea	10.30	19.50	—	29.80
2"	P1@1.00	Ea	8.25	24.40	—	32.65
2½"	P1@1.33	Ea	10.00	32.50	—	42.50
3"	P1@1.63	Ea	9.60	39.80	—	49.40
4"	P1@2.00	Ea	13.60	48.90	—	62.50
6"	ER@2.96	Ea	30.80	77.20	14.80	122.80
8"	ER@3.90	Ea	46.00	102.00	19.50	167.50
10"	ER@4.79	Ea	80.00	125.00	24.00	229.00
12"	ER@5.85	Ea	124.00	153.00	30.00	307.00
Schedule 80 carbon steel cap, butt-welded						
1/2"	P1@.180	Ea	6.70	4.40	—	11.10
3/4"	P1@.270	Ea	9.30	6.60	—	15.90
1"	P1@.360	Ea	4.35	8.80	—	13.15
1¼"	P1@.460	Ea	4.35	11.20	—	15.55
1½"	P1@.560	Ea	4.75	13.70	—	18.45
2"	P1@.740	Ea	4.10	18.10	—	22.20
2½"	P1@.920	Ea	4.60	22.50	—	27.10
3"	P1@1.11	Ea	4.90	27.10	—	32.00
4"	P1@1.50	Ea	6.55	36.70	—	43.25
6"	ER@2.24	Ea	12.10	58.40	11.20	81.70
8"	ER@2.68	Ea	18.80	69.90	13.40	102.10
10"	ER@3.36	Ea	28.00	87.60	16.80	132.40
12"	ER@4.00	Ea	37.00	104.00	20.00	161.00
Schedule 80 carbon steel union, butt-welded						
1/2"	P1@.300	Ea	5.64	7.33	—	12.97
3/4"	P1@.440	Ea	6.79	10.80	—	17.59
1"	P1@.590	Ea	8.87	14.40	—	23.27
1¼"	P1@.740	Ea	15.00	18.10	—	33.10
1½"	P1@.890	Ea	16.30	21.70	—	38.00
2"	P1@1.18	Ea	21.90	28.80	—	50.70
2½"	P1@1.49	Ea	47.60	36.40	—	84.00
3"	P1@1.77	Ea	59.60	43.30	—	102.90

Description	Craft@Hrs	Unit	Material $	Labor $	Equipment $	Total $
Carbon steel weldolet						
1/2"	P1@.440	Ea	7.34	10.80	—	18.14
3/4"	P1@.670	Ea	7.55	16.40	—	23.95
1"	P1@.890	Ea	8.23	21.70	—	29.93
1¼"	P1@1.11	Ea	10.50	27.10	—	37.60
1½"	P1@1.33	Ea	10.50	32.50	—	43.00
2"	P1@1.77	Ea	11.10	43.30	—	54.40
2½"	P1@2.22	Ea	25.00	54.30	—	79.30
3"	P1@2.66	Ea	25.65	65.00	—	90.65
4"	P1@3.55	Ea	57.00	86.80	—	143.80

Description	Craft@Hrs	Unit	Material $	Labor $	Equipment $	Total $
Carbon steel threadolet						
1/2"	P1@.300	Ea	6.20	7.33	—	13.53
3/4"	P1@.440	Ea	6.85	10.80	—	17.65
1"	P1@.590	Ea	8.45	14.40	—	22.85
1¼"	P1@.740	Ea	38.00	18.10	—	56.10
1½"	P1@.890	Ea	38.00	21.70	—	59.70
2"	P1@1.18	Ea	50.00	28.80	—	78.80

Description	Craft@Hrs	Unit	Material $	Labor $	Equipment $	Total $
300# forged steel slip-on companion flange, welding type						
2½"	P1@.810	Ea	21.20	19.80	—	41.00
3"	P1@.980	Ea	21.20	23.90	—	45.10
4"	P1@1.30	Ea	31.20	31.80	—	63.00
6"	ER@1.95	Ea	56.50	50.90	9.75	117.15
8"	ER@2.35	Ea	80.00	61.30	11.80	153.10
10"	ER@2.90	Ea	132.00	75.60	14.50	222.10
12"	ER@3.50	Ea	210.00	91.30	17.50	318.80

Description	Craft@Hrs	Unit	Material $	Labor $	Equipment $	Total $
Class 300 bronze body gate valve, threaded joints						
1/2"	P1@.210	Ea	14.20	5.13	—	19.33
3/4"	P1@.250	Ea	18.50	6.11	—	24.61
1"	P1@.300	Ea	22.45	7.33	—	29.78
1¼"	P1@.400	Ea	30.67	9.77	—	40.44
1½"	P1@.450	Ea	38.50	11.00	—	49.50
2"	P1@.500	Ea	56.60	12.20	—	68.80
2½"	P1@.750	Ea	115.00	18.30	—	133.30
3"	P1@.950	Ea	164.00	23.20	—	187.20

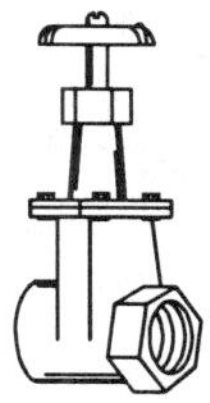

Carbon Steel, Schedule 80 with 300# Fittings & Butt-Welded Joints

Class 250 iron body gate valve, flanged

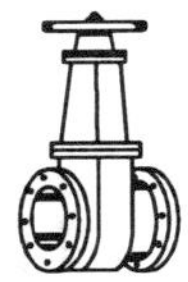

Description	Craft@Hrs	Unit	Material $	Labor $	Equipment $	Total $
2½"	P1@.600	Ea	369.00	14.70	—	383.70
3"	P1@.750	Ea	268.00	18.30	—	286.30
4"	P1@1.35	Ea	388.00	33.00	—	421.00
6"	ER@2.50	Ea	530.00	65.20	12.50	607.70
8"	ER@3.00	Ea	1,795.00	78.20	15.00	1,888.20
10"	ER@4.00	Ea	2,925.00	104.00	20.00	3,049.00
12"	ER@4.50	Ea	4,800.00	117.00	22.50	4,939.50

Class 300 bronze body globe valve, threaded

Description	Craft@Hrs	Unit	Material $	Labor $	Equipment $	Total $
1/2"	P1@.210	Ea	42.40	5.13	—	47.53
3/4"	P1@.250	Ea	53.65	6.11	—	59.76
1"	P1@.300	Ea	76.50	7.33	—	83.83
1¼"	P1@.400	Ea	110.10	9.77	—	119.87
1½"	P1@.450	Ea	148.00	11.00	—	159.00
2"	P1@.500	Ea	205.00	12.20	—	217.20

Class 250 iron body globe valve, flanged

Description	Craft@Hrs	Unit	Material $	Labor $	Equipment $	Total $
2"	P1@.500	Ea	310.00	12.20	—	322.20
2½"	P1@.600	Ea	386.50	14.70	—	401.20
3"	P1@.750	Ea	460.00	18.30	—	478.30
4"	P1@1.35	Ea	635.00	33.00	—	668.00
6"	P1@2.50	Ea	1,230.00	61.10	12.50	1,303.60
8"	P1@3.00	Ea	2,760.00	73.30	15.00	2,848.30

200 PSIG iron body butterfly valve, lug type

Description	Craft@Hrs	Unit	Material $	Labor $	Equipment $	Total $
2"	P1@.450	Ea	84.50	11.00	—	95.50
2½"	P1@.450	Ea	86.50	11.00	—	97.50
3"	P1@.550	Ea	91.70	13.40	—	105.10
4"	P1@.550	Ea	114.00	13.40	—	127.40
6"	ER@.800	Ea	185.00	20.90	4.00	209.90
8"	ER@.800	Ea	252.00	20.90	4.00	276.90
10"	ER@.900	Ea	350.00	23.50	4.50	378.00
12"	ER@1.00	Ea	460.00	26.10	5.00	491.10

200 PSIG iron body butterfly valve, wafer type

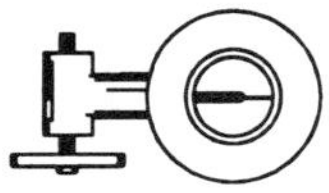

Description	Craft@Hrs	Unit	Material $	Labor $	Equipment $	Total $
2"	P1@.450	Ea	76.70	11.00	—	87.70
2½"	P1@.450	Ea	78.70	11.00	—	89.70
3"	P1@.550	Ea	84.50	13.40	—	97.90
4"	P1@.550	Ea	102.00	13.40	—	115.40
6"	ER@.800	Ea	170.00	20.90	4.00	194.90
8"	ER@.800	Ea	237.00	20.90	4.00	261.90
10"	ER@.900	Ea	332.00	23.50	4.50	360.00
12"	ER@1.00	Ea	436.00	26.10	5.00	467.10

Carbon Steel, Schedule 80 with 300# Fittings & Butt-Welded Joints

Class 300 bronze body swing check valve, threaded

Description	Craft@Hrs	Unit	Material $	Labor $	Equipment $	Total $
1/2"	P1@.210	Ea	54.00	5.13	—	59.13
3/4"	P1@.250	Ea	64.00	6.11	—	70.11
1"	P1@.300	Ea	98.00	7.33	—	105.33
1¼"	P1@.400	Ea	113.00	9.77	—	122.77
1½"	P1@.450	Ea	160.00	11.00	—	171.00
2"	P1@.500	Ea	225.00	12.20	—	237.20

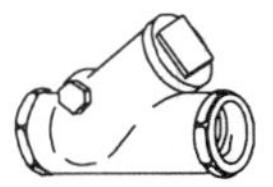

Class 250 iron body swing check valve, flanged

Description	Craft@Hrs	Unit	Material $	Labor $	Equipment $	Total $
2"	P1@.500	Ea	300.00	12.20	—	312.20
2½"	P1@.600	Ea	350.00	14.70	—	364.70
3"	P1@.750	Ea	435.00	18.30	—	453.30
4"	P1@1.35	Ea	535.00	33.00	—	568.00
6"	ER@2.50	Ea	1,025.00	65.20	12.50	1,102.70
8"	ER@3.00	Ea	1,950.00	78.20	15.00	2,043.20

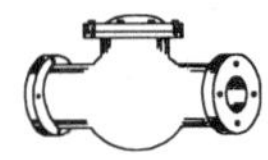

Class 250 iron body silent check valve, flanged

Description	Craft@Hrs	Unit	Material $	Labor $	Equipment $	Total $
2"	P1@.500	Ea	125.00	12.20	—	137.20
2½"	P1@.600	Ea	145.00	14.70	—	159.70
3"	P1@.750	Ea	160.00	18.30	—	178.30
4"	P1@1.35	Ea	220.00	33.00	—	253.00
6"	ER@2.50	Ea	385.00	65.20	12.50	462.70
8"	ER@3.00	Ea	700.00	78.20	15.00	793.20
10"	ER@4.00	Ea	1,100.00	104.00	20.00	1,224.00

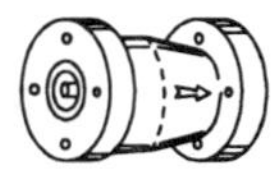

Class 250 bronze body strainer, threaded

Description	Craft@Hrs	Unit	Material $	Labor $	Equipment $	Total $
1/2"	P1@.210	Ea	16.50	5.13	—	21.63
3/4"	P1@.250	Ea	22.45	6.11	—	28.56
1"	P1@.300	Ea	28.50	7.33	—	35.83
1¼"	P1@.400	Ea	40.25	9.77	—	50.02
1½"	P1@.450	Ea	52.25	11.00	—	63.25
2"	P1@.500	Ea	90.00	12.20	—	102.20

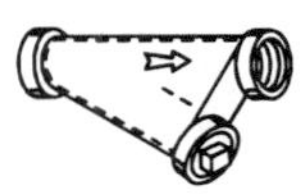

Class 250 iron body strainer, flanged

Description	Craft@Hrs	Unit	Material $	Labor $	Equipment $	Total $
2"	P1@.500	Ea	105.00	12.20	—	117.20
2½"	P1@.600	Ea	115.00	14.70	—	129.70
3"	P1@.750	Ea	136.00	18.30	—	154.30
4"	P1@1.35	Ea	240.00	33.00	—	273.00
6"	ER@2.50	Ea	385.00	65.20	12.50	462.70
8"	ER@3.00	Ea	705.00	78.20	15.00	798.20

Installation of 2-way control valve, threaded joints

Description	Craft@Hrs	Unit	Material $	Labor $	Equipment $	Total $
1/2"	P1@.210	Ea	—	5.13	—	5.13
3/4"	P1@.250	Ea	—	6.11	—	6.11
1"	P1@.300	Ea	—	7.33	—	7.33
1¼"	P1@.400	Ea	—	9.77	—	9.77
1½"	P1@.450	Ea	—	11.00	—	11.00
2"	P1@.500	Ea	—	12.20	—	12.20
2½"	P1@.830	Ea	—	20.30	—	20.30
3"	P1@.990	Ea	—	24.20	—	24.20

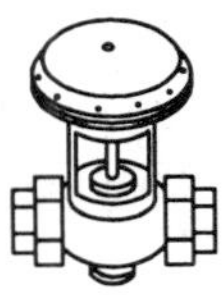

Carbon Steel, Schedule 80 with 300# Fittings & Butt-Welded Joints

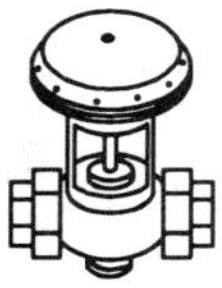

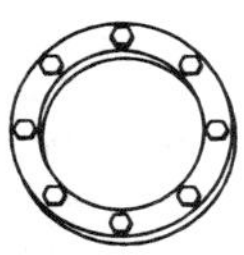

Description	Craft@Hrs	Unit	Material $	Labor $	Equipment $	Total $
Installation of 2-way control valve, flanged joints						
2"	P1@.500	Ea	—	12.20	—	12.20
2½"	P1@.600	Ea	—	14.70	—	14.70
3"	P1@.750	Ea	—	18.30	—	18.30
4"	P1@1.35	Ea	—	33.00	—	33.00
6"	P1@2.50	Ea	—	61.10	—	61.10
8"	P1@3.00	Ea	—	73.30	—	73.30
10"	P1@4.00	Ea	—	97.70	—	97.70
12"	P1@4.50	Ea	—	110.00	—	110.00
Installation of 3-way control valve, threaded joints						
1/2"	P1@.260	Ea	—	6.35	—	6.35
3/4"	P1@.365	Ea	—	8.92	—	8.92
1"	P1@.475	Ea	—	11.60	—	11.60
1¼"	P1@.575	Ea	—	14.10	—	14.10
1½"	P1@.680	Ea	—	16.60	—	16.60
2"	P1@.910	Ea	—	22.20	—	22.20
2½"	P1@1.12	Ea	—	27.40	—	27.40
3"	P1@1.33	Ea	—	32.50	—	32.50
Installation of 3-way control valve, flanged joints						
2"	P1@.910	Ea	—	22.20	—	22.20
2½"	P1@1.12	Ea	—	27.40	—	27.40
3"	P1@1.33	Ea	—	32.50	—	32.50
4"	P1@2.00	Ea	—	48.90	—	48.90
6"	P1@3.70	Ea	—	90.40	—	90.40
8"	P1@4.40	Ea	—	108.00	—	108.00
10"	P1@5.90	Ea	—	144.00	—	144.00
12"	P1@6.50	Ea	—	159.00	—	159.00
Bolt and gasket set						
2"	P1@.500	Ea	1.50	12.20	—	13.70
2½"	P1@.650	Ea	1.75	15.90	—	17.65
3"	P1@.750	Ea	3.00	18.30	—	21.30
4"	P1@1.00	Ea	5.00	24.40	—	29.40
6"	P1@1.20	Ea	8.25	29.30	—	37.55
8"	P1@1.25	Ea	9.35	30.50	—	39.85
10"	P1@1.70	Ea	16.00	41.50	—	57.50
12"	P1@2.20	Ea	18.00	53.80	—	71.80
Thermometer with well						
7"	P1@.250	Ea	15.00	6.11	—	21.11
9"	P1@.250	Ea	15.50	6.11	—	21.61

Description	Craft@Hrs	Unit	Material $	Labor $	Equipment $	Total $
Dial type pressure gauge						
2½"	P1@.200	Ea	15.50	4.89	—	20.39
3½"	P1@.200	Ea	20.50	4.89	—	25.39

Description	Craft@Hrs	Unit	Material $	Labor $	Equipment $	Total $
Pressure/temperature tap						
	P1@.150	Ea	7.30	3.67	—	10.97

Description	Craft@Hrs	Unit	Material $	Labor $	Equipment $	Total $
Hanger with swivel assembly						
1/2"	P1@.450	Ea	2.40	11.00	—	13.40
3/4"	P1@.450	Ea	2.65	11.00	—	13.65
1"	P1@.450	Ea	2.85	11.00	—	13.85
1½"	P1@.450	Ea	3.00	11.00	—	14.00
2"	P1@.450	Ea	3.90	11.00	—	14.90
2½"	P1@.450	Ea	5.00	11.00	—	16.00
3"	P1@.450	Ea	6.30	11.00	—	17.30
4"	P1@.450	Ea	10.75	11.00	—	21.75
5"	P1@.500	Ea	12.50	12.20	—	24.70
6"	P1@.500	Ea	14.00	12.20	—	26.20
8"	P1@.550	Ea	18.50	13.40	—	31.90
10"	P1@.650	Ea	25.50	15.90	—	41.40
12"	P1@.750	Ea	33.75	18.30	—	52.05

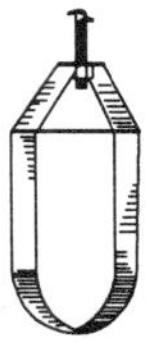

Description	Craft@Hrs	Unit	Material $	Labor $	Equipment $	Total $
Riser clamp						
1/2"	P1@.100	Ea	.95	2.44	—	3.39
3/4"	P1@.100	Ea	1.39	2.44	—	3.83
1"	P1@.100	Ea	1.41	2.44	—	3.85
1¼"	P1@.105	Ea	1.70	2.57	—	4.27
1½"	P1@.110	Ea	1.79	2.69	—	4.48
2"	P1@.115	Ea	1.89	2.81	—	4.70
2½"	P1@.120	Ea	2.00	2.93	—	4.93
3"	P1@.120	Ea	2.18	2.93	—	5.11
4"	P1@.125	Ea	2.75	3.05	—	5.80
5"	P1@.180	Ea	3.99	4.40	—	8.39
6"	P1@.200	Ea	4.79	4.89	—	9.68
8"	P1@.200	Ea	7.77	4.89	—	12.66
10"	P1@.250	Ea	11.60	6.11	—	17.71
12"	P1@.250	Ea	13.75	6.11	—	19.86

Carbon Steel, Schedule 80 with 300# M.I. Fittings & Threaded Joints

Schedule 80 carbon steel (ASTM A-53 and A-120) pipe with 300 pound malleable iron threaded fittings is commonly used for heating hot water, chilled water, steam, and steam condensate systems where operating pressures do not exceed 250 PSIG for pipe sizes 2 inches and smaller. For pipe sizes 2½ inches and larger, wrought steel fittings with welded joints should be used.

This section has been arranged to save the estimator's time by including all normally-used system components such as pipe, fittings, valves, hanger assemblies, riser clamps and miscellaneous items under one heading. Additional items can be found under "Plumbing and Piping Specialties." The cost estimates in this section are based on the conditions, limitations and wage rates described in the section "How to Use This Book" beginning on page 5.

Equipment cost, where shown, is $35 per hour for a 10-ton hydraulic truck-mounted crane.

Schedule 80 carbon steel pipe, threaded and coupled joints

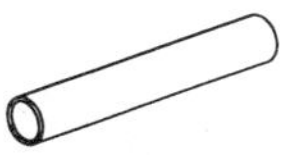

Description	Craft@Hrs	Unit	Material $	Labor $	Equipment $	Total $
1/2"	P1@.070	LF	.81	1.71	—	2.52
3/4"	P1@.080	LF	1.00	1.95	—	2.95
1"	P1@.100	LF	1.38	2.44	—	3.82
1¼"	P1@.110	LF	1.80	2.69	—	4.49
1½"	P1@.120	LF	2.13	2.93	—	5.06
2"	P1@.130	LF	2.90	3.18	—	6.08
2½"	P1@.190	LF	4.58	4.64	—	9.22
3"	P1@.220	LF	6.00	5.38	—	11.38
4"	P1@.290	LF	8.67	7.09	—	15.76

300# malleable iron 45 degree ell, threaded joints

Description	Craft@Hrs	Unit	Material $	Labor $	Equipment $	Total $
1/2"	P1@.132	Ea	5.78	3.23	—	9.01
3/4"	P1@.143	Ea	6.41	3.49	—	9.90
1"	P1@.198	Ea	7.12	4.84	—	11.96
1¼"	P1@.264	Ea	11.10	6.45	—	17.55
1½"	P1@.330	Ea	14.40	8.06	—	22.46
2"	P1@.418	Ea	21.40	10.20	—	31.60
2½"	P1@.561	Ea	33.80	13.70	—	47.50
3"	P1@.704	Ea	44.90	17.20	—	62.10
4"	P1@.935	Ea	87.00	22.80	—	109.80

300# malleable iron 90 degree ell, threaded joints

Description	Craft@Hrs	Unit	Material $	Labor $	Equipment $	Total $
1/2"	P1@.132	Ea	4.45	3.23	—	7.68
3/4"	P1@.143	Ea	5.14	3.49	—	8.63
1"	P1@.198	Ea	6.56	4.84	—	11.40
1¼"	P1@.264	Ea	9.10	6.45	—	15.55
1½"	P1@.330	Ea	11.00	8.06	—	19.06
2"	P1@.418	Ea	15.30	10.20	—	25.50
2½"	P1@.561	Ea	29.20	13.70	—	42.90
3"	P1@.704	Ea	40.00	17.20	—	57.20
4"	P1@.935	Ea	80.00	22.80	—	102.80

Carbon Steel, Schedule 80 with 300# M.I. Fittings & Threaded Joints

300# malleable iron tee, threaded joints

Description	Craft@Hrs	Unit	Material $	Labor $	Equipment $	Total $
1/2"	P1@.198	Ea	6.25	4.84	—	11.09
3/4"	P1@.209	Ea	6.87	5.11	—	11.98
1"	P1@.253	Ea	8.25	6.18	—	14.43
1¼"	P1@.341	Ea	11.00	8.33	—	19.33
1½"	P1@.429	Ea	13.70	10.50	—	24.20
2"	P1@.539	Ea	19.90	13.20	—	33.10
2½"	P1@.726	Ea	38.00	17.70	—	55.70
3"	P1@.913	Ea	63.70	22.30	—	86.00
4"	P1@1.21	Ea	108.00	29.60	—	137.60

300# malleable iron reducing tee, threaded joints

Description	Craft@Hrs	Unit	Material $	Labor $	Equipment $	Total $
3/4" x 1/2"	P1@.205	Ea	8.60	5.01	—	13.61
1" x 1/2"	P1@.205	Ea	10.80	5.01	—	15.81
1" x 3/4"	P1@.205	Ea	10.80	5.01	—	15.81
1¼" x 1/2"	P1@.310	Ea	14.00	7.58	—	21.58
1¼" x 3/4"	P1@.310	Ea	14.00	7.58	—	21.58
1¼" x 1"	P1@.310	Ea	14.00	7.58	—	21.58
1½" x 1/2"	P1@.400	Ea	17.00	9.77	—	26.77
1½" x 3/4"	P1@.400	Ea	17.00	9.77	—	26.77
1½" x 1"	P1@.400	Ea	17.00	9.77	—	26.77
1½" x 1¼"	P1@.400	Ea	17.00	9.77	—	26.77
2" x 1"	P1@.500	Ea	25.50	12.20	—	37.70
2" x 1¼"	P1@.500	Ea	25.50	12.20	—	37.70
2" x 1½"	P1@.500	Ea	25.50	12.20	—	37.70
3" x 1¼"	P1@.830	Ea	133.00	20.30	—	153.30
3" x 1½"	P1@.830	Ea	133.00	20.30	—	153.30
3" x 2"	P1@.830	Ea	133.00	20.30	—	153.30

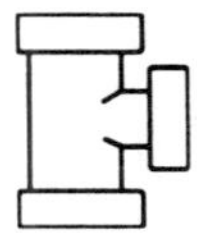

300# malleable iron reducer, threaded joints

Description	Craft@Hrs	Unit	Material $	Labor $	Equipment $	Total $
3/4" x 1/2"	P1@.135	Ea	5.10	3.30	—	8.40
1" x 1/2"	P1@.170	Ea	6.10	4.15	—	10.25
1" x 3/4"	P1@.170	Ea	6.10	4.15	—	10.25
1¼" x 3/4"	P1@.220	Ea	7.70	5.38	—	13.08
1¼" x 1"	P1@.220	Ea	7.70	5.38	—	13.08
1½" x 1"	P1@.285	Ea	10.50	6.96	—	17.46
1½" x 1¼"	P1@.285	Ea	10.50	6.96	—	17.46
2" x 1¼"	P1@.360	Ea	15.20	8.80	—	24.00
2" x 1½"	P1@.360	Ea	15.20	8.80	—	24.00
3" x 2"	P1@.550	Ea	37.00	13.40	—	50.40
3" x 2½"	P1@.550	Ea	37.00	13.40	—	50.40

Carbon Steel, Schedule 80 with 300# M.I. Fittings & Threaded Joints

300# malleable iron cross, threaded joints

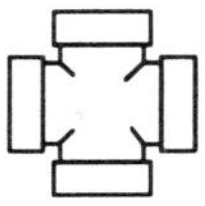

Description	Craft@Hrs	Unit	Material $	Labor $	Equipment $	Total $
1/2"	P1@.310	Ea	15.00	7.58	—	22.58
3/4"	P1@.350	Ea	16.30	8.55	—	24.85
1"	P1@.400	Ea	18.50	9.77	—	28.27
1¼"	P1@.530	Ea	23.00	13.00	—	36.00
1½"	P1@.660	Ea	30.00	16.10	—	46.10
2"	P1@.770	Ea	43.00	18.80	—	61.80

300# malleable iron cap, threaded joint

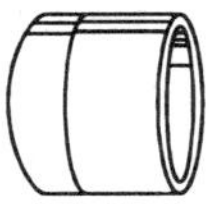

Description	Craft@Hrs	Unit	Material $	Labor $	Equipment $	Total $
1/2"	P1@.100	Ea	3.00	2.44	—	5.44
3/4"	P1@.110	Ea	4.00	2.69	—	6.69
1"	P1@.155	Ea	5.15	3.79	—	8.94
1¼"	P1@.200	Ea	7.00	4.89	—	11.89
1½"	P1@.250	Ea	8.30	6.11	—	14.41
2"	P1@.320	Ea	11.50	7.82	—	19.32
3"	P1@.530	Ea	24.00	13.00	—	37.00
4"	P1@.705	Ea	44.00	17.20	—	61.20

300# malleable iron plug, threaded joint

Description	Craft@Hrs	Unit	Material $	Labor $	Equipment $	Total $
1/2"	P1@.090	Ea	.74	2.20	—	2.94
3/4"	P1@.100	Ea	1.00	2.44	—	3.44
1"	P1@.140	Ea	1.20	3.42	—	4.62
1¼"	P1@.180	Ea	1.30	4.40	—	5.70
1½"	P1@.230	Ea	1.65	5.62	—	7.27
2"	P1@.290	Ea	2.55	7.09	—	9.64
2½"	P1@.380	Ea	3.60	9.29	—	12.89
3"	P1@.480	Ea	6.00	11.70	—	17.70
4"	P1@.640	Ea	17.00	15.60	—	32.60

300# malleable iron union, threaded joints

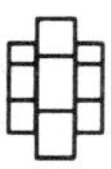

Description	Craft@Hrs	Unit	Material $	Labor $	Equipment $	Total $
1/2"	P1@.150	Ea	5.50	3.67	—	9.17
3/4"	P1@.165	Ea	6.20	4.03	—	10.23
1"	P1@.230	Ea	8.10	5.62	—	13.72
1¼"	P1@.310	Ea	12.50	7.58	—	20.08
1½"	P1@.400	Ea	13.50	9.77	—	23.27
2"	P1@.500	Ea	17.00	12.20	—	29.20
2½"	P1@.670	Ea	38.00	16.40	—	54.40
3"	P1@.840	Ea	64.00	20.50	—	84.50
4"	P1@.960	Ea	180.00	23.50	—	203.50

Carbon Steel, Schedule 80 with 300# M.I. Fittings & Threaded Joints

300# malleable iron coupling, threaded joints

Description	Craft@Hrs	Unit	Material $	Labor $	Equipment $	Total $
1/2"	P1@.132	Ea	3.35	3.23	—	6.58
3/4"	P1@.143	Ea	3.90	3.49	—	7.39
1"	P1@.198	Ea	4.60	4.84	—	9.44
1¼"	P1@.264	Ea	5.50	6.45	—	11.95
1½"	P1@.330	Ea	8.25	8.06	—	16.31
2"	P1@.418	Ea	11.50	10.20	—	21.70
2½"	P1@.561	Ea	14.70	13.70	—	28.40
3"	P1@.704	Ea	24.00	17.20	—	41.20
4"	P1@.935	Ea	57.00	22.80	—	79.80

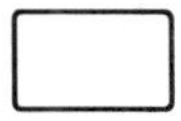

Class 300 bronze body gate valve, threaded joints

Description	Craft@Hrs	Unit	Material $	Labor $	Equipment $	Total $
1/2"	P1@.210	Ea	14.20	5.13	—	19.33
3/4"	P1@.250	Ea	18.50	6.11	—	24.61
1"	P1@.300	Ea	22.45	7.33	—	29.78
1¼"	P1@.400	Ea	30.67	9.77	—	40.44
1½"	P1@.450	Ea	38.50	11.00	—	49.50
2"	P1@.500	Ea	56.60	12.20	—	68.80

Class 250 iron body gate valve, flanged joints

Description	Craft@Hrs	Unit	Material $	Labor $	Equipment $	Total $
2"	P1@.500	Ea	369.00	12.20	—	381.20
2½"	P1@.600	Ea	268.00	14.70	—	282.70
3"	P1@.750	Ea	338.00	18.30	—	356.30
4"	P1@1.35	Ea	570.00	33.00	—	603.00

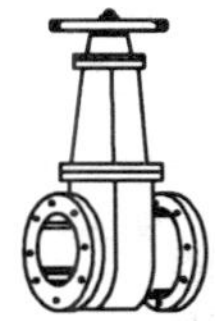

Class 300 bronze body globe valve, threaded joints

Description	Craft@Hrs	Unit	Material $	Labor $	Equipment $	Total $
1/2"	P1@.210	Ea	42.40	5.13	—	47.53
3/4"	P1@.250	Ea	53.65	6.11	—	59.76
1"	P1@.300	Ea	76.50	7.33	—	83.83
1¼"	P1@.400	Ea	110.10	9.77	—	119.87
1½"	P1@.450	Ea	148.00	11.00	—	159.00
2"	P1@.500	Ea	205.00	12.20	—	217.20

Class 250 iron body globe valve, flanged joints

Description	Craft@Hrs	Unit	Material $	Labor $	Equipment $	Total $
2"	P1@.500	Ea	310.00	12.20	—	322.20
2½"	P1@.600	Ea	386.00	14.70	—	400.70
3"	P1@.750	Ea	460.00	18.30	—	478.30
4"	P1@1.35	Ea	635.70	33.00	—	668.70

200 PSIG iron body butterfly valve, lug type

Description	Craft@Hrs	Unit	Material $	Labor $	Equipment $	Total $
2"	P1@.450	Ea	84.50	11.00	—	95.50
2½"	P1@.450	Ea	86.50	11.00	—	97.50
3"	P1@.550	Ea	91.70	13.40	—	105.10
4"	P1@.550	Ea	114.00	13.40	—	127.40

Carbon Steel, Schedule 80 with 300# M.I. Fittings & Threaded Joints

Description	Craft@Hrs	Unit	Material $	Labor $	Equipment $	Total $
200 PSIG iron body butterfly valve, wafer type 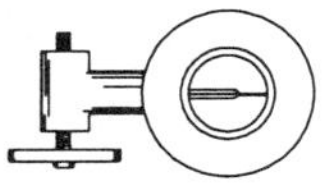						
2"	P1@.450	Ea	76.70	11.00	—	87.70
2½"	P1@.450	Ea	78.70	11.00	—	89.70
3"	P1@.550	Ea	84.50	13.40	—	97.90
4"	P1@.550	Ea	102.00	13.40	—	115.40
Class 150 bronze body 2-piece ball valve, threaded joints 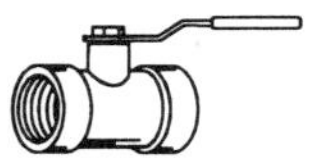						
1/2"	P1@.210	Ea	5.27	5.13	—	10.40
3/4"	P1@.250	Ea	8.61	6.11	—	14.72
1"	P1@.300	Ea	10.90	7.33	—	18.23
1¼"	P1@.400	Ea	13.70	9.77	—	23.47
1½"	P1@.450	Ea	18.50	11.00	—	29.50
2"	P1@.500	Ea	23.30	12.20	—	35.50
Class 300 bronze body swing check valve, threaded joints 						
1/2"	P1@.210	Ea	54.00	5.13	—	59.13
3/4"	P1@.250	Ea	64.00	6.11	—	70.11
1"	P1@.300	Ea	98.00	7.33	—	105.33
1¼"	P1@.400	Ea	113.00	9.77	—	122.77
1½"	P1@.450	Ea	160.00	11.00	—	171.00
2"	P1@.500	Ea	225.00	12.20	—	237.20
Class 250 iron body swing check valve, flanged joints 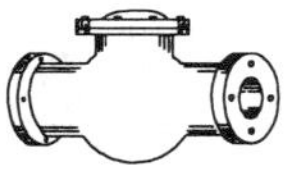						
2"	P1@.500	Ea	300.00	12.20	—	312.20
2½"	P1@.600	Ea	350.00	14.70	—	364.70
3"	P1@.750	Ea	435.00	18.30	—	453.30
4"	P1@1.35	Ea	535.00	33.00	—	568.00
Class 250 iron body silent check valve, wafer type 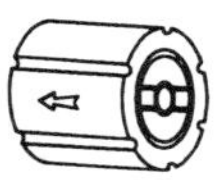						
2"	P1@.500	Ea	125.00	12.20	—	137.20
2½"	P1@.600	Ea	145.00	14.70	—	159.70
3"	P1@.750	Ea	160.00	18.30	—	178.30
4"	P1@1.35	Ea	220.00	33.00	—	253.00
Class 250 bronze body wye strainer, threaded joints 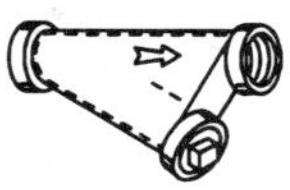						
1/2"	P1@.210	Ea	16.50	5.13	—	21.63
3/4"	P1@.250	Ea	22.45	6.11	—	28.56
1"	P1@.300	Ea	28.50	7.33	—	35.83
1¼"	P1@.400	Ea	40.25	9.77	—	50.02
1½"	P1@.450	Ea	52.25	11.00	—	63.25
2"	P1@.500	Ea	90.00	12.20	—	102.20

Description	Craft@Hrs	Unit	Material $	Labor $	Equipment $	Total $

Class 250 iron body strainer, flanged ends

Description	Craft@Hrs	Unit	Material $	Labor $	Equipment $	Total $
2"	P1@.500	Ea	105.00	12.20	—	117.20
2½"	P1@.600	Ea	115.00	14.70	—	129.70
3"	P1@.750	Ea	136.00	18.30	—	154.30
4"	P1@1.35	Ea	240.00	33.00	—	273.00

Installation of 2-way control valve, threaded joints

Description	Craft@Hrs	Unit	Material $	Labor $	Equipment $	Total $
1/2"	P1@.210	Ea	—	5.13	—	5.13
3/4"	P1@.275	Ea	—	6.72	—	6.72
1"	P1@.350	Ea	—	8.55	—	8.55
1¼"	P1@.430	Ea	—	10.50	—	10.50
1½"	P1@.505	Ea	—	12.30	—	12.30
2"	P1@.675	Ea	—	16.50	—	16.50
2½"	P1@.830	Ea	—	20.30	—	20.30
3"	P1@.990	Ea	—	24.20	—	24.20

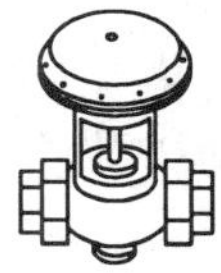

Installation of 3-way control valve, threaded joints

Description	Craft@Hrs	Unit	Material $	Labor $	Equipment $	Total $
1/2"	P1@.260	Ea	—	6.35	—	6.35
3/4"	P1@.365	Ea	—	8.92	—	8.92
1"	P1@.475	Ea	—	11.60	—	11.60
1¼"	P1@.575	Ea	—	14.10	—	14.10
1½"	P1@.680	Ea	—	16.60	—	16.60
2"	P1@.910	Ea	—	22.20	—	22.20
2½"	P1@1.12	Ea	—	27.40	—	27.40
3"	P1@1.33	Ea	—	32.50	—	32.50

Threaded companion flange, 300# malleable iron

Description	Craft@Hrs	Unit	Material $	Labor $	Equipment $	Total $
2"	P1@.290	Ea	40.50	7.09	—	47.59
2½"	P1@.380	Ea	55.00	9.29	—	64.29
3"	P1@.460	Ea	52.50	11.20	—	63.70
4"	P1@.600	Ea	98.00	14.70	—	112.70

Bolt and gasket sets

Description	Craft@Hrs	Unit	Material $	Labor $	Equipment $	Total $
2"	P1@.500	Ea	1.50	12.20	—	13.70
2½"	P1@.650	Ea	1.75	15.90	—	17.65
3"	P1@.750	Ea	3.00	18.30	—	21.30
4"	P1@1.00	Ea	5.00	24.40	—	29.40

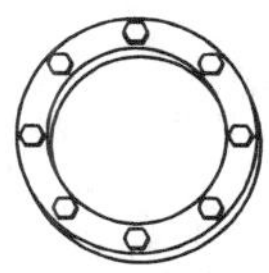

Carbon Steel, Schedule 80 with 300# M.I. Fittings & Threaded Joints

Description	Craft@Hrs	Unit	Material $	Labor $	Equipment $	Total $
Thermometer with well						
7"	P1@.250	Ea	15.00	6.11	—	21.11
9"	P1@.250	Ea	15.50	6.11	—	21.61

Description	Craft@Hrs	Unit	Material $	Labor $	Equipment $	Total $
Dial type pressure gauge						
2½"	P1@.200	Ea	15.50	4.89	—	20.39
3½"	P1@.200	Ea	20.50	4.89	—	25.39

Description	Craft@Hrs	Unit	Material $	Labor $	Equipment $	Total $
Pressure/temperature tap						
	P1@.150	Ea	7.30	3.67	—	10.97

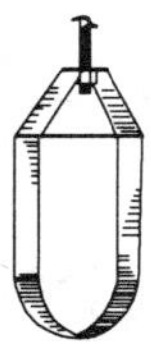

Description	Craft@Hrs	Unit	Material $	Labor $	Equipment $	Total $
Hanger with swivel assembly						
1/2"	P1@.450	Ea	1.75	11.00	—	12.75
3/4"	P1@.450	Ea	1.85	11.00	—	12.85
1"	P1@.450	Ea	1.95	11.00	—	12.95
1¼"	P1@.450	Ea	2.05	11.00	—	13.05
1½"	P1@.450	Ea	2.35	11.00	—	13.35
2"	P1@.450	Ea	2.60	11.00	—	13.60
2½"	P1@.450	Ea	3.35	11.00	—	14.35
3"	P1@.450	Ea	4.15	11.00	—	15.15
4"	P1@.450	Ea	6.40	11.00	—	17.40

Description	Craft@Hrs	Unit	Material $	Labor $	Equipment $	Total $
Riser clamp						
1/2"	P1@.100	Ea	.95	2.44	—	3.39
3/4"	P1@.100	Ea	1.39	2.44	—	3.83
1"	P1@.100	Ea	1.41	2.44	—	3.85
1¼"	P1@.105	Ea	1.70	2.57	—	4.27
1½"	P1@.110	Ea	1.79	2.69	—	4.48
2"	P1@.115	Ea	1.89	2.81	—	4.70
2½"	P1@.120	Ea	2.00	2.93	—	4.93
3"	P1@.120	Ea	2.18	2.93	—	5.11
4"	P1@.125	Ea	2.75	3.05	—	5.80

Schedule 160 carbon steel (ASTM A-106) pipe with steel fittings is commonly used for steam, steam condensate and re-circulating water systems.

This section has been arranged to save the estimator's time by including all normally-used system components such as pipe, fittings, valves, hanger assemblies, riser clamps and miscellaneous items under one heading. Additional items can be found under "Plumbing and Piping Specialties." The cost estimates in this section are based on the conditions, limitations and wage rates described in the section "How to Use This Book" beginning on page 5.

Schedule 160 carbon steel plain end pipe

Description	Craft@Hrs	Unit	Material $	Labor $	Equipment $	Total $
1/2"	P1@.090	LF	2.27	2.20	—	4.47
3/4"	P1@.100	LF	2.62	2.44	—	5.06
1"	P1@.120	LF	3.74	2.93	—	6.67
1¼"	P1@.130	LF	4.39	3.18	—	7.57
1½"	P1@.140	LF	6.83	3.42	—	10.25
2"	P1@.190	LF	5.65	4.64	—	10.29
3"	P1@.310	LF	10.90	7.58	—	18.48
4"	P1@.380	LF	17.10	9.29	—	26.39

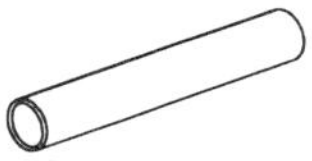

3,000# carbon steel 45 degree ell, threaded joints

Description	Craft@Hrs	Unit	Material $	Labor $	Equipment $	Total $
1/2"	P1@.132	Ea	4.99	3.23	—	8.22
3/4"	P1@.143	Ea	5.72	3.49	—	9.21
1"	P1@.198	Ea	7.84	4.84	—	12.68
1¼"	P1@.264	Ea	10.80	6.45	—	17.25
1½"	P1@.330	Ea	16.30	8.06	—	24.36
2"	P1@.418	Ea	22.50	10.20	—	32.70
2½"	P1@.561	Ea	55.50	13.70	—	69.20
3"	P1@.704	Ea	90.20	17.20	—	107.40
4"	P1@.935	Ea	165.00	22.80	—	187.80

3,000# carbon steel 45 degree ell, socket-welded joints

Description	Craft@Hrs	Unit	Material $	Labor $	Equipment $	Total $
1/2"	P1@.300	Ea	4.05	7.33	—	11.38
3/4"	P1@.440	Ea	4.89	10.80	—	15.69
1"	P1@.590	Ea	6.39	14.40	—	20.79
1¼"	P1@.740	Ea	8.64	18.10	—	26.74
1½"	P1@.890	Ea	10.50	21.70	—	32.20
2"	P1@1.18	Ea	18.30	28.80	—	47.10
2½"	P1@1.49	Ea	47.30	36.40	—	83.70
3"	P1@1.77	Ea	79.40	43.30	—	122.70
4"	P1@2.37	Ea	157.00	57.90	—	214.90

Carbon Steel, Schedule 160 with 3,000-6,000# Fittings

6,000# carbon steel 45 degree ell, threaded joints

Description	Craft@Hrs	Unit	Material $	Labor $	Equipment $	Total $
1/2"	P1@.132	Ea	8.38	3.23	—	11.61
3/4"	P1@.143	Ea	11.50	3.49	—	14.99
1"	P1@.198	Ea	13.00	4.84	—	17.84
1¼"	P1@.264	Ea	34.30	6.45	—	40.75
1½"	P1@.330	Ea	34.30	8.06	—	42.36
2"	P1@.418	Ea	47.90	10.20	—	58.10
2½"	P1@.561	Ea	182.00	13.70	—	195.70
3"	P1@.704	Ea	187.00	17.20	—	204.20

6,000# carbon steel 45 degree ell, socket-welded joints

Description	Craft@Hrs	Unit	Material $	Labor $	Equipment $	Total $
1/2"	P1@.300	Ea	6.58	7.33	—	13.91
3/4"	P1@.440	Ea	7.59	10.80	—	18.39
1"	P1@.590	Ea	10.20	14.40	—	24.60
1¼"	P1@.740	Ea	36.60	18.10	—	54.70
1½"	P1@.890	Ea	24.80	21.70	—	46.50
2"	P1@1.18	Ea	30.70	28.80	—	59.50
2½"	P1@1.49	Ea	122.00	36.40	—	158.40
3"	P1@1.77	Ea	146.00	43.30	—	189.30
4"	P1@2.37	Ea	170.00	57.90	—	227.90

3,000# carbon steel 90 degree ell, threaded joints

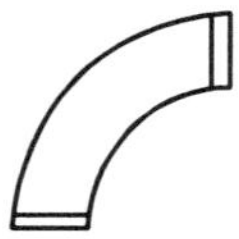

Description	Craft@Hrs	Unit	Material $	Labor $	Equipment $	Total $
1/2"	P1@.132	Ea	3.91	3.23	—	7.14
3/4"	P1@.143	Ea	4.68	3.49	—	8.17
1"	P1@.198	Ea	6.79	4.84	—	11.63
1¼"	P1@.264	Ea	11.50	6.45	—	17.95
1½"	P1@.330	Ea	17.10	8.06	—	25.16
2"	P1@.418	Ea	20.70	10.20	—	30.90
2½"	P1@.561	Ea	51.50	13.70	—	65.20
3"	P1@.704	Ea	78.30	17.20	—	95.50
4"	P1@.935	Ea	171.00	22.80	—	193.80

3,000# carbon steel 90 degree ell, socket-welded joints

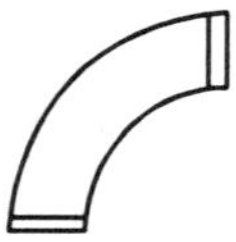

Description	Craft@Hrs	Unit	Material $	Labor $	Equipment $	Total $
1/2"	P1@.300	Ea	3.86	7.33	—	11.19
3/4"	P1@.440	Ea	3.98	10.80	—	14.78
1"	P1@.590	Ea	5.13	14.40	—	19.53
1¼"	P1@.740	Ea	8.45	18.10	—	26.55
1½"	P1@.890	Ea	10.70	21.70	—	32.40
2"	P1@1.18	Ea	16.30	28.80	—	45.10
2½"	P1@1.49	Ea	39.40	36.40	—	75.80
3"	P1@1.77	Ea	65.40	43.30	—	108.70
4"	P1@2.37	Ea	190.00	57.90	—	247.90

6,000# carbon steel 90 degree ell, threaded joints

Description	Craft@Hrs	Unit	Material $	Labor $	Equipment $	Total $
1/2"	P1@.132	Ea	6.34	3.23	—	9.57
3/4"	P1@.143	Ea	8.08	3.49	—	11.57
1"	P1@.198	Ea	10.40	4.84	—	15.24
1¼"	P1@.264	Ea	17.40	6.45	—	23.85
1½"	P1@.330	Ea	27.10	8.06	—	35.16
2"	P1@.418	Ea	47.50	10.20	—	57.70
2½"	P1@.561	Ea	70.10	13.70	—	83.80
3"	P1@.704	Ea	160.00	17.20	—	177.20

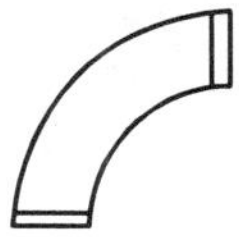

6,000# carbon steel 90 degree ell, socket-welded joints

Description	Craft@Hrs	Unit	Material $	Labor $	Equipment $	Total $
1/2"	P1@.300	Ea	5.36	7.33	—	12.69
3/4"	P1@.440	Ea	6.27	10.80	—	17.07
1"	P1@.590	Ea	8.62	14.40	—	23.02
1¼"	P1@.740	Ea	12.70	18.10	—	30.80
1½"	P1@.890	Ea	19.20	21.70	—	40.90
2"	P1@1.18	Ea	28.10	28.80	—	56.90
2½"	P1@1.49	Ea	58.60	36.40	—	95.00
3"	P1@1.77	Ea	103.00	43.30	—	146.30
4"	P1@2.37	Ea	228.00	57.90	—	285.90

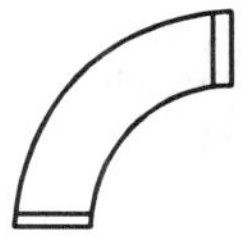

3,000# carbon steel tee, threaded joints

Description	Craft@Hrs	Unit	Material $	Labor $	Equipment $	Total $
1/2"	P1@.198	Ea	5.15	4.84	—	9.99
3/4"	P1@.209	Ea	7.33	5.11	—	12.44
1"	P1@.253	Ea	9.32	6.18	—	15.50
1¼"	P1@.341	Ea	15.50	8.33	—	23.83
1½"	P1@.429	Ea	21.20	10.50	—	31.70
2"	P1@.539	Ea	28.20	13.20	—	41.40
2½"	P1@.726	Ea	68.00	17.70	—	85.70
3"	P1@.913	Ea	128.00	22.30	—	150.30
4"	P1@1.21	Ea	211.00	29.60	—	240.60

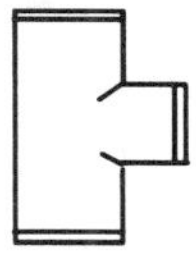

3,000# carbon steel tee, socket-welded joints

Description	Craft@Hrs	Unit	Material $	Labor $	Equipment $	Total $
1/2"	P1@.440	Ea	4.45	10.80	—	15.25
3/4"	P1@.670	Ea	5.60	16.40	—	22.00
1"	P1@.890	Ea	7.66	21.70	—	29.36
1¼"	P1@1.11	Ea	10.40	27.10	—	37.50
1½"	P1@1.33	Ea	16.20	32.50	—	48.70
2"	P1@1.77	Ea	22.70	43.30	—	66.00
2½"	P1@2.22	Ea	54.80	54.30	—	109.10
3"	P1@2.66	Ea	130.00	65.00	—	195.00
4"	P1@3.55	Ea	206.00	86.80	—	292.80

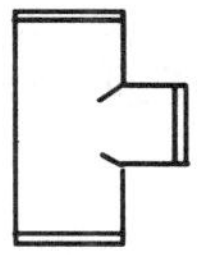

Carbon Steel, Schedule 160 with 3,000-6,000# Fittings

6,000# carbon steel tee, threaded joints

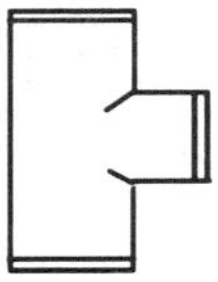

Description	Craft@Hrs	Unit	Material $	Labor $	Equipment $	Total $
1/2"	P1@.198	Ea	8.17	4.84	—	13.01
3/4"	P1@.209	Ea	11.10	5.11	—	16.21
1"	P1@.253	Ea	18.30	6.18	—	24.48
1¼"	P1@.341	Ea	22.90	8.33	—	31.23
1½"	P1@.429	Ea	31.10	10.50	—	41.60
2"	P1@.539	Ea	60.60	13.20	—	73.80
2½"	P1@.726	Ea	219.00	17.70	—	236.70
3"	P1@.913	Ea	220.00	22.30	—	242.30

6,000# carbon steel tee, socket-welded joints

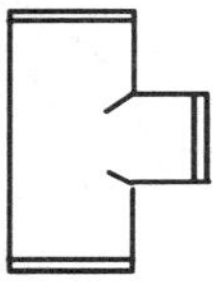

Description	Craft@Hrs	Unit	Material $	Labor $	Equipment $	Total $
1/2"	P1@.440	Ea	7.10	10.80	—	17.90
3/4"	P1@.670	Ea	9.28	16.40	—	25.68
1"	P1@.890	Ea	11.60	21.70	—	33.30
1¼"	P1@1.11	Ea	33.50	27.10	—	60.60
1½"	P1@1.33	Ea	30.00	32.50	—	62.50
2"	P1@1.77	Ea	38.00	43.30	—	81.30
2½"	P1@2.22	Ea	81.20	54.30	—	135.50
3"	P1@2.66	Ea	135.00	65.00	—	200.00
4"	P1@3.55	Ea	255.00	86.80	—	341.80

3,000# carbon steel union, threaded joints

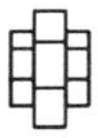

Description	Craft@Hrs	Unit	Material $	Labor $	Equipment $	Total $
1/2"	P1@.150	Ea	5.64	3.67	—	9.31
3/4"	P1@.165	Ea	6.79	4.03	—	10.82
1"	P1@.230	Ea	8.87	5.62	—	14.49
1¼"	P1@.310	Ea	15.00	7.58	—	22.58
1½"	P1@.400	Ea	16.30	9.77	—	26.07
2"	P1@.500	Ea	21.90	12.20	—	34.10
2½"	P1@.670	Ea	47.60	16.40	—	64.00
3"	P1@.840	Ea	59.60	20.50	—	80.10

3,000# carbon steel union, socket-welded joints

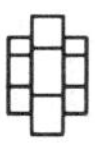

Description	Craft@Hrs	Unit	Material $	Labor $	Equipment $	Total $
1/2"	P1@.300	Ea	6.80	7.33	—	14.13
3/4"	P1@.440	Ea	7.73	10.80	—	18.53
1"	P1@.590	Ea	9.96	14.40	—	24.36
1¼"	P1@.740	Ea	16.00	18.10	—	34.10
1½"	P1@.890	Ea	16.50	21.70	—	38.20
2"	P1@1.18	Ea	23.60	28.80	—	52.40
2½"	P1@1.49	Ea	54.70	36.40	—	91.10
3"	P1@1.77	Ea	69.90	43.30	—	113.20

6,000# carbon steel union, threaded joints

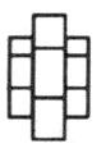

Description	Craft@Hrs	Unit	Material $	Labor $	Equipment $	Total $
1/2"	P1@.150	Ea	13.30	3.67	—	16.97
3/4"	P1@.165	Ea	15.10	4.03	—	19.13
1"	P1@.230	Ea	19.80	5.62	—	25.42
1¼"	P1@.310	Ea	32.30	7.58	—	39.88
1½"	P1@.400	Ea	39.90	9.77	—	49.67
2"	P1@.500	Ea	57.10	12.20	—	69.30

6,000# carbon steel union, socket-welded joints

Description	Craft@Hrs	Unit	Material $	Labor $	Equipment $	Total $
1/2"	P1@.300	Ea	13.70	7.33	—	21.03
3/4"	P1@.440	Ea	15.60	10.80	—	26.40
1"	P1@.590	Ea	23.10	14.40	—	37.50
1¼"	P1@.740	Ea	35.00	18.10	—	53.10
1½"	P1@.890	Ea	41.10	21.70	—	62.80
2"	P1@1.18	Ea	62.30	28.80	—	91.10

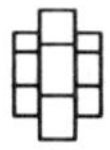

3,000# carbon steel reducer, socket-welded joints

Description	Craft@Hrs	Unit	Material $	Labor $	Equipment $	Total $
3/4"	P1@.360	Ea	3.67	8.80	—	12.47
1"	P1@.500	Ea	4.47	12.20	—	16.67
1¼"	P1@.665	Ea	5.19	16.30	—	21.49
1½"	P1@.800	Ea	6.55	19.50	—	26.05
2"	P1@1.00	Ea	8.88	24.40	—	33.28
2½"	P1@1.33	Ea	61.60	32.50	—	94.10
3"	P1@1.63	Ea	76.90	39.80	—	116.70

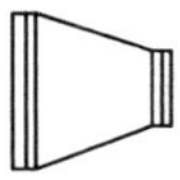

6,000# carbon steel reducer, socket-welded joints

Description	Craft@Hrs	Unit	Material $	Labor $	Equipment $	Total $
3/4"	P1@.360	Ea	14.10	8.80	—	22.90
1"	P1@.500	Ea	14.70	12.20	—	26.90
1¼"	P1@.665	Ea	31.40	16.30	—	47.70
1½"	P1@.800	Ea	31.40	19.50	—	50.90
2"	P1@1.00	Ea	41.70	24.40	—	66.10

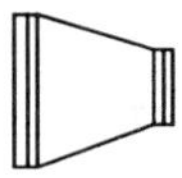

3,000# carbon steel cap, socket-welded joint

Description	Craft@Hrs	Unit	Material $	Labor $	Equipment $	Total $
1/2"	P1@.180	Ea	2.20	4.40	—	6.60
3/4"	P1@.270	Ea	2.56	6.60	—	9.16
1"	P1@.360	Ea	3.93	8.80	—	12.73
1¼"	P1@.460	Ea	4.59	11.20	—	15.79
1½"	P1@.560	Ea	6.26	13.70	—	19.96
2"	P1@.740	Ea	9.75	18.10	—	27.85
2½"	P1@.920	Ea	52.60	22.50	—	75.10
3"	P1@1.11	Ea	54.30	27.10	—	81.40
4"	P1@1.50	Ea	81.60	36.70	—	118.30

6,000# carbon steel cap, socket-welded joint

Description	Craft@Hrs	Unit	Material $	Labor $	Equipment $	Total $
1/2"	P1@.180	Ea	4.23	4.40	—	8.63
3/4"	P1@.270	Ea	6.15	6.60	—	12.75
1"	P1@.360	Ea	6.54	8.80	—	15.34
1¼"	P1@.460	Ea	15.20	11.20	—	26.40
1½"	P1@.560	Ea	19.80	13.70	—	33.50
2"	P1@.740	Ea	29.40	18.10	—	47.50

Carbon Steel, Schedule 160 with 3,000-6,000# Fittings

3,000# carbon steel weldolet

Description	Craft@Hrs	Unit	Material $	Labor $	Equipment $	Total $
1/2"	P1@.580	Ea	7.80	14.20	—	22.00
3/4"	P1@.890	Ea	8.02	21.70	—	29.72
1"	P1@1.18	Ea	8.72	28.80	—	37.52
1¼"	P1@1.48	Ea	11.10	36.20	—	47.30
1½"	P1@1.75	Ea	11.10	42.80	—	53.90
2"	P1@2.35	Ea	11.80	57.40	—	69.20
2½"	P1@2.90	Ea	26.60	70.90	—	97.50
3"	P1@3.50	Ea	27.30	85.50	—	112.80
4"	P1@4.75	Ea	34.10	116.00	—	150.10

6,000# carbon steel weldolet

Description	Craft@Hrs	Unit	Material $	Labor $	Equipment $	Total $
1/2"	P1@.690	Ea	34.10	16.90	—	51.00
3/4"	P1@1.05	Ea	35.50	25.70	—	61.20
1"	P1@1.40	Ea	37.20	34.20	—	71.40
1¼"	P1@1.75	Ea	44.70	42.80	—	87.50
1½"	P1@2.10	Ea	50.40	51.30	—	101.70
2"	P1@2.80	Ea	58.10	68.40	—	126.50
2½"	P1@3.50	Ea	85.20	85.50	—	170.70
3"	P1@4.20	Ea	95.10	103.00	—	198.10
4"	P1@5.70	Ea	161.00	139.00	—	300.00

3,000# carbon steel threadolet

Description	Craft@Hrs	Unit	Material $	Labor $	Equipment $	Total $
1/2"	P1@.400	Ea	6.20	9.77	—	15.97
3/4"	P1@.580	Ea	6.85	14.20	—	21.05
1"	P1@.790	Ea	8.45	19.30	—	27.75
1¼"	P1@.980	Ea	38.00	23.90	—	61.90
1½"	P1@1.20	Ea	38.00	29.30	—	67.30
2"	P1@1.60	Ea	50.00	39.10	—	89.10

6,000# carbon steel threadolet

Description	Craft@Hrs	Unit	Material $	Labor $	Equipment $	Total $
1/2"	P1@.480	Ea	9.50	11.70	—	21.20
3/4"	P1@.690	Ea	10.25	16.90	—	27.15
1"	P1@.950	Ea	14.75	23.20	—	37.95
1¼"	P1@1.20	Ea	55.00	29.30	—	84.30
1½"	P1@1.45	Ea	55.00	35.40	—	90.40
2"	P1@1.90	Ea	90.00	46.40	—	136.40

Class 300 forged steel companion flange, weld neck

Description	Craft@Hrs	Unit	Material $	Labor $	Equipment $	Total $
1/2"	P1@.240	Ea	10.98	5.86	—	16.84
3/4"	P1@.320	Ea	10.98	7.82	—	18.80
1"	P1@.340	Ea	10.98	8.31	—	19.29
1¼"	P1@.425	Ea	14.35	10.40	—	24.75
1½"	P1@.560	Ea	14.35	13.70	—	28.05
2"	P1@.740	Ea	15.60	18.10	—	33.70
2½"	P1@.810	Ea	22.95	19.80	—	42.75
3"	P1@.980	Ea	23.50	23.90	—	47.40
4"	P1@1.30	Ea	34.50	31.80	—	66.30

Class 300 forged steel companion flange, slip on

Description	Craft@Hrs	Unit	Material $	Labor $	Equipment $	Total $
1/2"	P1@.240	Ea	8.80	5.86	—	14.66
3/4"	P1@.320	Ea	8.80	7.82	—	16.62
1"	P1@.340	Ea	8.80	8.31	—	17.11
1½"	P1@.425	Ea	13.00	10.40	—	23.40
2"	P1@.560	Ea	12.50	13.70	—	26.20
2½"	P1@.740	Ea	21.20	18.10	—	39.30
3"	P1@.810	Ea	21.20	19.80	—	41.00
4"	P1@.980	Ea	31.20	23.90	—	55.10

Class 300 cast steel gate valve, flanged

Description	Craft@Hrs	Unit	Material $	Labor $	Equipment $	Total $
2"	P1@.500	Ea	537.00	12.20	—	549.20
2½"	P1@.600	Ea	725.00	14.70	—	739.70
3"	P1@.750	Ea	725.00	18.30	—	743.30
4"	P1@1.35	Ea	1,011.00	33.00	—	1,044.00

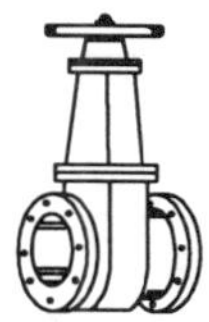

Class 600 cast steel gate valve, flanged

Description	Craft@Hrs	Unit	Material $	Labor $	Equipment $	Total $
2"	P1@.500	Ea	892.00	12.20	—	904.20
2½"	P1@.600	Ea	1,270.00	14.70	—	1,284.70
3"	P1@.750	Ea	1,270.00	18.30	—	1,288.30
4"	P1@1.35	Ea	2,050.00	33.00	—	2,083.00

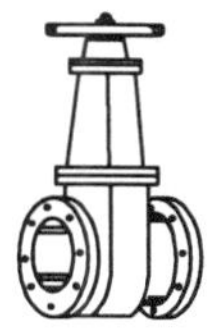

Class 300 cast steel body globe valve, flanged

Description	Craft@Hrs	Unit	Material $	Labor $	Equipment $	Total $
2"	P1@.500	Ea	716.00	12.20	—	728.20
2½"	P1@.600	Ea	966.00	14.70	—	980.70
3"	P1@.750	Ea	966.00	18.30	—	984.30
4"	P1@1.35	Ea	1,340.00	33.00	—	1,373.00

Class 600 cast steel body globe valve, flanged

Description	Craft@Hrs	Unit	Material $	Labor $	Equipment $	Total $
2"	P1@.500	Ea	1,002.00	12.20	—	1,014.20
2½"	P1@.600	Ea	1,510.00	14.70	—	1,524.70
3"	P1@.750	Ea	1,510.00	18.30	—	1,528.30
4"	P1@1.35	Ea	2,430.00	33.00	—	2,463.00

Carbon Steel, Schedule 160 with 3,000-6,000# Fittings

Class 300 cast steel body swing check valve, flanged

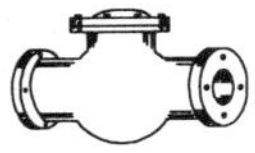

Description	Craft@Hrs	Unit	Material $	Labor $	Equipment $	Total $
2"	P1@.500	Ea	419.00	12.20	—	431.20
2½"	P1@.600	Ea	629.00	14.70	—	643.70
3"	P1@.750	Ea	629.00	18.30	—	647.30
4"	P1@1.35	Ea	894.00	33.00	—	927.00

Class 600 cast steel body swing check valve, flanged

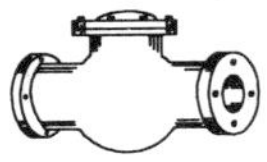

Description	Craft@Hrs	Unit	Material $	Labor $	Equipment $	Total $
2"	P1@.500	Ea	643.00	12.20	—	655.20
2½"	P1@.600	Ea	1,010.00	14.70	—	1,024.70
3"	P1@.750	Ea	1,010.00	18.30	—	1,028.30
4"	P1@1.35	Ea	1,600.00	33.00	—	1,633.00

Installation of cast steel body 2-way control valve, threaded joints

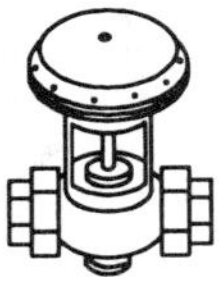

Description	Craft@Hrs	Unit	Material $	Labor $	Equipment $	Total $
1/2"	P1@.210	Ea	—	5.13	—	5.13
3/4"	P1@.250	Ea	—	6.11	—	6.11
1"	P1@.300	Ea	—	7.33	—	7.33
1¼"	P1@.400	Ea	—	9.77	—	9.77
1½"	P1@.450	Ea	—	11.00	—	11.00
2"	P1@.500	Ea	—	12.20	—	12.20
2½"	P1@.830	Ea	—	20.30	—	20.30
3"	P1@.990	Ea	—	24.20	—	24.20

Installation of cast steel body 2-way control valve, flanged joints

Description	Craft@Hrs	Unit	Material $	Labor $	Equipment $	Total $
2"	P1@.500	Ea	—	12.20	—	12.20
2½"	P1@.600	Ea	—	14.70	—	14.70
3"	P1@.750	Ea	—	18.30	—	18.30
4"	P1@1.35	Ea	—	33.00	—	33.00

Installation of cast steel body 3-way control valve, threaded joints

Description	Craft@Hrs	Unit	Material $	Labor $	Equipment $	Total $
1/2"	P1@.260	Ea	—	6.35	—	6.35
3/4"	P1@.365	Ea	—	8.92	—	8.92
1"	P1@.475	Ea	—	11.60	—	11.60
1¼"	P1@.575	Ea	—	14.10	—	14.10
1½"	P1@.680	Ea	—	16.60	—	16.60
2"	P1@.910	Ea	—	22.20	—	22.20
2½"	P1@1.12	Ea	—	27.40	—	27.40
3"	P1@1.33	Ea	—	32.50	—	32.50

Installation of cast steel body 3-way control valve, flanged joints

Description	Craft@Hrs	Unit	Material $	Labor $	Equipment $	Total $
2"	P1@.910	Ea	—	22.20	—	22.20
2½"	P1@1.12	Ea	—	27.40	—	27.40
3"	P1@1.33	Ea	—	32.50	—	32.50
4"	P1@2.00	Ea	—	48.90	—	48.90

Description	Craft@Hrs	Unit	Material $	Labor $	Equipment $	Total $
300# bolt and gasket set, ring face						
2"	P1@.500	Ea	3.00	12.20	—	15.20
2½"	P1@.650	Ea	3.50	15.90	—	19.40
3"	P1@.750	Ea	6.00	18.30	—	24.30
4"	P1@1.00	Ea	10.00	24.40	—	34.40
300# bolt and gasket set, full face						
2"	P1@.500	Ea	6.00	12.20	—	18.20
2½"	P1@.650	Ea	7.00	15.90	—	22.90
3"	P1@.750	Ea	12.00	18.30	—	30.30
4"	P1@1.00	Ea	20.00	24.40	—	44.40
Thermometer with well						
7"	P1@.250	Ea	15.00	6.11	—	21.11
9"	P1@.250	Ea	15.50	6.11	—	21.61
Dial type pressure gauge						
2½"	P1@.200	Ea	15.50	4.89	—	20.39
3½"	P1@.200	Ea	20.50	4.89	—	25.39
Pressure/temperature tap						
	P1@.150	Ea	7.30	3.67	—	10.97
Hanger with swivel assembly						
1/2"	P1@.450	Ea	2.40	11.00	—	13.40
3/4"	P1@.450	Ea	2.65	11.00	—	13.65
1"	P1@.450	Ea	2.75	11.00	—	13.75
1¼"	P1@.450	Ea	3.00	11.00	—	14.00
1½"	P1@.450	Ea	3.90	11.00	—	14.90
2"	P1@.450	Ea	5.00	11.00	—	16.00
2½"	P1@.450	Ea	6.30	11.00	—	17.30
3"	P1@.450	Ea	10.75	11.00	—	21.75
4"	P1@.450	Ea	14.00	11.00	—	25.00
Riser clamp						
1/2"	P1@.100	Ea	.95	2.44	—	3.39
3/4"	P1@.100	Ea	1.39	2.44	—	3.83
1"	P1@.100	Ea	1.41	2.44	—	3.85
1¼"	P1@.105	Ea	1.70	2.57	—	4.27
1½"	P1@.110	Ea	1.79	2.69	—	4.48
2"	P1@.115	Ea	1.89	2.81	—	4.70
2½"	P1@.120	Ea	2.00	2.93	—	4.93
3"	P1@.120	Ea	2.18	2.93	—	5.11
4"	P1@.125	Ea	2.75	3.05	—	5.80

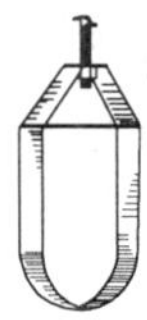

Carbon Steel, Schedule 40 with Roll-Grooved Joints

Schedule 40 carbon steel (ASTM A-53 and A-120) roll-grooved pipe with factory-grooved malleable iron, ductile iron, or steel fittings is commonly used for heating hot water, chilled water and condenser water systems.

Consult the manufacturers for maximum operating temperature/pressure ratings for the various combinations of pipe and fittings.

This section has been arranged to save the estimator's time by including all normally-used system components such as pipe, fittings, valves, hanger assemblies, riser clamps and miscellaneous items under one heading. Additional items can be found under "Plumbing and Piping Specialties." The cost estimates in this section are based on the conditions, limitations and wage rates described in the section "How to Use This Book" beginning on page 5.

Equipment cost, where shown, is $35 per hour for a 10-ton hydraulic truck-mounted crane.

Schedule 40 carbon steel pipe with roll-grooved joints

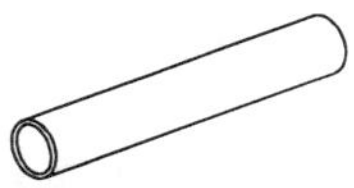

Description	Craft@Hrs	Unit	Material $	Labor $	Equipment $	Total $
2"	P1@.100	LF	1.75	2.44	—	4.19
2½"	P1@.115	LF	2.74	2.81	—	5.55
3"	P1@.135	LF	3.73	3.30	—	7.03
4"	P1@.190	LF	5.40	4.64	—	10.04
5"	ER@.240	LF	9.25	6.26	1.20	16.71
6"	ER@.315	LF	13.95	8.22	1.58	23.75
8"	ER@.370	LF	14.00	9.65	1.85	25.50
10"	ER@.460	LF	19.70	12.00	2.30	34.00
12"	ER@.550	LF	24.10	14.30	2.75	41.15

Schedule 40 carbon steel 45 degree ell with roll-grooved joints

Description	Craft@Hrs	Unit	Material $	Labor $	Equipment $	Total $
2"	P1@.170	Ea	7.85	4.15	—	12.00
2½"	P1@.185	Ea	10.65	4.52	—	15.17
3"	P1@.200	Ea	14.10	4.89	—	18.99
4"	P1@.230	Ea	20.70	5.62	—	26.32
5"	ER@.250	Ea	50.00	6.52	1.25	57.77
6"	ER@.350	Ea	59.00	9.13	1.75	69.88
8"	ER@.500	Ea	122.00	13.00	2.50	137.50
10"	ER@.750	Ea	166.00	19.60	3.75	189.35
12"	ER@1.00	Ea	290.00	26.10	5.00	321.10

Schedule 40 carbon steel 90 degree ell with roll-grooved joints

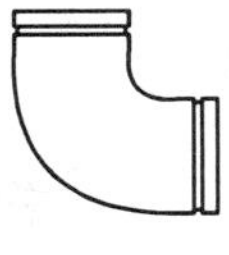

Description	Craft@Hrs	Unit	Material $	Labor $	Equipment $	Total $
2"	P1@.170	Ea	7.85	4.15	—	12.00
2½"	P1@.185	Ea	10.65	4.52	—	15.17
3"	P1@.200	Ea	14.10	4.89	—	18.99
4"	P1@.230	Ea	20.70	5.62	—	26.32
5"	ER@.250	Ea	50.00	6.52	1.25	57.77
6"	ER@.350	Ea	59.00	9.13	1.75	69.88
8"	ER@.500	Ea	122.00	13.00	2.50	137.50
10"	ER@.750	Ea	190.00	19.60	3.75	213.35
12"	ER@1.00	Ea	307.00	26.10	5.00	338.10

Description	Craft@Hrs	Unit	Material $	Labor $	Equipment $	Total $
Schedule 40 carbon steel tee with roll-grooved joints						
2"	P1@.200	Ea	12.00	4.89	—	16.89
2½"	P1@.210	Ea	16.40	5.13	—	21.53
3"	P1@.230	Ea	23.00	5.62	—	28.62
4"	P1@.250	Ea	35.00	6.11	—	41.11
5"	ER@.300	Ea	83.00	7.82	1.50	92.32
6"	ER@.500	Ea	96.00	13.00	2.50	111.50
8"	ER@.750	Ea	210.00	19.60	3.75	233.35
10"	ER@1.00	Ea	375.00	26.10	5.00	406.10
12"	ER@1.50	Ea	520.00	39.10	7.50	566.60
Schedule 40 carbon steel reducing tee with roll-grooved joints						
3" x 2"	P1@.230	Ea	31.00	5.62	—	36.62
4" x 2"	P1@.240	Ea	42.00	5.86	—	47.86
4" x 2½"	P1@.245	Ea	42.00	5.99	—	47.99
4" x 3"	P1@.250	Ea	42.00	6.11	—	48.11
5" x 2"	P1@.260	Ea	77.00	6.35	—	83.35
5" x 3"	P1@.280	Ea	84.00	6.84	—	90.84
5" x 4"	P1@.300	Ea	91.00	7.33	—	98.33
6" x 3"	ER@.400	Ea	100.00	10.40	2.00	112.40
6" x 4"	ER@.450	Ea	100.00	11.70	2.25	113.95
6" x 5"	ER@.500	Ea	100.00	13.00	2.50	115.50
8" x 3"	ER@.600	Ea	145.00	15.60	3.00	163.60
8" x 4"	ER@.650	Ea	210.00	17.00	3.25	230.25
8" x 5"	ER@.700	Ea	210.00	18.30	3.50	231.80
8" x 6"	ER@.750	Ea	210.00	19.60	3.75	233.35
10" x 4"	ER@.800	Ea	225.00	20.90	4.00	249.90
10" x 5"	ER@.900	Ea	230.00	23.50	4.50	258.00
10" x 6"	ER@1.00	Ea	230.00	26.10	5.00	261.10
10" x 8"	ER@1.10	Ea	230.00	28.70	5.50	264.20
12" x 6"	ER@1.20	Ea	335.00	31.30	6.00	372.30
12" x 8"	ER@1.35	Ea	345.00	35.20	6.75	386.95
12" x 10"	ER@1.50	Ea	360.00	39.10	7.50	406.60
Schedule 40 carbon steel male adapter						
2"	P1@.170	Ea	5.10	4.15	—	9.25
2½"	P1@.185	Ea	6.00	4.52	—	10.52
3"	P1@.200	Ea	7.50	4.89	—	12.39
4"	P1@.230	Ea	12.70	5.62	—	18.32
5"	ER@.290	Ea	26.50	7.56	1.45	35.51
6"	ER@.400	Ea	63.00	10.40	2.00	75.40
Schedule 40 carbon steel female adapter						
2"	P1@.170	Ea	11.50	4.15	—	15.65
3"	P1@.200	Ea	18.00	4.89	—	22.89
4"	P1@.230	Ea	25.00	5.62	—	30.62

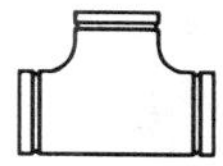

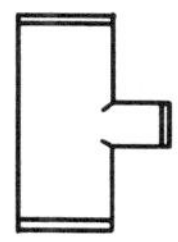

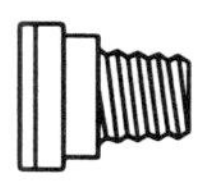

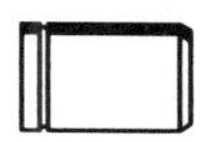

Carbon Steel, Schedule 40 with Roll-Grooved Joints

Schedule 40 carbon steel reducer, roll-grooved joints

Description	Craft@Hrs	Unit	Material $	Labor $	Equipment $	Total $
2"	P1@.160	Ea	10.80	3.91	—	14.71
3"	P1@.170	Ea	15.00	4.15	—	19.15
4"	P1@.210	Ea	18.25	5.13	—	23.38
5"	ER@.260	Ea	25.40	6.78	1.30	33.48
6"	ER@.310	Ea	29.50	8.09	1.55	39.14
8"	ER@.450	Ea	77.00	11.70	2.25	90.95
10"	ER@.690	Ea	135.00	18.00	3.45	156.45
12"	ER@.920	Ea	234.00	24.00	4.60	262.60

Schedule 40 carbon steel cap, roll-grooved joint

Description	Craft@Hrs	Unit	Material $	Labor $	Equipment $	Total $
2"	P1@.110	Ea	5.60	2.69	—	8.29
2½"	P1@.120	Ea	8.75	2.93	—	11.68
3"	P1@.130	Ea	8.75	3.18	—	11.93
4"	P1@.140	Ea	9.50	3.42	—	12.92
5"	ER@.150	Ea	22.00	3.91	.75	26.66
6"	ER@.170	Ea	23.00	4.43	.85	28.28
8"	ER@.200	Ea	44.00	5.22	1.00	50.22
10"	ER@.225	Ea	80.00	5.87	1.13	87.00
12"	ER@.250	Ea	130.00	6.52	1.25	137.77

Flange with gasket, grooved joint

Description	Craft@Hrs	Unit	Material $	Labor $	Equipment $	Total $
2"	P1@.280	Ea	16.75	6.84	—	23.59
2½"	P1@.340	Ea	19.85	8.31	—	28.16
3"	P1@.390	Ea	22.50	9.53	—	32.03
4"	P1@.560	Ea	32.50	13.70	—	46.20
6"	ER@.780	Ea	40.50	20.30	3.90	64.70
8"	ER@1.10	Ea	47.85	28.70	5.50	82.05
10"	ER@1.40	Ea	60.50	36.50	7.00	104.00
12"	ER@1.70	Ea	83.80	44.30	8.50	136.60

Roll-grooved coupling with gasket

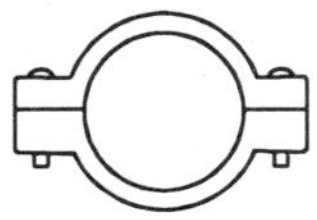

Description	Craft@Hrs	Unit	Material $	Labor $	Equipment $	Total $
2"	P1@.300	Ea	8.70	7.33	—	16.03
2½"	P1@.350	Ea	10.40	8.55	—	18.95
3"	P1@.400	Ea	11.50	9.77	—	21.27
4"	P1@.500	Ea	16.70	12.20	—	28.90
5"	ER@.600	Ea	25.70	15.60	3.00	44.30
6"	ER@.700	Ea	30.50	18.30	3.50	52.30
8"	ER@.900	Ea	50.00	23.50	4.50	78.00
10"	ER@1.10	Ea	70.00	28.70	5.50	104.20
12"	ER@1.35	Ea	80.00	35.20	6.75	121.95

Description	Craft@Hrs	Unit	Material $	Labor $	Equipment $	Total $
Class 125 iron body gate valve, flanged ends						
2"	P1@.500	Ea	101.00	12.20	—	113.20
2½"	P1@.600	Ea	137.00	14.70	—	151.70
3"	P1@.750	Ea	148.00	18.30	—	166.30
4"	P1@1.35	Ea	218.00	33.00	—	251.00
6"	ER@2.50	Ea	417.00	65.20	12.50	494.70
8"	ER@3.00	Ea	680.00	78.20	15.00	773.20
10"	ER@4.00	Ea	1,120.00	104.00	20.00	1,244.00
12"	ER@4.50	Ea	1,540.00	117.00	22.25	1,679.25

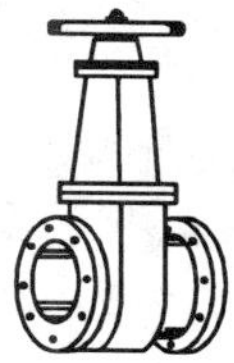

Description	Craft@Hrs	Unit	Material $	Labor $	Equipment $	Total $
Class 125 iron body globe valve, flanged joints						
2"	P1@.500	Ea	142.00	12.20	—	154.20
2½"	P1@.600	Ea	220.00	14.70	—	234.70
3"	P1@.750	Ea	264.00	18.30	—	282.30
4"	P1@1.35	Ea	352.00	33.00	—	385.00
6"	ER@2.50	Ea	619.00	65.20	12.50	696.70
8"	ER@3.00	Ea	952.00	78.20	15.00	1,045.20
10"	ER@4.00	Ea	1,450.00	104.00	20.00	1,574.00

Description	Craft@Hrs	Unit	Material $	Labor $	Equipment $	Total $
200 PSIG iron body butterfly valve, lug type						
2"	P1@.450	Ea	84.50	11.00	—	95.50
2½"	P1@.450	Ea	86.50	11.00	—	97.50
3"	P1@.550	Ea	91.70	13.40	—	105.10
4"	P1@.550	Ea	114.00	13.40	—	127.40
6"	ER@.800	Ea	185.00	20.90	4.00	209.90
8"	ER@.800	Ea	252.00	20.90	4.00	276.90
10"	ER@.900	Ea	350.00	23.50	4.50	378.00
12"	ER@1.00	Ea	460.00	26.10	5.00	491.10

Description	Craft@Hrs	Unit	Material $	Labor $	Equipment $	Total $
200 PSIG iron body butterfly valve, wafer type						
2"	P1@.450	Ea	76.70	11.00	—	87.70
2½"	P1@.450	Ea	78.70	11.00	—	89.70
3"	P1@.550	Ea	84.50	13.40	—	97.90
4"	P1@.550	Ea	102.00	13.40	—	115.40
6"	ER@.800	Ea	170.00	20.90	4.00	194.90
8"	ER@.800	Ea	237.00	20.90	4.00	261.90
10"	ER@.900	Ea	332.00	23.50	4.50	360.00
12"	ER@1.00	Ea	436.00	26.10	5.00	471.10

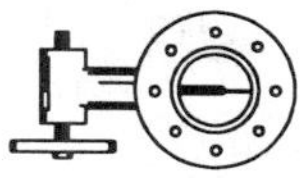

Carbon Steel, Schedule 40 with Roll-Grooved Joints

Class 125 iron body swing check valve, flanged ends

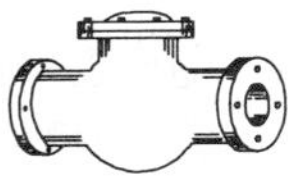

Description	Craft@Hrs	Unit	Material $	Labor $	Equipment $	Total $
2"	P1@.500	Ea	115.00	12.20	—	127.20
2½"	P1@.600	Ea	148.00	14.70	—	162.70
3"	P1@.750	Ea	183.00	18.30	—	201.30
4"	P1@1.35	Ea	271.00	33.00	—	304.00
6"	ER@2.50	Ea	526.00	65.20	12.50	603.70
8"	ER@3.00	Ea	938.00	78.20	15.00	1,031.20
10"	ER@4.00	Ea	1,535.00	104.00	20.00	1,659.00
12"	ER@4.50	Ea	2,100.00	117.00	22.25	2,239.25

Class 125 iron body silent check valve, flanged joints

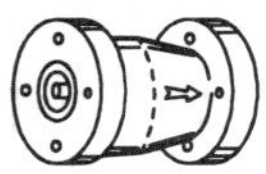

Description	Craft@Hrs	Unit	Material $	Labor $	Equipment $	Total $
2"	P1@.500	Ea	85.00	12.20	—	97.20
2½"	P1@.600	Ea	97.00	14.70	—	111.70
3"	P1@.750	Ea	110.00	18.30	—	128.30
4"	P1@1.35	Ea	145.00	33.00	—	178.00
5"	ER@2.00	Ea	210.00	52.20	10.00	262.20
6"	ER@2.50	Ea	260.00	65.20	12.50	337.70
8"	ER@3.00	Ea	470.00	78.20	15.00	563.20
10"	ER@4.00	Ea	730.00	104.00	20.00	854.00

Class 125 iron body strainer, flanged

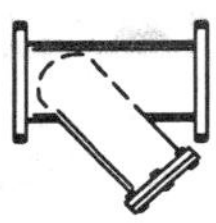

Description	Craft@Hrs	Unit	Material $	Labor $	Equipment $	Total $
2"	P1@.500	Ea	62.00	12.20	—	74.20
2½"	P1@.600	Ea	69.00	14.70	—	83.70
3"	P1@.750	Ea	80.00	18.30	—	98.30
4"	P1@1.35	Ea	137.00	33.00	—	170.00
6"	ER@2.50	Ea	277.00	65.20	12.50	354.70
8"	ER@3.00	Ea	470.00	78.20	15.00	563.20

Installation of 2-way control valve, flanged joints

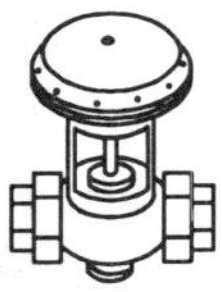

Description	Craft@Hrs	Unit	Material $	Labor $	Equipment $	Total $
2"	P1@.500	Ea	—	12.20	—	12.20
2½"	P1@.600	Ea	—	14.70	—	14.70
3"	P1@.750	Ea	—	18.30	—	18.30
4"	P1@1.35	Ea	—	33.00	—	33.00
6"	P1@2.50	Ea	—	61.10	—	61.10
8"	P1@3.00	Ea	—	73.30	—	73.30
10"	P1@4.00	Ea	—	97.70	—	97.70
12"	P1@4.50	Ea	—	110.00	—	110.00

Installation of 3-way control valve, flanged joints

Description	Craft@Hrs	Unit	Material $	Labor $	Equipment $	Total $
2"	P1@.910	Ea	—	22.20	—	22.20
2½"	P1@1.12	Ea	—	27.40	—	27.40
3"	P1@1.33	Ea	—	32.50	—	32.50
4"	P1@2.00	Ea	—	48.90	—	48.90
6"	P1@3.70	Ea	—	90.40	—	90.40
8"	P1@4.40	Ea	—	108.00	—	108.00
10"	P1@5.90	Ea	—	144.00	—	144.00
12"	P1@6.50	Ea	—	159.00	—	159.00

Description	Craft@Hrs	Unit	Material $	Labor $	Equipment $	Total $
Bolt and gasket sets						
2"	P1@.500	Ea	1.50	12.20	—	13.70
2½"	P1@.650	Ea	1.75	15.90	—	17.65
3"	P1@.750	Ea	3.00	18.30	—	21.30
4"	P1@1.00	Ea	5.00	24.40	—	29.40
6"	P1@1.20	Ea	8.25	29.30	—	37.55
8"	P1@1.25	Ea	9.35	30.50	—	39.85
10"	P1@1.70	Ea	16.00	41.50	—	57.50
12"	P1@2.20	Ea	18.00	53.80	—	71.80
Thermometer with well						
7"	P1@.250	Ea	15.00	6.11	—	21.11
9"	P1@.250	Ea	15.50	6.11	—	21.61
Dial type pressure gauge						
2½"	P1@.200	Ea	15.50	4.89	—	20.39
3½"	P1@.200	Ea	20.50	4.89	—	25.39
Pressure/temperature tap						
	P1@.150	Ea	7.30	3.67	—	10.97
Hanger with swivel assembly						
2"	P1@.450	Ea	2.60	11.00	—	13.60
2½"	P1@.450	Ea	3.35	11.00	—	14.35
3"	P1@.450	Ea	4.15	11.00	—	15.15
4"	P1@.450	Ea	6.40	11.00	—	17.40
5"	P1@.500	Ea	7.15	12.20	—	19.35
6"	P1@.500	Ea	9.40	12.20	—	21.60
8"	P1@.550	Ea	12.25	13.40	—	25.65
10"	P1@.650	Ea	16.95	15.90	—	32.85
12"	P1@.750	Ea	22.50	18.30	—	40.80
Riser clamp						
2"	P1@.115	Ea	1.89	2.81	—	4.70
2½"	P1@.120	Ea	2.00	2.93	—	4.93
3"	P1@.120	Ea	2.18	2.93	—	5.11
4"	P1@.125	Ea	2.75	3.05	—	5.80
5"	P1@.180	Ea	3.99	4.40	—	8.39
6"	P1@.200	Ea	4.79	4.89	—	9.68
8"	P1@.200	Ea	7.77	4.89	—	12.66
10"	P1@.250	Ea	11.60	6.11	—	17.71
12"	P1@.250	Ea	13.75	6.11	—	19.86

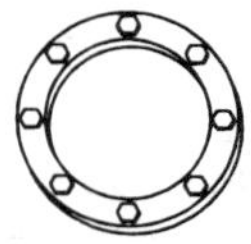

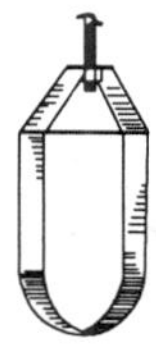

Carbon Steel, Schedule 10 with Roll-Grooved Joints

Schedule 10 carbon steel (A-106) roll-grooved pipe with factory-grooved malleable iron, ductile iron, or steel fittings is widely used for fire protection systems inside of buildings. Because of its relatively low cost, it is also gaining acceptance for use in closed heating hot water and chilled water systems.

Consult the manufacturers for maximum operating temperature/pressure ratings for the various combinations of pipe and fittings.

This section has been arranged to save the estimator's time by including all normally-used system components such as pipe, fittings, hanger assemblies and riser clamps under one heading. Additional items can be found under "Plumbing and Piping Specialties." The cost estimates in this section are based on the conditions, limitations and wage rates described in the section "How to Use This Book" beginning on page 5.

Equipment cost, where shown, is $35 per hour for a 10-ton hydraulic truck-mounted crane.

Schedule 10 carbon steel pipe, roll-grooved joints

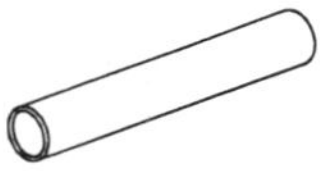

Description	Craft@Hrs	Unit	Material $	Labor $	Equipment $	Total $
2"	P1@.080	LF	1.46	1.95	—	3.41
2½"	P1@.100	LF	2.00	2.44	—	4.44
3"	P1@.115	LF	2.42	2.81	—	5.23
4"	P1@.160	LF	3.20	3.91	—	7.11
5"	ER@.200	LF	5.05	5.22	1.00	11.27
6"	ER@.260	LF	5.35	6.78	1.30	13.43
8"	ER@.300	LF	7.35	7.82	1.50	16.67
10"	ER@.380	LF	10.05	9.91	1.90	21.86
12"	ER@.450	LF	11.00	11.70	2.25	24.95

Schedule 10 carbon steel 45 degree ell, roll-grooved joints

Description	Craft@Hrs	Unit	Material $	Labor $	Equipment $	Total $
2"	P1@.170	Ea	7.85	4.15	—	12.00
2½"	P1@.185	Ea	10.65	4.52	—	15.17
3"	P1@.200	Ea	14.10	4.89	—	18.99
4"	P1@.230	Ea	20.70	5.62	—	26.32
5"	ER@.250	Ea	50.00	6.52	1.25	57.77
6"	ER@.350	Ea	59.00	9.13	1.75	69.88
8"	ER@.500	Ea	122.00	13.00	2.50	137.50
10"	ER@.750	Ea	166.00	19.60	3.75	189.35
12"	ER@1.00	Ea	290.00	26.10	5.00	321.10

Schedule 10 carbon steel 90 degree ell, roll-grooved joints

Description	Craft@Hrs	Unit	Material $	Labor $	Equipment $	Total $
2"	P1@.170	Ea	7.85	4.15	—	12.00
2½"	P1@.185	Ea	10.65	4.52	—	15.17
3"	P1@.200	Ea	14.10	4.89	—	18.99
4"	P1@.230	Ea	20.70	5.62	—	26.32
5"	ER@.250	Ea	50.00	6.52	1.25	57.77
6"	ER@.350	Ea	59.00	9.13	1.75	69.88
8"	ER@.500	Ea	122.00	13.00	2.50	137.50
10"	ER@.750	Ea	190.00	19.60	3.75	213.35
12"	ER@1.00	Ea	307.00	26.10	5.00	338.10

Schedule 10 carbon steel tee, roll-grooved joints

Description	Craft@Hrs	Unit	Material $	Labor $	Equipment $	Total $
2"	P1@.200	Ea	12.00	4.89	—	16.89
2½"	P1@.210	Ea	16.40	5.13	—	21.53
3"	P1@.230	Ea	23.00	5.62	—	28.62
4"	P1@.250	Ea	35.00	6.11	—	41.11
5"	ER@.300	Ea	83.00	7.82	1.50	92.32
6"	ER@.500	Ea	96.00	13.00	2.50	111.50
8"	ER@.750	Ea	210.00	19.60	3.75	233.35
10"	ER@1.00	Ea	375.00	26.10	5.00	406.10
12"	ER@1.50	Ea	520.00	39.10	7.50	566.60

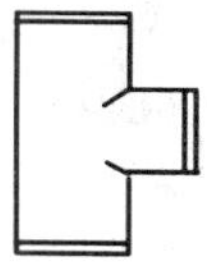

Schedule 10 carbon steel reducing tee, roll-grooved joints

Description	Craft@Hrs	Unit	Material $	Labor $	Equipment $	Total $
3" x 2"	P1@.230	Ea	31.00	5.62	—	36.62
4" x 2"	P1@.240	Ea	42.00	5.86	—	47.86
4" x 2½"	P1@.245	Ea	42.00	5.99	—	47.99
4" x 3"	P1@.250	Ea	42.00	6.11	—	48.11
5" x 2"	P1@.260	Ea	77.00	6.35	—	83.35
5" x 3"	P1@.280	Ea	84.00	6.84	—	90.84
5" x 4"	P1@.300	Ea	91.00	7.33	—	98.33
6" x 3"	ER@.400	Ea	100.00	10.40	2.00	112.40
6" x 4"	ER@.450	Ea	100.00	11.70	2.25	113.95
6" x 5"	ER@.500	Ea	100.00	13.00	2.50	115.50
8" x 3"	ER@.600	Ea	145.00	15.60	3.00	163.60
8" x 4"	ER@.650	Ea	210.00	17.00	3.25	230.25
8" x 5"	ER@.700	Ea	210.00	18.30	3.50	231.80
8" x 6"	ER@.750	Ea	210.00	19.60	3.75	233.35
10" x 4"	ER@.800	Ea	225.00	20.90	4.00	249.90
10" x 5"	ER@.900	Ea	230.00	23.50	4.50	258.00
10" x 6"	ER@1.00	Ea	230.00	26.10	5.00	261.10
10" x 8"	ER@1.10	Ea	230.00	28.70	5.50	264.20
12" x 6"	ER@1.20	Ea	335.00	31.30	6.00	372.30
12" x 8"	ER@1.35	Ea	345.00	35.20	6.25	386.45
12" x 10"	ER@1.50	Ea	360.00	39.10	7.50	406.60

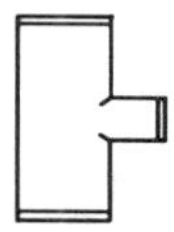

Schedule 10 carbon steel male adapter

Description	Craft@Hrs	Unit	Material $	Labor $	Equipment $	Total $
2"	P1@.170	Ea	5.10	4.15	—	9.25
2½"	P1@.185	Ea	6.00	4.52	—	10.52
3"	P1@.200	Ea	7.50	4.89	—	12.39
4"	P1@.230	Ea	12.70	5.62	—	18.32
5"	ER@.290	Ea	26.50	7.56	1.45	35.51
6"	ER@.400	Ea	63.00	10.40	2.00	75.40

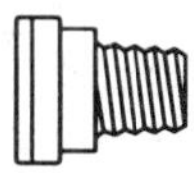

Schedule 10 carbon steel female adapter

Description	Craft@Hrs	Unit	Material $	Labor $	Equipment $	Total $
2"	P1@.170	Ea	11.50	4.15	—	15.65
3"	P1@.200	Ea	18.00	4.89	—	22.89
4"	P1@.230	Ea	25.00	5.62	—	30.62

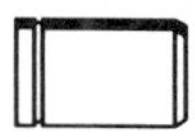

Carbon Steel, Schedule 10 with Roll-Grooved Joints

Schedule 10 carbon steel reducer, roll-grooved joints

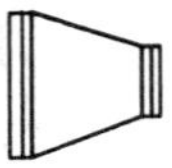

Description	Craft@Hrs	Unit	Material $	Labor $	Equipment $	Total $
2"	P1@.160	Ea	10.80	3.91	—	14.71
3"	P1@.170	Ea	15.00	4.15	—	19.15
4"	P1@.210	Ea	18.25	5.13	—	23.38
5"	ER@.260	Ea	25.40	6.78	1.30	33.48
6"	ER@.310	Ea	29.50	8.09	1.55	39.14
8"	ER@.450	Ea	77.00	11.70	2.25	90.95
10"	ER@.690	Ea	135.00	18.00	3.45	156.45
12"	ER@.920	Ea	234.00	24.00	4.60	262.60

Schedule 10 carbon steel cap, roll-grooved joints

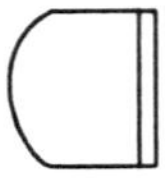

Description	Craft@Hrs	Unit	Material $	Labor $	Equipment $	Total $
2"	P1@.110	Ea	5.60	2.69	—	8.29
2½"	P1@.120	Ea	8.75	2.93	—	11.68
3"	P1@.130	Ea	8.75	3.18	—	11.93
4"	P1@.140	Ea	9.50	3.42	—	12.92
5"	ER@.150	Ea	22.00	3.91	.75	26.66
6"	ER@.170	Ea	23.00	4.43	.85	28.28
8"	ER@.200	Ea	44.00	5.22	1.00	50.22
10"	ER@.225	Ea	80.00	5.87	1.13	87.00
12"	ER@.250	Ea	130.00	6.52	1.25	137.77

Flange with gasket, grooved joint

Description	Craft@Hrs	Unit	Material $	Labor $	Equipment $	Total $
2"	P1@.280	Ea	16.75	6.84	—	23.59
2½"	P1@.340	Ea	19.85	8.31	—	28.16
3"	P1@.390	Ea	22.50	9.53	—	32.03
4"	P1@.560	Ea	32.50	13.70	—	46.20
6"	ER@.780	Ea	40.50	20.30	3.90	64.70
8"	ER@1.10	Ea	47.85	28.70	5.50	82.05
10"	ER@1.40	Ea	60.50	36.50	7.00	104.00
12"	ER@1.70	Ea	83.80	44.30	8.50	136.60

Roll-grooved coupling with gasket

Description	Craft@Hrs	Unit	Material $	Labor $	Equipment $	Total $
2"	P1@.300	Ea	8.70	7.33	—	16.03
2½"	P1@.350	Ea	10.40	8.55	—	18.95
3"	P1@.400	Ea	11.50	9.77	—	21.27
4"	P1@.500	Ea	16.70	12.20	—	28.90
5"	ER@.600	Ea	25.70	15.60	3.00	44.30
6"	ER@.700	Ea	30.50	18.30	3.50	52.30
8"	ER@.900	Ea	50.00	23.50	4.50	78.00
10"	ER@1.10	Ea	70.00	28.70	5.50	104.20
12"	ER@1.35	Ea	80.00	35.20	6.75	121.95

Description	Craft@Hrs	Unit	Material $	Labor $	Equipment $	Total $
Class 125 iron body gate valve, flanged joints						
2"	P1@.500	Ea	101.00	12.20	—	113.20
2½"	P1@.600	Ea	137.00	14.70	—	151.70
3"	P1@.750	Ea	148.00	18.30	—	166.30
4"	P1@1.35	Ea	218.00	33.00	—	251.00
5"	ER@2.00	Ea	417.00	52.20	10.00	479.20
6"	ER@2.50	Ea	417.00	65.20	12.50	494.70
8"	ER@3.00	Ea	680.00	78.20	15.00	773.20
10"	ER@4.00	Ea	1,120.00	104.00	20.00	1,244.00
12"	ER@4.50	Ea	1,540.00	117.00	22.50	1,679.50
Class 125 iron body globe valve, flanged joints						
2"	P1@.500	Ea	142.00	12.20	—	154.20
2½"	P1@.600	Ea	220.00	14.70	—	234.70
3"	P1@.750	Ea	264.00	18.30	—	282.30
4"	P1@1.35	Ea	352.00	33.00	—	385.00
5"	ER@2.00	Ea	619.00	52.20	10.00	681.20
6"	ER@2.50	Ea	619.00	65.20	12.50	696.70
8"	ER@3.00	Ea	952.00	78.20	15.00	1,045.20
10"	ER@4.00	Ea	1,450.00	104.00	20.00	1,574.00
200 PSIG iron body butterfly valve, lug type						
2"	P1@.450	Ea	84.50	11.00	—	95.50
2½"	P1@.450	Ea	86.50	11.00	—	97.50
3"	P1@.550	Ea	91.70	13.40	—	105.10
4"	P1@.550	Ea	114.00	13.40	—	127.40
5"	ER@.800	Ea	151.00	20.90	4.00	175.90
6"	ER@.800	Ea	185.00	20.90	4.00	209.90
8"	ER@.800	Ea	252.00	20.90	4.00	276.90
10"	ER@.900	Ea	350.00	23.50	4.50	378.00
12"	ER@1.00	Ea	460.00	26.10	5.00	491.10
200 PSIG iron body butterfly valve, wafer type						
2"	P1@.450	Ea	76.70	11.00	—	87.70
2½"	P1@.450	Ea	78.70	11.00	—	89.70
3"	P1@.550	Ea	84.50	13.40	—	97.90
4"	P1@.550	Ea	102.00	13.40	—	115.40
5"	ER@.800	Ea	137.00	20.90	4.00	161.90
6"	ER@.800	Ea	170.00	20.90	4.00	194.90
8"	ER@.800	Ea	237.00	20.90	4.00	261.90
10"	ER@.900	Ea	332.00	23.50	4.50	360.00
12"	ER@1.00	Ea	436.00	26.10	5.00	467.10

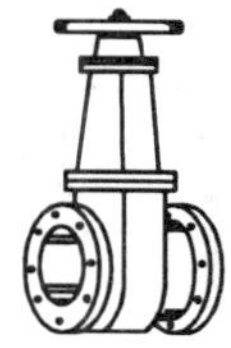

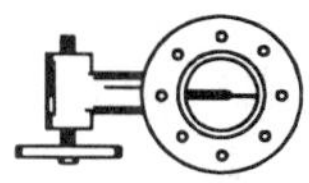

Carbon Steel, Schedule 10 with Roll-Grooved Joints

Description	Craft@Hrs	Unit	Material $	Labor $	Equipment $	Total $

Class 125 iron body swing check valve, flanged joints

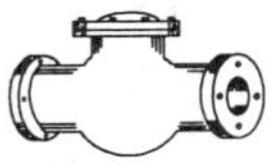

Description	Craft@Hrs	Unit	Material $	Labor $	Equipment $	Total $
2"	P1@.500	Ea	115.00	12.20	—	127.20
2½"	P1@.600	Ea	148.00	14.70	—	162.70
3"	P1@.750	Ea	183.00	18.30	—	201.30
4"	P1@1.35	Ea	271.00	33.00	—	304.00
5"	ER@2.00	Ea	522.00	52.20	10.00	584.20
6"	ER@2.50	Ea	526.00	65.20	12.50	603.70
8"	ER@3.00	Ea	938.00	78.20	15.00	1,031.20
10"	ER@4.00	Ea	1,538.00	104.00	20.00	1,662.00
12"	ER@4.50	Ea	2,100.00	117.00	22.50	2,239.50

Class 125 iron body silent check valve, flanged joints

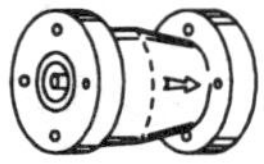

Description	Craft@Hrs	Unit	Material $	Labor $	Equipment $	Total $
2"	P1@.500	Ea	85.00	12.20	—	97.20
2½"	P1@.600	Ea	97.00	14.70	—	111.70
3"	P1@.750	Ea	110.00	18.30	—	128.30
4"	P1@1.35	Ea	145.00	33.00	—	178.00
5"	ER@2.00	Ea	210.00	52.20	10.00	272.20
6"	ER@2.50	Ea	260.00	65.20	12.50	337.70
8"	ER@3.00	Ea	470.00	78.20	15.00	563.20
10"	ER@4.00	Ea	730.00	104.00	20.00	854.00

Class 125 iron body strainer, flanged

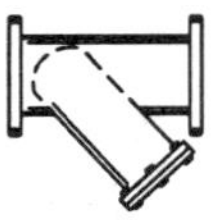

Description	Craft@Hrs	Unit	Material $	Labor $	Equipment $	Total $
2"	P1@.500	Ea	62.00	12.20	—	74.20
2½"	P1@.600	Ea	69.00	14.70	—	83.70
3"	P1@.750	Ea	80.00	18.30	—	98.30
4"	P1@1.35	Ea	137.00	33.00	—	170.00
5"	P1@2.00	Ea	277.00	48.90	10.00	335.90
6"	P1@2.50	Ea	277.00	61.10	12.50	350.60
8"	P1@3.00	Ea	470.00	73.30	15.00	558.30

Installation of 2-way control valve, flanged joints

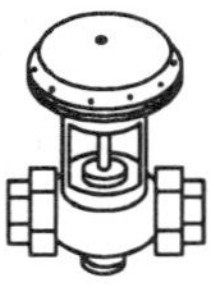

Description	Craft@Hrs	Unit	Material $	Labor $	Equipment $	Total $
2"	P1@.500	Ea	—	12.20	—	12.20
2½"	P1@.600	Ea	—	14.70	—	14.70
3"	P1@.750	Ea	—	18.30	—	18.30
4"	P1@1.35	Ea	—	33.00	—	33.00
6"	P1@2.50	Ea	—	61.10	—	61.10
8"	P1@3.00	Ea	—	73.30	—	73.30
10"	P1@4.00	Ea	—	97.70	—	97.70
12"	P1@4.50	Ea	—	110.00	—	110.00

Installation of 3-way control valve, flanged joints

Description	Craft@Hrs	Unit	Material $	Labor $	Equipment $	Total $
2"	P1@.910	Ea	—	22.20	—	22.20
2½"	P1@1.12	Ea	—	27.40	—	27.40
3"	P1@1.33	Ea	—	32.50	—	32.50
4"	P1@2.00	Ea	—	48.90	—	48.90
6"	P1@3.70	Ea	—	90.40	—	90.40
8"	P1@4.40	Ea	—	108.00	—	108.00
10"	P1@5.90	Ea	—	144.00	—	144.00
12"	P1@6.50	Ea	—	159.00	—	159.00

Bolt and gasket sets

Description	Craft@Hrs	Unit	Material $	Labor $	Equipment $	Total $
2"	P1@.500	Ea	1.50	12.20	—	13.70
2½"	P1@.650	Ea	1.75	15.90	—	17.65
3"	P1@.750	Ea	3.00	18.30	—	21.30
4"	P1@1.00	Ea	5.00	24.40	—	29.40
5"	P1@1.10	Ea	8.25	26.90	—	35.15
6"	P1@1.20	Ea	8.25	29.30	—	37.55
8"	P1@1.25	Ea	9.35	30.50	—	39.85
10"	P1@1.70	Ea	16.00	41.50	—	57.50
12"	P1@2.20	Ea	18.00	53.80	—	71.80

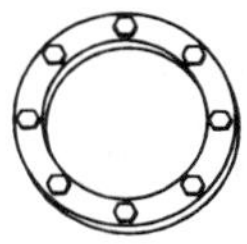

Thermometer with well

Description	Craft@Hrs	Unit	Material $	Labor $	Equipment $	Total $
7"	P1@.250	Ea	15.00	6.11	—	21.11
9"	P1@.250	Ea	15.50	6.11	—	21.61

Dial type pressure gauge

Description	Craft@Hrs	Unit	Material $	Labor $	Equipment $	Total $
2½"	P1@.200	Ea	15.50	4.89	—	20.39
3½"	P1@.200	Ea	20.50	4.89	—	25.39

Pressure/temperature tap

Description	Craft@Hrs	Unit	Material $	Labor $	Equipment $	Total $
	P1@.150	Ea	7.30	3.67	—	10.97

Hanger with swivel assembly

Description	Craft@Hrs	Unit	Material $	Labor $	Equipment $	Total $
2"	P1@.450	Ea	2.60	11.00	—	13.60
2½"	P1@.450	Ea	3.35	11.00	—	14.35
3"	P1@.450	Ea	4.15	11.00	—	15.15
4"	P1@.450	Ea	6.40	11.00	—	17.40
5"	P1@.500	Ea	7.15	12.20	—	19.35
6"	P1@.500	Ea	9.40	12.20	—	21.60
8"	P1@.550	Ea	12.25	13.40	—	25.65
10"	P1@.650	Ea	16.95	15.90	—	32.85
12"	P1@.750	Ea	22.50	18.30	—	40.80

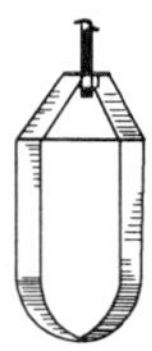

Carbon Steel, Schedule 10 with Roll-Grooved Joints

Description	Craft@Hrs	Unit	Material $	Labor $	Equipment $	Total $
Riser clamp						
2"	P1@.115	Ea	1.89	2.81	—	4.70
2½"	P1@.120	Ea	2.00	2.93	—	4.93
3"	P1@.120	Ea	2.18	2.93	—	5.11
4"	P1@.125	Ea	2.75	3.05	—	5.80
5"	P1@.180	Ea	3.99	4.40	—	8.39
6"	P1@.200	Ea	4.79	4.89	—	9.68
8"	P1@.200	Ea	7.77	4.89	—	12.66
10"	P1@.250	Ea	11.60	6.11	—	17.71
12"	P1@.250	Ea	13.75	6.11	—	19.86

Schedule 40 carbon steel (ASTM A-53 and A-120) cut-grooved pipe with factory-grooved malleable iron, ductile iron or steel fittings is commonly used for heating hot water, chilled water and condenser water systems.

Consult the manufacturers for maximum operating temperature/pressure ratings for various combinations of pipe and fittings.

This section has been arranged to save the estimator's time by including all normally-used system components such as pipe, fittings, valves, hanger assemblies, riser clamps and miscellaneous items under one heading. Additional items can be found under "Plumbing and Piping Specialties." The cost estimates in this section are based on the conditions, limitations and wage rates described in the section "How to Use This Book" beginning on page 5.

Equipment cost, where shown, is $35 per hour for a 10-ton hydraulic truck-mounted crane.

Schedule 40 carbon steel pipe with cut-grooved joints

Description	Craft@Hrs	Unit	Material $	Labor $	Equipment $	Total $
2"	P1@.110	LF	1.75	2.69	—	4.44
2½"	P1@.130	LF	2.74	3.18	—	5.92
3"	P1@.150	LF	3.73	3.67	—	7.40
4"	P1@.210	LF	5.40	5.13	—	10.53
5"	ER@.270	LF	9.25	7.04	1.35	17.64
6"	ER@.350	LF	14.00	9.13	1.75	24.88
8"	ER@.410	LF	14.00	10.70	2.05	26.75
10"	ER@.510	LF	19.70	13.30	2.55	35.55
12"	ER@.610	LF	24.10	15.90	3.05	43.05

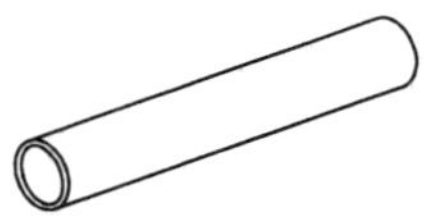

Schedule 40 carbon steel 45 degree ell with cut-grooved joints

Description	Craft@Hrs	Unit	Material $	Labor $	Equipment $	Total $
2"	P1@.170	Ea	7.85	4.15	—	12.00
2½"	P1@.185	Ea	10.70	4.52	—	15.22
3"	P1@.200	Ea	14.10	4.89	—	18.99
4"	P1@.230	Ea	20.70	5.62	—	26.32
5"	ER@.250	Ea	50.00	6.52	1.25	57.77
6"	ER@.350	Ea	59.00	9.13	1.75	69.88
8"	ER@.500	Ea	122.00	13.00	2.50	137.50
10"	ER@.750	Ea	166.00	19.60	3.75	189.35
12"	ER@1.00	Ea	290.00	26.10	5.00	321.10

Schedule 40 carbon steel 90 degree ell with cut-grooved joints

Description	Craft@Hrs	Unit	Material $	Labor $	Equipment $	Total $
2"	P1@.170	Ea	7.85	4.15	—	12.00
2½"	P1@.185	Ea	10.70	4.52	—	15.22
3"	P1@.200	Ea	14.10	4.89	—	18.99
4"	P1@.230	Ea	20.70	5.62	—	26.32
5"	ER@.250	Ea	50.00	6.52	1.25	57.77
6"	ER@.350	Ea	59.00	9.13	1.75	69.88
8"	ER@.500	Ea	122.00	13.00	2.50	137.50
10"	ER@.750	Ea	190.00	19.60	3.75	213.35
12"	ER@1.00	Ea	520.00	26.10	5.00	551.10

Carbon Steel, Schedule 40 with Cut-Grooved Joints

Schedule 40 carbon steel tee with cut-grooved joints

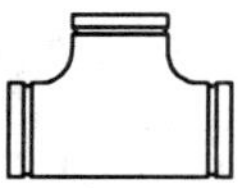

Description	Craft@Hrs	Unit	Material $	Labor $	Equipment $	Total $
2"	P1@.200	Ea	12.00	4.89	—	16.89
2½"	P1@.210	Ea	16.40	5.13	—	21.53
3"	P1@.230	Ea	23.00	5.62	—	28.62
4"	P1@.250	Ea	35.00	6.11	—	41.11
5"	ER@.300	Ea	83.00	7.82	1.50	92.32
6"	ER@.500	Ea	96.00	13.00	2.50	111.50
8"	ER@.750	Ea	210.00	19.60	3.75	233.35
10"	ER@1.00	Ea	375.00	26.10	5.00	406.10
12"	ER@1.50	Ea	520.00	39.10	7.50	566.60

Schedule 40 carbon steel reducing tee with cut-grooved joints

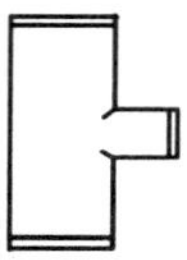

Description	Craft@Hrs	Unit	Material $	Labor $	Equipment $	Total $
3" x 2"	P1@.230	Ea	31.00	5.62	—	36.62
4" x 2"	P1@.240	Ea	42.00	5.86	—	47.86
4" x 2½"	P1@.245	Ea	42.00	5.99	—	47.99
4" x 3"	P1@.250	Ea	42.00	6.11	—	48.11
5" x 2"	P1@.260	Ea	77.00	6.35	—	83.35
5" x 3"	P1@.280	Ea	84.00	6.84	—	90.84
5" x 4"	P1@.300	Ea	91.00	7.33	—	98.33
6" x 3"	ER@.400	Ea	100.00	10.40	2.00	112.40
6" x 4"	ER@.450	Ea	100.00	11.70	2.25	113.95
6" x 5"	ER@.500	Ea	100.00	13.00	2.50	115.50
8" x 3"	ER@.600	Ea	145.00	15.60	3.00	163.60
8" x 4"	ER@.650	Ea	210.00	17.00	3.25	230.25
8" x 5"	ER@.700	Ea	210.00	18.30	3.50	231.80
8" x 6"	ER@.750	Ea	210.00	19.60	3.75	233.35
10" x 4"	ER@.800	Ea	225.00	20.90	4.00	249.90
10" x 5"	ER@.900	Ea	230.00	23.50	4.50	258.00
10" x 6"	ER@1.00	Ea	230.00	26.10	5.00	261.10
10" x 8"	ER@1.10	Ea	230.00	28.70	5.50	264.20
12" x 6"	ER@1.20	Ea	335.00	31.30	6.00	372.30
12" x 8"	ER@1.35	Ea	345.00	35.20	6.75	386.95
12" x 10"	ER@1.50	Ea	360.00	39.10	7.50	406.60

Schedule 40 carbon steel male adapter

Description	Craft@Hrs	Unit	Material $	Labor $	Equipment $	Total $
2"	P1@.170	Ea	5.10	4.15	—	9.25
2½"	P1@.185	Ea	6.00	4.52	—	10.52
3"	P1@.200	Ea	7.50	4.89	—	12.39
4"	P1@.230	Ea	12.65	5.62	—	18.27
5"	ER@.290	Ea	26.50	7.56	1.45	35.51
6"	ER@.400	Ea	63.00	10.40	2.00	75.40

Schedule 40 carbon steel female adapter

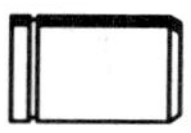

Description	Craft@Hrs	Unit	Material $	Labor $	Equipment $	Total $
2"	P1@.170	Ea	11.50	4.15	—	15.65
3"	P1@.200	Ea	18.00	4.89	—	22.89
4"	P1@.230	Ea	25.00	5.62	—	30.62

Description	Craft@Hrs	Unit	Material $	Labor $	Equipment $	Total $
Schedule 40 carbon steel reducer						
2"	P1@.160	Ea	10.80	3.91	—	14.71
3"	P1@.170	Ea	15.00	4.15	—	19.15
4"	P1@.210	Ea	18.25	5.13	—	23.38
5"	ER@.260	Ea	25.40	6.78	1.30	33.48
6"	ER@.310	Ea	29.50	8.09	1.55	39.14
8"	ER@.450	Ea	77.00	11.70	2.25	90.95
10"	ER@.690	Ea	135.00	18.00	3.45	156.45
12"	ER@.920	Ea	234.00	24.00	4.60	262.60
Schedule 40 carbon steel cap						
2"	P1@.110	Ea	5.60	2.69	—	8.29
2½"	P1@.120	Ea	8.75	2.93	—	11.68
3"	P1@.130	Ea	8.75	3.18	—	11.93
4"	P1@.140	Ea	9.50	3.42	—	12.92
5"	ER@.150	Ea	22.00	3.91	.75	26.66
6"	ER@.170	Ea	23.00	4.43	.85	28.28
8"	ER@.200	Ea	44.00	5.22	1.00	50.22
10"	ER@.225	Ea	80.00	5.87	1.13	87.00
12"	ER@.250	Ea	130.00	6.52	1.25	137.77
Flange with gasket, grooved joint						
2"	P1@.280	Ea	16.75	6.84	—	23.59
2½"	P1@.340	Ea	19.85	8.31	—	28.16
3"	P1@.390	Ea	22.50	9.53	—	32.03
4"	P1@.560	Ea	32.50	13.70	—	46.20
6"	ER@.780	Ea	40.50	20.30	3.90	64.70
8"	ER@1.10	Ea	47.85	28.70	5.50	82.05
10"	ER@1.40	Ea	60.50	36.50	7.00	104.00
12"	ER@1.70	Ea	83.80	44.30	8.50	136.60
Cut grooved joint coupling with gasket						
2"	P1@.300	Ea	8.70	7.33	—	16.03
2½"	P1@.350	Ea	10.40	8.55	—	18.95
3"	P1@.400	Ea	11.50	9.77	—	21.27
4"	P1@.500	Ea	16.70	12.20	—	28.90
5"	ER@.600	Ea	25.70	15.60	3.00	44.30
6"	ER@.700	Ea	30.50	18.30	3.50	52.30
8"	ER@.900	Ea	50.00	23.50	4.50	78.00
10"	ER@1.10	Ea	70.00	28.70	5.50	104.20
12"	ER@1.35	Ea	80.00	35.20	6.75	121.95

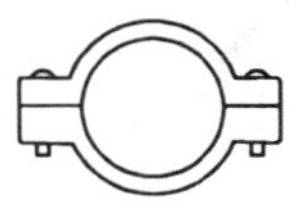

Description	Craft@Hrs	Unit	Material $	Labor $	Equipment $	Total $

Class 125 iron body gate valve, flanged ends

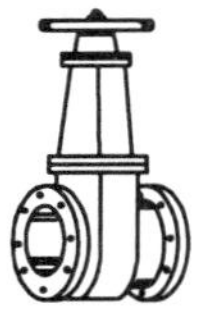

Description	Craft@Hrs	Unit	Material $	Labor $	Equipment $	Total $
2"	P1@.500	Ea	101.00	12.20	—	113.20
2½"	P1@.600	Ea	137.00	14.70	—	151.70
3"	P1@.750	Ea	148.00	18.30	—	166.30
4"	P1@1.35	Ea	218.00	33.00	—	251.00
6"	ER@2.50	Ea	417.00	65.20	12.50	494.70
8"	ER@3.00	Ea	680.00	78.20	15.00	773.20
10"	ER@4.00	Ea	1,120.00	104.00	20.00	1,244.00
12"	ER@4.50	Ea	1,540.00	117.00	22.50	1,679.50

Class 125 iron body globe valve, flanged joints

Description	Craft@Hrs	Unit	Material $	Labor $	Equipment $	Total $
2"	P1@.500	Ea	142.00	12.20	—	154.20
2½"	P1@.600	Ea	220.00	14.70	—	234.70
3"	P1@.750	Ea	264.00	18.30	—	282.30
4"	P1@1.35	Ea	352.00	33.00	—	385.00
6"	ER@2.50	Ea	619.00	65.20	12.50	696.70
8"	ER@3.00	Ea	952.00	78.20	15.00	1,045.20
10"	ER@4.00	Ea	1,450.00	104.00	20.00	1,574.00

200 PSIG iron body butterfly valve, lug type

Description	Craft@Hrs	Unit	Material $	Labor $	Equipment $	Total $
2"	P1@.450	Ea	84.50	11.00	—	95.50
2½"	P1@.450	Ea	86.50	11.00	—	97.50
3"	P1@.550	Ea	91.70	13.40	—	105.10
4"	P1@.550	Ea	114.00	13.40	—	127.40
6"	ER@.800	Ea	185.00	20.90	4.00	209.90
8"	ER@.800	Ea	252.00	20.90	4.00	276.90
10"	ER@.900	Ea	350.00	23.50	4.50	378.00
12"	ER@1.00	Ea	460.00	26.10	5.00	491.10

200 PSIG iron body butterfly valve, wafer type

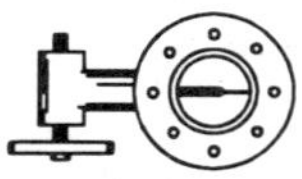

Description	Craft@Hrs	Unit	Material $	Labor $	Equipment $	Total $
2"	P1@.450	Ea	76.70	11.00	—	87.70
2½"	P1@.450	Ea	78.70	11.00	—	89.70
3"	P1@.550	Ea	84.50	13.40	—	97.90
4"	P1@.550	Ea	102.00	13.40	—	115.40
6"	ER@.800	Ea	170.00	20.90	4.00	194.90
8"	ER@.800	Ea	237.00	20.90	4.00	261.90
10"	ER@.900	Ea	332.00	23.50	4.50	360.00
12"	ER@1.00	Ea	436.00	26.10	5.00	467.10

Description	Craft@Hrs	Unit	Material $	Labor $	Equipment $	Total $

Class 125 iron body swing check valve, flanged ends

Description	Craft@Hrs	Unit	Material $	Labor $	Equipment $	Total $
2"	P1@.500	Ea	115.00	12.20	—	127.20
2½"	P1@.600	Ea	148.00	14.70	—	162.70
3"	P1@.750	Ea	183.00	18.30	—	201.30
4"	P1@1.35	Ea	271.00	33.00	—	304.00
6"	ER@2.50	Ea	526.00	65.20	12.50	603.70
8"	ER@3.00	Ea	938.00	78.20	15.00	1,031.20
10"	ER@4.00	Ea	1,535.00	104.00	20.00	1,659.00
12"	ER@4.50	Ea	2,100.00	117.00	22.50	2,239.50

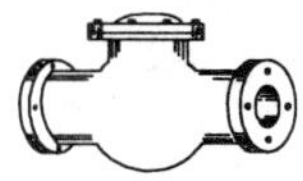

Class 125 iron body silent check valve, flanged joints

Description	Craft@Hrs	Unit	Material $	Labor $	Equipment $	Total $
2"	P1@.500	Ea	85.00	12.20	—	97.20
2½"	P1@.600	Ea	97.00	14.70	—	111.70
3"	P1@.750	Ea	110.00	18.30	—	128.30
4"	P1@1.35	Ea	145.00	33.00	—	178.00
5"	ER@2.00	Ea	210.00	52.20	10.00	272.20
6"	ER@2.50	Ea	260.00	65.20	12.50	337.70
8"	ER@3.00	Ea	470.00	78.20	15.00	563.20
10"	ER@4.00	Ea	730.00	104.00	20.00	854.00

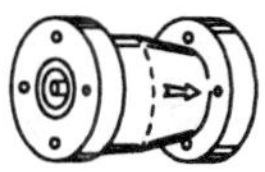

Class 125 iron body strainer, flanged

Description	Craft@Hrs	Unit	Material $	Labor $	Equipment $	Total $
2"	P1@.500	Ea	62.00	12.20	—	74.20
2½"	P1@.600	Ea	69.00	14.70	—	83.70
3"	P1@.750	Ea	80.00	18.30	—	98.30
4"	P1@1.35	Ea	137.00	33.00	—	170.00
6"	ER@2.50	Ea	277.00	65.20	12.50	354.70
8"	ER@3.00	Ea	470.00	78.20	15.00	563.20

Installation of 2-way control valve, flanged joints

Description	Craft@Hrs	Unit	Material $	Labor $	Equipment $	Total $
2"	P1@.500	Ea	—	12.20	—	12.20
2½"	P1@.600	Ea	—	14.70	—	14.70
3"	P1@.750	Ea	—	18.30	—	18.30
4"	P1@1.35	Ea	—	33.00	—	33.00
6"	P1@2.50	Ea	—	61.10	—	61.10
8"	P1@3.00	Ea	—	73.30	—	73.30
10"	P1@4.00	Ea	—	97.70	—	97.70
12"	P1@4.50	Ea	—	110.00	—	110.00

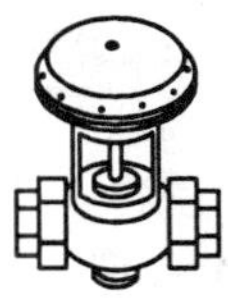

Carbon Steel, Schedule 40 with Cut-Grooved Joints

Installation of 3-way control valve, flanged joints

Description	Craft@Hrs	Unit	Material $	Labor $	Equipment $	Total $
2"	P1@.910	Ea	—	22.20	—	22.20
2½"	P1@1.12	Ea	—	27.40	—	27.40
3"	P1@1.33	Ea	—	32.50	—	32.50
4"	P1@2.00	Ea	—	48.90	—	48.90
6"	P1@3.70	Ea	—	90.40	—	90.40
8"	P1@4.40	Ea	—	108.00	—	108.00
10"	P1@5.90	Ea	—	144.00	—	144.00
12"	P1@6.50	Ea	—	159.00	—	159.00

Bolt and gasket sets

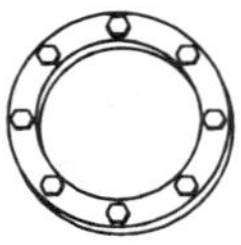

Description	Craft@Hrs	Unit	Material $	Labor $	Equipment $	Total $
2"	P1@.500	Ea	1.50	12.20	—	13.70
2½"	P1@.650	Ea	1.75	15.90	—	17.65
3"	P1@.750	Ea	3.00	18.30	—	21.30
4"	P1@1.00	Ea	5.00	24.40	—	29.40
6"	P1@1.20	Ea	8.25	29.30	—	37.55
8"	P1@1.25	Ea	9.35	30.50	—	39.85
10"	P1@1.70	Ea	16.00	41.50	—	57.50
12"	P1@2.20	Ea	18.00	53.80	—	71.80

Thermometer with well

Description	Craft@Hrs	Unit	Material $	Labor $	Equipment $	Total $
7"	P1@.250	Ea	15.00	6.11	—	21.11
9"	P1@.250	Ea	15.50	6.11	—	21.61

Dial type pressure gauge

Description	Craft@Hrs	Unit	Material $	Labor $	Equipment $	Total $
2½"	P1@.200	Ea	15.50	4.89	—	20.39
3½"	P1@.200	Ea	20.50	4.89	—	25.39

Pressure/temperature tap

Description	Craft@Hrs	Unit	Material $	Labor $	Equipment $	Total $
	P1@.150	Ea	7.30	3.67	—	10.97

Hanger with swivel assembly

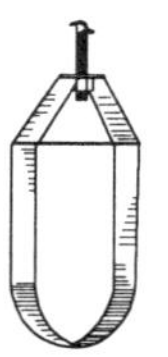

Description	Craft@Hrs	Unit	Material $	Labor $	Equipment $	Total $
2"	P1@.450	Ea	2.60	11.00	—	13.60
2½"	P1@.450	Ea	3.35	11.00	—	14.35
3"	P1@.450	Ea	4.15	11.00	—	15.15
4"	P1@.450	Ea	6.40	11.00	—	17.40
5"	P1@.500	Ea	7.15	12.20	—	19.35
6"	P1@.500	Ea	9.40	12.20	—	21.60
8"	P1@.550	Ea	12.25	13.40	—	25.65
10"	P1@.650	Ea	16.95	15.90	—	32.85
12"	P1@.750	Ea	22.50	18.30	—	40.80

Description	Craft@Hrs	Unit	Material $	Labor $	Equipment $	Total $

Riser clamp

Description	Craft@Hrs	Unit	Material $	Labor $	Equipment $	Total $
2"	P1@.115	Ea	1.89	2.81	—	4.70
2½"	P1@.120	Ea	2.00	2.93	—	4.93
3"	P1@.120	Ea	2.18	2.93	—	5.11
4"	P1@.125	Ea	2.75	3.05	—	5.80
5"	P1@.180	Ea	3.99	4.40	—	8.39
6"	P1@.200	Ea	4.79	4.89	—	9.68
8"	P1@.200	Ea	7.77	4.89	—	12.66
10"	P1@.250	Ea	11.60	6.11	—	17.71
12"	P1@.250	Ea	13.75	6.11	—	19.86

Galvanized steel pipe sleeve

Description	Craft@Hrs	Unit	Material $	Labor $	Equipment $	Total $
2"	P1@.130	Ea	2.55	3.18	—	5.73
2½"	P1@.150	Ea	2.59	3.67	—	6.26
3"	P1@.180	Ea	2.66	4.40	—	7.06
4"	P1@.220	Ea	3.02	5.38	—	8.40
5"	P1@.250	Ea	3.75	6.11	—	9.86
6"	P1@.270	Ea	4.05	6.60	—	10.65
8"	P1@.270	Ea	4.66	6.60	—	11.26
10"	P1@.290	Ea	5.43	7.09	—	12.52
12"	P1@.310	Ea	7.36	7.58	—	14.94
14"	P1@.330	Ea	7.36	8.06	—	15.42

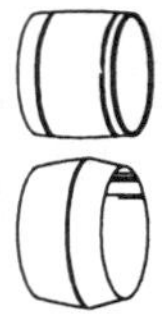

Description	Craft@Hrs	Unit	Material $	Labor $	Equipment $	Total $

Packaged self-contained roof-top DX air conditioning unit, gas heating. Set in place only. Make additional allowances for gas and electrical connections. Cooling loads based on 400 cfm per ton cooling. Cost includes basic single zone controls.

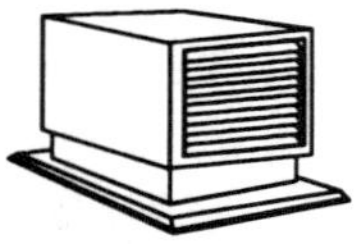

Description	Craft@Hrs	Unit	Material $	Labor $	Equipment $	Total $
2-T, 800 CFM	SN@5.00	Ea	2,500.00	129.00	25.00	2,654.00
2½-T, 1,000 CFM	SN@6.00	Ea	3,000.00	155.00	30.00	3,185.00
3-T, 1,200 CFM	SN@7.00	Ea	3,600.00	180.00	35.00	3,815.00
4-T, 1,600 CFM	SN@8.00	Ea	4,400.00	206.00	40.00	4,646.00
5-T, 2,000 CFM	SN@9.00	Ea	5,500.00	232.00	45.00	5,777.00
7½-T, 3,000 CFM	SN@10.0	Ea	7,500.00	258.00	50.00	7,808.00
10-T, 4,000 CFM	SN@12.0	Ea	9,500.00	309.00	60.00	9,869.00
12-T, 5,000 CFM	SN@14.0	Ea	11,340.00	361.00	70.00	11,771.00
15-T, 6,000 CFM	SN@15.0	Ea	14,000.00	386.00	75.00	14,461.00
20-T, 8,000 CFM	SN@16.0	Ea	18,500.00	412.00	80.00	18,992.00
25-T, 10,000 CFM	SN@17.0	Ea	22,500.00	438.00	85.00	23,023.00
30-T, 12,000 CFM	SN@19.0	Ea	25,500.00	489.00	95.00	26,084.00
40-T, 16,000 CFM	SN@22.0	Ea	32,000.00	567.00	110.00	32,677.00

Roof-top DX air conditioning unit, hot water coil. Set in place only. Make additional allowances for hot water coil and electrical connections. Cooling loads based on 400 cfm per ton cooling. Cost includes basic single zone controls.

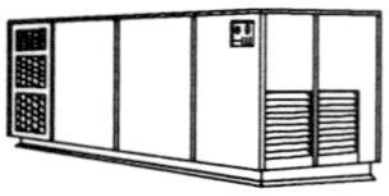

Description	Craft@Hrs	Unit	Material $	Labor $	Equipment $	Total $
2-T, 800 CFM	SN@5.00	Ea	2,450.00	129.00	25.00	2,604.00
2½-T, 1,000 CFM	SN@6.00	Ea	2,950.00	155.00	30.00	3,135.00
3-T, 1,200 CFM	SN@7.00	Ea	3,525.00	180.00	35.00	3,740.00
4-T, 1,600 CFM	SN@8.00	Ea	4,300.00	206.00	40.00	4,546.00
5-T, 2,000 CFM	SN@9.00	Ea	5,375.00	232.00	45.00	5,652.00
7½-T, 3,000 CFM	SN@10.0	Ea	8,050.00	258.00	50.00	8,358.00
10-T, 4,000 CFM	SN@12.0	Ea	9,250.00	309.00	60.00	9,619.00
12-T, 5,000 CFM	SN@14.0	Ea	11,050.00	361.00	70.00	11,481.00
15-T, 6,000 CFM	SN@15.0	Ea	13,650.00	386.00	75.00	14,111.00
20-T, 8,000 CFM	SN@16.0	Ea	18,000.00	412.00	80.00	18,492.00
25-T, 10,000 CFM	SN@17.0	Ea	21,875.00	438.00	85.00	22,398.00
30-T, 12,000 CFM	SN@19.0	Ea	25,500.00	489.00	95.00	26,084.00
40-T, 16,000 CFM	SN@22.0	Ea	32,000.00	567.00	110.00	32,677.00

Description	Craft@Hrs	Unit	Material $	Labor $	Equipment $	Total $

Packaged air handler with chilled water and hot water/steam coil. Costs shown based on 400 cfm per ton cooling. Unit includes insulated single wall casing, fan section, cooling coil section, heating coil section, mixing plenum section, bag filter section, fan motor, variable pitch drive, vibration isolators and drain pan. Set in place only. Make additional allowances for coil connections, controls, motor starters and power wiring. (12000 btus equals 1 ton cooling) Use these costs for preliminary estimates

Description	Craft@Hrs	Unit	Material $	Labor $	Equipment $	Total $
3-T, 1,200 CFM	SN@4.00	Ea	2,070.00	103.00	20.00	2,193.00
4-T, 1,600 CFM	SN@5.50	Ea	2,730.00	142.00	27.50	2,899.50
5-T, 2,000 CFM	SN@7.00	Ea	3,250.00	180.00	35.00	3,465.00
7½-T, 3,000 CFM	SN@9.00	Ea	4,580.00	232.00	45.00	4,857.00
10-T, 4,000 CFM	SN@11.0	Ea	5,810.00	283.00	55.00	6,148.00
12-T, 5,000 CFM	SN@13.0	Ea	6,860.00	335.00	65.00	7,260.00
15-T, 6,000 CFM	SN@14.0	Ea	7,660.00	361.00	70.00	8,091.00
20-T, 8,000 CFM	SN@15.0	Ea	9,150.00	386.00	75.00	9,611.00
25-T, 10,000 CFM	SN@16.0	Ea	9,010.00	412.00	80.00	9,502.00
30-T, 12,000 CFM	SN@18.0	Ea	11,500.00	464.00	90.00	12,054.00
40-T, 16,000 CFM	SN@21.0	Ea	14,300.00	541.00	105.00	14,946.00

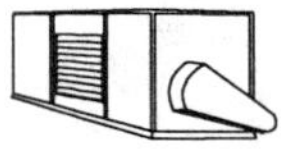

Air handling unit accessories and options. Use these costs for preliminary estimates.

Description	Craft@Hrs	Unit	Material $	Labor $	Equipment $	Total $
DDC controls per zone complete	SN@2.75	Ea	750.00	70.80	105.00	925.80
Electric controls per zone complete	SN@2.75	Ea	400.00	70.80	105.00	575.80
Pneumatic controls per zone complete	SN@2.75	Ea	375.00	70.80	105.00	550.80
Variable speed drive 5 HP	SN@4.00	Ea	3,000.00	103.00	105.00	3,208.00
Variable speed drive 7.5HP	SN@4.00	Ea	3,500.00	103.00	105.00	3,708.00
Variable speed drive 10 HP	SN@6.00	Ea	4,000.00	155.00	105.00	4,260.00
Variable speed drive 15 HP	SN@8.00	Ea	4,500.00	206.00	105.00	4,811.00
Variable speed drive 20 HP	SN@12.0	Ea	5,000.00	309.00	105.00	5,414.00
Variable speed drive 25 HP	SN@16.0	Ea	5,500.00	412.00	105.00	6,017.00
Variable speed drive 30 HP	SN@18.0	Ea	6,500.00	464.00	105.00	7,069.00
Variable speed drive 40 HP	SN@22.0	Ea	8,000.00	567.00	105.00	8,672.00
Variable speed drive 50 HP	SN@26.0	Ea	9,500.00	670.00	105.00	10,275.00
Variable Inlet vanes 1000-1500 CFM	SN@4.00	Ea	150.00	103.00	105.00	358.00
Variable Inlet vanes 1600-2500 CFM	SN@4.00	Ea	325.00	103.00	105.00	533.00
Variable Inlet vanes 2600-5000 CFM	SN@6.00	Ea	900.00	155.00	105.00	1,160.00
Variable Inlet vanes 6000-10,000 CFM	SN@8.00	Ea	1,500.00	206.00	105.00	1,811.00
Variable Inlet vanes 11,000-20,000 CFM	SN@10.0	Ea	2,500.00	258.00	105.00	2,863.00

Engineering drawings of coil piping details have a bad reputation among HVAC contractors and estimators. They're notorious for what they leave out. They'll rarely show more than one coil bank, no matter how big the system is. Furthermore, the drawings hardly ever call out sizes for either the piping or the control valves. Don't be taken in by the apparent simplicity of the system as shown in these drawings. It's likely to be only the tip of the iceberg. For example, unless the air handling capacity of the system is less than 16,000 CFM the single coil bank shown won't be adequate. Add one or two more coil banks and you're looking at a lot more piping — and a more complex system that takes longer to install.

You probably haven't even decided which equipment supplier to use. This is hardly the time for you to start researching heating and cooling coils. Nevertheless, you need better, more complete and realistic data to come up with a competitive estimate.

We'll see how and where to track down the hard data that you have to take the time to find. I'll also pass along a few tips on estimating water coil piping. They'll help you avoid leaving something out of your estimates. That's a real pitfall for any beginner. It's those little (but essential) items, so easily overlooked, that are so deadly to a profit margin. Finally, at the end of this section there are three pages of diagrams with tables. The data given there, combined with the data you collected earlier, forms the basis for informed guesswork.

The Hard Data

There are two things you absolutely must know to estimate water coil piping. First, the size of the branch piping to the coils. Second, the CFM rating of the air handling units for the system.

To find the pipe sizes, look at either the floor plans or the details for the equipment room. They'll list the sizes of the branch run-out pipes. Once you know them, you can make a good guess at the right size for the control valves. (See the diagrams and tables on pages 244, 245 and 246.) The only information you need is the CFM ratings of the air handling units and the branch piping sizes to the coils. If the system's capacity is over 16,000 CFM, you need two or three coil banks.

A Few Tips on Estimating Water Coil Piping

1) Coil connection sizes seldom match branch pipe sizes. Be sure to include the reducing fittings you'll need in any estimate.

2) Two-way and three-way control valves are both usually one pipe size smaller than the pipe where they're installed. That means you'll need either two or three reducing fittings per control valve. Be sure you include their cost in your estimates.

Using the Diagrams

In the following diagrams, for clarity some items are not included. These items are: balance valves, shut-off valves, reducers, strainers, gauges and gauge taps. Any details you need about these items for your estimate are in the engineer's coil piping details.

Water Coil Piping

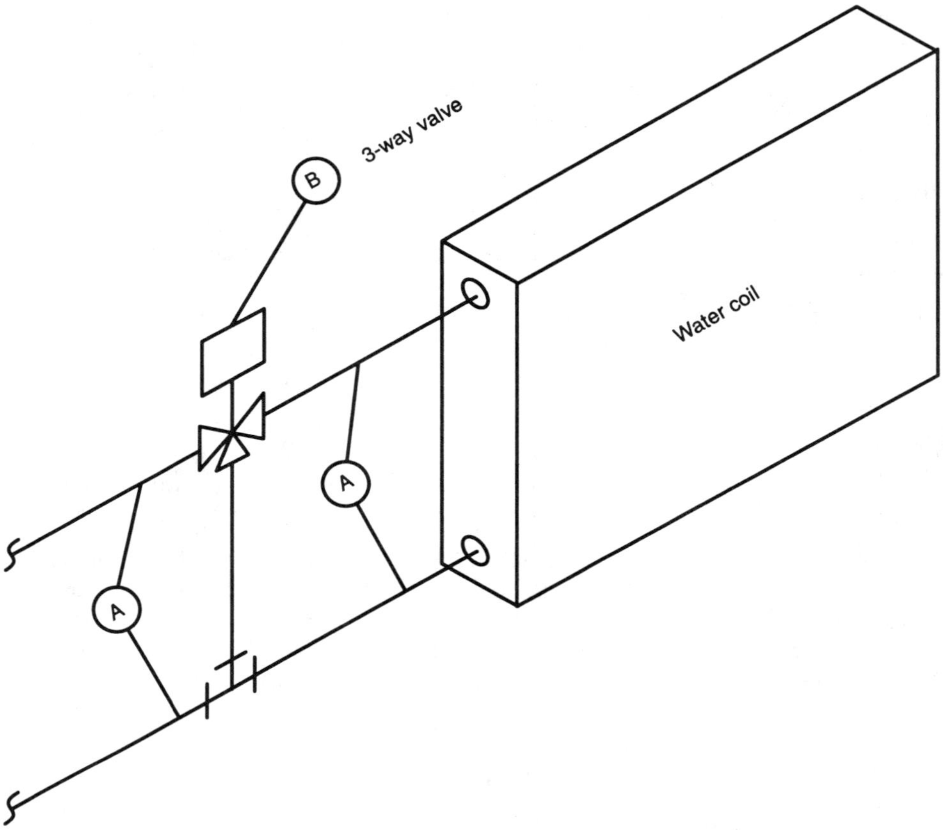

If Ⓐ is:	Then Ⓑ is:
1"	¾"
1¼"	1"
1½"	1¼"
2"	1½"
2½"	2"
3"	2½"

Typical water coil piping for A.H. units up to 16,000 CFM

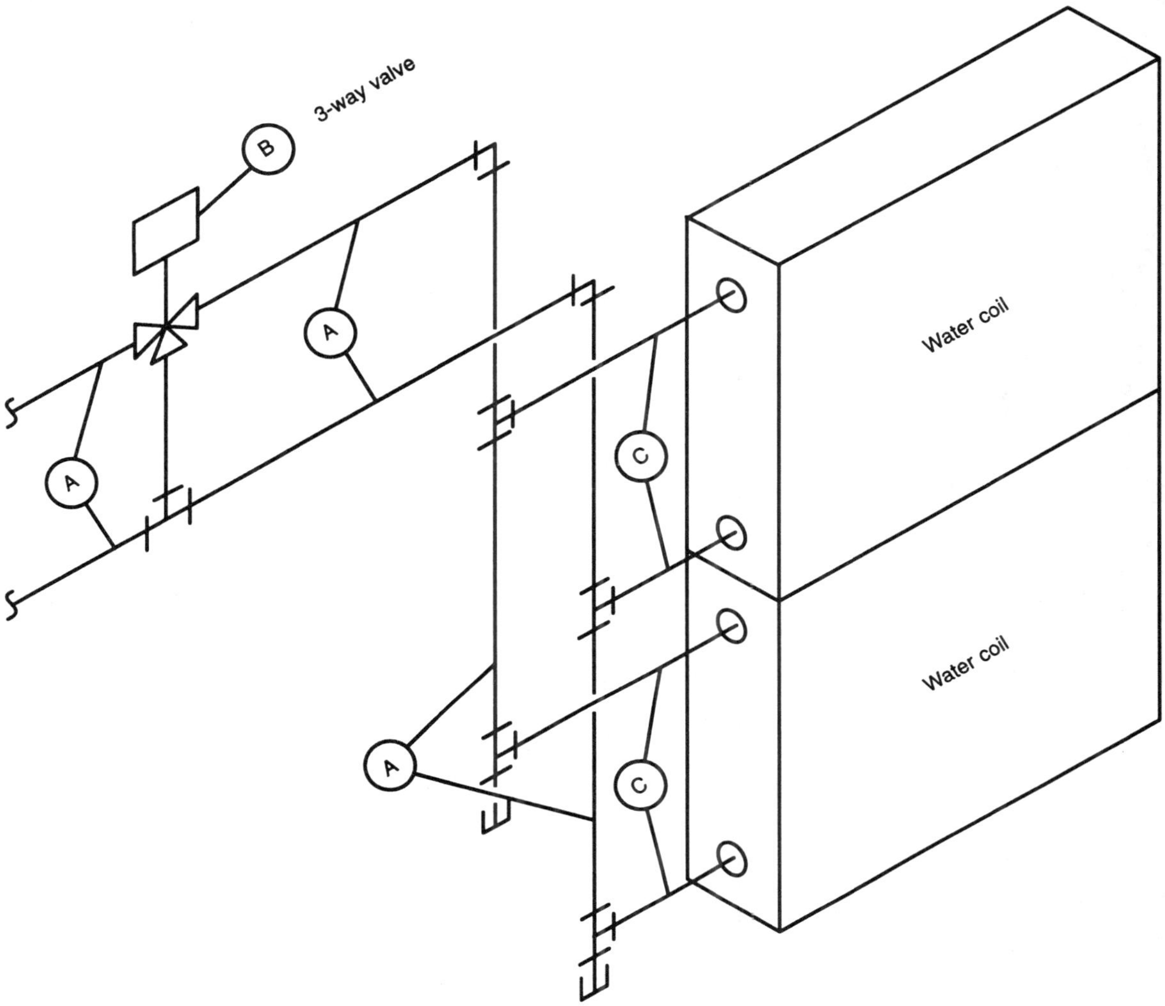

If (A) is:	Then (B) is:	And (C) is:
2½"	2"	2"
3"	2½"	2½"
4"	3"	3"

Typical water coil piping for A.H. units from 16,000 to 26,000 CFM

Water Coil Piping

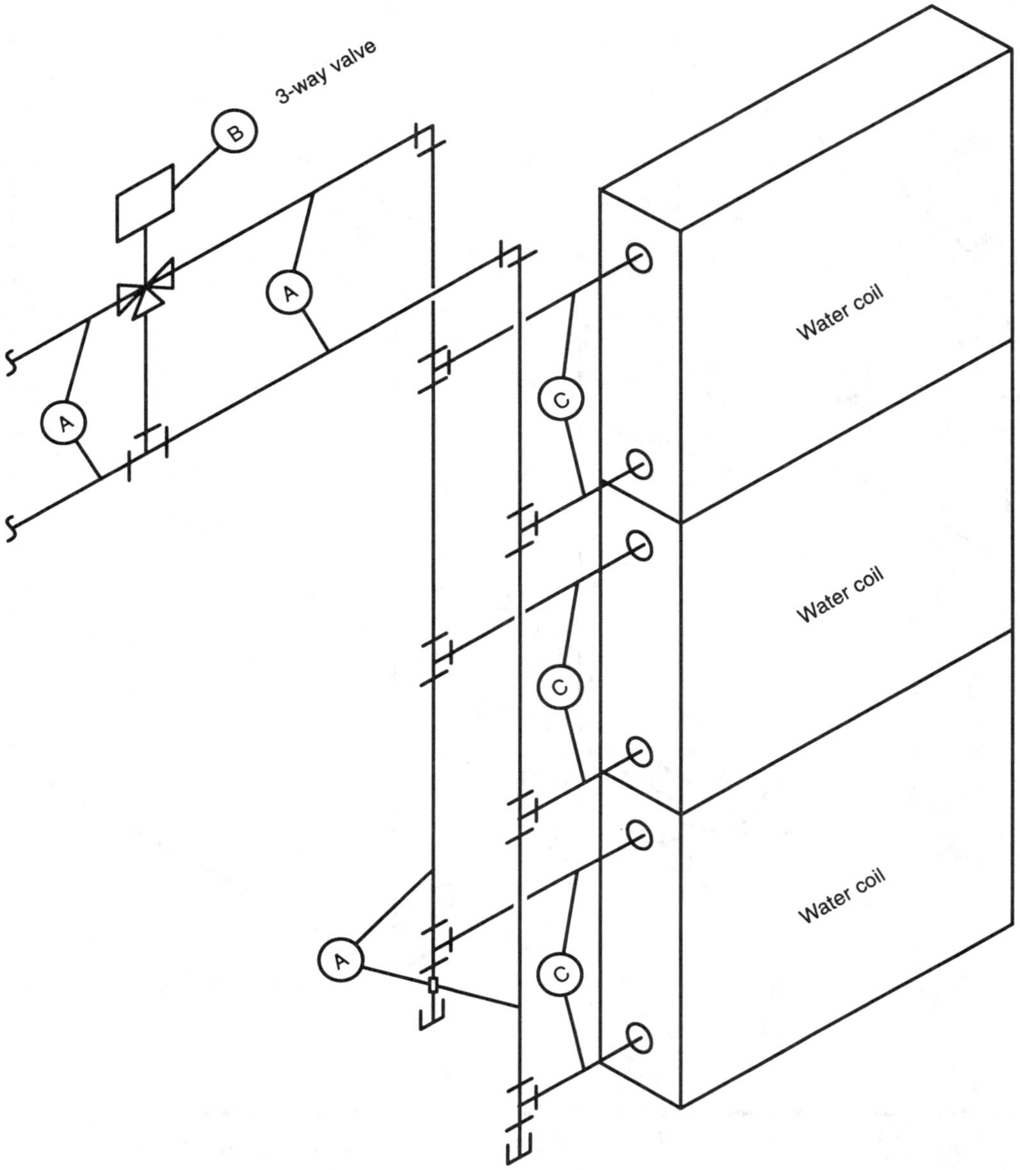

If Ⓐ is:	Then Ⓑ is:	And Ⓒ is:
4"	3"	2½"
6"	4"	3"
8"	6"	4"
10"	8"	6"

Typical water coil piping for A.H. units over 26,000 CFM

Description	Craft@Hrs	Unit	Material $	Labor $	Equipment $	Total $

Air handling unit coil connection, one row coil bank, non-regulated flow. Connection assembly includes pipe, fittings, pipe insulation, valves, gauges, thermometers and vents.

Description	Craft@Hrs	Unit	Material $	Labor $	Equipment $	Total $
1½" supply	SN@14.0	Ea	950.00	361.00	33.60	1,344.60
2" supply	SN@18.0	Ea	1,175.00	464.00	33.60	1,672.60
2½" supply	SN@28.0	Ea	1,550.92	721.00	33.60	2,305.52
3" supply	SN@31.5	Ea	1,703.47	811.00	37.80	2,552.27
4" supply	SN@34.0	Ea	1,830.60	876.00	40.80	2,747.40
6" supply	SN@41.0	Ea	2,644.20	1,060.00	49.20	3,753.40

Air handling unit coil connection, one row coil bank, 2-way control valve design. Connection assembly includes pipe and fittings, pipe insulation, valves, gauges, thermometers and vents.

Description	Craft@Hrs	Unit	Material $	Labor $	Equipment $	Total $
1½" supply	SN@16.5	Ea	1,075.00	425.00	33.60	1,533.60
2" supply	SN@21.0	Ea	1,390.00	541.00	33.60	1,964.60
2½" supply	SN@33.6	Ea	1,678.05	865.00	40.32	2,583.37
3" supply	SN@36.0	Ea	1,805.17	927.00	43.20	2,775.37
4" supply	SN@43.8	Ea	2,196.72	1,130.00	52.56	3,379.28
6" supply	SN@54.0	Ea	3,508.65	1,390.00	64.80	4,963.45

Air handling unit coil connection, one row coil bank, 3-way control valve design. Connection assembly includes pipe and fittings, pipe insulation, valves, gauges, thermometers and vents.

Description	Craft@Hrs	Unit	Material $	Labor $	Equipment $	Total $
1½" supply	SN@18.0	Ea	1,250.00	464.00	33.60	1,747.60
2" supply	SN@26.0	Ea	1,525.00	670.00	33.60	2,228.60
2½" supply	SN@41.5	Ea	1,932.30	1,070.00	49.80	3,052.10
3" supply	SN@43.8	Ea	2,135.70	1,130.00	52.56	3,318.26
4" supply	SN@54.6	Ea	2,745.90	1,410.00	65.52	4,221.42
6" supply	SN@61.4	Ea	3,915.45	1,580.00	73.68	5,569.13

Air handling unit coil connection, two row coil bank, non-regulated flow. Connection assembly includes pipe and fittings, pipe insulation, valves, gauges, thermometers and vents.

Description	Craft@Hrs	Unit	Material $	Labor $	Equipment $	Total $
2½" supply	SN@48.0	Ea	2,339.10	1,240.00	57.60	3,636.70
3" supply	SN@51.5	Ea	2,847.60	1,330.00	61.80	4,239.40
4" supply	SN@63.6	Ea	3,762.90	1,640.00	76.32	5,479.22
6" supply	SN@79.2	Ea	5,491.80	2,040.00	95.04	7,626.84

Air handling unit coil connection, two row coil bank, 2-way control valve design. Connection assembly includes pipe and fittings, pipe insulation, valves, gauges, thermometers and vents.

Description	Craft@Hrs	Unit	Material $	Labor $	Equipment $	Total $
2½" supply	SN@54.0	Ea	2,618.78	1,390.00	64.80	4,073.58
3" supply	SN@58.5	Ea	3,300.16	1,510.00	70.20	4,880.36
4" supply	SN@67.0	Ea	4,052.74	1,730.00	80.40	5,863.14
6" supply	SN@83.5	Ea	5,593.50	2,150.00	100.20	7,843.70

Air Handling Unit Coil Connections

Description	Craft@Hrs	Unit	Material $	Labor $	Equipment $	Total $

Air handling unit coil connection, two row coil bank, 3-way control valve design. Connection assembly includes pipe and fittings, pipe insulation, valves, gauges, thermometers and vents.

Description	Craft@Hrs	Unit	Material $	Labor $	Equipment $	Total $
2½" supply	SN@58.8	Ea	2,740.81	1,510.00	70.56	4,321.37
3" supply	SN@64.0	Ea	3,457.80	1,650.00	76.80	5,184.60
4" supply	SN@72.0	Ea	4,388.35	1,850.00	86.40	6,324.75
6" supply	SN@87.0	Ea	5,817.24	2,240.00	104.40	8,161.64

Air handling unit coil connection, three row coil bank, non-regulated flow. Connection assembly includes pipe and fittings, pipe insulation, valves, gauges, thermometers and vents.

Description	Craft@Hrs	Unit	Material $	Labor $	Equipment $	Total $
4" supply	SN@72.0	Ea	3,966.30	1,850.00	86.40	5,902.70
6" supply	SN@94.0	Ea	6,915.60	2,420.00	112.80	9,448.40
8" supply	SN@120.	Ea	9,356.40	3,090.00	144.00	12,590.40

Air handling unit coil connection, three row coil bank, 2-way control valve design. Connection assembly icludes pipe and fittings, pipe insulation, valves, gauges, thermometers and vents.

Description	Craft@Hrs	Unit	Material $	Labor $	Equipment $	Total $
4" supply	SN@78.0	Ea	4,169.70	2,010.00	93.60	6,273.30
6" supply	SN@105.	Ea	7,424.10	2,700.00	126.00	10,250.10
8" supply	SN@132.	Ea	10,576.80	3,400.00	158.40	14,135.20

Air handling unit coil connection, three row coil bank, 3-way control valve design. Connection assembly includes pipe and fittings, pipe insulation, valves, gauges, thermometers and vents.

Description	Craft@Hrs	Unit	Material $	Labor $	Equipment $	Total $
4" supply	SN@82.0	Ea	4,881.60	2,110.00	98.40	7,090.00
6" supply	SN@112.	Ea	7,780.05	2,880.00	134.40	10,794.45
8" supply	SN@140.	Ea	11,797.20	3,610.00	168.00	15,575.20

Description	Craft@Hrs	Unit	Material $	Labor $	Equipment $	Total $

Residential gas-fired upflow furnace. Set in place only. Make additional allowances for gas and electrical connections. Costs shown per thousand Btu per hour (MBH) Mid-efficiency, 80% AFUE.

Description	Craft@Hrs	Unit	Material $	Labor $	Equipment $	Total $
50 MBH input	SN@2.00	Ea	515.00	51.50	6.00	572.50
75 MBH input	SN@2.00	Ea	675.00	51.50	7.50	734.00
100 MBH input	SN@2.00	Ea	780.00	51.50	9.00	840.50
125 MBH input	SN@2.00	Ea	890.00	51.50	10.00	951.50

Residential combination gas fired furnace and air conditioning unit with remote exterior condenser/compressor. Set in place only. Make additional allowances for gas, electrical and DX connections. Costs shown per thousand Btu per hour (MBH) Mid-efficiency, 80% AFUE

Description	Craft@Hrs	Unit	Material $	Labor $	Equipment $	Total $
30 MBH, 1.5 tons	SN@2.50	Ea	1,450.00	64.40	12.50	1,526.90
36 MBH, 2 tons	SN@2.75	Ea	1,730.00	70.80	13.75	1,814.55
42 MBH, 2 tons	SN@3.00	Ea	1,930.00	77.30	15.00	2,022.30
48 MBH, 2.5 tons	SN@3.50	Ea	2,260.00	90.10	17.50	2,367.60
60 MBH, 3 tons	SN@4.00	Ea	2,690.00	103.00	20.00	2,813.00

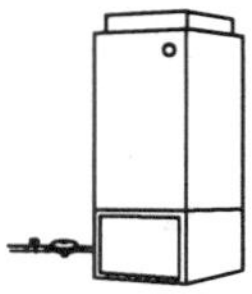

Commercial Fans and Blowers

Centrifugal air-foil wheel blower, with motor, drive and isolation, set in place only.

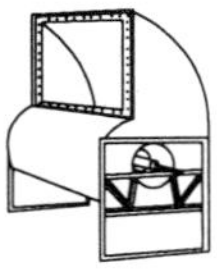

Description	Craft@Hrs	Unit	Material $	Labor $	Equipment $	Total $
1/3 HP, 12" dia.	SN@2.50	Ea	731.00	64.40	12.50	807.90
1/2 HP, 15" dia.	SN@4.50	Ea	1,300.00	116.00	22.50	1,438.50
1 HP, 18" dia.	SN@7.00	Ea	1,470.00	180.00	35.00	1,685.00
2 HP, 18" dia.	SN@7.50	Ea	1,750.00	193.00	37.50	1,980.50
3 HP, 27" dia.	SN@9.00	Ea	2,770.00	232.00	45.00	3,047.00
5 HP, 30" dia.	SN@11.0	Ea	3,290.00	283.00	55.00	3,628.00
7½ HP, 36" dia.	SN@12.0	Ea	3,960.00	309.00	60.00	4,329.00
10 HP, 40" dia.	SN@14.0	Ea	5,520.00	361.00	70.00	5,951.00
15 HP, 44" dia.	SN@17.0	Ea	7,310.00	438.00	85.00	7,833.00
20 HP, 60" dia.	SN@20.0	Ea	8,350.00	515.00	100.00	8,965.00

Utility centrifugal blower, with motor, drive and isolation, set in place only.

Description	Craft@Hrs	Unit	Material $	Labor $	Equipment $	Total $
1/3 HP, 10" dia.	SN@2.50	Ea	479.00	64.40	12.50	555.90
1/2 HP, 12" dia.	SN@4.50	Ea	628.00	116.00	22.50	766.50
1 HP, 15" dia.	SN@6.50	Ea	1,020.00	167.00	32.50	1,219.50
2 HP, 18" dia.	SN@7.50	Ea	1,160.00	193.00	37.50	1,390.50
3 HP, 24" dia.	SN@9.00	Ea	1,660.00	232.00	45.00	1,937.00
5 HP, 27" dia.	SN@11.0	Ea	1,910.00	283.00	55.00	2,248.00
7½ HP, 36" dia.	SN@12.0	Ea	3,090.00	309.00	60.00	3,459.00
10 HP, 40" dia.	SN@14.0	Ea	3,720.00	361.00	70.00	4,151.00
15 HP, 44" dia.	SN@17.0	Ea	4,030.00	438.00	85.00	4,553.00

Vane-axial fan, set in place only.

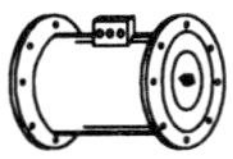

Description	Craft@Hrs	Unit	Material $	Labor $	Equipment $	Total $
30 HP, 36" dia.	SN@16.0	Ea	10,000.00	412.00	80.00	10,492.00
50 HP, 42" dia.	SN@24.0	Ea	11,200.00	618.00	120.00	11,938.00
100 HP, 48" dia.	SN@32.0	Ea	14,000.00	824.00	160.00	14,984.00
200 HP, 54" dia.	SN@40.0	Ea	19,600.00	1,030.00	200.00	20,830.00

Tube-axial fan, set in place only.

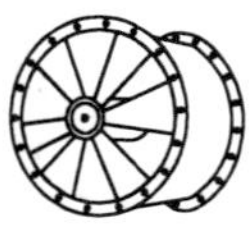

Description	Craft@Hrs	Unit	Material $	Labor $	Equipment $	Total $
15 HP, 24" dia.	SN@10.0	Ea	2,970.00	258.00	50.00	3,278.00
25 HP, 30" dia.	SN@13.0	Ea	4,600.00	335.00	65.00	5,000.00
30 HP, 36" dia.	SN@16.0	Ea	6,110.00	412.00	80.00	6,602.00
75 HP, 40" dia.	SN@22.0	Ea	9,170.00	567.00	110.00	9,847.00
100 HP, 48" dia.	SN@32.0	Ea	14,100.00	824.00	160.00	15,084.00

Roof exhaust fan, set in place only. Make additional allowances for roof curb.

Description	Craft@Hrs	Unit	Material $	Labor $	Equipment $	Total $
1/4 HP, 14" dia.	SN@2.50	Ea	772.00	64.40	12.50	848.90
1/3 HP, 22" dia.	SN@2.90	Ea	950.00	74.70	14.50	1,039.20
1/2 HP, 24" dia.	SN@3.50	Ea	1,080.00	90.10	17.50	1,187.60
1 HP, 30" dia.	SN@4.50	Ea	1,480.00	116.00	22.50	1,618.50
2 HP, 36" dia.	SN@5.00	Ea	2,140.00	129.00	25.00	2,294.00
3 HP, 40" dia.	SN@6.00	Ea	3,210.00	155.00	30.00	3,395.00
5 HP, 48" dia.	SN@7.00	Ea	4,440.00	180.00	35.00	4,655.00

Unitary or factory-packaged air conditioning units are often too small to meet the heating and cooling needs of a large single-zone space. In such cases, engineers design what's called a "built-up" system. They're custom fitted to the client's needs and the space. A typical "built-up" system includes the following equipment:

- centrifugal blowers or axial-flow fans
- damper assemblies
- coil banks
- filter banks

Once the system is assembled, it is enclosed in a shop-fabricated housing or casing. The illustration on page 262 shows a standard apparatus housing. The walls and roof are shop-built panels. These panels consist of a layer of thermal insulation covered on one or both sides with sheets of 18 gauge galvanized steel. An interior framework of steel angles or channels supports the panels. Seal all mating surfaces to minimize air leakage, using gaskets, caulking or sealant.

Apparatus Housings

There are two types of panels used in building apparatus housings: "double-skin" panels and "single-skin" panels. Double-skin panels have a layer of thermal insulation material sandwiched between two sheets of 18 gauge galvanized steel. Single-skin panels also have a layer of insulation. However, only the outside surface is clad in 18 gauge steel sheeting. On the inside the panel insulation is exposed.

The costs of constructing and installing apparatus housings vary greatly. This is due to the wide range of possible conditions and needs that you face on each job. Some of the many factors that influence costs here include the following:

- internal operating pressures
- external wind loads
- seismic requirements
- physical locations

That's why I recommend that you always prepare a detailed cost estimate. Then, you can take all the particulars into account. However, for budget estimating purposes only, use the following cost data:

Single-skin panels	$8.05 per SF*
Double-skin panels	$13.70 per SF*
Access doors (20" x 60")	$280.00 each

*Total area of walls and roof, less doors.

When you're estimating material costs for apparatus housings, remember to add in an extra 25 percent to cover the following costs: waste, seams, gaskets, sealants and miscellaneous assembly hardware.

Labor for Shop-Fabrication and Field Assembly of Apparatus Housings. The costs listed are manhours per square foot (SF) of panel. To find the total SF of panel, add up areas for all the panels.

Description	Shop Labor (MH per SF)	Field Labor (MH per SF)
Single-skin panels	.040	.250
Double-skin panels	.085	.300
Add for access doors	.400	240

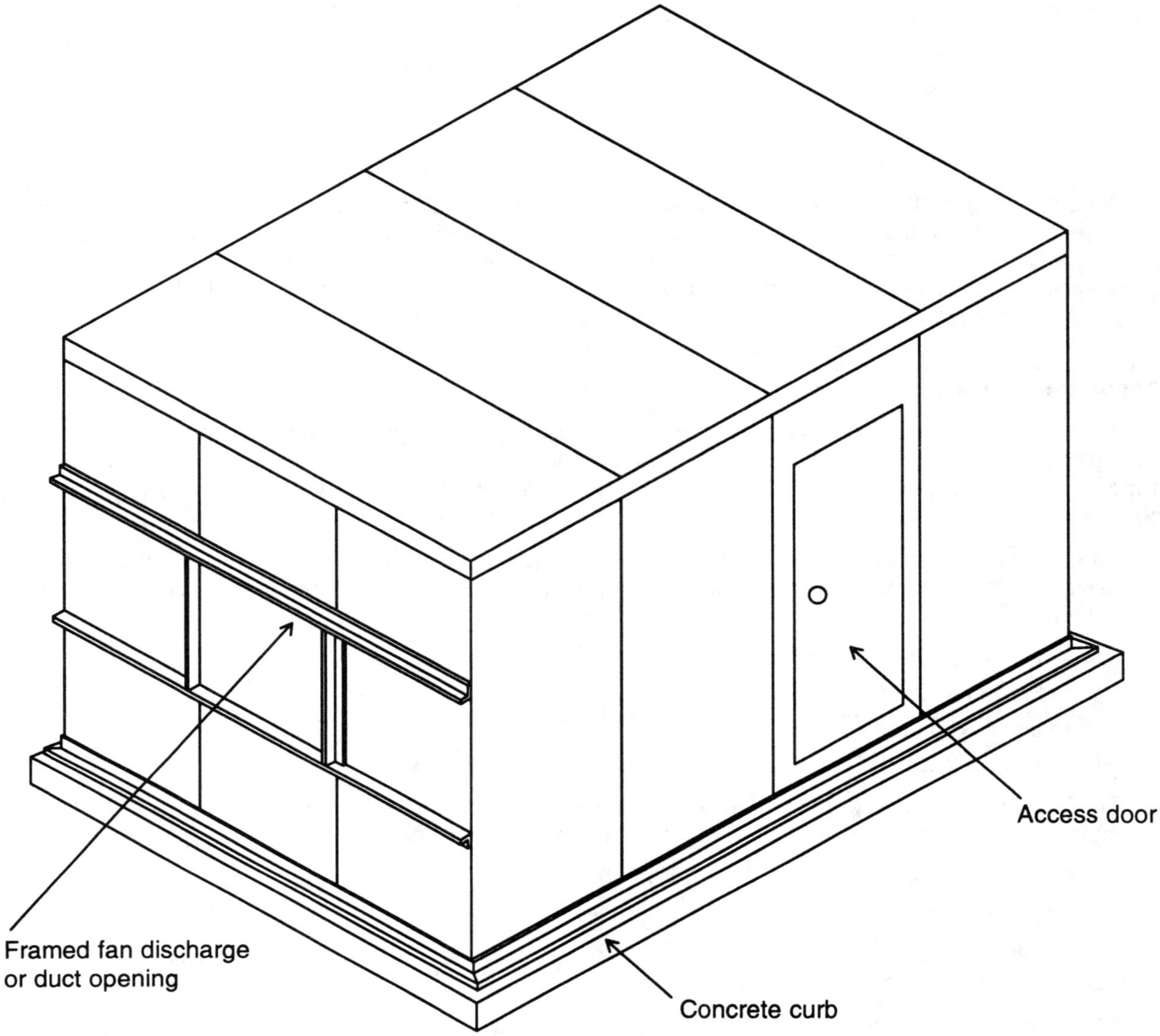

Typical apparatus housing

Supply air side-wall register, fixed pattern, with damper. Steel construction.

Description	Craft@Hrs	Unit	Material $	Labor $	Equipment $	Total $
8" x 4"	S2@.250	Ea	4.74	6.04	—	10.78
8" x 6"	S2@.250	Ea	4.92	6.04	—	10.96
10" x 4"	S2@.250	Ea	5.34	6.04	—	11.38
10" x 6"	S2@.250	Ea	9.10	6.04	—	15.14
10" x 8"	S2@.250	Ea	8.63	6.04	—	14.67
12" x 6"	S2@.250	Ea	9.19	6.04	—	15.23
12" x 8"	S2@.250	Ea	7.51	6.04	—	13.55
14" x 6"	S2@.250	Ea	7.51	6.04	—	13.55
14" x 8"	S2@.250	Ea	8.58	6.04	—	14.62

Return air side-wall grille, fixed pattern, without damper. Steel construction.

Description	Craft@Hrs	Unit	Material $	Labor $	Equipment $	Total $
14" x 6"	S2@.250	Ea	4.08	6.04	—	10.12
14" x 8"	S2@.250	Ea	3.96	6.04	—	10.00
16" x 16"	S2@.250	Ea	11.90	6.04	—	17.94
20" x 10"	S2@.300	Ea	17.70	7.24	—	24.94
20" x 16"	S2@.300	Ea	19.50	7.24	—	26.74
20" x 20"	S2@.300	Ea	25.80	7.24	—	33.04
24" x 24"	S2@.350	Ea	33.60	8.45	—	42.05

Supply air double deflection side-wall register, adjustable pattern. Aluminum construction.

Description	Craft@Hrs	Unit	Material $	Labor $	Equipment $	Total $
8" x 4"	S2@.250	Ea	15.20	6.04	—	21.24
8" x 6"	S2@.250	Ea	17.00	6.04	—	23.04
10" x 4"	S2@.250	Ea	19.70	6.04	—	25.74
10" x 6"	S2@.250	Ea	20.20	6.04	—	26.24
10" x 8"	S2@.250	Ea	21.60	6.04	—	27.64
12" x 6"	S2@.250	Ea	22.30	6.04	—	28.34
12" x 8"	S2@.250	Ea	25.10	6.04	—	31.14
12" x 10"	S2@.250	Ea	28.40	6.04	—	34.44
14" x 6"	S2@.250	Ea	28.40	6.04	—	34.44
14" x 8"	S2@.250	Ea	28.80	6.04	—	34.84
14" x 10"	S2@.250	Ea	30.70	6.04	—	36.74
16" x 8"	S2@.250	Ea	30.70	6.04	—	36.74
16" x 10"	S2@.250	Ea	33.60	6.04	—	39.64
18" x 8"	S2@.300	Ea	34.90	7.24	—	42.14
18" x 10"	S2@.300	Ea	37.40	7.24	—	44.64
18" x 12"	S2@.300	Ea	39.50	7.24	—	46.74

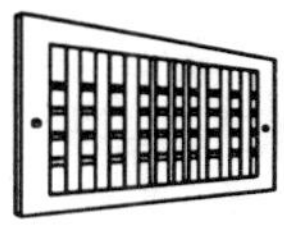

Description	Craft@Hrs	Unit	Material $	Labor $	Equipment $	Total $

Return air single deflection side-wall grille, adjustable pattern. Aluminum construction.

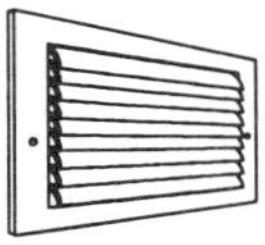

Description	Craft@Hrs	Unit	Material $	Labor $	Equipment $	Total $
8" x 4"	S2@.250	Ea	13.90	6.04	—	19.94
8" x 6"	S2@.250	Ea	15.20	6.04	—	21.24
10" x 4"	S2@.250	Ea	16.80	6.04	—	22.84
10" x 6"	S2@.250	Ea	18.20	6.04	—	24.24
10" x 8"	S2@.250	Ea	19.40	6.04	—	25.44
12" x 6"	S2@.250	Ea	20.10	6.04	—	26.14
12" x 8"	S2@.250	Ea	22.60	6.04	—	28.64
12" x 10"	S2@.250	Ea	25.40	6.04	—	31.44
14" x 6"	S2@.250	Ea	25.40	6.04	—	31.44
14" x 8"	S2@.250	Ea	25.90	6.04	—	31.94
14" x 10"	S2@.250	Ea	27.40	6.04	—	33.44
16" x 8"	S2@.250	Ea	27.40	6.04	—	33.44
16" x 10"	S2@.300	Ea	30.20	7.24	—	37.44
18" x 8"	S2@.300	Ea	31.50	7.24	—	38.74
18" x 10"	S2@.300	Ea	33.80	7.24	—	41.04
18" x 12"	S2@.300	Ea	35.60	7.24	—	42.84

4-way adjustable pattern ceiling diffuser 24" x 24" with collar and volume damper. Steel construction.

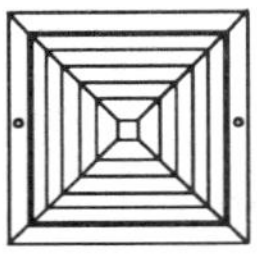

Description	Craft@Hrs	Unit	Material $	Labor $	Equipment $	Total $
6" neck	S2@.350	Ea	40.00	8.45	—	48.45
8" neck	S2@.350	Ea	42.00	8.45	—	50.45
10" neck	S2@.350	Ea	45.00	8.45	—	53.45
12" neck	S2@.350	Ea	47.00	8.45	—	55.45
14" neck	S2@.400	Ea	56.00	9.66	—	65.66

Fixed pattern ceiling diffuser 24" x 24" with collar, without volume damper. Steel construction.

Description	Craft@Hrs	Unit	Material $	Labor $	Equipment $	Total $
6" neck	S2@.350	Ea	24.00	8.45	—	32.45
8" neck	S2@.350	Ea	26.00	8.45	—	34.45
10" neck	S2@.350	Ea	29.00	8.45	—	37.45
12" neck	S2@.350	Ea	31.00	8.45	—	39.45
14" neck	S2@.350	Ea	40.00	8.45	—	48.45

Perforated ceiling diffuser 24" x 24" fixed pattern with collar, without volume damper. Steel construction.

Description	Craft@Hrs	Unit	Material $	Labor $	Equipment $	Total $
6" neck	S2@.350	Ea	32.00	8.45	—	40.45
8" neck	S2@.350	Ea	33.25	8.45	—	41.70
10" neck	S2@.350	Ea	35.50	8.45	—	43.95
12" neck	S2@.350	Ea	38.00	8.45	—	46.45
14" neck	S2@.400	Ea	42.50	9.66	—	52.16
16" neck	S2@.720	Ea	72.30	17.40	—	89.70

Description	Craft@Hrs	Unit	Material $	Labor $	Equipment $	Total $

Round 3-cone, 2-position ceiling diffuser with collar, without volume damper. Steel construction.

Description	Craft@Hrs	Unit	Material $	Labor $	Equipment $	Total $
6"	S2@.350	Ea	33.00	8.45	—	41.45
8"	S2@.350	Ea	38.00	8.45	—	46.45
10"	S2@.350	Ea	46.00	8.45	—	54.45
12"	S2@.350	Ea	54.00	8.45	—	62.45
14"	S2@.350	Ea	73.00	8.45	—	81.45

Round adjustable ceiling diffuser with collar and volume damper. Steel construction.

Description	Craft@Hrs	Unit	Material $	Labor $	Equipment $	Total $
6"	S2@.350	Ea	50.00	8.45	—	58.45
8"	S2@.350	Ea	59.00	8.45	—	67.45
10"	S2@.350	Ea	46.00	8.45	—	54.45
12"	S2@.350	Ea	73.00	8.45	—	81.45
14"	S2@.350	Ea	136.00	8.45	—	144.45
16"	S2@.450	Ea	155.00	10.90	—	165.90
18"	S2@.450	Ea	182.00	10.90	—	192.90
20"	S2@.550	Ea	215.00	13.30	—	228.30
24"	S2@.650	Ea	295.00	15.70	—	310.70

Terminal Units (VAV)

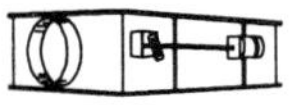

Variable-air volume terminal unit with analog electric controls, thermostat, actuator and controller

Description	Craft@Hrs	Unit	Material $	Labor $	Equipment $	Total $
200- 400 CFM	SN@2.00	Ea	255.00	51.50	10.00	316.50
400- 600 CFM	SN@2.00	Ea	285.00	51.50	12.00	348.50
600- 800 CFM	SN@2.50	Ea	300.00	64.40	14.00	378.40
800-1,000 CFM	SN@2.75	Ea	325.00	70.80	16.00	411.80
1,000-1,500 CFM	SN@3.00	Ea	475.00	77.30	18.00	570.30
1,500-2,000 CFM	SN@3.50	Ea	600.00	90.10	20.00	710.10

Variable-air volume terminal reheat unit with analog electric controls, thermostat, actuator, controller and a two row hot water coil

Description	Craft@Hrs	Unit	Material $	Labor $	Equipment $	Total $
200- 400 CFM	SN@2.25	Ea	295.00	57.90	12.50	365.40
400- 600 CFM	SN@2.25	Ea	325.00	57.90	14.50	397.40
600- 800 CFM	SN@2.75	Ea	375.00	70.80	16.50	462.30
800-1,000 CFM	SN@3.00	Ea	450.00	77.30	18.50	545.80
1,000-1,500 CFM	SN@3.50	Ea	600.00	90.10	20.50	710.60
1,500-2,000 CFM	SN@4.00	Ea	750.00	103.00	22.50	875.50

Description	Craft@Hrs	Unit	Material $	Labor $	Equipment $	Total $
Opposed-blade damper for rectangular ductwork						
12" x 6"	S2@.600	Ea	26.20	14.50	—	40.70
12" x 10"	S2@.650	Ea	30.90	15.70	—	46.60
12" x 12"	S2@.700	Ea	38.00	16.90	—	54.90
18" x 12"	S2@.800	Ea	43.80	19.30	—	63.10
18" x 18"	S2@.850	Ea	48.60	20.50	—	69.10
24" x 12"	S2@.850	Ea	51.10	20.50	—	71.60
24" x 18"	S2@.900	Ea	55.70	21.70	—	77.40
24" x 24"	S2@1.00	Ea	62.80	24.10	—	86.90
30" x 12"	S2@.900	Ea	58.10	21.70	—	79.80
30" x 18"	S2@1.00	Ea	65.20	24.10	—	89.30
30" x 24"	S2@1.10	Ea	73.60	26.60	—	100.20
30" x 30"	S2@1.20	Ea	79.50	29.00	—	108.50
36" x 18"	S2@1.10	Ea	76.00	26.60	—	102.60
36" x 36"	S2@1.40	Ea	104.00	33.80	—	137.80
42" x 24"	S2@1.30	Ea	106.00	31.40	—	137.40
42" x 42"	S2@1.50	Ea	146.00	36.20	—	182.20
48" x 24"	S2@1.40	Ea	109.00	33.80	—	142.80
48" x 48"	S2@1.60	Ea	152.00	38.60	—	190.60
54" x 24"	S2@1.45	Ea	140.00	35.00	—	175.00
54" x 54"	S2@1.70	Ea	175.00	41.10	—	216.10
Single-blade damper for rectangular ductwork						
12" x 6"	S2@.600	Ea	17.30	14.50	—	31.80
12" x 8"	S2@.625	Ea	18.90	15.10	—	34.00
12" x 10"	S2@.650	Ea	21.20	15.70	—	36.90
12" x 12"	S2@.700	Ea	24.90	16.90	—	41.80
18" x 12"	S2@.800	Ea	27.60	19.30	—	46.90
18" x 18"	S2@.850	Ea	31.40	20.50	—	51.90
24" x 12"	S2@.850	Ea	33.20	20.50	—	53.70
24" x 18"	S2@.900	Ea	34.50	21.70	—	56.20
24" x 24"	S2@1.00	Ea	38.00	24.10	—	62.10
Single-blade butterfly damper for round ductwork						
6"	S2@.600	Ea	17.30	14.50	—	31.80
8"	S2@.550	Ea	16.20	13.30	—	29.50
10"	S2@.600	Ea	18.80	14.50	—	33.30
12"	S2@.650	Ea	21.30	15.70	—	37.00
14"	S2@.700	Ea	24.30	16.90	—	41.20
16"	S2@.765	Ea	26.90	18.50	—	45.40
18"	S2@.835	Ea	29.60	20.20	—	49.80
20"	S2@.900	Ea	33.30	21.70	—	55.00

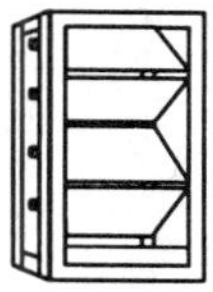

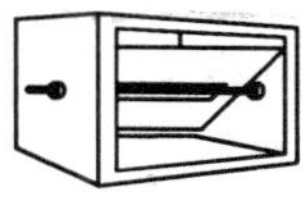

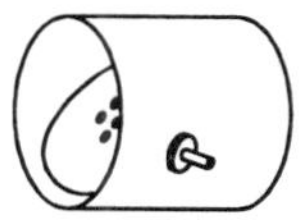

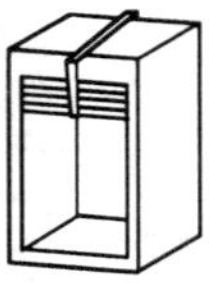

Curtain-type duct fire damper, 1½ hour rating, U.L. label

Description	Craft@Hrs	Unit	Material $	Labor $	Equipment $	Total $
12" x 6"	S2@1.00	Ea	27.50	24.10	—	51.60
12" x 8"	S2@1.00	Ea	29.60	24.10	—	53.70
12" x 10"	S2@1.05	Ea	32.40	25.40	—	57.80
12" x 12"	S2@1.10	Ea	35.00	26.60	—	61.60
18" x 12"	S2@1.20	Ea	48.90	29.00	—	77.90
18" x 18"	S2@1.30	Ea	52.90	31.40	—	84.30
24" x 12"	S2@1.40	Ea	59.10	33.80	—	92.90
24" x 18"	S2@1.50	Ea	67.20	36.20	—	103.40
24" x 24"	S2@1.60	Ea	80.20	38.60	—	118.80
30" x 12"	S2@1.50	Ea	68.10	36.20	—	104.30
30" x 18"	S2@1.80	Ea	75.60	43.50	—	119.10
30" x 24"	S2@2.00	Ea	99.80	48.30	—	148.10
30" x 30"	S2@2.30	Ea	108.00	55.50	—	163.50
36" x 18"	S2@2.00	Ea	86.10	48.30	—	134.40
36" x 36"	S2@2.40	Ea	129.00	58.00	—	187.00
42" x 24"	S2@2.40	Ea	112.00	58.00	—	170.00
42" x 42"	S2@2.90	Ea	162.00	70.00	—	232.00
48" x 24"	S2@2.90	Ea	121.00	70.00	—	191.00
48" x 48"	S2@3.50	Ea	202.00	84.50	—	286.50
54" x 24"	S2@3.00	Ea	131.00	72.40	—	203.40
54" x 54"	S2@3.80	Ea	234.00	91.80	—	325.80

Correction factors:

(1) For 22 gauge break-away connection, multiply material cost by 1.30

(2) For blades out of air stream cap, multiply material cost by 1.25

Single skin 2" fixed position turning vanes for duct

Description	Craft@Hrs	Unit	Material $	Labor $	Equipment $	Total $
12" x 6"	S2@.280	Ea	3.20	6.76	—	9.96
12" x 8"	S2@.285	Ea	3.97	6.88	—	10.85
12" x 10"	S2@.290	Ea	5.14	7.00	—	12.14
12" x 12"	S2@.300	Ea	6.35	7.24	—	13.59
18" x 12"	S2@.375	Ea	9.56	9.06	—	18.62
18" x 18"	S2@.450	Ea	11.60	10.90	—	22.50
24" x 12"	S2@.450	Ea	12.70	10.90	—	23.60
24" x 18"	S2@.500	Ea	13.90	12.10	—	26.00
24" x 24"	S2@.525	Ea	15.70	12.70	—	28.40
30" x 12"	S2@.550	Ea	15.90	13.30	—	29.20
30" x 18"	S2@.750	Ea	23.00	18.10	—	41.10
30" x 24"	S2@.800	Ea	30.60	19.30	—	49.90
30" x 30"	S2@.900	Ea	38.70	21.70	—	60.40
36" x 18"	S2@.750	Ea	27.40	18.10	—	45.50
36" x 36"	S2@.800	Ea	55.50	19.30	—	74.80
42" x 24"	S2@.900	Ea	43.10	21.70	—	64.80
42" x 42"	S2@1.00	Ea	75.70	24.10	—	99.80
48" x 24"	S2@1.10	Ea	48.90	26.60	—	75.50
48" x 48"	S2@1.50	Ea	97.90	36.20	—	134.10
54" x 24"	S2@1.15	Ea	54.80	27.80	—	82.60
54" x 54"	S2@1.25	Ea	124.00	30.20	—	154.20
60" x 24"	S2@1.30	Ea	61.70	31.40	—	93.10
60" x 36"	S2@1.40	Ea	93.10	33.80	—	126.90
60" x 60"	S2@1.50	Ea	153.00	36.20	—	189.20
72" x 36"	S2@1.65	Ea	112.00	39.80	—	151.80
72" x 54"	S2@2.40	Ea	165.00	58.00	—	223.00

Note: Labor includes cost to cut, assemble and install vanes and rails. For 4-inch air-foil vanes, multiply costs by 1.80 and labor costs by .90

Description	Craft@Hrs	Unit	Material $	Labor $	Equipment $	Total $
Spin-in galvanized steel collars without volume damper						
4"	S2@.200	Ea	2.45	4.83	—	7.28
5"	S2@.200	Ea	2.75	4.83	—	7.58
6"	S2@.200	Ea	3.04	4.83	—	7.87
7"	S2@.200	Ea	3.33	4.83	—	8.16
8"	S2@.250	Ea	3.86	6.04	—	9.90
9"	S2@.250	Ea	4.26	6.04	—	10.30
10"	S2@.300	Ea	4.61	7.24	—	11.85
12"	S2@.350	Ea	5.48	8.45	—	13.93
14"	S2@.400	Ea	6.58	9.66	—	16.24
16"	S2@.450	Ea	7.80	10.90	—	18.70
18"	S2@.500	Ea	9.73	12.10	—	21.83

Description	Craft@Hrs	Unit	Material $	Labor $	Equipment $	Total $
Spin-in galvanized steel collar with volume damper						
4"	S2@.250	Ea	8.25	6.04	—	14.29
5"	S2@.250	Ea	8.38	6.04	—	14.42
6"	S2@.250	Ea	8.49	6.04	—	14.53
7"	S2@.250	Ea	8.61	6.04	—	14.65
8"	S2@.300	Ea	8.79	7.24	—	16.03
9"	S2@.300	Ea	8.91	7.24	—	16.15
10"	S2@.350	Ea	9.08	8.45	—	17.53
12"	S2@.400	Ea	9.80	9.66	—	19.46
14"	S2@.450	Ea	10.50	10.90	—	21.40
16"	S2@.500	Ea	11.10	12.10	—	23.20
18"	S2@.550	Ea	11.90	13.30	—	25.20

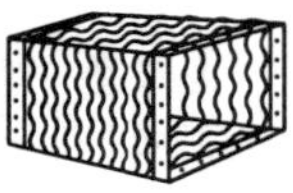

Description	Craft@Hrs	Unit	Material $	Labor $	Equipment $	Total $
Duct-to-equipment flexible connection						
18" x 12"	S2@.650	Ea	7.49	15.70	—	23.19
18" x 18"	S2@.850	Ea	8.99	20.50	—	29.49
24" x 12"	S2@.850	Ea	8.99	20.50	—	29.49
24" x 18"	S2@1.00	Ea	10.50	24.10	—	34.60
24" x 24"	S2@1.30	Ea	12.00	31.40	—	43.40
30" x 12"	S2@1.00	Ea	10.50	24.10	—	34.60
30" x 18"	S2@1.30	Ea	12.00	31.40	—	43.40
30" x 30"	S2@1.60	Ea	15.00	38.60	—	53.60
36" x 12"	S2@1.30	Ea	12.00	31.40	—	43.40
36" x 24"	S2@1.60	Ea	15.00	38.60	—	53.60
36" x 36"	S2@1.80	Ea	17.90	43.50	—	61.40
48" x 24"	S2@1.80	Ea	17.90	43.50	—	61.40
48" x 48"	S2@2.00	Ea	23.70	48.30	—	72.00

The costs for fabricating and installing galvanized steel ductwork are usually based on the total weights of the duct and fittings. Because of cost differences, fitting weights should be kept separate from straight duct weights.

Round spiral duct: Normally-used duct gauge/size relationships for low-pressure systems, up to 2 inches static pressure, are as follows:

Duct Diameter (inches)	U.S. Standard Gauge
Up to 12	26
13 through 24	24
26 through 36	22

Rectangular duct: Normally-used duct gauge/size relationships for low-pressure systems, up to 2 inches static pressure, are as follows:

Duct Size (inches)	U.S. Standard Gauge
Up to 12	26
13 through 30	24
31 through 54	22
55 through 72	20

The fitting weights in the following tables are in accordance with the above gauges.

Other material costs: All weights in the tables are net weights; they do not include bracing, cleats, scrap, hangers, end closures, sealants and miscellaneous hardware. Add 15% to calculated duct/fitting weights to cover these items if ductwork is purchased. Add 20% if ducting is manufactured in contractor's shop. The additional 5% covers the costs for scrap which is already included in an outside vendor's price.

Ductwork costs: Typical costs for purchased ductwork are as follows:

Straight duct, less than 1,000 pounds per order	$1.10/lb.
Straight duct, over 1,000 pounds per order	$ .93/lb.
Fittings, less than 1,000 pounds per order	$2.21/lb
Fittings, over 1,000 pounds per order	$1.93/lb.
Lined ducts and fittings, add	$ .50/lb.*
Delivery costs	$ .05/lb.

**Duct sizes shown on drawings are always inside, or net, dimensions and must be increased in size when estimating lined ductwork.*

If ductwork is manufactured in the contractor's fabrication shop, the above costs can be reduced by about 20%, depending on the shop's rate of productivity.

Galvanized Steel Ductwork

Installation costs: An average crew can install approximately 25 pounds of duct and fittings per manhour under normal conditions. See "Applying Correction Factors" on page 6 for situations that do not conform to the definition of a standard labor unit.

EXAMPLE:

What is the cost to furnish and install 2,585 pounds of unlined duct and 560 pounds of unlined fittings? The ductwork will be purchased from an outside vendor. A 15% allowance for miscellaneous material is included in the weights.

Material:	
2,585 pounds of straight duct x $.93/lb.	= $2,404.05
560 pounds of fittings x $2.21/lb.	= 1,237.60
Delivery cost: (2,585 + 560) x $.05/lb.	= 157.25
Total material cost	= $3,798.90*
Labor:	
$\frac{2{,}585 + 560}{25} = \mathit{125.8\ MH};\ \ \mathit{125.8\ MH} \times \$24.15$	= $3,038.07
Total installed cost: $3,798.90 + $3,038.07	= $6,836.97*

Sales tax not included.

Weights of Galvanized Steel Spiral Duct (pounds per LF)					
Diameter (inches)	U.S. Standard Gauge				
	26	24	22	20	18
3	.76	1.01	1.23	1.46	—
4	1.02	1.35	1.64	1.94	—
5	1.28	1.69	2.06	2.42	—
6	1.54	2.03	2.47	2.91	3.88
7	1.79	2.37	2.88	3.40	4.52
8	2.05	2.71	3.29	3.86	5.17
9	2.31	3.05	3.71	4.37	5.82
10	2.57	3.39	4.12	4.86	6.47
12	3.08	4.07	4.95	5.83	7.75
14	3.59	4.74	5.77	6.81	9.05
16	4.11	5.42	6.60	7.78	10.34
18	4.63	6.10	7.43	8.76	11.64
20	5.15	6.78	8.25	9.73	12.93
22	5.65	7.46	9.08	10.71	14.93
24	6.16	8.14	9.91	11.68	15.52
26	6.67	8.82	10.73	12.66	16.82
28	7.18	9.50	11.56	13.63	18.11
30	7.71	10.18	12.38	14.60	19.41
32	8.22	10.84	13.21	15.58	20.71
34	—	11.52	14.05	16.55	22.00
36	—	12.20	14.90	17.53	23.29

Galvanized Steel Round Spiral Fittings

Weights of Galvanized Steel Round Spiral Fittings (pounds per piece)				
Diameter (inches)	**90° elbow**	**45° elbow**	**Coupling**	**Reducer**
3	1.3	1.0	0.5	1.0
4	2.2	1.3	0.6	1.2
5	3.3	1.9	0.7	1.4
6	4.3	2.5	0.9	1.8
7	5.8	3.3	1.0	2.0
8	7.3	4.3	1.2	2.4
9	8.8	5.3	2.6	3.3
10	11.8	7.5	2.9	4.4
12	16.3	10.0	3.5	5.3
14	22.0	13.0	4.1	6.2
16	28.3	15.8	4.6	6.9
18	34.5	19.0	5.2	7.8
20	41.5	23.5	5.8	8.7
22	48.3	27.5	6.4	9.6
24	57.5	32.0	6.9	10.3
26	68.8	37.5	7.5	11.2
28	76.8	42.8	8.1	12.1
30	87.0	48.0	8.7	13.0
32	99.5	55.0	9.2	13.8
34	112.0	61.0	9.8	14.7
36	161.0	89.5	10.4	15.6

Largest run diam. (inches)	Weights of Galvanized Steel Round Spiral Fittings (pounds per piece) Tee with reducing run and branch Branch diameter (inches)													
	3	**4**	**5**	**6**	**7**	**8**	**9**	**10**	**12**	**14**	**16**	**18**	**20**	**22**
4	2.3													
5	2.8	3.0												
6	3.2	3.5	3.7											
7	3.7	3.9	4.1	4.5										
8	4.1	4.4	4.7	5.0	5.3									
9	4.5	4.8	5.1	5.5	5.8	6.2								
10	6.4	6.7	7.1	7.5	7.8	8.2	8.6							
12	7.7	8.1	8.5	8.9	9.4	9.9	10.3	10.7						
14	8.9	9.4	10.0	10.5	11.0	11.5	12.1	12.6	13.6					
16	10.2	10.8	11.4	12.0	12.5	13.1	13.7	14.3	15.5	16.7				
18	11.5	12.1	12.7	13.5	14.1	14.8	15.5	16.1	17.5	18.8	19.2			
20	12.7	13.4	14.2	14.9	17.7	16.4	17.2	17.9	19.4	20.9	22.4	23.9		
22	14.0	14.8	15.6	16.4	17.2	18.1	18.9	19.7	21.3	23.0	24.6	26.2	27.9	
24	15.3	16.1	17.0	17.8	18.7	19.6	20.5	21.4	23.2	25.0	26.8	28.5	30.3	32.1
26	16.5	17.5	18.5	19.5	20.4	21.3	22.3	23.3	25.2	26.2	29.1	31.1	33.0	34.9
28	17.7	18.7	19.8	20.8	21.9	22.9	23.9	25.0	27.0	29.2	31.2	33.2	35.4	37.4
30	18.9	20.0	21.1	22.2	23.3	24.5	25.5	26.7	28.9	31.1	33.3	35.5	37.7	40.0
32	20.4	21.6	22.6	24.0	25.2	26.4	27.6	28.8	31.2	33.6	36.0	38.4	40.8	42.2
34	21.6	22.9	24.2	25.4	26.7	28.0	29.3	30.5	33.1	35.6	38.1	40.7	42.2	45.6
36	29.6	31.4	33.1	34.8	36.6	38.6	40.1	41.8	45.3	48.8	52.2	55.7	59.2	62.7

Run diam. (inches)	Weights of Galvanized Steel Round Spiral Fittings (pounds per piece) Cross with reducing run and branch Branch diameter (inches)													
	3	4	5	6	7	8	9	10	12	14	16	18	20	22
4	2.6													
5	3.2	3.6												
6	3.8	4.2	4.6											
7	4.4	4.9	5.4	5.8										
8	4.9	5.4	6.0	6.5	7.0									
9	5.4	6.0	6.6	7.2	7.7	8.3								
10	7.6	8.2	8.9	9.5	10.2	10.8	11.4							
12	8.8	9.6	10.3	11.0	11.7	12.5	13.2	13.8						
14	10.1	10.9	11.7	12.6	13.4	14.2	15.1	15.9	17.5					
16	11.3	12.2	13.1	14.0	15.0	15.9	16.8	17.7	19.5	21.4				
18	12.5	13.5	14.5	15.5	16.5	17.5	18.5	19.5	21.5	23.5	25.5			
20	13.8	15.0	16.1	17.2	18.2	19.4	20.6	21.7	23.9	26.2	28.4	30.6		
22	15.2	16.4	17.7	18.9	20.1	21.4	22.6	23.8	26.2	28.7	31.2	33.6	36.1	
24	16.6	17.9	19.3	20.6	22.0	23.3	24.6	26.0	28.7	31.5	34.1	36.7	39.4	42.1
26	18.0	19.4	20.9	22.3	23.8	25.2	26.7	28.2	31.1	33.9	36.9	39.8	42.7	45.6
28	19.3	20.8	22.4	24.0	25.5	27.1	28.6	30.2	32.3	36.4	39.6	42.7	45.8	48.9
30	20.6	22.3	23.9	25.6	27.3	28.9	30.6	32.3	35.6	38.9	42.3	45.6	49.9	52.2
32	22.2	24.0	25.8	27.6	29.4	31.2	33.0	34.8	38.4	42.0	45.6	49.1	52.8	56.5
34	23.5	25.4	27.3	29.2	31.1	33.1	34.9	36.9	40.7	44.5	48.3	52.1	55.9	59.8
36	32.2	34.8	37.4	40.0	42.8	45.3	48.0	50.5	55.7	61.0	66.2	71.5	76.7	82.0

Weights of 26 Gauge Galvanized Steel Rectangular Duct (pounds per LF)									
Size (inches)	4	5	6	7	8	9	10	11	12
4	1.21								
5	1.36	1.51							
6	1.51	1.66	1.81						
7	1.66	1.81	1.96	2.11					
8	1.81	1.96	2.11	2.27	2.42				
9	1.96	2.11	2.27	2.42	2.57	2.72			
10	2.11	2.27	2.42	2.57	2.72	2.87	3.02		
11	2.27	2.42	2.57	2.72	2.87	3.02	3.17	3.32	
12	2.42	2.57	2.72	2.87	3.02	3.17	3.32	3.47	3.62

Galvanized Steel Rectangular Ductwork

Weights of 24 Gauge Galvanized Steel Rectangular Duct (pounds per LF)									
Size (inches)	14	16	18	20	22	24	26	28	30
6	3.85	4.24	4.62	5.01	5.40	5.78	6.17	6.55	6.94
8	4.24	4.62	5.01	5.40	5.78	6.17	6.55	6.94	7.32
10	4.62	5.01	5.40	5.78	6.17	6.55	6.94	7.32	7.71
12	5.01	5.40	5.78	6.17	6.55	6.94	7.32	7.71	8.09
14	5.40	5.78	6.17	6.55	6.94	7.32	7.71	8.09	8.48
16	5.78	6.17	6.55	6.94	7.32	7.71	8.09	8.48	8.86
18	6.17	6.55	6.94	7.32	7.71	8.09	8.48	8.86	9.25
20	6.55	6.94	7.32	7.71	8.09	8.48	8.86	9.25	9.64
22	6.94	7.32	7.71	8.09	8.48	8.86	9.25	9.64	10.00
24	7.32	7.71	8.09	8.48	8.86	9.25	9.64	10.00	10.40
26	7.71	8.09	8.48	8.86	9.25	9.64	10.00	10.40	10.80
28	8.09	8.48	8.86	9.25	9.64	10.00	10.40	10.80	11.20
30	8.48	8.86	9.25	9.64	10.00	10.40	10.80	11.20	11.60

Weights of 22 Gauge Galvanized Steel Rectangular Duct (pounds per LF)												
Size (inches)	32	34	36	38	40	42	44	46	48	50	52	54
8	9.37	9.84	10.3	10.8	11.3	11.7	12.2	12.7	13.1	13.6	14.1	14.5
10	9.84	10.3	10.8	11.3	11.7	12.2	12.7	13.1	13.6	14.1	14.5	15.0
12	10.3	10.8	11.3	11.7	12.2	12.7	13.1	13.6	14.1	14.5	15.0	15.5
14	10.8	11.3	11.7	12.2	12.7	13.1	13.6	14.1	14.5	15.0	15.5	15.9
16	11.3	11.7	12.2	12.7	13.1	13.6	14.1	14.5	15.0	15.5	15.9	16.4
18	11.7	12.2	12.7	13.1	13.6	14.1	14.5	15.0	15.5	15.9	16.4	16.9
20	12.2	12.7	13.1	13.6	14.1	14.5	15.0	15.5	15.9	16.4	16.9	17.3
22	12.7	13.1	13.6	14.1	14.5	15.0	15.5	15.9	16.4	16.9	17.3	17.8
24	13.1	13.6	14.1	14.5	15.0	15.5	15.9	16.4	16.9	17.3	17.8	18.3
26	13.6	14.1	14.5	15.0	15.5	15.9	16.4	16.9	17.3	17.8	18.3	18.7
28	14.1	14.5	15.0	15.5	15.9	16.4	16.9	17.3	17.8	18.3	18.7	19.2
30	14.5	15.0	15.5	15.9	16.4	16.9	17.3	17.8	18.3	18.7	19.2	19.7
32	15.0	15.5	15.9	16.4	16.9	17.3	17.8	18.3	18.7	19.2	19.7	20.2
34	15.5	15.9	16.4	16.9	17.3	17.8	18.3	18.7	19.2	19.7	20.2	20.6
36	15.9	16.4	16.9	17.3	17.8	18.3	18.7	19.2	19.7	20.2	20.6	21.1
38	16.4	16.9	17.3	17.8	18.3	18.7	19.2	19.7	20.2	20.6	21.1	21.6
40	16.9	17.3	17.8	18.3	18.7	19.2	19.7	20.2	20.6	21.1	21.6	22.0
42	17.3	17.8	18.3	18.7	19.2	19.7	20.2	20.6	21.1	21.6	22.0	22.5
44	17.8	18.3	18.7	19.2	19.7	20.2	20.6	21.1	21.6	22.0	22.5	23.0
46	18.3	18.7	19.2	19.7	20.2	20.6	21.1	21.6	22.0	22.5	23.0	23.4
48	18.7	19.2	19.7	20.2	20.6	21.1	21.6	22.0	22.5	23.0	23.4	23.9
50	19.2	19.7	20.2	20.6	21.1	21.6	22.0	22.5	23.0	23.4	23.9	24.4
52	19.7	20.2	20.6	21.1	21.6	22.0	22.5	23.0	23.4	23.9	24.4	24.8
54	20.2	20.6	21.1	21.6	22.0	22.5	23.0	23.4	23.9	24.4	24.8	25.3

Weights of 20 Gauge Galvanized Steel Rectangular Duct (pounds per LF)									
Size (inches)	56	58	60	62	64	66	68	70	72
18	20.4	21.0	21.5	22.1	22.6	23.2	23.7	24.3	24.8
20	21.0	21.5	22.1	22.6	23.2	23.7	24.3	24.8	25.4
22	21.5	22.1	22.6	23.2	23.7	24.3	24.8	25.4	25.9
24	22.1	22.6	23.2	23.7	24.3	24.8	25.4	25.9	26.5
26	22.6	23.2	23.7	24.3	24.8	25.4	25.9	26.5	27.0
28	23.2	23.7	24.3	24.8	25.4	25.9	26.5	27.0	27.6
30	23.7	24.3	24.8	25.4	25.9	26.5	27.0	27.6	28.2
32	24.3	24.8	25.4	25.9	26.5	27.0	27.6	28.2	28.7
34	24.8	25.4	25.9	26.5	27.0	27.6	28.2	28.7	29.3
36	25.4	25.9	26.5	27.0	27.6	28.2	28.7	29.3	29.8
38	25.9	26.5	27.0	27.6	28.2	28.7	29.3	29.8	30.4
40	26.5	27.0	27.6	28.2	28.7	29.3	29.8	30.4	30.9
42	27.0	27.6	28.2	28.7	29.3	29.8	30.4	30.9	31.5
44	27.6	28.2	28.7	29.3	29.8	30.4	30.9	31.5	32.0
46	28.2	28.7	29.3	29.8	30.4	30.9	31.5	32.0	32.6
48	28.7	29.3	29.8	30.4	30.9	31.5	32.0	32.6	33.1
50	29.3	29.8	30.4	30.9	31.5	32.0	32.6	33.1	33.7
52	29.8	30.4	30.9	31.5	32.0	32.6	33.1	33.7	34.2
54	30.4	30.9	31.5	32.0	32.6	33.1	33.7	34.2	34.8
56	30.9	31.5	32.0	32.6	33.1	33.7	34.2	34.8	35.3
58	31.5	32.0	32.6	33.1	33.7	34.2	34.8	35.3	35.9
60	32.0	32.6	33.1	33.7	34.2	34.8	35.3	35.9	36.4
62	32.6	33.1	33.7	34.2	34.8	35.3	35.9	36.4	37.0
64	33.1	33.7	34.2	34.8	35.3	35.9	36.4	37.0	37.5
66	33.7	34.2	34.8	35.3	35.9	36.4	37.0	37.5	38.1
68	34.2	34.8	35.3	35.9	36.4	37.0	37.5	38.1	38.6
70	34.8	35.3	35.9	36.4	37.0	37.5	38.1	38.6	39.2
72	35.3	35.9	36.4	37.0	37.5	38.1	38.6	39.2	39.7

Weights of Galvanized Steel Rectangular 90 Degree Elbow (pounds per piece)									
Size (inches)	4	6	8	10	12	14	16	18	20
4	1.05	1.45	1.95	2.60	3.30	5.30	6.50	7.80	9.25
6	1.45	1.95	2.60	3.30	3.70	5.80	7.75	8.75	9.75
8	1.95	2.60	3.30	3.70	4.25	6.50	8.50	9.75	10.4
10	2.60	3.30	3.70	4.25	4.85	7.30	9.50	10.7	11.2
12	3.30	3.70	4.25	4.85	5.60	8.20	10.5	11.7	12.2
14	5.30	5.80	6.50	7.30	8.20	9.25	11.5	12.6	13.2
16	6.50	7.50	8.50	9.50	10.5	11.5	12.5	13.4	14.4
18	7.80	8.75	9.75	10.7	11.7	12.6	13.4	14.5	15.7
20	9.25	9.75	10.4	11.2	12.2	13.8	14.4	15.7	17.2
22	10.8	11.8	12.8	13.8	14.8	25.8	16.8	17.8	18.8
24	12.6	13.6	14.6	15.6	16.6	17.6	18.6	19.6	20.5
26	14.4	15.4	16.4	17.4	18.4	19.4	20.4	21.4	22.3
28	16.4	17.4	18.4	19.4	20.4	21.4	22.4	23.4	24.3
30	18.5	19.5	20.5	21.5	22.5	23.5	24.5	25.5	26.4
32	24.3	25.7	27.0	28.4	29.7	31.1	32.4	33.7	35.0
34	28.2	29.4	30.6	31.8	33.0	34.3	35.5	36.7	37.9
36	31.3	32.5	33.7	34.9	36.1	37.3	38.5	39.7	40.9
38	34.5	35.7	36.9	38.1	39.3	40.5	41.7	42.9	44.1
40	37.9	39.1	40.3	41.5	42.7	43.9	45.1	46.3	47.5
42	41.4	42.6	43.8	45.0	46.2	47.4	48.6	49.8	51.0
44	45.1	46.4	47.6	48.8	50.0	51.2	52.4	53.6	54.8
46	49.0	50.2	51.4	52.6	53.8	55.0	56.2	57.4	58.6
48	53.0	54.2	55.4	56.6	57.8	59.0	60.2	61.4	62.6
50	57.2	58.5	59.7	60.9	62.1	63.3	64.5	65.7	66.9

Note: Elbow weights do not include turning vanes.

Weights of Galvanized Steel Rectangular 90 Degree Elbows (pounds per piece)									
Size (inches)	**22**	**24**	**26**	**28**	**30**	**32**	**34**	**36**	**38**
22	19.6	21.9	24.1	26.3	28.5	37.0	40.8	44.5	48.3
24	21.9	24.2	26.4	28.6	30.9	39.5	43.6	47.2	50.8
26	24.1	26.4	28.7	30.9	33.2	42.0	46.4	49.8	53.3
28	26.3	28.6	30.9	33.3	35.6	44.5	49.2	52.5	55.8
30	28.5	30.9	33.2	35.6	37.9	47.0	52.0	55.1	58.3
32	37.0	39.5	42.0	44.5	47.0	49.5	52.0	57.8	60.8
34	40.8	43.6	46.4	49.2	52.0	54.8	57.6	60.5	63.3
36	44.5	47.2	49.8	52.5	55.1	57.8	60.5	63.1	65.8
38	48.3	50.8	53.3	55.8	58.3	60.8	63.3	65.8	68.3
40	52.1	54.6	57.1	59.6	62.1	64.6	67.1	69.6	72.1
42	55.9	58.4	60.9	63.4	65.9	68.4	70.9	73.4	75.9
44	59.6	62.1	64.6	67.1	69.6	72.1	74.6	77.1	79.6
46	63.4	65.9	68.4	70.9	73.4	75.9	78.4	80.9	83.4
48	67.2	69.7	72.2	74.7	77.2	79.7	82.2	84.7	87.2
50	70.9	73.4	75.9	78.4	80.9	83.4	85.9	88.4	90.9
52	74.8	77.3	79.8	82.3	84.8	87.3	89.8	92.3	94.7
54	78.5	81.0	83.5	86.0	88.5	91.0	93.5	96.0	98.5
56	97.2	100	103	106	109	112	115	118	121
58	103	106	109	112	115	118	121	124	127
60	110	113	116	119	122	125	128	131	134
62	116	119	122	125	128	131	134	137	140
64	122	125	129	132	135	138	141	144	147
66	128	131	135	138	141	144	148	151	154
68	135	138	142	145	148	151	154	157	160
70	141	144	148	151	154	157	160	163	166
72	147	150	154	157	160	163	166	169	172

Note: Elbow weights do not include turning vanes.

Weights of Galvanized Steel Rectangular 90 Degree Elbows (pounds per piece)									
Size (inches)	**40**	**42**	**44**	**46**	**48**	**50**	**52**	**54**	**56**
40	74.6	78.4	82.1	85.9	89.7	93.4	97.2	102	125
42	78.4	82.3	86.2	90.1	94.0	97.9	102	106	130
44	82.1	86.2	90.3	94.4	98.5	102	106	110	135
46	85.9	90.1	94.4	98.7	102	106	110	113	140
48	89.7	94.0	98.5	102	106	110	113	117	144
50	93.4	97.9	102	106	110	113	117	122	149
52	97.2	102	106	110	113	117	122	126	154
54	102	106	110	113	117	122	126	131	159
56	126	130	134	139	145	150	155	160	165
58	132	137	142	147	151	156	161	166	170
60	137	142	147	152	157	162	167	172	176
62	143	148	153	158	163	168	173	178	182
64	150	155	160	165	169	174	179	184	188
66	156	161	166	171	176	181	186	191	195
68	163	168	173	178	182	187	192	197	201
70	170	175	180	185	189	194	199	204	208
72	176	181	185	190	194	199	204	210	215

Note: Elbow weights do not include turning vanes.

Galvanized Steel Rectangular 90 Degree Elbows

Weights of Galvanized Steel Rectangular 90 Degree Elbows (pounds per piece)								
Size (inches)	**58**	**60**	**62**	**64**	**66**	**68**	**70**	**72**
58	176	182	187	194	200	207	214	221
60	182	187	194	200	207	214	221	227
62	187	194	200	207	214	221	227	233
64	194	200	207	214	221	227	233	239
66	200	207	214	221	227	233	239	246
68	207	214	221	227	233	239	246	252
70	214	221	227	233	239	246	252	259
72	221	227	233	239	246	252	259	266

Note: Elbow weights do not include turning vanes.

Weights of Galvanized Steel Rectangular Drops and Tap-In Tees (pounds per piece)									
Size (inches)	**6**	**8**	**10**	**12**	**16**	**20**	**24**	**28**	**32**
6	1.81	2.11	2.42	2.72	4.24	5.01	5.78	6.55	8.90
8	2.11	2.42	2.72	3.02	4.62	5.40	6.17	6.94	9.37
10	2.42	2.72	3.02	3.32	5.01	5.78	6.55	7.32	9.84
12	2.72	3.02	3.32	3.62	5.40	6.17	6.94	7.71	10.3
16	4.24	4.62	5.01	5.40	6.17	6.94	7.71	8.48	11.3
20	5.01	5.40	5.78	6.17	6.94	7.71	8.48	9.25	12.2
24	5.78	6.17	6.55	6.94	7.71	8.48	9.25	10.0	13.1
28	6.55	6.94	7.32	7.71	8.48	9.25	10.0	10.8	14.1
32	8.90	9.37	9.84	10.3	11.3	12.2	13.1	14.1	15.0
36	9.84	10.3	10.8	11.3	12.2	13.1	14.1	15.0	15.9
40	10.8	11.3	11.7	12.2	13.1	14.1	15.0	15.9	16.9
44	11.7	12.2	12.7	13.1	14.1	15.0	15.9	16.9	17.8
48	12.7	13.1	13.6	14.1	15.0	15.9	16.9	17.8	18.7
52	13.6	14.1	14.5	15.0	15.9	16.9	17.8	18.7	19.7
56	16.0	17.6	18.2	18.8	19.6	21.0	22.1	23.2	24.3
60	18.7	19.0	19.3	19.9	21.0	22.1	23.2	24.3	25.4

Note: Weights of drops and tap-in tees do not include splitter dampers.

Costs for Steel Spiral Duct. Costs per linear foot of duct, based on quantities less than 1,000 pounds. Add 50% to the material cost for lined duct. These costs include delivery, typical bracing, cleats, scrap, hangers, end closures, sealants and miscellaneous hardware.

Galvanized steel spiral duct 26 gauge

Description	Craft@Hrs	Unit	Material $	Labor $	Equipment $	Total $
3"	S2@.030	LF	.89	.72	—	1.61
4"	S2@.041	LF	1.19	.99	—	2.18
5"	S2@.051	LF	1.49	1.23	—	2.72
6"	S2@.061	LF	1.78	1.47	—	3.25
7"	S2@.072	LF	2.07	1.74	—	3.81
8"	S2@.082	LF	2.38	1.98	—	4.36
9"	S2@.092	LF	2.68	2.22	—	4.90
10"	S2@.103	LF	2.98	2.49	—	5.47
12"	S2@.123	LF	3.57	2.97	—	6.54
14" - 16"	S2@.174	LF	4.46	4.20	—	8.66
18" - 20"	S2@.195	LF	5.67	4.71	—	10.38
22" - 24"	S2@.236	LF	6.83	5.70	—	12.53
26" - 28"	S2@.277	LF	8.01	6.69	—	14.70
30" - 32"	S2@.318	LF	9.21	7.68	—	16.89

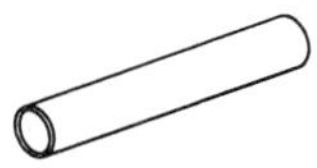

Galvanized steel spiral duct 24 gauge

Description	Craft@Hrs	Unit	Material $	Labor $	Equipment $	Total $
3"	S2@.041	LF	1.18	.99	—	2.17
4"	S2@.054	LF	1.56	1.30	—	2.86
5"	S2@.067	LF	1.96	1.62	—	3.58
6"	S2@.081	LF	2.35	1.96	—	4.31
7"	S2@.095	LF	2.74	2.29	—	5.03
8"	S2@.108	LF	3.14	2.61	—	5.75
9"	S2@.122	LF	3.53	2.95	—	6.48
10"	S2@.135	LF	3.92	3.26	—	7.18
12"	S2@.163	LF	4.72	3.94	—	8.66
14" - 16"	S2@.203	LF	5.88	4.90	—	10.78
18" - 20"	S2@.257	LF	7.46	6.21	—	13.67
22" - 24"	S2@.312	LF	9.02	7.53	—	16.55
26" - 28"	S2@.366	LF	10.60	8.84	—	19.44
30" - 32"	S2@.419	LF	12.10	10.10	—	22.20
34" - 36"	S2@.474	LF	13.80	11.40	—	25.20

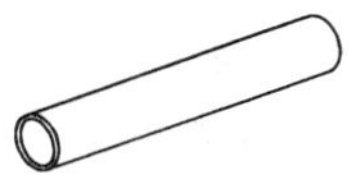

Galvanized Steel Spiral Duct

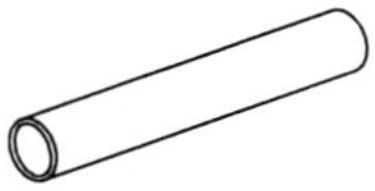

Description	Craft@Hrs	Unit	Material $	Labor $	Equipment $	Total $
Galvanized steel spiral duct 22 gauge						
3"	S2@.049	LF	1.43	1.18	—	2.61
4"	S2@.065	LF	1.91	1.57	—	3.48
5"	S2@.082	LF	2.39	1.98	—	4.37
6"	S2@.099	LF	2.86	2.39	—	5.25
7"	S2@.115	LF	3.33	2.78	—	6.11
8"	S2@.132	LF	3.81	3.19	—	7.00
9"	S2@.148	LF	4.29	3.57	—	7.86
10"	S2@.165	LF	4.77	3.98	—	8.75
12"	S2@.198	LF	5.73	4.78	—	10.51
14" - 16"	S2@.247	LF	7.15	5.96	—	13.11
18" - 20"	S2@.313	LF	9.07	7.56	—	16.63
22" - 24"	S2@.380	LF	11.00	9.18	—	20.18
26" - 28"	S2@.444	LF	12.90	10.70	—	23.60
30" - 32"	S2@.491	LF	14.80	11.90	—	26.70
34" - 36"	S2@.576	LF	16.80	13.90	—	30.70

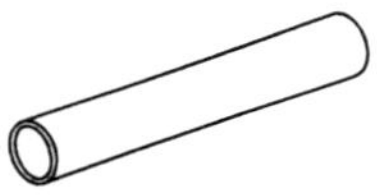

Description	Craft@Hrs	Unit	Material $	Labor $	Equipment $	Total $
Galvanized steel spiral duct 20 gauge						
3"	S2@.058	LF	1.70	1.40	—	3.10
4"	S2@.078	LF	2.25	1.88	—	4.13
5"	S2@.097	LF	2.81	2.34	—	5.15
6"	S2@.116	LF	3.37	2.80	—	6.17
7"	S2@.136	LF	3.94	3.28	—	7.22
8"	S2@.154	LF	4.47	3.72	—	8.19
9"	S2@.175	LF	5.05	4.23	—	9.28
10"	S2@.194	LF	5.62	4.68	—	10.30
12"	S2@.233	LF	6.75	5.63	—	12.38
14" - 16"	S2@.291	LF	8.44	7.03	—	15.47
18" - 20"	S2@.369	LF	10.70	8.91	—	19.61
22" - 24"	S2@.446	LF	12.90	10.80	—	23.70
26" - 28"	S2@.525	LF	15.20	12.70	—	27.90
30" - 32"	S2@.604	LF	17.40	14.60	—	32.00
34" - 36"	S2@.681	LF	19.70	16.40	—	36.10

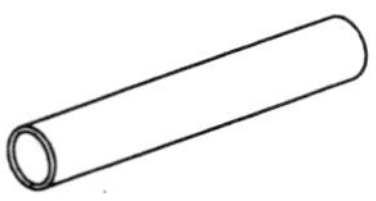

Description	Craft@Hrs	Unit	Material $	Labor $	Equipment $	Total $
Galvanized steel spiral duct 18 gauge						
6"	S2@.155	LF	4.49	3.74	—	8.23
7"	S2@.181	LF	5.23	4.37	—	9.60
8"	S2@.207	LF	5.98	5.00	—	10.98
9"	S2@.233	LF	6.74	5.63	—	12.37
10"	S2@.259	LF	7.49	6.25	—	13.74
12"	S2@.310	LF	8.97	7.49	—	16.46
14" - 16"	S2@.387	LF	11.20	9.35	—	20.55
18" - 20"	S2@.491	LF	14.20	11.90	—	26.10
22" - 24"	S2@.608	LF	17.60	14.70	—	32.30
26" - 28"	S2@.696	LF	20.20	16.80	—	37.00
30" - 32"	S2@.803	LF	23.20	19.40	—	42.60
34" - 36"	S2@.905	LF	26.20	21.90	—	48.10

Costs for Steel Spiral Duct Fittings. Costs per fitting, based on quantities less than 1,000 pounds. For quantities over 1,000 pounds, deduct 15% from material costs. For lined duct add 50%. These costs include delivery, typical bracing, cleats, scrap, hangers, end closures, sealants and miscellaneous hardware.

Description	Craft@Hrs	Unit	Material $	Labor $	Equipment $	Total $
Galvanized steel spiral duct 90 degree elbows						
3"	S2@.052	Ea	3.00	1.26	—	4.26
4"	S2@.088	Ea	5.05	2.13	—	7.18
5"	S2@.132	Ea	7.59	3.19	—	10.78
6"	S2@.172	Ea	9.90	4.15	—	14.05
7"	S2@.232	Ea	13.30	5.60	—	18.90
8"	S2@.292	Ea	16.80	7.05	—	23.85
9"	S2@.352	Ea	20.30	8.50	—	28.80
10"	S2@.470	Ea	27.10	11.30	—	38.40
12"	S2@.653	Ea	37.50	15.80	—	53.30
14" - 16"	S2@1.01	Ea	57.70	24.40	—	82.10
18" - 20"	S2@1.52	Ea	87.40	36.70	—	124.10
22" - 24"	S2@2.11	Ea	121.00	51.00	—	172.00
26" - 28"	S2@2.91	Ea	168.00	70.30	—	238.30
30" - 32"	S2@3.73	Ea	215.00	90.10	—	305.10
34" - 36"	S2@5.47	Ea	314.00	132.00	—	446.00

Description	Craft@Hrs	Unit	Material $	Labor $	Equipment $	Total $
Galvanized steel spiral duct 45 degree elbow						
3"	S2@.040	Ea	2.30	.97	—	3.27
4"	S2@.052	Ea	3.00	1.26	—	4.26
5"	S2@.076	Ea	4.36	1.84	—	6.20
6"	S2@.100	Ea	5.75	2.41	—	8.16
7"	S2@.132	Ea	7.59	3.19	—	10.78
8"	S2@.172	Ea	9.90	4.15	—	14.05
9"	S2@.212	Ea	12.10	5.12	—	17.22
10"	S2@.300	Ea	17.20	7.24	—	24.44
12"	S2@.400	Ea	23.00	9.66	—	32.66
14" - 16"	S2@.576	Ea	33.00	13.90	—	46.90
18" - 20"	S2@.850	Ea	49.00	20.50	—	69.50
22" - 24"	S2@1.19	Ea	68.40	28.70	—	97.10
26" - 28"	S2@1.61	Ea	92.30	38.90	—	131.20
30" - 32"	S2@2.06	Ea	118.00	49.70	—	167.70
34" - 36"	S2@3.01	Ea	173.00	72.70	—	245.70

Galvanized Steel Spiral Duct Fittings

Galvanized steel spiral duct coupling

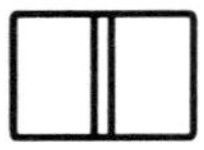

Description	Craft@Hrs	Unit	Material $	Labor $	Equipment $	Total $
3"	S2@.020	Ea	1.15	.48	—	1.63
4"	S2@.024	Ea	1.39	.58	—	1.97
5"	S2@.028	Ea	1.60	.68	—	2.28
6"	S2@.036	Ea	2.06	.87	—	2.93
7"	S2@.040	Ea	2.30	.97	—	3.27
8"	S2@.048	Ea	2.75	1.16	—	3.91
9"	S2@.104	Ea	5.98	2.51	—	8.49
10"	S2@.116	Ea	6.67	2.80	—	9.47
12"	S2@.140	Ea	8.05	3.38	—	11.43
14" - 16"	S2@.174	Ea	10.00	4.20	—	14.20
18" - 20"	S2@.220	Ea	12.60	5.31	—	17.91
22" - 24"	S2@.266	Ea	15.30	6.42	—	21.72
26" - 28"	S2@.312	Ea	17.90	7.53	—	25.43
30" - 32"	S2@.357	Ea	20.60	8.62	—	29.22
34" - 36"	S2@.403	Ea	23.20	9.73	—	32.93

Galvanized steel spiral duct with reducer

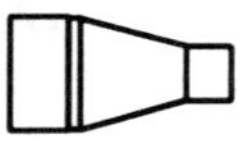

Description	Craft@Hrs	Unit	Material $	Labor $	Equipment $	Total $
3"	S2@.040	Ea	2.30	.97	—	3.27
4"	S2@.048	Ea	2.75	1.16	—	3.91
5"	S2@.056	Ea	3.22	1.35	—	4.57
6"	S2@.072	Ea	4.15	1.74	—	5.89
7"	S2@.080	Ea	4.59	1.93	—	6.52
8"	S2@.096	Ea	5.52	2.32	—	7.84
9"	S2@.132	Ea	7.59	3.19	—	10.78
10"	S2@.176	Ea	10.10	4.25	—	14.35
12"	S2@.212	Ea	12.10	5.12	—	17.22
14" - 16"	S2@.262	Ea	15.00	6.33	—	21.33
18" - 20"	S2@.330	Ea	19.00	7.97	—	26.97
22" - 24"	S2@.398	Ea	22.90	9.61	—	32.51
26" - 28"	S2@.465	Ea	26.80	11.20	—	38.00
30" - 32"	S2@.536	Ea	30.70	12.90	—	43.60
34" - 36"	S2@.606	Ea	34.80	14.60	—	49.40

Costs for Steel Spiral Tees with Reducing Branch. Costs per tee, based on quantities less than 1,000 pounds. Deduct 15% from the material cost for quantities over 1,000 pounds. Add 50% to the material cost for lined duct. These costs include delivery, typical bracing, cleats, scrap, hangers, end closures, sealants and miscellaneous hardware.

Galvanized steel spiral duct tees with 3" reducing branch

Description	Craft@Hrs	Unit	Material $	Labor $	Equipment $	Total $
4"	S2@.109	Ea	3.91	2.63	—	6.54
5"	S2@.112	Ea	4.59	2.70	—	7.29
6"	S2@.128	Ea	5.29	3.09	—	8.38
7"	S2@.148	Ea	5.98	3.57	—	9.55
8"	S2@.164	Ea	6.67	3.96	—	10.63
9"	S2@.180	Ea	7.12	4.35	—	11.47
10"	S2@.256	Ea	11.20	6.18	—	17.38
12"	S2@.308	Ea	13.30	7.44	—	20.74
14" - 16"	S2@.382	Ea	16.80	9.22	—	26.02
18" - 20"	S2@.485	Ea	21.20	11.70	—	32.90
22" - 24"	S2@.585	Ea	25.70	14.10	—	39.80
26" - 28"	S2@.683	Ea	30.00	16.50	—	46.50
30" - 32"	S2@.786	Ea	34.50	19.00	—	53.50
34" - 36"	S2@1.02	Ea	44.90	24.60	—	69.50

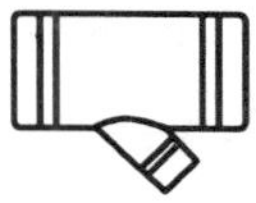

Galvanized steel spiral duct tees with 4" reducing branch

Description	Craft@Hrs	Unit	Material $	Labor $	Equipment $	Total $
5"	S2@.120	Ea	5.05	2.90	—	7.95
6"	S2@.140	Ea	5.98	3.38	—	9.36
7"	S2@.156	Ea	6.45	3.77	—	10.22
8"	S2@.176	Ea	7.35	4.25	—	11.60
9"	S2@.192	Ea	7.81	4.64	—	12.45
10"	S2@.268	Ea	11.90	6.47	—	18.37
12"	S2@.324	Ea	14.50	7.82	—	22.32
14" - 16"	S2@.403	Ea	18.00	9.73	—	27.73
18" - 20"	S2@.508	Ea	22.80	12.30	—	35.10
22" - 24"	S2@.619	Ea	27.50	14.90	—	42.40
26" - 28"	S2@.724	Ea	32.30	17.50	—	49.80
30" - 32"	S2@.830	Ea	37.10	20.00	—	57.10
34" - 36"	S2@1.08	Ea	48.40	26.10	—	74.50

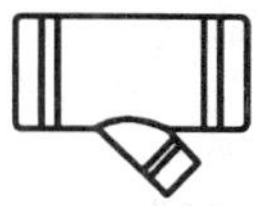

Galvanized steel spiral duct tees with 5" reducing branch

Description	Craft@Hrs	Unit	Material $	Labor $	Equipment $	Total $
6"	S2@.148	Ea	6.45	3.57	—	10.02
7"	S2@.164	Ea	7.12	3.96	—	11.08
8"	S2@.188	Ea	8.05	4.54	—	12.59
9"	S2@.204	Ea	8.74	4.93	—	13.67
10"	S2@.284	Ea	12.90	6.86	—	19.76
12"	S2@.340	Ea	15.40	8.21	—	23.61
14" - 16"	S2@.428	Ea	19.40	10.30	—	29.70
18" - 20"	S2@.538	Ea	24.60	13.00	—	37.60
22" - 24"	S2@.651	Ea	29.50	15.70	—	45.20
26" - 28"	S2@.764	Ea	34.60	18.40	—	53.00
30" - 32"	S2@.873	Ea	39.80	21.10	—	60.90
34" - 36"	S2@1.15	Ea	52.00	27.80	—	79.80

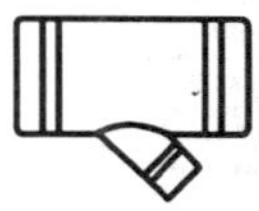

Galvanized Steel Spiral Tees

Galvanized steel spiral duct tees with 6" reducing branch

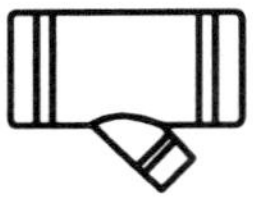

Description	Craft@Hrs	Unit	Material $	Labor $	Equipment $	Total $
7"	S2@.180	Ea	8.05	4.35	—	12.40
8"	S2@.200	Ea	8.74	4.83	—	13.57
9"	S2@.220	Ea	9.43	5.31	—	14.74
10"	S2@.300	Ea	13.80	7.24	—	21.04
12"	S2@.356	Ea	16.60	8.60	—	25.20
14" - 16"	S2@.449	Ea	20.70	10.80	—	31.50
18" - 20"	S2@.566	Ea	26.10	13.70	—	39.80
22" - 24"	S2@.683	Ea	31.40	16.50	—	47.90
26" - 28"	S2@.807	Ea	36.90	19.50	—	56.40
30" - 32"	S2@.924	Ea	42.40	22.30	—	64.70
34" - 36"	S2@1.20	Ea	55.20	29.00	—	84.20

Galvanized steel spiral duct tees with 7" reducing branch

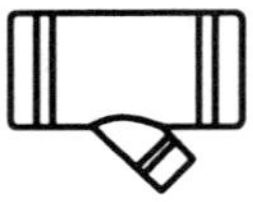

Description	Craft@Hrs	Unit	Material $	Labor $	Equipment $	Total $
8"	S2@.164	Ea	9.43	3.96	—	13.39
9"	S2@.180	Ea	10.30	4.35	—	14.65
10"	S2@.252	Ea	14.50	6.09	—	20.59
12"	S2@.304	Ea	17.50	7.34	—	24.84
14" - 16"	S2@.381	Ea	21.90	9.20	—	31.10
18" - 20"	S2@.482	Ea	27.60	11.60	—	39.20
22" - 24"	S2@.579	Ea	33.20	14.00	—	47.20
26" - 28"	S2@.683	Ea	39.40	16.50	—	55.90
30" - 32"	S2@.784	Ea	45.00	18.90	—	63.90
34" - 36"	S2@1.02	Ea	58.80	24.60	—	83.40

Galvanized steel spiral duct tees with 8" reducing branch

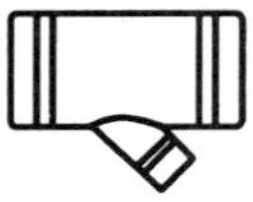

Description	Craft@Hrs	Unit	Material $	Labor $	Equipment $	Total $
9"	S2@.228	Ea	13.10	5.51	—	18.61
10"	S2@.268	Ea	15.40	6.47	—	21.87
12"	S2@.324	Ea	18.60	7.82	—	26.42
14" - 16"	S2@.401	Ea	23.10	9.68	—	32.78
18" - 20"	S2@.508	Ea	29.30	12.30	—	41.60
22" - 24"	S2@.617	Ea	35.30	14.90	—	50.20
26" - 28"	S2@.722	Ea	41.60	17.40	—	59.00
30" - 32"	S2@.830	Ea	47.80	20.00	—	67.80
34" - 36"	S2@1.08	Ea	62.20	26.10	—	88.30

Galvanized steel spiral duct tees with 9" reducing branch

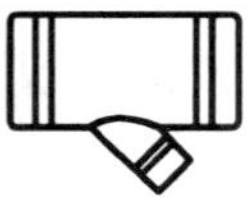

Description	Craft@Hrs	Unit	Material $	Labor $	Equipment $	Total $
10"	S2@.284	Ea	16.40	6.86	—	23.26
12"	S2@.340	Ea	19.50	8.21	—	27.71
14" - 16"	S2@.426	Ea	24.60	10.30	—	34.90
18" - 20"	S2@.540	Ea	30.90	13.00	—	43.90
22" - 24"	S2@.649	Ea	37.30	15.70	—	53.00
26" - 28"	S2@.760	Ea	43.70	18.40	—	62.10
30" - 32"	S2@.875	Ea	50.30	21.10	—	71.40
34" - 36"	S2@1.15	Ea	65.80	27.80	—	93.60

Description	Craft@Hrs	Unit	Material $	Labor $	Equipment $	Total $
Galvanized steel spiral duct tees with 10" reducing branch						
12"	S2@.360	Ea	20.70	8.69	—	29.39
14" - 16"	S2@.447	Ea	25.70	10.80	—	36.50
18" - 20"	S2@.564	Ea	32.50	13.60	—	46.10
22" - 24"	S2@.683	Ea	39.40	16.50	—	55.90
26" - 28"	S2@.805	Ea	46.10	19.40	—	65.50
30" - 32"	S2@.924	Ea	53.10	22.30	—	75.40
34" - 36"	S2@1.20	Ea	69.20	29.00	—	98.20
Galvanized steel spiral duct tees with 12" reducing branch						
14" - 16"	S2@.572	Ea	28.30	13.80	—	42.10
18" - 20"	S2@.681	Ea	35.80	16.40	—	52.20
22" - 24"	S2@.820	Ea	43.30	19.80	—	63.10
26" - 28"	S2@.965	Ea	50.60	23.30	—	73.90
30" - 32"	S2@1.11	Ea	58.30	26.80	—	85.10
34" - 36"	S2@1.44	Ea	76.10	34.80	—	110.90
Galvanized steel spiral duct tees with 14" reducing branch						
14" - 16"	S2@.572	Ea	32.80	13.80	—	46.60
18" - 20"	S2@.681	Ea	39.10	16.40	—	55.50
22" - 24"	S2@.820	Ea	47.20	19.80	—	67.00
26" - 28"	S2@.965	Ea	55.50	23.30	—	78.80
30" - 32"	S2@1.11	Ea	63.60	26.80	—	90.40
34" - 36"	S2@1.44	Ea	83.00	34.80	—	117.80
Galvanized steel spiral duct tees with 16" reducing branch						
18" - 20"	S2@.739	Ea	42.40	17.80	—	60.20
22" - 24"	S2@.888	Ea	51.10	21.40	—	72.50
26" - 28"	S2@1.04	Ea	60.00	25.10	—	85.10
30" - 32"	S2@1.20	Ea	69.00	29.00	—	98.00
34" - 36"	S2@1.56	Ea	89.90	37.70	—	127.60
Galvanized steel spiral duct tees with 18" reducing branch						
18" - 20"	S2@.837	Ea	48.00	20.20	—	68.20
22" - 24"	S2@.954	Ea	55.00	23.00	—	78.00
26" - 28"	S2@1.12	Ea	64.60	27.00	—	91.60
30" - 32"	S2@1.29	Ea	74.30	31.20	—	105.50
34" - 36"	S2@1.68	Ea	96.80	40.60	—	137.40
Galvanized steel spiral duct tees with 20" reducing branch						
22" - 24"	S2@1.02	Ea	59.00	24.60	—	83.60
26" - 28"	S2@1.20	Ea	69.30	29.00	—	98.30
30" - 32"	S2@1.38	Ea	79.50	33.30	—	112.80
34" - 36"	S2@1.81	Ea	104.00	43.70	—	147.70
Galvanized steel spiral duct tees with 22" reducing branch						
22" - 24"	S2@1.57	Ea	65.40	37.90	—	103.30
26" - 28"	S2@1.32	Ea	76.10	31.90	—	108.00
30" - 32"	S2@1.48	Ea	84.90	35.70	—	120.60
34" - 36"	S2@1.92	Ea	111.00	46.40	—	157.40

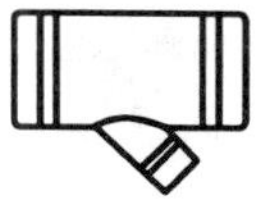
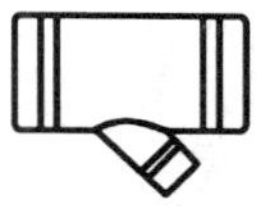
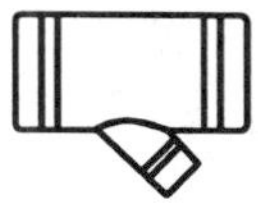
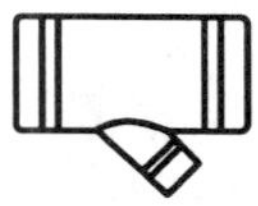

Galvanized Steel Spiral Tees

Costs for Steel Spiral Tees with Reducing Run and Branch. Costs per tee, based on quantities less than 1,000 pounds. Deduct 15% from the material cost for quantities over 1,000 pounds. Add 50% to the material cost for lined duct. These costs include delivery, typical bracing, cleats, scrap, hangers, end closures, sealants and miscellaneous hardware.

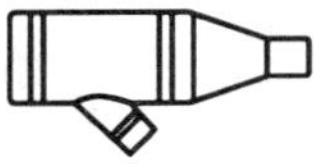

Galvanized steel spiral duct tees with 3" run and branch

Description	Craft@Hrs	Unit	Material $	Labor $	Equipment $	Total $
4"	S2@.109	Ea	6.31	2.63	—	8.94
5"	S2@.112	Ea	6.45	2.70	—	9.15
6"	S2@.128	Ea	7.35	3.09	—	10.44
7"	S2@.148	Ea	8.50	3.57	—	12.07
8"	S2@.164	Ea	9.43	3.96	—	13.39
9"	S2@.180	Ea	10.30	4.35	—	14.65
10"	S2@.256	Ea	14.70	6.18	—	20.88
12"	S2@.308	Ea	17.70	7.44	—	25.14
14" - 16"	S2@.382	Ea	21.90	9.22	—	31.12
18" - 20"	S2@.485	Ea	27.90	11.70	—	39.60
22" - 24"	S2@.585	Ea	33.70	14.10	—	47.80
26" - 28"	S2@.683	Ea	39.40	16.50	—	55.90
30" - 32"	S2@.786	Ea	45.20	19.00	—	64.20
34" - 36"	S2@1.02	Ea	58.80	24.60	—	83.40

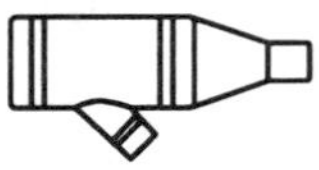

Galvanized steel spiral duct tees with 4" run and branch

Description	Craft@Hrs	Unit	Material $	Labor $	Equipment $	Total $
5"	S2@.120	Ea	6.90	2.90	—	9.80
6"	S2@.140	Ea	8.05	3.38	—	11.43
7"	S2@.156	Ea	8.97	3.77	—	12.74
8"	S2@.176	Ea	10.10	4.25	—	14.35
9"	S2@.192	Ea	11.00	4.64	—	15.64
10"	S2@.268	Ea	15.40	6.47	—	21.87
12"	S2@.324	Ea	18.60	7.82	—	26.42
14" - 16"	S2@.403	Ea	23.20	9.73	—	32.93
18" - 20"	S2@.508	Ea	29.30	12.30	—	41.60
22" - 24"	S2@.619	Ea	35.40	14.90	—	50.30
26" - 28"	S2@.724	Ea	41.70	17.50	—	59.20
30" - 32"	S2@.830	Ea	47.80	20.00	—	67.80
34" - 36"	S2@1.08	Ea	62.40	26.10	—	88.50

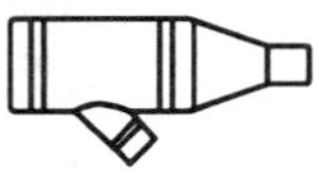

Galvanized steel spiral duct tees with 5" run and branch

Description	Craft@Hrs	Unit	Material $	Labor $	Equipment $	Total $
6"	S2@.148	Ea	8.50	3.57	—	12.07
7"	S2@.164	Ea	9.43	3.96	—	13.39
8"	S2@.188	Ea	10.80	4.54	—	15.34
9"	S2@.204	Ea	11.70	4.93	—	16.63
10"	S2@.284	Ea	16.40	6.86	—	23.26
12"	S2@.340	Ea	19.50	8.21	—	27.71
14" - 16"	S2@.428	Ea	24.60	10.30	—	34.90
18" - 20"	S2@.538	Ea	30.80	13.00	—	43.80
22" - 24"	S2@.651	Ea	37.50	15.70	—	53.20
26" - 28"	S2@.764	Ea	44.10	18.40	—	62.50
30" - 32"	S2@.873	Ea	50.20	21.10	—	71.30
34" - 36"	S2@1.15	Ea	65.80	27.80	—	93.60

Description	Craft@Hrs	Unit	Material $	Labor $	Equipment $	Total $
Galvanized steel spiral duct tees with 6" run and branch						
7"	S2@.180	Ea	10.30	4.35	—	14.65
8"	S2@.200	Ea	11.50	4.83	—	16.33
9"	S2@.220	Ea	12.60	5.31	—	17.91
10"	S2@.300	Ea	17.20	7.24	—	24.44
12"	S2@.356	Ea	20.50	8.60	—	29.10
14" - 16"	S2@.449	Ea	25.80	10.80	—	36.60
18" - 20"	S2@.566	Ea	32.60	13.70	—	46.30
22" - 24"	S2@.683	Ea	39.40	16.50	—	55.90
26" - 28"	S2@.807	Ea	46.20	19.50	—	65.70
30" - 32"	S2@.924	Ea	53.10	22.30	—	75.40
34" - 36"	S2@1.20	Ea	69.20	29.00	—	98.20
Galvanized steel spiral duct tees with 7" run and branch						
8"	S2@.212	Ea	12.10	5.12	—	17.22
9"	S2@.232	Ea	13.30	5.60	—	18.90
10"	S2@.312	Ea	17.90	7.53	—	25.43
12"	S2@.376	Ea	21.60	9.08	—	30.68
14" - 16"	S2@.470	Ea	27.00	11.30	—	38.30
18" - 20"	S2@.636	Ea	36.60	15.40	—	52.00
22" - 24"	S2@.717	Ea	41.40	17.30	—	58.70
26" - 28"	S2@.845	Ea	48.50	20.40	—	68.90
30" - 32"	S2@.969	Ea	55.70	23.40	—	79.10
34" - 36"	S2@1.27	Ea	72.70	30.70	—	103.40
Galvanized steel spiral duct tees with 8" run and branch						
9"	S2@.248	Ea	14.20	5.99	—	20.19
10"	S2@.328	Ea	18.90	7.92	—	26.82
12"	S2@.396	Ea	22.70	9.56	—	32.26
14" - 16"	S2@.493	Ea	28.20	11.90	—	40.10
18" - 20"	S2@.623	Ea	35.80	15.00	—	50.80
22" - 24"	S2@.751	Ea	43.30	18.10	—	61.40
26" - 28"	S2@.882	Ea	50.70	21.30	—	72.00
30" - 32"	S2@1.02	Ea	58.30	24.60	—	82.90
34" - 36"	S2@1.33	Ea	76.50	32.10	—	108.60
Galvanized steel spiral duct tees with 9" run and branch						
10"	S2@.344	Ea	19.70	8.31	—	28.01
12"	S2@.412	Ea	23.60	9.95	—	33.55
14" - 16"	S2@.515	Ea	29.60	12.40	—	42.00
18" - 20"	S2@.653	Ea	37.60	15.80	—	53.40
22" - 24"	S2@.779	Ea	45.30	18.80	—	64.10
26" - 28"	S2@.924	Ea	53.00	22.30	—	75.30
30" - 32"	S2@1.06	Ea	60.90	25.60	—	86.50
34" - 36"	S2@1.39	Ea	79.70	33.60	—	113.30

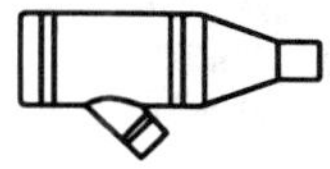

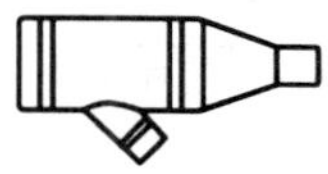

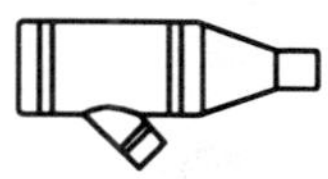

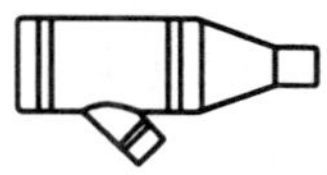

Galvanized Steel Spiral Tees

Galvanized steel spiral duct tees with 10" run and branch

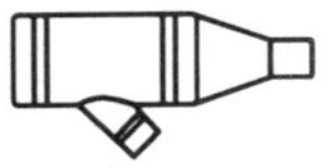

Description	Craft@Hrs	Unit	Material $	Labor $	Equipment $	Total $
12"	S2@.427	Ea	12.30	10.30	—	22.60
14" - 16"	S2@.538	Ea	15.50	13.00	—	28.50
18" - 20"	S2@.681	Ea	39.00	16.40	—	55.40
22" - 24"	S2@.820	Ea	47.20	19.80	—	67.00
26" - 28"	S2@.965	Ea	55.40	23.30	—	78.70
30" - 32"	S2@1.11	Ea	63.60	26.80	—	90.40
34" - 36"	S2@1.44	Ea	82.90	34.80	—	117.70

Galvanized steel spiral duct tees with 12" run and branch

Description	Craft@Hrs	Unit	Material $	Labor $	Equipment $	Total $
14" - 16"	S2@.581	Ea	33.30	14.00	—	47.30
18" - 20"	S2@.739	Ea	42.40	17.80	—	60.20
22" - 24"	S2@.888	Ea	51.10	21.40	—	72.50
26" - 28"	S2@1.04	Ea	59.90	25.10	—	85.00
30" - 32"	S2@1.20	Ea	69.00	29.00	—	98.00
34" - 36"	S2@1.57	Ea	90.00	37.90	—	127.90

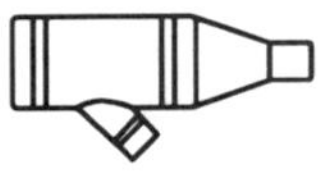

Galvanized steel spiral duct tees with 14" run and branch

Description	Craft@Hrs	Unit	Material $	Labor $	Equipment $	Total $
14" - 16"	S2@.666	Ea	38.30	16.10	—	54.40
18" - 20"	S2@.794	Ea	45.60	19.20	—	64.80
22" - 24"	S2@.959	Ea	55.10	23.20	—	78.30
26" - 28"	S2@1.11	Ea	63.50	26.80	—	90.30
30" - 32"	S2@1.29	Ea	74.30	31.20	—	105.50
34" - 36"	S2@1.69	Ea	96.80	40.80	—	137.60

Galvanized steel spiral duct tees with 16" run and branch

Description	Craft@Hrs	Unit	Material $	Labor $	Equipment $	Total $
18" - 20"	S2@.833	Ea	47.80	20.10	—	67.90
22" - 24"	S2@1.03	Ea	59.20	24.90	—	84.10
26" - 28"	S2@1.20	Ea	69.30	29.00	—	98.30
30" - 32"	S2@1.39	Ea	79.70	33.60	—	113.30
34" - 36"	S2@1.81	Ea	104.00	43.70	—	147.70

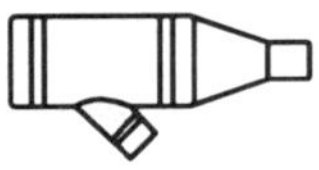

Galvanized steel spiral duct tees with 18" run and branch

Description	Craft@Hrs	Unit	Material $	Labor $	Equipment $	Total $
18" - 20"	S2@.956	Ea	55.00	23.10	—	78.10
22" - 24"	S2@1.09	Ea	62.90	26.30	—	89.20
26" - 28"	S2@1.29	Ea	74.00	31.20	—	105.20
30" - 32"	S2@1.48	Ea	84.90	35.70	—	120.60
34" - 36"	S2@1.93	Ea	111.00	46.60	—	157.60

Galvanized steel spiral duct tees with 20" run and branch

Description	Craft@Hrs	Unit	Material $	Labor $	Equipment $	Total $
22" - 24"	S2@1.16	Ea	66.90	28.00	—	94.90
26" - 28"	S2@1.37	Ea	78.50	33.10	—	111.60
30" - 32"	S2@1.57	Ea	90.20	37.90	—	128.10
34" - 36"	S2@2.03	Ea	117.00	49.00	—	166.00

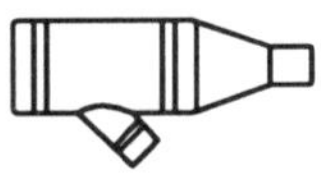

Galvanized steel spiral duct tees with 22" run and branch

Description	Craft@Hrs	Unit	Material $	Labor $	Equipment $	Total $
22" - 24"	S2@1.28	Ea	73.70	30.90	—	104.60
26" - 28"	S2@1.45	Ea	83.00	35.00	—	118.00
30" - 32"	S2@1.64	Ea	94.50	39.60	—	134.10
34" - 36"	S2@2.16	Ea	124.00	52.20	—	176.20

Costs for Steel Spiral Crosses with Reducing Branches. Costs per cross, based on quantities less than 1,000 pounds. Deduct 15% from the material cost for quantities over 1,000 pounds. Add 50% to the material cost for lined duct. These costs include delivery, typical bracing, cleats, scrap, hangers, end closures, sealants and miscellaneous hardware.

Galvanized steel spiral duct crosses with 3" branch

Description	Craft@Hrs	Unit	Material $	Labor $	Equipment $	Total $
4"	S2@.104	Ea	5.98	2.51	—	8.49
5"	S2@.128	Ea	7.35	3.09	—	10.44
6"	S2@.152	Ea	8.74	3.67	—	12.41
7"	S2@.176	Ea	10.10	4.25	—	14.35
8"	S2@.196	Ea	11.20	4.73	—	15.93
9"	S2@.216	Ea	12.30	5.22	—	17.52
10"	S2@.304	Ea	17.50	7.34	—	24.84
12"	S2@.352	Ea	20.30	8.50	—	28.80
14" - 16"	S2@.428	Ea	24.70	10.30	—	35.00
18" - 20"	S2@.525	Ea	30.20	12.70	—	42.90
22" - 24"	S2@.634	Ea	36.60	15.30	—	51.90
26" - 28"	S2@.745	Ea	42.90	18.00	—	60.90
30" - 32"	S2@.856	Ea	49.20	20.70	—	69.90
34" - 36"	S2@1.11	Ea	64.00	26.80	—	90.80

Galvanized steel spiral duct crosses with 4" branch

Description	Craft@Hrs	Unit	Material $	Labor $	Equipment $	Total $
5"	S2@.144	Ea	8.28	3.48	—	11.76
6"	S2@.168	Ea	9.65	4.06	—	13.71
7"	S2@.196	Ea	11.20	4.73	—	15.93
8"	S2@.216	Ea	12.30	5.22	—	17.52
9"	S2@.240	Ea	13.80	5.80	—	19.60
10"	S2@.328	Ea	18.90	7.92	—	26.82
12"	S2@.384	Ea	22.20	9.27	—	31.47
14" - 16"	S2@.461	Ea	26.60	11.10	—	37.70
18" - 20"	S2@.568	Ea	32.70	13.70	—	46.40
22" - 24"	S2@.685	Ea	39.50	16.50	—	56.00
26" - 28"	S2@.805	Ea	46.10	19.40	—	65.50
30" - 32"	S2@.927	Ea	53.20	22.40	—	75.60
34" - 36"	S2@1.20	Ea	69.20	29.00	—	98.20

Galvanized steel spiral duct crosses with 5" branch

Description	Craft@Hrs	Unit	Material $	Labor $	Equipment $	Total $
6"	S2@.184	Ea	10.60	4.44	—	15.04
7"	S2@.216	Ea	12.30	5.22	—	17.52
8"	S2@.240	Ea	13.80	5.80	—	19.60
9"	S2@.264	Ea	15.20	6.38	—	21.58
10"	S2@.356	Ea	20.50	8.60	—	29.10
12"	S2@.412	Ea	23.60	9.95	—	33.55
14" - 16"	S2@.497	Ea	28.40	12.00	—	40.40
18" - 20"	S2@.613	Ea	35.10	14.80	—	49.90
22" - 24"	S2@.741	Ea	42.50	17.90	—	60.40
26" - 28"	S2@.867	Ea	49.70	20.90	—	70.60
30" - 32"	S2@.993	Ea	57.20	24.00	—	81.20
34" - 36"	S2@1.29	Ea	74.40	31.20	—	105.60

Galvanized Steel Spiral Crosses

Description	Craft@Hrs	Unit	Material $	Labor $	Equipment $	Total $
Galvanized steel spiral duct crosses with 6" branch						
7"	S2@.232	Ea	13.30	5.60	—	18.90
8"	S2@.260	Ea	14.90	6.28	—	21.18
9"	S2@.288	Ea	16.60	6.95	—	23.55
10"	S2@.380	Ea	21.80	9.18	—	30.98
12"	S2@.440	Ea	25.30	10.60	—	35.90
14" - 16"	S2@.532	Ea	30.50	12.80	—	43.30
18" - 20"	S2@.653	Ea	37.60	15.80	—	53.40
22" - 24"	S2@.790	Ea	45.40	19.10	—	64.50
26" - 28"	S2@.927	Ea	53.20	22.40	—	75.60
30" - 32"	S2@1.06	Ea	61.10	25.60	—	86.70
34" - 36"	S2@1.38	Ea	79.50	33.30	—	112.80
Galvanized steel spiral duct crosses with 7" branch						
8"	S2@.280	Ea	16.00	6.76	—	22.76
9"	S2@.308	Ea	17.70	7.44	—	25.14
10"	S2@.408	Ea	23.40	9.85	—	33.25
12"	S2@.470	Ea	26.90	11.30	—	38.20
14" - 16"	S2@.566	Ea	32.60	13.70	—	46.30
18" - 20"	S2@.692	Ea	39.90	16.70	—	56.60
22" - 24"	S2@.841	Ea	48.30	20.30	—	68.60
26" - 28"	S2@.986	Ea	56.70	23.80	—	80.50
30" - 32"	S2@1.13	Ea	65.20	27.30	—	92.50
34" - 36"	S2@1.48	Ea	84.90	35.70	—	120.60
Galvanized steel spiral duct crosses with 8" branch						
9"	S2@.310	Ea	19.10	7.49	—	26.59
10"	S2@.403	Ea	24.90	9.73	—	34.63
12"	S2@.470	Ea	28.60	11.30	—	39.90
14" - 16"	S2@.561	Ea	34.60	13.50	—	48.10
18" - 20"	S2@.690	Ea	42.40	16.70	—	59.10
22" - 24"	S2@.835	Ea	51.40	20.20	—	71.60
26" - 28"	S2@.980	Ea	60.10	23.70	—	83.80
30" - 32"	S2@1.12	Ea	69.10	27.00	—	96.10
34" - 36"	S2@1.47	Ea	90.10	35.50	—	125.60
Galvanized steel spiral duct crosses with 9" branch						
10"	S2@.427	Ea	26.10	10.30	—	36.40
12"	S2@.495	Ea	30.30	12.00	—	42.30
14" - 16"	S2@.596	Ea	36.70	14.40	—	51.10
18" - 20"	S2@.732	Ea	44.90	17.70	—	62.60
22" - 24"	S2@.882	Ea	54.20	21.30	—	75.50
26" - 28"	S2@1.04	Ea	63.50	25.10	—	88.60
30" - 32"	S2@1.19	Ea	73.10	28.70	—	101.80
34" - 36"	S2@1.55	Ea	95.30	37.40	—	132.70

Galvanized steel spiral duct crosses with 12" branch

Description	Craft@Hrs	Unit	Material $	Labor $	Equipment $	Total $
14" - 16"	S2@.692	Ea	42.50	16.70	—	59.20
18" - 20"	S2@.850	Ea	52.20	20.50	—	72.70
22" - 24"	S2@1.03	Ea	63.10	24.90	—	88.00
26" - 28"	S2@1.18	Ea	72.80	28.50	—	101.30
30" - 32"	S2@1.38	Ea	85.00	33.30	—	118.30
34" - 36"	S2@1.80	Ea	111.00	43.50	—	154.50

Galvanized steel spiral duct crosses with 14" branch

Description	Craft@Hrs	Unit	Material $	Labor $	Equipment $	Total $
14" - 16"	S2@1.81	Ea	21.60	43.70	—	65.30
18" - 20"	S2@2.11	Ea	25.10	51.00	—	76.10
22" - 24"	S2@2.55	Ea	30.50	61.60	—	92.10
26" - 28"	S2@2.98	Ea	35.60	72.00	—	107.60
30" - 32"	S2@3.43	Ea	40.90	82.80	—	123.70
34" - 36"	S2@4.47	Ea	53.40	108.00	—	161.40

Galvanized steel spiral duct crosses with 16" branch

Description	Craft@Hrs	Unit	Material $	Labor $	Equipment $	Total $
18" - 20"	S2@1.08	Ea	61.90	26.10	—	88.00
22" - 24"	S2@1.30	Ea	75.00	31.40	—	106.40
26" - 28"	S2@1.53	Ea	87.90	36.90	—	124.80
30" - 32"	S2@1.76	Ea	101.00	42.50	—	143.50
34" - 36"	S2@2.31	Ea	132.00	55.80	—	187.80

Galvanized steel spiral duct crosses with 18" branch

Description	Craft@Hrs	Unit	Material $	Labor $	Equipment $	Total $
18" - 20"	S2@1.22	Ea	70.30	29.50	—	99.80
22" - 24"	S2@1.40	Ea	80.70	33.80	—	114.50
26" - 28"	S2@1.65	Ea	94.80	39.80	—	134.60
30" - 32"	S2@1.89	Ea	109.00	45.60	—	154.60
34" - 36"	S2@2.47	Ea	142.00	59.60	—	201.60

Galvanized steel spiral duct crosses with 20" branch

Description	Craft@Hrs	Unit	Material $	Labor $	Equipment $	Total $
22" - 24"	S2@1.51	Ea	86.70	36.50	—	123.20
26" - 28"	S2@1.77	Ea	102.00	42.70	—	144.70
30" - 32"	S2@2.05	Ea	118.00	49.50	—	167.50
34" - 36"	S2@2.65	Ea	152.00	64.00	—	216.00

Galvanized steel spiral duct crosses with 22" branch

Description	Craft@Hrs	Unit	Material $	Labor $	Equipment $	Total $
22" - 24"	S2@1.68	Ea	96.80	40.60	—	137.40
26" - 28"	S2@1.89	Ea	109.00	45.60	—	154.60
30" - 32"	S2@2.17	Ea	124.00	52.40	—	176.40
34" - 36"	S2@2.84	Ea	164.00	68.60	—	232.60

Galvanized Steel Rectangular Ductwork

Costs for Steel 26 Gauge Rectangular Duct. Costs per linear foot of duct, based on quantities less than 1,000 pounds of prefabricated duct. Deduct 15% from the material cost for quantities over 1,000 pounds. Add 50% to the material cost for lined duct. These costs include delivery, typical bracing, cleats, scrap, hangers, end closures, sealants and miscellaneous hardware.

Galvanized steel 4" rectangular duct 26 gauge

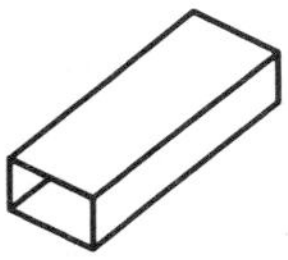

Description	Craft@Hrs	Unit	Material $	Labor $	Equipment $	Total $
4"	S2@.048	LF	1.41	1.16	—	2.57
5"	S2@.054	LF	1.57	1.30	—	2.87
6"	S2@.060	LF	1.75	1.45	—	3.20
7"	S2@.066	LF	1.93	1.59	—	3.52
8"	S2@.072	LF	2.10	1.74	—	3.84
9"	S2@.078	LF	2.27	1.88	—	4.15
10"	S2@.084	LF	2.45	2.03	—	4.48
11"	S2@.091	LF	2.64	2.20	—	4.84
12"	S2@.097	LF	2.81	2.34	—	5.15

Galvanized steel 5" rectangular duct 26 gauge

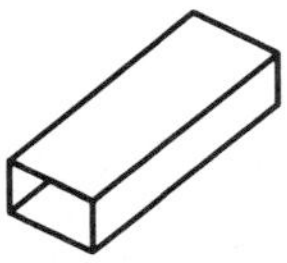

Description	Craft@Hrs	Unit	Material $	Labor $	Equipment $	Total $
5"	S2@.060	LF	1.75	1.45	—	3.20
6"	S2@.066	LF	1.93	1.59	—	3.52
7"	S2@.072	LF	2.10	1.74	—	3.84
8"	S2@.078	LF	2.27	1.88	—	4.15
9"	S2@.084	LF	2.45	2.03	—	4.48
10"	S2@.091	LF	2.64	2.20	—	4.84
11"	S2@.097	LF	2.81	2.34	—	5.15
12"	S2@.103	LF	2.98	2.49	—	5.47

Galvanized steel 6" rectangular duct 26 gauge

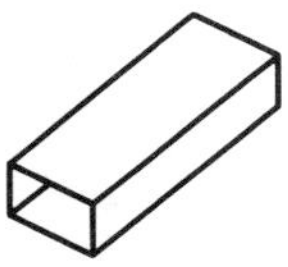

Description	Craft@Hrs	Unit	Material $	Labor $	Equipment $	Total $
6"	S2@.072	LF	2.10	1.74	—	3.84
7"	S2@.078	LF	2.27	1.88	—	4.15
8"	S2@.084	LF	2.45	2.03	—	4.48
9"	S2@.091	LF	2.64	2.20	—	4.84
10"	S2@.097	LF	2.81	2.34	—	5.15
11"	S2@.103	LF	2.98	2.49	—	5.47
12"	S2@.109	LF	3.15	2.63	—	5.78

Galvanized steel 7" rectangular duct 26 gauge

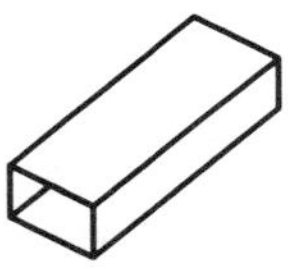

Description	Craft@Hrs	Unit	Material $	Labor $	Equipment $	Total $
7"	S2@.084	LF	2.45	2.03	—	4.48
8"	S2@.091	LF	2.64	2.20	—	4.84
9"	S2@.097	LF	2.81	2.34	—	5.15
10"	S2@.103	LF	2.98	2.49	—	5.47
11"	S2@.109	LF	3.15	2.63	—	5.78
12"	S2@.115	LF	3.32	2.78	—	6.10

Description	Craft@Hrs	Unit	Material $	Labor $	Equipment $	Total $

Galvanized steel 8" rectangular duct 26 gauge

Description	Craft@Hrs	Unit	Material $	Labor $	Equipment $	Total $
8"	S2@.097	LF	2.81	2.34	—	5.15
9"	S2@.103	LF	2.98	2.49	—	5.47
10"	S2@.109	LF	3.15	2.63	—	5.78
11"	S2@.115	LF	3.32	2.78	—	6.10
12"	S2@.121	LF	3.50	2.92	—	6.42

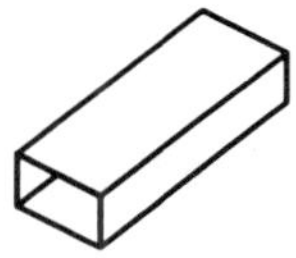

Galvanized steel 9" rectangular duct 26 gauge

Description	Craft@Hrs	Unit	Material $	Labor $	Equipment $	Total $
9"	S2@.109	LF	3.15	2.63	—	5.78
10"	S2@.115	LF	3.32	2.78	—	6.10
11"	S2@.121	LF	3.50	2.92	—	6.42
12"	S2@.127	LF	3.67	3.07	—	6.74

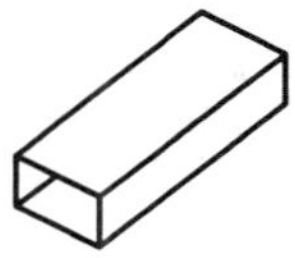

Galvanized steel 10" rectangular duct 26 gauge

Description	Craft@Hrs	Unit	Material $	Labor $	Equipment $	Total $
10"	S2@.129	LF	3.50	3.12	—	6.62
11"	S2@.127	LF	3.67	3.07	—	6.74
12"	S2@.133	LF	3.84	3.21	—	7.05

Galvanized steel 11" rectangular duct 26 gauge

Description	Craft@Hrs	Unit	Material $	Labor $	Equipment $	Total $
11"	S2@.133	LF	3.84	3.21	—	7.05
12"	S2@.139	LF	4.02	3.36	—	7.38

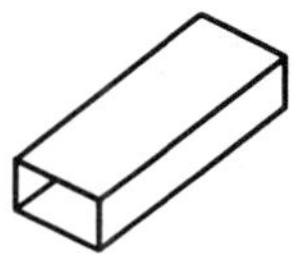

Galvanized steel 12" rectangular duct 26 gauge

Description	Craft@Hrs	Unit	Material $	Labor $	Equipment $	Total $
12"	S2@.145	LF	4.20	3.50	—	7.70

Galvanized Steel Rectangular Ductwork

Costs for Galvanized 24 Gauge Rectangular Duct. Costs per linear foot of duct, based on quantities less than 1,000 pounds of prefabricated duct. Deduct 15% from the material cost for over 1,000 pounds. Add 50% to the material cost for lined duct. These costs include delivery, typical bracing, cleats, scrap, hangers, end closures.

14" wide galvanized steel rectangular duct 24 gauge

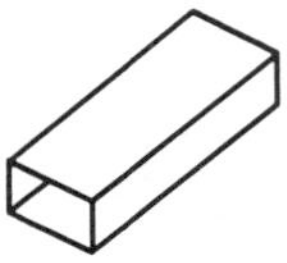

Description	Craft@Hrs	Unit	Material $	Labor $	Equipment $	Total $
6"	S2@.154	LF	4.50	3.72	—	8.22
8"	S2@.170	LF	4.96	4.11	—	9.07
10"	S2@.184	LF	5.40	4.44	—	9.84
12"	S2@.200	LF	5.86	4.83	—	10.69
14" - 16"	S2@.223	LF	6.53	5.39	—	11.92
18" - 20"	S2@.254	LF	7.42	6.13	—	13.55
22" - 24"	S2@.285	LF	8.33	6.88	—	15.21
26" - 28"	S2@.316	LF	9.23	7.63	—	16.86
30"	S2@.339	LF	9.90	8.19	—	18.09

16" wide galvanized steel rectangular duct 24 gauge

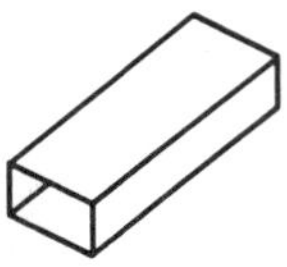

Description	Craft@Hrs	Unit	Material $	Labor $	Equipment $	Total $
6"	S2@.170	LF	4.96	4.11	—	9.07
8"	S2@.184	LF	5.40	4.44	—	9.84
10"	S2@.200	LF	5.86	4.83	—	10.69
12"	S2@.216	LF	6.30	5.22	—	11.52
14" - 16"	S2@.239	LF	6.98	5.77	—	12.75
18" - 20"	S2@.269	LF	7.87	6.50	—	14.37
22" - 24"	S2@.300	LF	8.78	7.24	—	16.02
26" - 28"	S2@.331	LF	9.67	7.99	—	17.66
30"	S2@.354	LF	10.40	8.55	—	18.95

18" wide galvanized steel rectangular duct 24 gauge

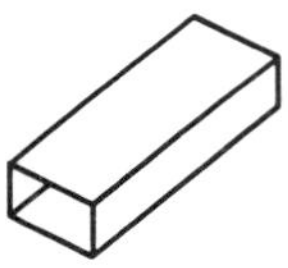

Description	Craft@Hrs	Unit	Material $	Labor $	Equipment $	Total $
6"	S2@.184	LF	5.40	4.44	—	9.84
8"	S2@.200	LF	5.86	4.83	—	10.69
10"	S2@.216	LF	6.30	5.22	—	11.52
12"	S2@.231	LF	6.75	5.58	—	12.33
14" - 16"	S2@.254	LF	7.42	6.13	—	13.55
18" - 20"	S2@.285	LF	8.33	6.88	—	15.21
22" - 24"	S2@.316	LF	9.23	7.63	—	16.86
26" - 28"	S2@.346	LF	10.10	8.36	—	18.46
30"	S2@.370	LF	10.80	8.93	—	19.73

Description	Craft@Hrs	Unit	Material $	Labor $	Equipment $	Total $
20" wide galvanized steel rectangular duct 24 gauge						
6"	S2@.200	LF	5.86	4.83	—	10.69
8"	S2@.216	LF	6.30	5.22	—	11.52
10"	S2@.231	LF	6.75	5.58	—	12.33
12"	S2@.247	LF	7.21	5.96	—	13.17
14" - 16"	S2@.269	LF	7.87	6.50	—	14.37
18" - 20"	S2@.300	LF	8.78	7.24	—	16.02
22" - 24"	S2@.331	LF	9.67	7.99	—	17.66
26" - 28"	S2@.362	LF	10.60	8.74	—	19.34
30"	S2@.385	LF	11.20	9.30	—	20.50
22" wide galvanized steel rectangular duct 24 gauge						
6"	S2@.216	LF	6.30	5.22	—	11.52
8"	S2@.231	LF	6.75	5.58	—	12.33
10"	S2@.247	LF	7.21	5.96	—	13.17
12"	S2@.262	LF	7.65	6.33	—	13.98
14" - 16"	S2@.285	LF	8.33	6.88	—	15.21
18" - 20"	S2@.316	LF	9.23	7.63	—	16.86
22" - 24"	S2@.346	LF	10.10	8.36	—	18.46
26" - 28"	S2@.377	LF	11.10	9.10	—	20.20
30"	S2@.400	LF	11.70	9.66	—	21.36
24" wide galvanized steel rectangular duct 24 gauge						
6"	S2@.231	LF	6.75	5.58	—	12.33
8"	S2@.247	LF	7.21	5.96	—	13.17
10"	S2@.262	LF	7.65	6.33	—	13.98
12"	S2@.278	LF	8.10	6.71	—	14.81
14" - 16"	S2@.300	LF	8.78	7.24	—	16.02
18" - 20"	S2@.331	LF	9.67	7.99	—	17.66
22" - 24"	S2@.362	LF	10.60	8.74	—	19.34
26" - 28"	S2@.392	LF	11.50	9.47	—	20.97
30"	S2@.415	LF	12.10	10.00	—	22.10
26" wide galvanized steel rectangular duct 24 gauge						
6"	S2@.247	LF	7.21	5.96	—	13.17
8"	S2@.262	LF	7.65	6.33	—	13.98
10"	S2@.278	LF	8.10	6.71	—	14.81
12"	S2@.292	LF	8.55	7.05	—	15.60
14" - 16"	S2@.316	LF	9.23	7.63	—	16.86
18" - 20"	S2@.346	LF	10.10	8.36	—	18.46
22" - 24"	S2@.377	LF	11.10	9.10	—	20.20
26" - 28"	S2@.407	LF	11.90	9.83	—	21.73
30"	S2@.431	LF	12.60	10.40	—	23.00

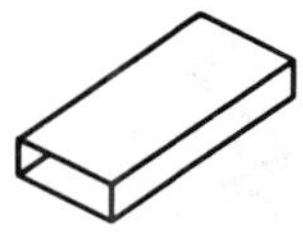

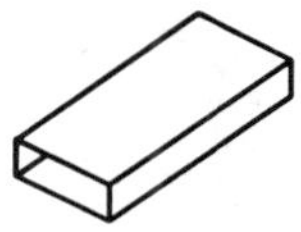

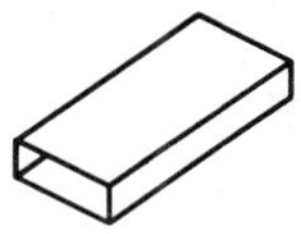

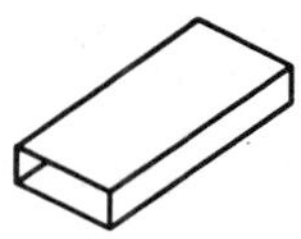

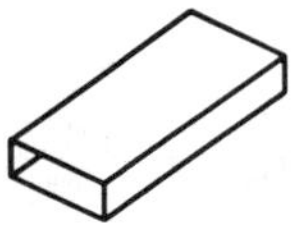

28" wide galvanized steel rectangular duct 24 gauge

Description	Craft@Hrs	Unit	Material $	Labor $	Equipment $	Total $
6"	S2@.262	LF	7.65	6.33	—	13.98
8"	S2@.278	LF	8.10	6.71	—	14.81
10"	S2@.292	LF	8.55	7.05	—	15.60
12"	S2@.308	LF	9.01	7.44	—	16.45
14" - 16"	S2@.331	LF	9.67	7.99	—	17.66
18" - 20"	S2@.362	LF	10.60	8.74	—	19.34
22" - 24"	S2@.392	LF	11.50	9.47	—	20.97
26" - 28"	S2@.423	LF	12.40	10.20	—	22.60
30"	S2@.448	LF	13.10	10.80	—	23.90

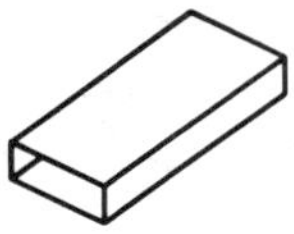

30" wide galvanized steel rectangular duct 24 gauge

Description	Craft@Hrs	Unit	Material $	Labor $	Equipment $	Total $
6"	S2@.278	LF	8.10	6.71	—	14.81
8"	S2@.292	LF	8.55	7.05	—	15.60
10"	S2@.308	LF	9.01	7.44	—	16.45
12"	S2@.323	LF	9.45	7.80	—	17.25
14" - 16"	S2@.346	LF	10.10	8.36	—	18.46
18" - 20"	S2@.377	LF	11.10	9.10	—	20.20
22" - 24"	S2@.407	LF	11.90	9.83	—	21.73
26" - 28"	S2@.440	LF	12.80	10.60	—	23.40
30"	S2@.465	LF	13.50	11.20	—	24.70

Costs for Galvanized 22 Gauge Rectangular Duct. Costs per linear foot of duct, based on quantities less than 1,000 pounds of prefabricated duct. Deduct 15% from the material cost for over 1,000 pounds. Add 50% to the material cost for lined duct. These costs include delivery, typical bracing, cleats, scrap, hangers, end closures.

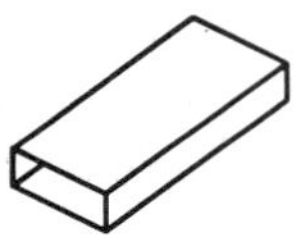

32" wide galvanized steel rectangular duct 22 gauge

Description	Craft@Hrs	Unit	Material $	Labor $	Equipment $	Total $
8"	S2@.374	LF	11.00	9.03	—	20.03
10"	S2@.393	LF	11.50	9.49	—	20.99
12"	S2@.412	LF	12.00	9.95	—	21.95
14" - 16"	S2@.442	LF	12.90	10.70	—	23.60
18" - 20"	S2@.478	LF	14.00	11.50	—	25.50
22" - 24"	S2@.517	LF	15.00	12.50	—	27.50
26" - 28"	S2@.553	LF	16.10	13.40	—	29.50
30" - 32"	S2@.589	LF	17.30	14.20	—	31.50
34" - 36"	S2@.628	LF	18.30	15.20	—	33.50
38" - 40"	S2@.664	LF	19.50	16.00	—	35.50
42" - 44"	S2@.702	LF	20.40	17.00	—	37.40
46" - 48"	S2@.739	LF	21.60	17.80	—	39.40
50" - 52"	S2@.777	LF	22.70	18.80	—	41.50
54"	S2@.807	LF	23.50	19.50	—	43.00

34" wide galvanized steel rectangular duct 22 gauge

Description	Craft@Hrs	Unit	Material $	Labor $	Equipment $	Total $
8"	S2@.393	LF	11.50	9.49	—	20.99
10"	S2@.412	LF	12.00	9.95	—	21.95
12"	S2@.431	LF	12.60	10.40	—	23.00
14" - 16"	S2@.461	LF	13.40	11.10	—	24.50
18" - 20"	S2@.497	LF	14.60	12.00	—	26.60
22" - 24"	S2@.534	LF	15.50	12.90	—	28.40
26" - 28"	S2@.572	LF	16.70	13.80	—	30.50
30" - 32"	S2@.608	LF	17.80	14.70	—	32.50
34" - 36"	S2@.645	LF	18.90	15.60	—	34.50
38" - 40"	S2@.683	LF	20.00	16.50	—	36.50
42" - 44"	S2@.722	LF	21.10	17.40	—	38.50
46" - 48"	S2@.758	LF	22.20	18.30	—	40.50
50" - 52"	S2@.796	LF	23.20	19.20	—	42.40
54"	S2@.824	LF	24.10	19.90	—	44.00

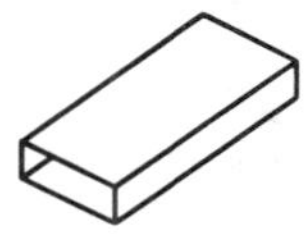

36" wide galvanized steel rectangular duct 22 gauge

Description	Craft@Hrs	Unit	Material $	Labor $	Equipment $	Total $
8"	S2@.412	LF	12.00	9.95	—	21.95
10"	S2@.431	LF	12.60	10.40	—	23.00
12"	S2@.453	LF	13.20	10.90	—	24.10
14" - 16"	S2@.478	LF	14.00	11.50	—	25.50
18" - 20"	S2@.517	LF	15.00	12.50	—	27.50
22" - 24"	S2@.553	LF	16.10	13.40	—	29.50
26" - 28"	S2@.589	LF	17.30	14.20	—	31.50
30" - 32"	S2@.628	LF	18.30	15.20	—	33.50
34" - 36"	S2@.664	LF	19.50	16.00	—	35.50
38" - 40"	S2@.702	LF	20.40	17.00	—	37.40
42" - 44"	S2@.739	LF	21.60	17.80	—	39.40
46" - 48"	S2@.777	LF	22.70	18.80	—	41.50
50" - 52"	S2@.816	LF	23.70	19.70	—	43.40
54"	S2@.841	LF	24.70	20.30	—	45.00

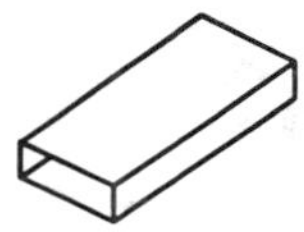

38" wide galvanized steel rectangular duct 22 gauge

Description	Craft@Hrs	Unit	Material $	Labor $	Equipment $	Total $
8"	S2@.431	LF	12.60	10.40	—	23.00
10"	S2@.453	LF	13.20	10.90	—	24.10
12"	S2@.470	LF	13.60	11.30	—	24.90
14" - 16"	S2@.497	LF	14.60	12.00	—	26.60
18" - 20"	S2@.534	LF	15.50	12.90	—	28.40
22" - 24"	S2@.572	LF	16.70	13.80	—	30.50
26" - 28"	S2@.608	LF	17.80	14.70	—	32.50
30" - 32"	S2@.645	LF	18.90	15.60	—	34.50
34" - 36"	S2@.683	LF	20.00	16.50	—	36.50
38" - 40"	S2@.722	LF	21.10	17.40	—	38.50
42" - 44"	S2@.758	LF	22.20	18.30	—	40.50
46" - 48"	S2@.796	LF	23.20	19.20	—	42.40
50" - 52"	S2@.833	LF	24.40	20.10	—	44.50
54"	S2@.863	LF	25.20	20.80	—	46.00

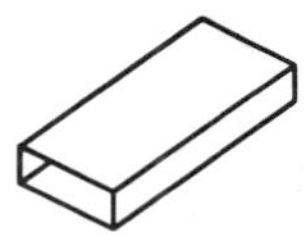

Galvanized Steel Rectangular Ductwork

40" wide galvanized steel rectangular duct 22 gauge

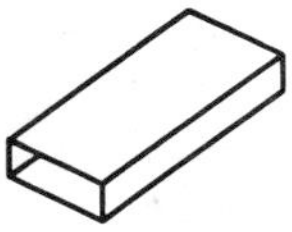

Description	Craft@Hrs	Unit	Material $	Labor $	Equipment $	Total $
8"	S2@.453	LF	13.20	10.90	—	24.10
10"	S2@.470	LF	13.60	11.30	—	24.90
12"	S2@.487	LF	14.30	11.80	—	26.10
14" - 16"	S2@.517	LF	15.00	12.50	—	27.50
18" - 20"	S2@.553	LF	16.10	13.40	—	29.50
22" - 24"	S2@.589	LF	17.30	14.20	—	31.50
26" - 28"	S2@.628	LF	18.30	15.20	—	33.50
30" - 32"	S2@.664	LF	19.50	16.00	—	35.50
34" - 36"	S2@.702	LF	20.40	17.00	—	37.40
38" - 40"	S2@.739	LF	21.60	17.80	—	39.40
42" - 44"	S2@.777	LF	22.70	18.80	—	41.50
46" - 48"	S2@.816	LF	23.70	19.70	—	43.40
50" - 52"	S2@.852	LF	24.90	20.60	—	45.50
54"	S2@.880	LF	25.70	21.30	—	47.00

42" wide galvanized steel rectangular duct 22 gauge

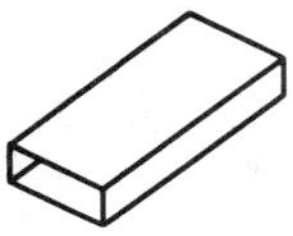

Description	Craft@Hrs	Unit	Material $	Labor $	Equipment $	Total $
8"	S2@.470	LF	13.60	11.30	—	24.90
10"	S2@.487	LF	14.30	11.80	—	26.10
12"	S2@.508	LF	14.80	12.30	—	27.10
14" - 16"	S2@.534	LF	15.50	12.90	—	28.40
18" - 20"	S2@.572	LF	16.70	13.80	—	30.50
22" - 24"	S2@.611	LF	17.80	14.80	—	32.60
26" - 28"	S2@.645	LF	18.90	15.60	—	34.50
30" - 32"	S2@.683	LF	20.00	16.50	—	36.50
34" - 36"	S2@.722	LF	21.10	17.40	—	38.50
38" - 40"	S2@.758	LF	22.20	18.30	—	40.50
42" - 44"	S2@.798	LF	23.20	19.30	—	42.50
46" - 48"	S2@.833	LF	24.40	20.10	—	44.50
50" - 52"	S2@.871	LF	25.50	21.00	—	46.50
54"	S2@.901	LF	26.20	21.80	—	48.00

44" wide galvanized steel rectangular duct 22 gauge

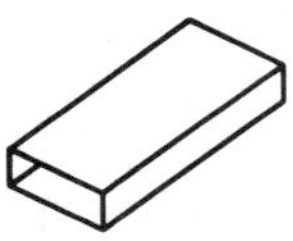

Description	Craft@Hrs	Unit	Material $	Labor $	Equipment $	Total $
8"	S2@.487	LF	14.30	11.80	—	26.10
10"	S2@.508	LF	14.80	12.30	—	27.10
12"	S2@.525	LF	15.30	12.70	—	28.00
14" - 16"	S2@.555	LF	16.10	13.40	—	29.50
18" - 20"	S2@.589	LF	17.30	14.20	—	31.50
22" - 24"	S2@.628	LF	18.30	15.20	—	33.50
26" - 28"	S2@.666	LF	19.50	16.10	—	35.60
30" - 32"	S2@.705	LF	20.40	17.00	—	37.40
34" - 36"	S2@.739	LF	21.60	17.80	—	39.40
38" - 40"	S2@.777	LF	22.70	18.80	—	41.50
42" - 44"	S2@.816	LF	23.70	19.70	—	43.40
46" - 48"	S2@.852	LF	24.90	20.60	—	45.50
50" - 52"	S2@.890	LF	26.00	21.50	—	47.50
54"	S2@.918	LF	26.80	22.20	—	49.00

46" wide galvanized steel rectangular duct 22 gauge

Description	Craft@Hrs	Unit	Material $	Labor $	Equipment $	Total $
8"	S2@.508	LF	14.80	12.30	—	27.10
10"	S2@.525	LF	15.30	12.70	—	28.00
12"	S2@.542	LF	15.90	13.10	—	29.00
14" - 16"	S2@.572	LF	16.70	13.80	—	30.50
18" - 20"	S2@.611	LF	17.80	14.80	—	32.60
22" - 24"	S2@.649	LF	18.90	15.70	—	34.60
26" - 28"	S2@.683	LF	20.00	16.50	—	36.50
30" - 32"	S2@.722	LF	21.10	17.40	—	38.50
34" - 36"	S2@.758	LF	22.20	18.30	—	40.50
38" - 40"	S2@.796	LF	23.20	19.20	—	42.40
42" - 44"	S2@.833	LF	24.40	20.10	—	44.50
46" - 48"	S2@.871	LF	25.50	21.00	—	46.50
50" - 52"	S2@.909	LF	26.60	22.00	—	48.60
54"	S2@.935	LF	27.40	22.60	—	50.00

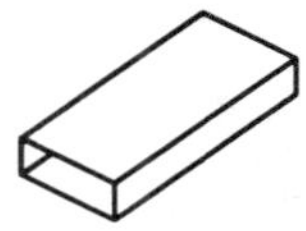

48" wide galvanized steel rectangular duct 22 gauge

Description	Craft@Hrs	Unit	Material $	Labor $	Equipment $	Total $
8"	S2@.525	LF	15.30	12.70	—	28.00
10"	S2@.542	LF	15.90	13.10	—	29.00
12"	S2@.564	LF	16.50	13.60	—	30.10
14" - 16"	S2@.589	LF	17.30	14.20	—	31.50
18" - 20"	S2@.628	LF	18.30	15.20	—	33.50
22" - 24"	S2@.664	LF	19.50	16.00	—	35.50
26" - 28"	S2@.702	LF	20.40	17.00	—	37.40
30" - 32"	S2@.739	LF	21.60	17.80	—	39.40
34" - 36"	S2@.777	LF	22.70	18.80	—	41.50
38" - 40"	S2@.816	LF	23.70	19.70	—	43.40
42" - 44"	S2@.852	LF	24.90	20.60	—	45.50
46" - 48"	S2@.890	LF	26.00	21.50	—	47.50
50" - 52"	S2@.927	LF	27.20	22.40	—	49.60
54"	S2@.956	LF	27.90	23.10	—	51.00

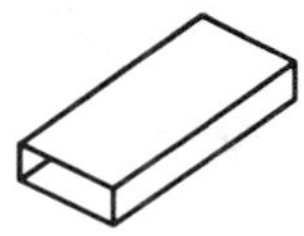

50" wide galvanized steel rectangular duct 22 gauge

Description	Craft@Hrs	Unit	Material $	Labor $	Equipment $	Total $
8"	S2@.542	LF	15.90	13.10	—	29.00
10"	S2@.564	LF	16.50	13.60	—	30.10
12"	S2@.581	LF	16.90	14.00	—	30.90
14" - 16"	S2@.611	LF	17.80	14.80	—	32.60
18" - 20"	S2@.645	LF	18.90	15.60	—	34.50
22" - 24"	S2@.683	LF	20.00	16.50	—	36.50
26" - 28"	S2@.722	LF	21.10	17.40	—	38.50
30" - 32"	S2@.758	LF	22.20	18.30	—	40.50
34" - 36"	S2@.798	LF	23.20	19.30	—	42.50
38" - 40"	S2@.833	LF	24.40	20.10	—	44.50
42" - 44"	S2@.871	LF	25.50	21.00	—	46.50
46" - 48"	S2@.909	LF	26.60	22.00	—	48.60
50" - 52"	S2@.946	LF	27.60	22.80	—	50.40
54"	S2@.974	LF	28.40	23.50	—	51.90

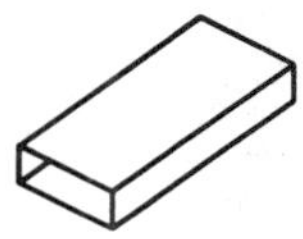

Galvanized Steel Rectangular Ductwork

52" wide galvanized steel rectangular duct 22 gauge

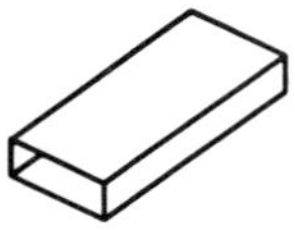

Description	Craft@Hrs	Unit	Material $	Labor $	Equipment $	Total $
8"	S2@.564	LF	16.50	13.60	—	30.10
10"	S2@.581	LF	16.90	14.00	—	30.90
12"	S2@.598	LF	17.50	14.40	—	31.90
14" - 16"	S2@.628	LF	18.30	15.20	—	33.50
18" - 20"	S2@.662	LF	19.50	16.00	—	35.50
22" - 24"	S2@.700	LF	20.40	16.90	—	37.30
26" - 28"	S2@.739	LF	21.60	17.80	—	39.40
30" - 32"	S2@.777	LF	22.70	18.80	—	41.50
34" - 36"	S2@.816	LF	23.70	19.70	—	43.40
38" - 40"	S2@.852	LF	24.90	20.60	—	45.50
42" - 44"	S2@.890	LF	26.00	21.50	—	47.50
46" - 48"	S2@.927	LF	27.20	22.40	—	49.60
50" - 52"	S2@.976	LF	28.20	23.60	—	51.80
54"	S2@.991	LF	28.90	23.90	—	52.80

54" wide galvanized steel rectangular duct 22 gauge

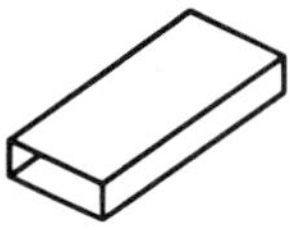

Description	Craft@Hrs	Unit	Material $	Labor $	Equipment $	Total $
8"	S2@.581	LF	16.90	14.00	—	30.90
10"	S2@.598	LF	17.50	14.40	—	31.90
12"	S2@.619	LF	18.10	14.90	—	33.00
14" - 16"	S2@.645	LF	18.90	15.60	—	34.50
18" - 20"	S2@.683	LF	20.00	16.50	—	36.50
22" - 24"	S2@.722	LF	21.10	17.40	—	38.50
26" - 28"	S2@.758	LF	22.20	18.30	—	40.50
30" - 32"	S2@.796	LF	23.20	19.20	—	42.40
34" - 36"	S2@.833	LF	24.40	20.10	—	44.50
38" - 40"	S2@.871	LF	25.50	21.00	—	46.50
42" - 44"	S2@.909	LF	26.60	22.00	—	48.60
46" - 48"	S2@.946	LF	27.60	22.80	—	50.40
50" - 52"	S2@.982	LF	28.60	23.70	—	52.30
54"	S2@1.01	LF	29.60	24.40	—	54.00

Costs for Galvanized 20 Gauge Rectangular Duct. Costs per linear foot of duct, based on quantities less than 1,000 pounds of prefabricated duct. Deduct 15% from the material cost for over 1,000 pounds. Add 50% to the material cost for lined duct. These costs include delivery, typical bracing, cleats, scrap, hangers, end closures, sealants and miscellaneous hardware. See installation costs that follow.

56" wide galvanized steel rectangular duct 20 gauge

Description	Craft@Hrs	Unit	Material $	Labor $	Equipment $	Total $
18" - 20"	S2@.828	LF	24.40	20.00	—	44.40
22" - 24"	S2@.871	LF	25.70	21.00	—	46.70
26" - 28"	S2@.916	LF	26.90	22.10	—	49.00
30" - 32"	S2@.959	LF	28.30	23.20	—	51.50
34" - 36"	S2@1.00	LF	29.60	24.10	—	53.70
38" - 40"	S2@1.05	LF	30.90	25.40	—	56.30
42" - 44"	S2@1.09	LF	32.20	26.30	—	58.50
46" - 48"	S2@1.14	LF	33.50	27.50	—	61.00
50" - 52"	S2@1.18	LF	34.80	28.50	—	63.30
54" - 56"	S2@1.23	LF	36.10	29.70	—	65.80
58" - 60"	S2@1.27	LF	37.40	30.70	—	68.10
62" - 64"	S2@1.31	LF	38.80	31.60	—	70.40
66" - 68"	S2@1.36	LF	39.90	32.80	—	72.70
70" - 72"	S2@1.40	LF	41.30	33.80	—	75.10

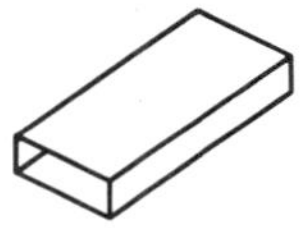

58" wide galvanized steel rectangular duct 20 gauge

Description	Craft@Hrs	Unit	Material $	Labor $	Equipment $	Total $
18" - 20"	S2@.850	LF	25.00	20.50	—	45.50
22" - 24"	S2@.895	LF	26.40	21.60	—	48.00
26" - 28"	S2@.937	LF	27.60	22.60	—	50.20
30" - 32"	S2@.980	LF	29.00	23.70	—	52.70
34" - 36"	S2@1.02	LF	30.20	24.60	—	54.80
38" - 40"	S2@1.07	LF	31.50	25.80	—	57.30
42" - 44"	S2@1.11	LF	32.90	26.80	—	59.70
46" - 48"	S2@1.16	LF	34.10	28.00	—	62.10
50" - 52"	S2@1.02	LF	35.50	24.60	—	60.10
54" - 56"	S2@1.25	LF	36.80	30.20	—	67.00
58" - 60"	S2@1.29	LF	38.00	31.20	—	69.20
62" - 64"	S2@1.33	LF	39.40	32.10	—	71.50
66" - 68"	S2@1.38	LF	40.70	33.30	—	74.00
70" - 72"	S2@1.42	LF	42.00	34.30	—	76.30

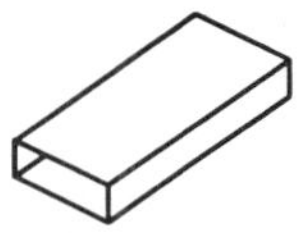

60" wide galvanized steel rectangular duct 20 gauge

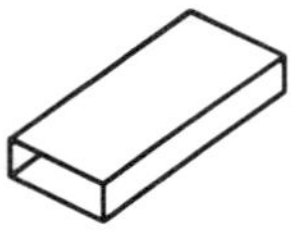

Description	Craft@Hrs	Unit	Material $	Labor $	Equipment $	Total $
18" - 20"	S2@.871	LF	25.70	21.00	—	46.70
22" - 24"	S2@.916	LF	26.90	22.10	—	49.00
26" - 28"	S2@.959	LF	28.30	23.20	—	51.50
30" - 32"	S2@1.00	LF	29.60	24.10	—	53.70
34" - 36"	S2@1.05	LF	30.90	25.40	—	56.30
38" - 40"	S2@1.09	LF	32.20	26.30	—	58.50
42" - 44"	S2@1.14	LF	33.50	27.50	—	61.00
46" - 48"	S2@1.18	LF	34.80	28.50	—	63.30
50" - 52"	S2@1.23	LF	36.10	29.70	—	65.80
54" - 56"	S2@1.27	LF	37.40	30.70	—	68.10
58" - 60"	S2@1.31	LF	38.80	31.60	—	70.40
62" - 64"	S2@1.36	LF	39.90	32.80	—	72.70
66" - 68"	S2@1.40	LF	41.30	33.80	—	75.10
70" - 72"	S2@1.45	LF	42.70	35.00	—	77.70

62" wide galvanized steel rectangular duct 20 gauge

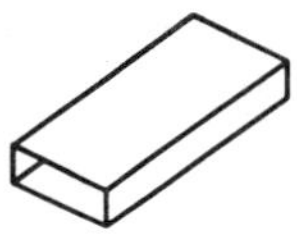

Description	Craft@Hrs	Unit	Material $	Labor $	Equipment $	Total $
18" - 20"	S2@.895	LF	26.40	21.60	—	48.00
22" - 24"	S2@.937	LF	27.60	22.60	—	50.20
26" - 28"	S2@.980	LF	29.00	23.70	—	52.70
30" - 32"	S2@1.02	LF	30.20	24.60	—	54.80
34" - 36"	S2@1.07	LF	31.50	25.80	—	57.30
38" - 40"	S2@1.11	LF	32.90	26.80	—	59.70
42" - 44"	S2@1.16	LF	34.10	28.00	—	62.10
46" - 48"	S2@1.20	LF	35.50	29.00	—	64.50
50" - 52"	S2@1.25	LF	36.80	30.20	—	67.00
54" - 56"	S2@1.29	LF	38.00	31.20	—	69.20
58" - 60"	S2@1.33	LF	39.40	32.10	—	71.50
62" - 64"	S2@1.38	LF	40.70	33.30	—	74.00
66" - 68"	S2@1.42	LF	42.00	34.30	—	76.30
70" - 72"	S2@1.47	LF	43.20	35.50	—	78.70

64" wide galvanized steel rectangular duct 20 gauge

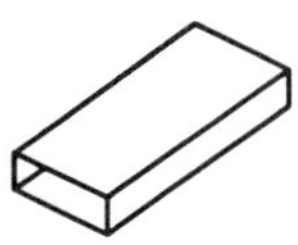

Description	Craft@Hrs	Unit	Material $	Labor $	Equipment $	Total $
18" - 20"	S2@.916	LF	26.90	22.10	—	49.00
22" - 24"	S2@.959	LF	28.30	23.20	—	51.50
26" - 28"	S2@1.00	LF	29.60	24.10	—	53.70
30" - 32"	S2@1.05	LF	30.90	25.40	—	56.30
34" - 36"	S2@1.09	LF	32.20	26.30	—	58.50
38" - 40"	S2@1.14	LF	33.50	27.50	—	61.00
42" - 44"	S2@1.18	LF	34.80	28.50	—	63.30
46" - 48"	S2@1.23	LF	36.10	29.70	—	65.80
50" - 52"	S2@1.27	LF	37.40	30.70	—	68.10
54" - 56"	S2@1.31	LF	38.80	31.60	—	70.40
58" - 60"	S2@1.36	LF	39.90	32.80	—	72.70
62" - 64"	S2@1.40	LF	41.30	33.80	—	75.10
66" - 68"	S2@1.45	LF	42.70	35.00	—	77.70
70" - 72"	S2@1.49	LF	43.90	36.00	—	79.90

Description	Craft@Hrs	Unit	Material $	Labor $	Equipment $	Total $

66" wide galvanized steel rectangular duct 20 gauge

Description	Craft@Hrs	Unit	Material $	Labor $	Equipment $	Total $
18" - 20"	S2@.937	LF	27.60	22.60	—	50.20
22" - 24"	S2@.980	LF	29.00	23.70	—	52.70
26" - 28"	S2@1.02	LF	30.20	24.60	—	54.80
30" - 32"	S2@1.07	LF	31.50	25.80	—	57.30
34" - 36"	S2@1.11	LF	32.90	26.80	—	59.70
38" - 40"	S2@1.16	LF	34.10	28.00	—	62.10
42" - 44"	S2@1.20	LF	35.50	29.00	—	64.50
46" - 48"	S2@1.25	LF	36.80	30.20	—	67.00
50" - 52"	S2@1.29	LF	38.00	31.20	—	69.20
54" - 56"	S2@1.33	LF	39.40	32.10	—	71.50
58" - 60"	S2@1.38	LF	40.70	33.30	—	74.00
62" - 64"	S2@1.42	LF	42.00	34.30	—	76.30
66" - 68"	S2@1.47	LF	43.20	35.50	—	78.70
70" - 72"	S2@1.51	LF	44.50	36.50	—	81.00

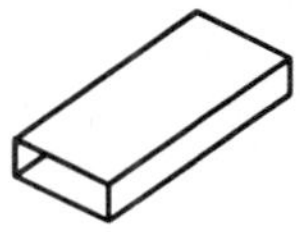

68" wide galvanized steel rectangular duct 20 gauge

Description	Craft@Hrs	Unit	Material $	Labor $	Equipment $	Total $
18" - 20"	S2@.959	LF	28.30	23.20	—	51.50
22" - 24"	S2@1.00	LF	29.60	24.10	—	53.70
26" - 28"	S2@1.05	LF	30.90	25.40	—	56.30
30" - 32"	S2@1.09	LF	32.20	26.30	—	58.50
34" - 36"	S2@1.14	LF	33.50	27.50	—	61.00
38" - 40"	S2@1.18	LF	34.80	28.50	—	63.30
42" - 44"	S2@1.23	LF	36.10	29.70	—	65.80
46" - 48"	S2@1.27	LF	37.40	30.70	—	68.10
50" - 52"	S2@1.31	LF	38.80	31.60	—	70.40
54" - 56"	S2@1.36	LF	39.90	32.80	—	72.70
58" - 60"	S2@1.40	LF	41.30	33.80	—	75.10
62" - 64"	S2@1.45	LF	42.70	35.00	—	77.70
66" - 68"	S2@1.49	LF	43.90	36.00	—	79.90
70" - 72"	S2@1.53	LF	45.30	36.90	—	82.20

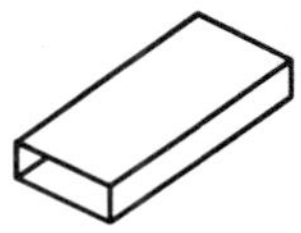

70" wide galvanized steel rectangular duct 20 gauge

Description	Craft@Hrs	Unit	Material $	Labor $	Equipment $	Total $
18" - 20"	S2@.980	LF	29.00	23.70	—	52.70
22" - 24"	S2@1.02	LF	30.20	24.60	—	54.80
26" - 28"	S2@1.07	LF	31.50	25.80	—	57.30
30" - 32"	S2@1.11	LF	32.90	26.80	—	59.70
34" - 36"	S2@1.16	LF	34.10	28.00	—	62.10
38" - 40"	S2@1.20	LF	35.50	29.00	—	64.50
42" - 44"	S2@1.25	LF	36.80	30.20	—	67.00
46" - 48"	S2@1.29	LF	38.00	31.20	—	69.20
50" - 52"	S2@1.33	LF	39.40	32.10	—	71.50
54" - 56"	S2@1.38	LF	40.70	33.30	—	74.00
58" - 60"	S2@1.42	LF	42.00	34.30	—	76.30
62" - 64"	S2@1.47	LF	43.20	35.50	—	78.70
66" - 68"	S2@1.51	LF	44.50	36.50	—	81.00
70" - 72"	S2@1.55	LF	45.80	37.40	—	83.20

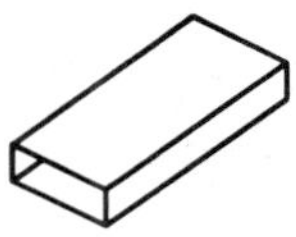

Galvanized Steel Rectangular Ductwork

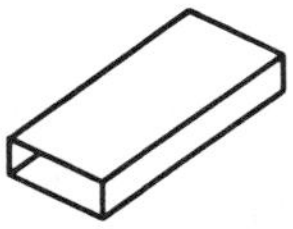

Description	Craft@Hrs	Unit	Material $	Labor $	Equipment $	Total $
72" wide galvanized steel rectangular duct 20 gauge						
18" - 20"	S2@1.00	LF	29.60	24.10	—	53.70
22" - 24"	S2@1.05	LF	30.90	25.40	—	56.30
26" - 28"	S2@1.09	LF	32.20	26.30	—	58.50
30" - 32"	S2@1.14	LF	33.50	27.50	—	61.00
34" - 36"	S2@1.18	LF	34.80	28.50	—	63.30
38" - 40"	S2@1.23	LF	36.10	29.70	—	65.80
42" - 44"	S2@1.27	LF	37.40	30.70	—	68.10
46" - 48"	S2@1.31	LF	38.80	31.60	—	70.40
50" - 52"	S2@1.36	LF	39.90	32.80	—	72.70
54" - 56"	S2@1.40	LF	41.30	33.80	—	75.10
58" - 60"	S2@1.45	LF	42.70	35.00	—	77.70
62" - 64"	S2@1.49	LF	43.90	36.00	—	79.90
66" - 68"	S2@1.53	LF	45.30	36.90	—	82.20
70" - 72"	S2@1.58	LF	46.40	38.20	—	84.60

Costs for Galvanized Rectangular 90 Degree Elbows. Costs per elbow, based on quantities less than 1,000 pounds. Deduct 15% from the material cost for over 1,000 pounds. Add 50% to the material cost for lined duct. These costs include delivery, typical bracing, cleats, scrap, hangers, end closures, sealants and miscellaneous hardware. For turning vanes, please see page 267.

4" galvanized steel rectangular duct 90 degree elbow

Description	Craft@Hrs	Unit	Material $	Labor $	Equipment $	Total $
4"	S2@.042	Ea	2.47	1.01	—	3.48
6"	S2@.058	Ea	3.40	1.40	—	4.80
8"	S2@.078	Ea	4.57	1.88	—	6.45
10"	S2@.104	Ea	6.09	2.51	—	8.60
12"	S2@.132	Ea	7.73	3.19	—	10.92
14" - 16"	S2@.236	Ea	13.80	5.70	—	19.50
18" - 20"	S2@.339	Ea	20.00	8.19	—	28.19
22" - 24"	S2@.468	Ea	27.40	11.30	—	38.70
26" - 28"	S2@.615	Ea	36.00	14.90	—	50.90
30" - 32"	S2@.854	Ea	50.20	20.60	—	70.80
34" - 36"	S2@1.19	Ea	69.80	28.70	—	98.50
38" - 40"	S2@1.45	Ea	84.80	35.00	—	119.80
42" - 44"	S2@1.73	Ea	101.00	41.80	—	142.80
46" - 48"	S2@2.04	Ea	119.00	49.30	—	168.30
50"	S2@2.28	Ea	134.00	55.10	—	189.10

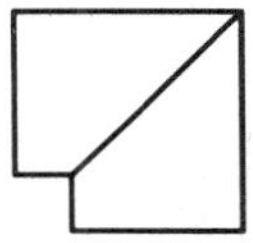

6" galvanized steel rectangular duct 90 degree elbow

Description	Craft@Hrs	Unit	Material $	Labor $	Equipment $	Total $
4"	S2@.058	Ea	3.40	1.40	—	4.80
6"	S2@.078	Ea	4.57	1.88	—	6.45
8"	S2@.104	Ea	6.09	2.51	—	8.60
10"	S2@.132	Ea	7.73	3.19	—	10.92
12"	S2@.148	Ea	8.67	3.57	—	12.24
14" - 16"	S2@.266	Ea	15.70	6.42	—	22.12
18" - 20"	S2@.370	Ea	21.60	8.93	—	30.53
22" - 24"	S2@.506	Ea	29.70	12.20	—	41.90
26" - 28"	S2@.655	Ea	38.30	15.80	—	54.10
30" - 32"	S2@.905	Ea	52.80	21.90	—	74.70
34" - 36"	S2@1.24	Ea	72.50	29.90	—	102.40
38" - 40"	S2@1.49	Ea	87.50	36.00	—	123.50
42" - 44"	S2@1.78	Ea	104.00	43.00	—	147.00
46" - 48"	S2@2.09	Ea	122.00	50.50	—	172.50
50"	S2@2.34	Ea	137.00	56.50	—	193.50

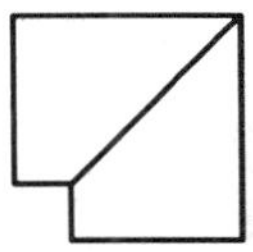

Galvanized Steel Rectangular Elbows

8" galvanized steel rectangular duct 90 degree elbow

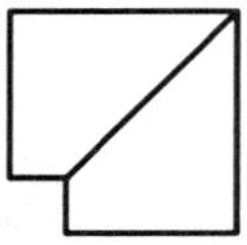

Description	Craft@Hrs	Unit	Material $	Labor $	Equipment $	Total $
4"	S2@.078	Ea	4.57	1.88	—	6.45
6"	S2@.104	Ea	6.09	2.51	—	8.60
8"	S2@.132	Ea	7.73	3.19	—	10.92
10"	S2@.148	Ea	8.67	3.57	—	12.24
12"	S2@.170	Ea	9.97	4.11	—	14.08
14" - 16"	S2@.300	Ea	17.50	7.24	—	24.74
18" - 20"	S2@.403	Ea	23.60	9.73	—	33.33
22" - 24"	S2@.549	Ea	32.10	13.30	—	45.40
26" - 28"	S2@.694	Ea	40.80	16.80	—	57.60
30" - 32"	S2@.950	Ea	55.70	22.90	—	78.60
34" - 36"	S2@1.28	Ea	75.30	30.90	—	106.20
38" - 40"	S2@1.54	Ea	90.30	37.20	—	127.50
42" - 44"	S2@1.83	Ea	107.00	44.20	—	151.20
46" - 48"	S2@2.13	Ea	125.00	51.40	—	176.40
50"	S2@2.39	Ea	141.00	57.70	—	198.70

10" galvanized steel rectangular duct 90 degree elbow

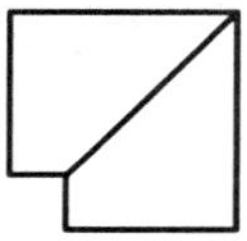

Description	Craft@Hrs	Unit	Material $	Labor $	Equipment $	Total $
4"	S2@.104	Ea	6.09	2.51	—	8.60
6"	S2@.132	Ea	7.73	3.19	—	10.92
8"	S2@.148	Ea	8.67	3.57	—	12.24
10"	S2@.170	Ea	9.97	4.11	—	14.08
12"	S2@.194	Ea	11.30	4.68	—	15.98
14" - 16"	S2@.336	Ea	19.60	8.11	—	27.71
18" - 20"	S2@.438	Ea	25.70	10.60	—	36.30
22" - 24"	S2@.587	Ea	34.50	14.20	—	48.70
26" - 28"	S2@.737	Ea	43.10	17.80	—	60.90
30" - 32"	S2@.997	Ea	58.40	24.10	—	82.50
34" - 36"	S2@1.33	Ea	78.20	32.10	—	110.30
38" - 40"	S2@1.59	Ea	93.30	38.40	—	131.70
42" - 44"	S2@1.87	Ea	110.00	45.20	—	155.20
46" - 48"	S2@2.18	Ea	128.00	52.60	—	180.60
50"	S2@2.43	Ea	143.00	58.70	—	201.70

12" galvanized steel rectangular duct 90 degree elbow

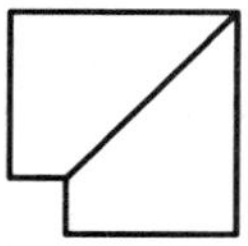

Description	Craft@Hrs	Unit	Material $	Labor $	Equipment $	Total $
4"	S2@.132	Ea	7.73	3.19	—	10.92
6"	S2@.148	Ea	8.67	3.57	—	12.24
8"	S2@.170	Ea	9.97	4.11	—	14.08
10"	S2@.194	Ea	11.30	4.68	—	15.98
12"	S2@.224	Ea	13.10	5.41	—	18.51
18" - 20"	S2@.478	Ea	27.90	11.50	—	39.40
22" - 24"	S2@.628	Ea	36.80	15.20	—	52.00
26" - 28"	S2@.775	Ea	45.50	18.70	—	64.20
30" - 32"	S2@1.04	Ea	61.00	25.10	—	86.10
34" - 36"	S2@1.38	Ea	80.90	33.30	—	114.20
38" - 40"	S2@1.64	Ea	96.10	39.60	—	135.70
42" - 44"	S2@1.92	Ea	112.00	46.40	—	158.40
46" - 48"	S2@2.23	Ea	131.00	53.90	—	184.90
50"	S2@2.48	Ea	145.00	59.90	—	204.90

14" galvanized steel rectangular duct 90 degree elbow

Description	Craft@Hrs	Unit	Material $	Labor $	Equipment $	Total $
4"	S2@.212	Ea	12.40	5.12	—	17.52
6"	S2@.232	Ea	13.60	5.60	—	19.20
8"	S2@.260	Ea	15.10	6.28	—	21.38
10"	S2@.292	Ea	17.00	7.05	—	24.05
12"	S2@.328	Ea	19.20	7.92	—	27.12
14" - 16"	S2@.415	Ea	24.40	10.00	—	34.40
18" - 20"	S2@.527	Ea	30.90	12.70	—	43.60
22" - 24"	S2@.867	Ea	39.10	20.90	—	60.00
26" - 28"	S2@.816	Ea	47.70	19.70	—	67.40
30" - 32"	S2@1.09	Ea	64.00	26.30	—	90.30
34" - 36"	S2@1.43	Ea	83.80	34.50	—	118.30
38" - 40"	S2@1.69	Ea	98.80	40.80	—	139.60
42" - 44"	S2@1.97	Ea	116.00	47.60	—	163.60
46" - 48"	S2@2.28	Ea	133.00	55.10	—	188.10
50"	S2@2.53	Ea	148.00	61.10	—	209.10

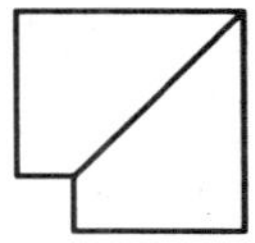

16" galvanized steel rectangular duct 90 degree elbow

Description	Craft@Hrs	Unit	Material $	Labor $	Equipment $	Total $
4"	S2@.260	Ea	15.10	6.28	—	21.38
6"	S2@.310	Ea	18.20	7.49	—	25.69
8"	S2@.340	Ea	19.90	8.21	—	28.11
10"	S2@.380	Ea	22.30	9.18	—	31.48
12"	S2@.420	Ea	24.60	10.10	—	34.70
14" - 16"	S2@.480	Ea	28.10	11.60	—	39.70
18" - 20"	S2@.555	Ea	32.60	13.40	—	46.00
22" - 24"	S2@.707	Ea	41.50	17.10	—	58.60
26" - 28"	S2@.856	Ea	50.10	20.70	—	70.80
30" - 32"	S2@1.14	Ea	66.70	27.50	—	94.20
34" - 36"	S2@1.48	Ea	86.70	35.70	—	122.40
38" - 40"	S2@1.73	Ea	102.00	41.80	—	143.80
42" - 44"	S2@2.02	Ea	118.00	48.80	—	166.80
46" - 48"	S2@2.32	Ea	137.00	56.00	—	193.00
50"	S2@2.58	Ea	150.00	62.30	—	212.30

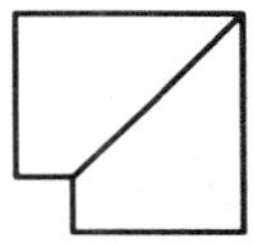

18" galvanized steel rectangular duct 90 degree elbow

Description	Craft@Hrs	Unit	Material $	Labor $	Equipment $	Total $
4"	S2@.312	Ea	18.30	7.53	—	25.83
6"	S2@.350	Ea	20.60	8.45	—	29.05
8"	S2@.390	Ea	22.90	9.42	—	32.32
10"	S2@.427	Ea	25.10	10.30	—	35.40
12"	S2@.470	Ea	27.40	11.30	—	38.70
14" - 16"	S2@.519	Ea	30.50	12.50	—	43.00
18" - 20"	S2@.604	Ea	35.40	14.60	—	50.00
22" - 24"	S2@.747	Ea	43.90	18.00	—	61.90
26" - 28"	S2@.895	Ea	52.40	21.60	—	74.00
30" - 32"	S2@1.18	Ea	69.30	28.50	—	97.80
34" - 36"	S2@1.53	Ea	89.50	36.90	—	126.40
38" - 40"	S2@1.78	Ea	104.00	43.00	—	147.00
42" - 44"	S2@2.07	Ea	121.00	50.00	—	171.00
46" - 48"	S2@2.37	Ea	140.00	57.20	—	197.20
50"	S2@2.63	Ea	154.00	63.50	—	217.50

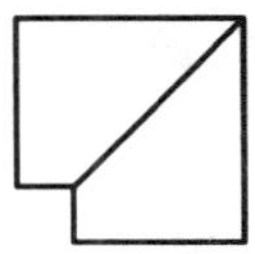

Galvanized Steel Rectangular Elbows

20" galvanized steel rectangular duct 90 degree elbow

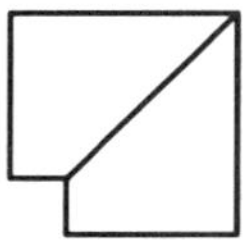

Description	Craft@Hrs	Unit	Material $	Labor $	Equipment $	Total $
4"	S2@.370	Ea	21.60	8.93	—	30.53
6"	S2@.390	Ea	22.90	9.42	—	32.32
8"	S2@.415	Ea	24.40	10.00	—	34.40
10"	S2@.448	Ea	26.30	10.80	—	37.10
12"	S2@.487	Ea	28.50	11.80	—	40.30
14" - 16"	S2@.553	Ea	32.40	13.40	—	45.80
18" - 20"	S2@.658	Ea	38.60	15.90	—	54.50
22" - 24"	S2@.786	Ea	46.00	19.00	—	65.00
26" - 28"	S2@.931	Ea	54.50	22.50	—	77.00
30" - 32"	S2@1.23	Ea	72.00	29.70	—	101.70
34" - 36"	S2@1.58	Ea	92.20	38.20	—	130.40
38" - 40"	S2@1.83	Ea	108.00	44.20	—	152.20
42" - 44"	S2@2.11	Ea	124.00	51.00	—	175.00
46" - 48"	S2@2.42	Ea	142.00	58.40	—	200.40
50"	S2@2.67	Ea	158.00	64.50	—	222.50

22" galvanized steel rectangular duct 90 degree elbow

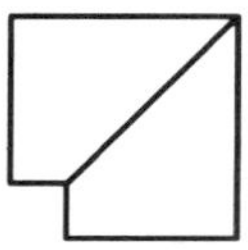

Description	Craft@Hrs	Unit	Material $	Labor $	Equipment $	Total $
22" - 24"	S2@.828	Ea	48.60	20.00	—	68.60
26" - 28"	S2@1.01	Ea	59.00	24.40	—	83.40
30" - 32"	S2@1.31	Ea	76.70	31.60	—	108.30
34" - 36"	S2@1.71	Ea	99.90	41.30	—	141.20
38" - 40"	S2@2.01	Ea	118.00	48.50	—	166.50
42" - 44"	S2@2.31	Ea	134.00	55.80	—	189.80
46" - 48"	S2@2.61	Ea	152.00	63.00	—	215.00
50" - 52"	S2@2.91	Ea	170.00	70.30	—	240.30
54" - 56"	S2@3.51	Ea	206.00	84.80	—	290.80
58" - 60"	S2@4.26	Ea	250.00	103.00	—	353.00
62" - 64"	S2@4.76	Ea	278.00	115.00	—	393.00
66" - 68"	S2@5.25	Ea	308.00	127.00	—	435.00
70" - 72"	S2@5.76	Ea	337.00	139.00	—	476.00

24" galvanized steel rectangular duct 90 degree elbow

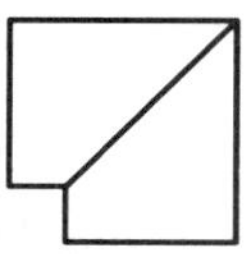

Description	Craft@Hrs	Unit	Material $	Labor $	Equipment $	Total $
22" - 24"	S2@.922	Ea	54.00	22.30	—	76.30
26" - 28"	S2@1.10	Ea	64.40	26.60	—	91.00
30" - 32"	S2@1.41	Ea	82.40	34.00	—	116.40
34" - 36"	S2@1.81	Ea	106.00	43.70	—	149.70
38" - 40"	S2@2.10	Ea	124.00	50.70	—	174.70
42" - 44"	S2@2.41	Ea	141.00	58.20	—	199.20
46" - 48"	S2@2.71	Ea	159.00	65.40	—	224.40
50" - 52"	S2@3.01	Ea	176.00	72.70	—	248.70
54" - 56"	S2@3.62	Ea	212.00	87.40	—	299.40
58" - 60"	S2@4.38	Ea	257.00	106.00	—	363.00
62" - 64"	S2@4.87	Ea	285.00	118.00	—	403.00
66" - 68"	S2@5.38	Ea	314.00	130.00	—	444.00
70" - 72"	S2@5.87	Ea	345.00	142.00	—	487.00

Description	Craft@Hrs	Unit	Material $	Labor $	Equipment $	Total $

26" galvanized steel rectangular duct 90 degree elbow

Description	Craft@Hrs	Unit	Material $	Labor $	Equipment $	Total $
22" - 24"	S2@1.01	Ea	59.10	24.40	—	83.50
26" - 28"	S2@1.19	Ea	69.90	28.70	—	98.60
30" - 32"	S2@1.50	Ea	88.10	36.20	—	124.30
34" - 36"	S2@1.92	Ea	112.00	46.40	—	158.40
38" - 40"	S2@2.21	Ea	129.00	53.40	—	182.40
42" - 44"	S2@2.51	Ea	147.00	60.60	—	207.60
46" - 48"	S2@2.81	Ea	164.00	67.90	—	231.90
50" - 52"	S2@3.11	Ea	183.00	75.10	—	258.10
54" - 56"	S2@3.73	Ea	219.00	90.10	—	309.10
58" - 60"	S2@4.50	Ea	264.00	109.00	—	373.00
62" - 64"	S2@5.02	Ea	293.00	121.00	—	414.00
66" - 68"	S2@5.53	Ea	325.00	134.00	—	459.00
70" - 72"	S2@6.04	Ea	354.00	146.00	—	500.00

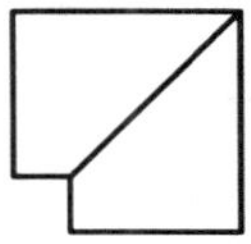

28" galvanized steel rectangular duct 90 degree elbow

Description	Craft@Hrs	Unit	Material $	Labor $	Equipment $	Total $
22" - 24"	S2@1.10	Ea	64.30	26.60	—	90.90
26" - 28"	S2@1.28	Ea	75.20	30.90	—	106.10
30" - 32"	S2@1.60	Ea	93.70	38.60	—	132.30
34" - 36"	S2@2.03	Ea	119.00	49.00	—	168.00
38" - 40"	S2@2.31	Ea	134.00	55.80	—	189.80
42" - 44"	S2@2.61	Ea	152.00	63.00	—	215.00
46" - 48"	S2@2.91	Ea	170.00	70.30	—	240.30
50" - 52"	S2@3.21	Ea	189.00	77.50	—	266.50
54" - 56"	S2@3.84	Ea	226.00	92.70	—	318.70
58" - 60"	S2@4.61	Ea	271.00	111.00	—	382.00
62" - 64"	S2@5.15	Ea	300.00	124.00	—	424.00
66" - 68"	S2@5.66	Ea	331.00	137.00	—	468.00
70" - 72"	S2@6.15	Ea	360.00	149.00	—	509.00

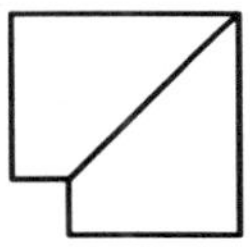

30" galvanized steel rectangular duct 90 degree elbow

Description	Craft@Hrs	Unit	Material $	Labor $	Equipment $	Total $
22" - 24"	S2@1.19	Ea	69.40	28.70	—	98.10
26" - 28"	S2@1.37	Ea	80.60	33.10	—	113.70
30" - 32"	S2@1.70	Ea	99.50	41.10	—	140.60
34" - 36"	S2@2.14	Ea	125.00	51.70	—	176.70
38" - 40"	S2@2.41	Ea	141.00	58.20	—	199.20
42" - 44"	S2@2.71	Ea	159.00	65.40	—	224.40
46" - 48"	S2@3.01	Ea	176.00	72.70	—	248.70
50" - 52"	S2@3.31	Ea	194.00	79.90	—	273.90
54" - 56"	S2@3.95	Ea	231.00	95.40	—	326.40
58" - 60"	S2@4.74	Ea	277.00	114.00	—	391.00
62" - 64"	S2@5.25	Ea	308.00	127.00	—	435.00
66" - 68"	S2@5.79	Ea	338.00	140.00	—	478.00
70" - 72"	S2@6.28	Ea	368.00	152.00	—	520.00

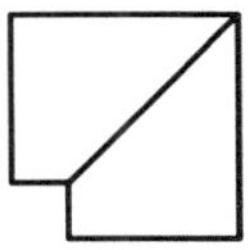

Galvanized Steel Rectangular Elbows

Description	Craft@Hrs	Unit	Material $	Labor $	Equipment $	Total $

32" galvanized steel rectangular duct 90 degree elbow

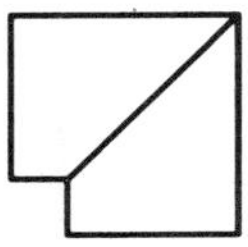

Description	Craft@Hrs	Unit	Material $	Labor $	Equipment $	Total $
22" - 24"	S2@1.53	Ea	89.60	36.90	—	126.50
26" - 28"	S2@1.73	Ea	101.00	41.80	—	142.80
30" - 32"	S2@1.93	Ea	113.00	46.60	—	159.60
34" - 36"	S2@2.25	Ea	132.00	54.30	—	186.30
38" - 40"	S2@2.51	Ea	147.00	60.60	—	207.60
42" - 44"	S2@2.81	Ea	164.00	67.90	—	231.90
46" - 48"	S2@3.11	Ea	182.00	75.10	—	257.10
50" - 52"	S2@3.41	Ea	200.00	82.30	—	282.30
54" - 56"	S2@4.06	Ea	237.00	98.00	—	335.00
58" - 60"	S2@4.85	Ea	284.00	117.00	—	401.00
62" - 64"	S2@5.38	Ea	314.00	130.00	—	444.00
66" - 68"	S2@5.89	Ea	346.00	142.00	—	488.00
70" - 72"	S2@6.40	Ea	375.00	155.00	—	530.00

34" galvanized steel rectangular duct 90 degree elbow

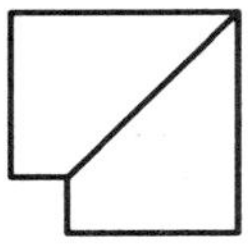

Description	Craft@Hrs	Unit	Material $	Labor $	Equipment $	Total $
22" - 24"	S2@1.67	Ea	98.00	40.30	—	138.30
26" - 28"	S2@1.91	Ea	112.00	46.10	—	158.10
30" - 32"	S2@2.12	Ea	124.00	51.20	—	175.20
34" - 36"	S2@2.36	Ea	138.00	57.00	—	195.00
38" - 40"	S2@2.60	Ea	152.00	62.80	—	214.80
42" - 44"	S2@2.91	Ea	170.00	70.30	—	240.30
46" - 48"	S2@3.21	Ea	187.00	77.50	—	264.50
50" - 52"	S2@3.51	Ea	206.00	84.80	—	290.80
54" - 56"	S2@4.17	Ea	245.00	101.00	—	346.00
58" - 60"	S2@4.97	Ea	292.00	120.00	—	412.00
62" - 64"	S2@5.49	Ea	322.00	133.00	—	455.00
66" - 68"	S2@6.04	Ea	354.00	146.00	—	500.00
70" - 72"	S2@6.51	Ea	381.00	157.00	—	538.00

36" galvanized steel rectangular duct 90 degree elbow

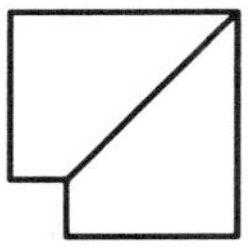

Description	Craft@Hrs	Unit	Material $	Labor $	Equipment $	Total $
22" - 24"	S2@1.83	Ea	108.00	44.20	—	152.20
26" - 28"	S2@2.04	Ea	120.00	49.30	—	169.30
30" - 32"	S2@2.26	Ea	132.00	54.60	—	186.60
34" - 36"	S2@2.47	Ea	145.00	59.60	—	204.60
38" - 40"	S2@2.70	Ea	159.00	65.20	—	224.20
42" - 44"	S2@3.01	Ea	176.00	72.70	—	248.70
46" - 48"	S2@3.31	Ea	194.00	79.90	—	273.90
50" - 52"	S2@3.61	Ea	211.00	87.20	—	298.20
54" - 56"	S2@4.27	Ea	250.00	103.00	—	353.00
58" - 60"	S2@5.10	Ea	298.00	123.00	—	421.00
62" - 64"	S2@5.61	Ea	329.00	135.00	—	464.00
66" - 68"	S2@6.15	Ea	360.00	149.00	—	509.00
70" - 72"	S2@6.64	Ea	389.00	160.00	—	549.00

38" galvanized steel rectangular duct 90 degree elbow

Description	Craft@Hrs	Unit	Material $	Labor $	Equipment $	Total $
22" - 24"	S2@1.98	Ea	116.00	47.80	—	163.80
26" - 28"	S2@2.18	Ea	128.00	52.60	—	180.60
30" - 32"	S2@2.38	Ea	140.00	57.50	—	197.50
34" - 36"	S2@2.58	Ea	151.00	62.30	—	213.30
38" - 40"	S2@2.81	Ea	164.00	67.90	—	231.90
42" - 44"	S2@3.11	Ea	182.00	75.10	—	257.10
46" - 48"	S2@3.41	Ea	200.00	82.30	—	282.30
50" - 52"	S2@3.71	Ea	217.00	89.60	—	306.60
54" - 56"	S2@4.38	Ea	257.00	106.00	—	363.00
58" - 60"	S2@5.21	Ea	306.00	126.00	—	432.00
62" - 64"	S2@5.74	Ea	335.00	139.00	—	474.00
66" - 68"	S2@6.28	Ea	368.00	152.00	—	520.00
70" - 72"	S2@6.75	Ea	396.00	163.00	—	559.00

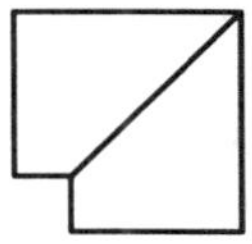

40" galvanized steel rectangular duct 90 degree elbow

Description	Craft@Hrs	Unit	Material $	Labor $	Equipment $	Total $
40" - 42"	S2@3.06	Ea	180.00	73.90	—	253.90
44" - 46"	S2@3.36	Ea	196.00	81.10	—	277.10
48" - 50"	S2@3.66	Ea	214.00	88.40	—	302.40
52" - 54"	S2@3.98	Ea	233.00	96.10	—	329.10
56" - 58"	S2@5.17	Ea	302.00	125.00	—	427.00
60" - 62"	S2@5.59	Ea	328.00	135.00	—	463.00
64" - 66"	S2@6.11	Ea	358.00	148.00	—	506.00
68" - 70"	S2@6.66	Ea	391.00	161.00	—	552.00
72"	S2@7.05	Ea	412.00	170.00	—	582.00

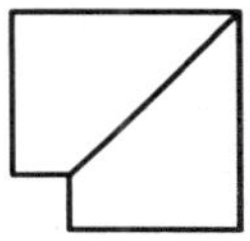

42" galvanized steel rectangular duct 90 degree elbow

Description	Craft@Hrs	Unit	Material $	Labor $	Equipment $	Total $
40" - 42"	S2@3.21	Ea	189.00	77.50	—	266.50
44" - 46"	S2@3.52	Ea	207.00	85.00	—	292.00
48" - 50"	S2@3.83	Ea	226.00	92.50	—	318.50
52" - 54"	S2@4.16	Ea	244.00	100.00	—	344.00
56" - 58"	S2@5.34	Ea	312.00	129.00	—	441.00
60" - 62"	S2@5.81	Ea	339.00	140.00	—	479.00
64" - 66"	S2@6.32	Ea	371.00	153.00	—	524.00
68" - 70"	S2@6.85	Ea	402.00	165.00	—	567.00
72"	S2@7.22	Ea	424.00	174.00	—	598.00

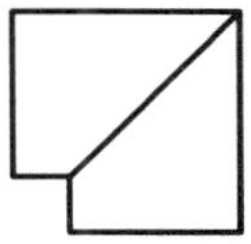

44" galvanized steel rectangular duct 90 degree elbow

Description	Craft@Hrs	Unit	Material $	Labor $	Equipment $	Total $
40" - 42"	S2@3.36	Ea	196.00	81.10	—	277.10
44" - 46"	S2@3.69	Ea	216.00	89.10	—	305.10
48" - 50"	S2@4.01	Ea	234.00	96.80	—	330.80
52" - 54"	S2@4.32	Ea	253.00	104.00	—	357.00
56" - 58"	S2@5.51	Ea	324.00	133.00	—	457.00
60" - 62"	S2@6.00	Ea	351.00	145.00	—	496.00
64" - 66"	S2@6.51	Ea	381.00	157.00	—	538.00
68" - 70"	S2@7.05	Ea	414.00	170.00	—	584.00
72"	S2@7.39	Ea	433.00	178.00	—	611.00

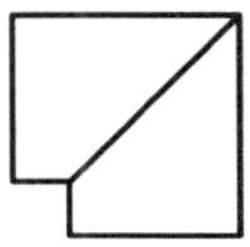

Galvanized Steel Rectangular Elbows

46" galvanized steel rectangular duct 90 degree elbow

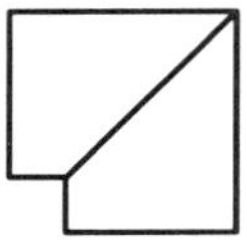

Description	Craft@Hrs	Unit	Material $	Labor $	Equipment $	Total $
40" - 42"	S2@3.52	Ea	207.00	85.00	—	292.00
44" - 46"	S2@3.86	Ea	227.00	93.20	—	320.20
48" - 50"	S2@4.16	Ea	244.00	100.00	—	344.00
52" - 54"	S2@4.32	Ea	262.00	104.00	—	366.00
56" - 58"	S2@5.72	Ea	334.00	138.00	—	472.00
60" - 62"	S2@6.19	Ea	362.00	149.00	—	511.00
64" - 66"	S2@6.70	Ea	394.00	162.00	—	556.00
68" - 70"	S2@7.26	Ea	425.00	175.00	—	600.00
72"	S2@7.60	Ea	445.00	184.00	—	629.00

48" galvanized steel rectangular duct 90 degree elbow

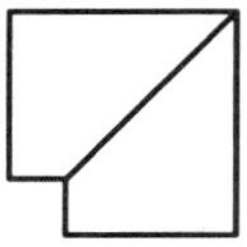

Description	Craft@Hrs	Unit	Material $	Labor $	Equipment $	Total $
40" - 42"	S2@3.67	Ea	215.00	88.60	—	303.60
44" - 46"	S2@4.01	Ea	234.00	96.80	—	330.80
48" - 50"	S2@4.32	Ea	253.00	104.00	—	357.00
52" - 54"	S2@4.61	Ea	269.00	111.00	—	380.00
56" - 58"	S2@5.91	Ea	347.00	143.00	—	490.00
60" - 62"	S2@6.40	Ea	375.00	155.00	—	530.00
64" - 66"	S2@6.90	Ea	404.00	167.00	—	571.00
68" - 70"	S2@7.41	Ea	435.00	179.00	—	614.00
72"	S2@7.77	Ea	455.00	188.00	—	643.00

50" galvanized steel rectangular duct 90 degree elbow

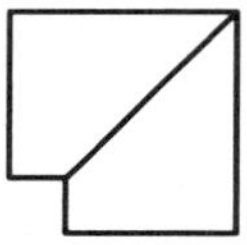

Description	Craft@Hrs	Unit	Material $	Labor $	Equipment $	Total $
40" - 42"	S2@3.82	Ea	225.00	92.20	—	317.20
44" - 46"	S2@4.16	Ea	244.00	100.00	—	344.00
48" - 50"	S2@4.46	Ea	262.00	108.00	—	370.00
52" - 54"	S2@4.78	Ea	279.00	115.00	—	394.00
56" - 58"	S2@6.11	Ea	358.00	148.00	—	506.00
60" - 62"	S2@6.60	Ea	387.00	159.00	—	546.00
64" - 66"	S2@7.09	Ea	416.00	171.00	—	587.00
68" - 70"	S2@7.62	Ea	446.00	184.00	—	630.00
72"	S2@7.94	Ea	465.00	192.00	—	657.00

52" galvanized steel rectangular duct 90 degree elbow

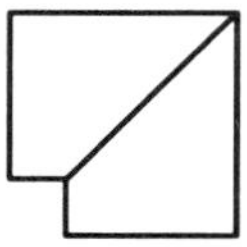

Description	Craft@Hrs	Unit	Material $	Labor $	Equipment $	Total $
40" - 42"	S2@3.98	Ea	233.00	96.10	—	329.10
44" - 46"	S2@4.32	Ea	253.00	104.00	—	357.00
48" - 50"	S2@4.61	Ea	269.00	111.00	—	380.00
52" - 54"	S2@4.95	Ea	290.00	120.00	—	410.00
56" - 58"	S2@6.32	Ea	371.00	153.00	—	524.00
60" - 62"	S2@6.79	Ea	398.00	164.00	—	562.00
64" - 66"	S2@7.30	Ea	428.00	176.00	—	604.00
68" - 70"	S2@7.81	Ea	458.00	189.00	—	647.00
72"	S2@8.16	Ea	477.00	197.00	—	674.00

54" galvanized steel rectangular duct 90 degree elbow

Description	Craft@Hrs	Unit	Material $	Labor $	Equipment $	Total $
40" - 42"	S2@4.16	Ea	244.00	100.00	—	344.00
44" - 46"	S2@4.46	Ea	262.00	108.00	—	370.00
48" - 50"	S2@4.78	Ea	279.00	115.00	—	394.00
52" - 54"	S2@5.15	Ea	300.00	124.00	—	424.00
56" - 58"	S2@6.51	Ea	381.00	157.00	—	538.00
60" - 62"	S2@7.00	Ea	410.00	169.00	—	579.00
64" - 66"	S2@7.49	Ea	440.00	181.00	—	621.00
68" - 70"	S2@8.01	Ea	470.00	193.00	—	663.00
72"	S2@8.41	Ea	492.00	203.00	—	695.00

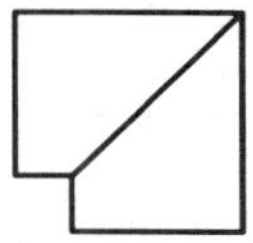

56" galvanized steel rectangular duct 90 degree elbow

Description	Craft@Hrs	Unit	Material $	Labor $	Equipment $	Total $
40" - 42"	S2@5.10	Ea	298.00	123.00	—	421.00
44" - 46"	S2@5.49	Ea	322.00	133.00	—	455.00
48" - 50"	S2@5.85	Ea	344.00	141.00	—	485.00
52" - 54"	S2@6.26	Ea	367.00	151.00	—	518.00
56" - 58"	S2@6.68	Ea	393.00	161.00	—	554.00
60" - 62"	S2@7.15	Ea	420.00	173.00	—	593.00
64" - 66"	S2@7.66	Ea	449.00	185.00	—	634.00
68" - 70"	S2@8.18	Ea	478.00	198.00	—	676.00
72"	S2@8.58	Ea	503.00	207.00	—	710.00

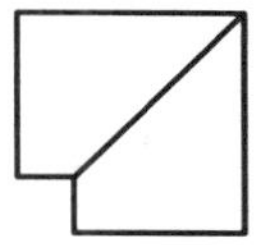

58" galvanized steel rectangular duct 90 degree elbow

Description	Craft@Hrs	Unit	Material $	Labor $	Equipment $	Total $
58" - 60"	S2@7.15	Ea	420.00	173.00	—	593.00
62" - 64"	S2@7.62	Ea	446.00	184.00	—	630.00
66" - 68"	S2@8.13	Ea	476.00	196.00	—	672.00
70" - 72"	S2@8.69	Ea	510.00	210.00	—	720.00

60" galvanized steel rectangular duct 90 degree elbow

Description	Craft@Hrs	Unit	Material $	Labor $	Equipment $	Total $
58" - 60"	S2@7.37	Ea	432.00	178.00	—	610.00
62" - 64"	S2@7.88	Ea	461.00	190.00	—	651.00
66" - 68"	S2@8.41	Ea	493.00	203.00	—	696.00
70" - 72"	S2@8.95	Ea	524.00	216.00	—	740.00

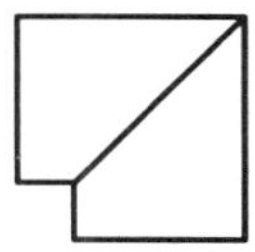

62" galvanized steel rectangular duct 90 degree elbow

Description	Craft@Hrs	Unit	Material $	Labor $	Equipment $	Total $
58" - 60"	S2@7.62	Ea	446.00	184.00	—	630.00
62" - 64"	S2@8.13	Ea	476.00	196.00	—	672.00
66" - 68"	S2@8.69	Ea	510.00	210.00	—	720.00
70" - 72"	S2@9.18	Ea	539.00	222.00	—	761.00

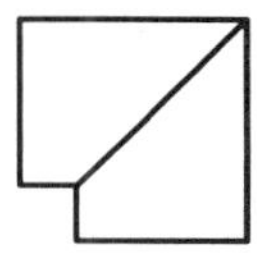

64" galvanized steel rectangular duct 90 degree elbow

Description	Craft@Hrs	Unit	Material $	Labor $	Equipment $	Total $
58" - 60"	S2@7.88	Ea	461.00	190.00	—	651.00
62" - 64"	S2@8.41	Ea	493.00	203.00	—	696.00
66" - 68"	S2@8.95	Ea	524.00	216.00	—	740.00
70" - 72"	S2@9.44	Ea	553.00	228.00	—	781.00

Galvanized Steel Rectangular Elbows

66" galvanized steel rectangular duct 90 degree elbow

Description	Craft@Hrs	Unit	Material $	Labor $	Equipment $	Total $
58" - 60"	S2@8.13	Ea	476.00	196.00	—	672.00
62" - 64"	S2@8.69	Ea	510.00	210.00	—	720.00
66" - 68"	S2@9.18	Ea	539.00	222.00	—	761.00
70" - 72"	S2@9.69	Ea	568.00	234.00	—	802.00

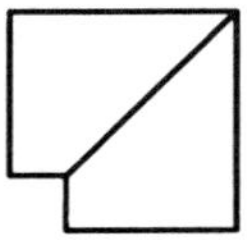

68" galvanized steel rectangular duct 90 degree elbow

Description	Craft@Hrs	Unit	Material $	Labor $	Equipment $	Total $
58" - 60"	S2@8.41	Ea	493.00	203.00	—	696.00
62" - 64"	S2@8.95	Ea	524.00	216.00	—	740.00
66" - 68"	S2@9.44	Ea	553.00	228.00	—	781.00
70" - 72"	S2@9.95	Ea	584.00	240.00	—	824.00

70" galvanized steel rectangular duct 90 degree elbow

Description	Craft@Hrs	Unit	Material $	Labor $	Equipment $	Total $
58" - 60"	S2@8.69	Ea	510.00	210.00	—	720.00
62" - 64"	S2@9.18	Ea	539.00	222.00	—	761.00
66" - 68"	S2@9.69	Ea	568.00	234.00	—	802.00
70" - 72"	S2@10.2	Ea	599.00	246.00	—	845.00

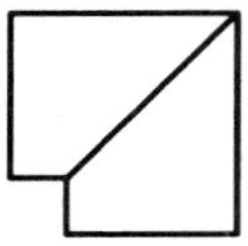

72" galvanized steel rectangular duct 90 degree elbow

Description	Craft@Hrs	Unit	Material $	Labor $	Equipment $	Total $
58" - 60"	S2@8.95	Ea	524.00	216.00	—	740.00
62" - 64"	S2@9.44	Ea	553.00	228.00	—	781.00
66" - 68"	S2@9.95	Ea	584.00	240.00	—	824.00
70" - 72"	S2@10.5	Ea	616.00	254.00	—	870.00

Costs for Galvanized Rectangular Drops and Tap-in Tees. Costs per drop or tap-in tee, based on quantities less than 1,000 pounds. Deduct 15% from the material cost for over 1,000 pounds. Add 50% to the material cost for the lined duct. These costs include delivery, typical bracing, cleats, scrap, hangers, end closures, sealants and miscellaneous hardware. For turning vanes, please see page 259.

6" wide galvanized steel rectangular duct drops and tap-in tees

Description	Craft@Hrs	Unit	Material $	Labor $	Equipment $	Total $
6"	S2@.072	Ea	4.61	1.74	—	6.35
8"	S2@.084	Ea	5.38	2.03	—	7.41
10"	S2@.097	Ea	6.18	2.34	—	8.52
12" - 16"	S2@.139	Ea	8.87	3.36	—	12.23
20" - 24"	S2@.216	Ea	13.70	5.22	—	18.92
28" - 32"	S2@.309	Ea	19.60	7.46	—	27.06
36" - 40"	S2@.412	Ea	26.30	9.95	—	36.25
44" - 48"	S2@.489	Ea	31.10	11.80	—	42.90
52" - 56"	S2@.591	Ea	37.70	14.30	—	52.00
60"	S2@.747	Ea	47.60	18.00	—	65.60

8" wide galvanized steel rectangular duct drops and tap-in tees

Description	Craft@Hrs	Unit	Material $	Labor $	Equipment $	Total $
6"	S2@.084	Ea	5.38	2.03	—	7.41
8"	S2@.097	Ea	6.18	2.34	—	8.52
10"	S2@.109	Ea	6.93	2.63	—	9.56
12" - 16"	S2@.152	Ea	9.73	3.67	—	13.40
20" - 24"	S2@.231	Ea	14.70	5.58	—	20.28
28" - 32"	S2@.326	Ea	20.80	7.87	—	28.67
36" - 40"	S2@.432	Ea	27.50	10.40	—	37.90
44" - 48"	S2@.506	Ea	32.20	12.20	—	44.40
52" - 56"	S2@.634	Ea	40.40	15.30	—	55.70
60"	S2@.760	Ea	48.40	18.40	—	66.80

10" wide galvanized steel rectangular duct drops and tap-in tees

Description	Craft@Hrs	Unit	Material $	Labor $	Equipment $	Total $
6"	S2@.097	Ea	6.18	2.34	—	8.52
8"	S2@.109	Ea	6.93	2.63	—	9.56
10"	S2@.121	Ea	7.69	2.92	—	10.61
12" - 16"	S2@.167	Ea	10.60	4.03	—	14.63
20" - 24"	S2@.246	Ea	15.80	5.94	—	21.74
28" - 32"	S2@.343	Ea	21.90	8.28	—	30.18
36" - 40"	S2@.450	Ea	28.60	10.90	—	39.50
44" - 48"	S2@.525	Ea	33.40	12.70	—	46.10
52" - 56"	S2@.653	Ea	41.60	15.80	—	57.40
60"	S2@.773	Ea	49.20	18.70	—	67.90

Galvanized Steel Rectangular Drops and Tees

12" wide galvanized steel rectangular duct drops and tap-in tees

Description	Craft@Hrs	Unit	Material $	Labor $	Equipment $	Total $
6"	S2@.109	Ea	6.93	2.63	—	9.56
8"	S2@.121	Ea	7.69	2.92	—	10.61
10"	S2@.133	Ea	8.46	3.21	—	11.67
12" - 16"	S2@.180	Ea	11.50	4.35	—	15.85
20" - 24"	S2@.262	Ea	16.70	6.33	—	23.03
28" - 32"	S2@.360	Ea	22.90	8.69	—	31.59
36" - 40"	S2@.470	Ea	29.80	11.30	—	41.10
44" - 48"	S2@.544	Ea	34.60	13.10	—	47.70
52" - 56"	S2@.675	Ea	43.00	16.30	—	59.30
60"	S2@.794	Ea	50.60	19.20	—	69.80

16" wide galvanized steel rectangular duct drops and tap-in tees

Description	Craft@Hrs	Unit	Material $	Labor $	Equipment $	Total $
6"	S2@.170	Ea	10.80	4.11	—	14.91
8"	S2@.184	Ea	11.80	4.44	—	16.24
10"	S2@.200	Ea	12.80	4.83	—	17.63
12" - 16"	S2@.231	Ea	14.70	5.58	—	20.28
20" - 24"	S2@.293	Ea	18.60	7.08	—	25.68
28" - 32"	S2@.396	Ea	25.10	9.56	—	34.66
36" - 40"	S2@.506	Ea	32.20	12.20	—	44.40
44" - 48"	S2@.581	Ea	37.10	14.00	—	51.10
52" - 56"	S2@.709	Ea	45.30	17.10	—	62.40
60"	S2@.841	Ea	53.50	20.30	—	73.80

20" wide galvanized steel rectangular duct drops and tap-in tees

Description	Craft@Hrs	Unit	Material $	Labor $	Equipment $	Total $
6"	S2@.200	Ea	12.80	4.83	—	17.63
8"	S2@.216	Ea	13.70	5.22	—	18.92
10"	S2@.231	Ea	14.70	5.58	—	20.28
12" - 16"	S2@.262	Ea	16.70	6.33	—	23.03
20" - 24"	S2@.324	Ea	20.70	7.82	—	28.52
28" - 32"	S2@.428	Ea	27.30	10.30	—	37.60
36" - 40"	S2@.544	Ea	34.60	13.10	—	47.70
44" - 48"	S2@.617	Ea	39.30	14.90	—	54.20
52" - 56"	S2@.758	Ea	48.20	18.30	—	66.50
60"	S2@.884	Ea	56.20	21.30	—	77.50

Description	Craft@Hrs	Unit	Material $	Labor $	Equipment $	Total $
24" wide galvanized steel rectangular duct drops and tap-in tees						
6"	S2@.231	Ea	14.70	5.58	—	20.28
8"	S2@.247	Ea	15.80	5.96	—	21.76
10"	S2@.262	Ea	16.70	6.33	—	23.03
12" - 16"	S2@.293	Ea	18.60	7.08	—	25.68
20" - 24"	S2@.354	Ea	22.60	8.55	—	31.15
28" - 32"	S2@.462	Ea	29.20	11.20	—	40.40
36" - 40"	S2@.581	Ea	37.10	14.00	—	51.10
44" - 48"	S2@.655	Ea	41.70	15.80	—	57.50
52" - 56"	S2@.798	Ea	50.70	19.30	—	70.00
60"	S2@.927	Ea	59.00	22.40	—	81.40
28" wide galvanized steel rectangular duct drops and tap-in tees						
6"	S2@.262	Ea	16.70	6.33	—	23.03
8"	S2@.278	Ea	17.60	6.71	—	24.31
10"	S2@.292	Ea	18.60	7.05	—	25.65
12" - 16"	S2@.324	Ea	20.70	7.82	—	28.52
20" - 24"	S2@.385	Ea	24.50	9.30	—	33.80
28" - 32"	S2@.497	Ea	31.60	12.00	—	43.60
36" - 40"	S2@.617	Ea	39.30	14.90	—	54.20
44" - 48"	S2@.694	Ea	44.10	16.80	—	60.90
52" - 56"	S2@.837	Ea	53.40	20.20	—	73.60
60"	S2@.969	Ea	61.90	23.40	—	85.30
32" wide galvanized steel rectangular duct drops and tap-in tees						
6"	S2@.356	Ea	22.70	8.60	—	31.30
8"	S2@.374	Ea	24.00	9.03	—	33.03
10"	S2@.393	Ea	25.00	9.49	—	34.49
12" - 16"	S2@.432	Ea	27.50	10.40	—	37.90
20" - 24"	S2@.506	Ea	32.20	12.20	—	44.40
28" - 32"	S2@.581	Ea	37.10	14.00	—	51.10
36" - 40"	S2@.655	Ea	41.70	15.80	—	57.50
44" - 48"	S2@.730	Ea	46.40	17.60	—	64.00
52" - 56"	S2@.877	Ea	56.00	21.20	—	77.20
60"	S2@1.02	Ea	64.60	24.60	—	89.20

Galvanized Steel Round Ductwork

Description	Craft@Hrs	Unit	Material $	Labor $	Equipment $	Total $

Galvanized steel round duct snap-lock with slip joints

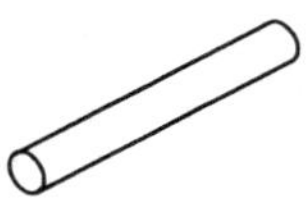

Description	Craft@Hrs	Unit	Material $	Labor $	Equipment $	Total $
3"	S2@.080	LF	1.04	1.93	—	2.97
4"	S2@.090	LF	1.07	2.17	—	3.24
5"	S2@.095	LF	1.15	2.29	—	3.44
6"	S2@.100	LF	1.33	2.41	—	3.74
7"	S2@.110	LF	1.65	2.66	—	4.31
8"	S2@.120	LF	1.69	2.90	—	4.59
9"	S2@.140	LF	1.78	3.38	—	5.16
10"	S2@.160	LF	1.86	3.86	—	5.72

Galvanized steel round duct 90 degree adjustable elbow

Description	Craft@Hrs	Unit	Material $	Labor $	Equipment $	Total $
3"	S2@.250	Ea	10.70	6.04	—	16.74
4"	S2@.300	Ea	14.60	7.24	—	21.84
5"	S2@.350	Ea	17.90	8.45	—	26.35
6"	S2@.400	Ea	23.30	9.66	—	32.96
7"	S2@.450	Ea	26.70	10.90	—	37.60
8"	S2@.500	Ea	30.00	12.10	—	42.10
9"	S2@.600	Ea	32.40	14.50	—	46.90
10"	S2@.700	Ea	35.30	16.90	—	52.20

Galvanized steel round duct 45 degree adjustable elbow

Description	Craft@Hrs	Unit	Material $	Labor $	Equipment $	Total $
3"	S2@.250	Ea	6.72	6.04	—	12.76
4"	S2@.300	Ea	8.00	7.24	—	15.24
5"	S2@.350	Ea	10.20	8.45	—	18.65
6"	S2@.400	Ea	13.50	9.66	—	23.16
7"	S2@.450	Ea	15.20	10.90	—	26.10
8"	S2@.500	Ea	16.90	12.10	—	29.00
9"	S2@.600	Ea	18.40	14.50	—	32.90
10"	S2@.700	Ea	20.20	16.90	—	37.10

Standard fiberglass ductwork, or duct board, should not be used in systems where air velocities exceed 2,500 feet per minute, or where static pressures are higher than 2 inches of water gauge.

Additionally, it should not be used for the following applications:

1) Equipment rooms
2) Underground or outdoors
3) Final connections to air handling equipment
4) Plenums or casings
5) Fume, heat or moisture exhaust systems

The costs for manufacturing and installing fiberglass ductwork are based on the net areas of the duct and fittings.

Here are approximate current material costs, per square foot:

Fiberglass Type	Up to 2,500 SF Material $/SF	Over 2,500 SF Material $/SF
475	1.06	.96
800	1.16	1.06
1400	1.41	1.36

Additional costs: Approximately 25 cents per square foot should be added for the cost of tie rods, hangers, staples, tape and waste.

Type 475 fiberglass board is used almost exclusively for residential and small commercial air systems, while Types 800 and 1400 are used in larger commercial and industrial systems.

Fiberglass Ductwork Fabrication Labor*
Size ranges (inches)

Activity	Up through 16		17 through 32		33 through 48	
	SF/Hour Output	Labor $/SF	SF/Hour Output	Labor $/SF	SF/Hour Output	Labor $/SF
Duct only	60	.46	58	.48	56	.49
Fittings only	40	.70	38	.74	35	.79
Duct & 15% fittings	55	.51	53	.52	50	.55
Duct & 25% fittings	53	.53	51	.54	48	.58
Duct & 35% fittings	51	.54	49	.58	46	.60

**Labor rates are based on using a standard grooving machine and include all tie-rod reinforcing.*

Labor correction factors:

1) Hand grooving in lieu of machine grooving: 1.45
2) Use of an auto-closer: .75
3) Tie-rods enclosed in conduit: 1.25
4) T-bar or channel reinforcing in lieu of tie-rods: 1.15

Description	Craft@Hrs	Unit	Material $	Labor $	Equipment $	Total $
Fiberglass Ductwork Insulation						
12" (2.00 SF/LF)	S2@.070	LF	—	1.69	—	1.69
14" (2.35 SF/LF)	S2@.082	LF	—	1.98	—	1.98
16" (2.65 SF/LF)	S2@.093	LF	—	2.25	—	2.25
18" (3.00 SF/LF)	S2@.105	LF	—	2.54	—	2.54
20" (3.35 SF/LF)	S2@.117	LF	—	2.83	—	2.83
22" (3.65 SF/LF)	S2@.128	LF	—	3.09	—	3.09
24" (4.00 SF/LF)	S2@.140	LF	—	3.38	—	3.38
26" (4.35 SF/LF)	S2@.152	LF	—	3.67	—	3.67
28" (4.65 SF/LF)	S2@.163	LF	—	3.94	—	3.94
30" (5.00 SF/LF)	S2@.175	LF	—	4.23	—	4.23
32" (5.35 SF/LF)	S2@.187	LF	—	4.52	—	4.52
34" (5.65 SF/LF)	S2@.198	LF	—	4.78	—	4.78
36" (6.00 SF/LF)	S2@.210	LF	—	5.07	—	5.07
38" (6.35 SF/LF)	S2@.222	LF	—	5.36	—	5.36
40" (6.65 SF/LF)	S2@.232	LF	—	5.60	—	5.60
42" (7.00 SF/LF)	S2@.245	LF	—	5.92	—	5.92
44" (7.35 SF/LF)	S2@.257	LF	—	6.21	—	6.21
46" (7.65 SF/LF)	S2@.268	LF	—	6.47	—	6.47
48" (8.00 SF/LF)	S2@.280	LF	—	6.76	—	6.76
50" (8.35 SF/LF)	S2@.292	LF	—	7.05	—	7.05
52" (8.65 SF/LF)	S2@.303	LF	—	7.32	—	7.32
54" (9.00 SF/LF)	S2@.315	LF	—	7.61	—	7.61
56" (9.35 SF/LF)	S2@.327	LF	—	7.90	—	7.90
58" (9.65 SF/LF)	S2@.338	LF	—	8.16	—	8.16
60" (10.0 SF/LF)	S2@.350	LF	—	8.45	—	8.45
64" (10.6 SF/LF)	S2@.373	LF	—	9.01	—	9.01
68" (11.4 SF/LF)	S2@.397	LF	—	9.59	—	9.59
72" (12.0 SF/LF)	S2@.420	LF	—	10.10	—	10.10
76" (12.6 SF/LF)	S2@.443	LF	—	10.70	—	10.70
80" (13.4 SF/LF)	S2@.467	LF	—	11.30	—	11.30
84" (14.0 SF/LF)	S2@.490	LF	—	11.80	—	11.80
88" (14.6 SF/LF)	S2@.513	LF	—	12.40	—	12.40
92" (15.4 SF/LF)	S2@.537	LF	—	13.00	—	13.00
96" (16.0 SF/LF)	S2@.560	LF	—	13.50	—	13.50

For estimating purposes, when measuring the length of ductwork, include the fitting. Then apply the labor per run foot to the total.

1¼" thick wire reinforced flexible fiberglass duct with vinyl cover

Description	Craft@Hrs	Unit	Material $	Labor $	Equipment $	Total $
4"	S2@.060	LF	1.75	1.45	—	3.20
5"	S2@.070	LF	1.81	1.69	—	3.50
6"	S2@.080	LF	1.99	1.93	—	3.92
8"	S2@.090	LF	2.52	2.17	—	4.69
9"	S2@.100	LF	2.70	2.41	—	5.11
10"	S2@.120	LF	2.82	2.90	—	5.72
12"	S2@.140	LF	3.64	3.38	—	7.02
14"	S2@.160	LF	4.36	3.86	—	8.22
16"	S2@.180	LF	5.06	4.35	—	9.41
18"	S2@.200	LF	5.75	4.83	—	10.58

Description	Craft@Hrs	Unit	Material $	Labor $	Equipment $	Total $

Fiberglass Pipe Insulation with AP-T Plus (all purpose self sealing) Jacket. By nominal pipe diameter. R factor equals 2.56 at 300 degrees F. These costs do not include scaffolding, add for same, if required. Also see fittings, flanges and valves at the end of this section.

Jacketed fiberglass insulation for 1/2" pipe

Description	Craft@Hrs	Unit	Material $	Labor $	Equipment $	Total $
1/2" thick	P1@.038	LF	.77	.93	—	1.70
1" thick	P1@.038	LF	.91	.93	—	1.84
1½" thick	P1@.040	LF	1.86	.98	—	2.84

Jacketed fiberglass insulation for 3/4" pipe

Description	Craft@Hrs	Unit	Material $	Labor $	Equipment $	Total $
1/2" thick	P1@.038	LF	.84	.93	—	1.77
1" thick	P1@.038	LF	1.04	.93	—	1.97
1½" thick	P1@.040	LF	1.93	.98	—	2.91

Jacketed fiberglass insulation for 1" pipe

Description	Craft@Hrs	Unit	Material $	Labor $	Equipment $	Total $
1/2" thick	P1@.040	LF	.90	.98	—	1.88
1" thick	P1@.040	LF	1.20	.98	—	2.18
1½" thick	P1@.042	LF	2.05	1.03	—	3.08

Jacketed fiberglass insulation for 1¼" pipe

Description	Craft@Hrs	Unit	Material $	Labor $	Equipment $	Total $
1/2" thick	P1@.040	LF	1.00	.98	—	1.98
1" thick	P1@.040	LF	1.21	.98	—	2.19
1½" thick	P1@.042	LF	2.18	1.03	—	3.21

Jacketed fiberglass insulation for 1½" pipe

Description	Craft@Hrs	Unit	Material $	Labor $	Equipment $	Total $
1/2" thick	P1@.042	LF	1.07	1.03	—	2.10
1" thick	P1@.042	LF	1.36	1.03	—	2.39
1½" thick	P1@.044	LF	2.35	1.08	—	3.43

Jacketed fiberglass insulation for 2" pipe

Description	Craft@Hrs	Unit	Material $	Labor $	Equipment $	Total $
1/2" thick	P1@.044	LF	1.14	1.08	—	2.22
1" thick	P1@.044	LF	1.56	1.08	—	2.64
1½" thick	P1@.046	LF	2.52	1.12	—	3.64

Fiberglass Pipe Insulation

Description	Craft@Hrs	Unit	Material $	Labor $	Equipment $	Total $
Jacketed fiberglass insulation for 2½" pipe						
1" thick	P1@.047	LF	1.21	1.15	—	2.36
1½" thick	P1@.048	LF	2.13	1.17	—	3.30
2" thick	P1@.050	LF	3.47	1.22	—	4.69
Jacketed fiberglass insulation for 3" pipe						
1" thick	P1@.051	LF	.84	1.25	—	2.09
1½" thick	P1@.054	LF	2.31	1.32	—	3.63
2" thick	P1@.056	LF	3.80	1.37	—	5.17
Jacketed fiberglass insulation for 4" pipe						
1" thick	P1@.063	LF	1.04	1.54	—	2.58
1½" thick	P1@.066	LF	2.66	1.61	—	4.27
2" thick	P1@.069	LF	4.45	1.69	—	6.14
Jacketed fiberglass insulation for 6" pipe						
1" thick	P1@.077	LF	2.43	1.88	—	4.31
1½" thick	P1@.080	LF	3.34	1.95	—	5.29
2" thick	P1@.085	LF	5.61	2.08	—	7.69
Jacketed fiberglass insulation for 8" pipe						
1" thick	P1@.117	LF	3.34	2.86	—	6.20
1½" thick	P1@.123	LF	4.09	3.01	—	7.10
2" thick	P1@.129	LF	6.60	3.15	—	9.75
Jacketed fiberglass insulation for 10" pipe						
1" thick	P1@.140	LF	3.94	3.42	—	7.36
1½" thick	P1@.147	LF	5.01	3.59	—	8.60
2" thick	P1@.154	LF	7.92	3.76	—	11.68
Jacketed fiberglass insulation for 12" pipe						
1" thick	P1@.155	LF	4.55	3.79	—	8.34
1½" thick	P1@.163	LF	5.61	3.98	—	9.59
2" thick	P1@.171	LF	8.72	4.18	—	12.90
Add for .016" aluminum jacket Per SF						
Per SF of surface	P1@.035	SF	.65	.86	—	1.51

Pipe fittings and flanges

For each fitting or flange use the cost for 3 LF of pipe of the same pipe size

Valves

Body only: for each valve body use the cost for 5 LF of pipe of the same pipe size

Body and bonnet or yoke: for each valve use the cost for 10 LF of pipe of the same pipe size

Flanged valves: add the cost for of flanges per above

Description	Craft@Hrs	Unit	Material $	Labor $	Equipment $	Total $

Calcium Silicate Pipe Insulation with .016" Aluminum Jacket. By nominal pipe diameter. R factor equals 2.20 at 300 degrees F. Manufactured by Pabco. These costs do not include scaffolding, add for same, if required. Also see fittings, flanges and valves at the end of this section.

Jacketed calcium silicate insulation for 6" pipe

Description	Craft@Hrs	Unit	Material $	Labor $	Equipment $	Total $
2" thick	P1@.183	LF	9.77	4.47	—	14.24
4" thick	P1@.218	LF	21.40	5.33	—	26.73
6" thick	P1@.288	LF	30.00	7.04	—	37.04

Jacketed calcium silicate insulation for 8" pipe

Description	Craft@Hrs	Unit	Material $	Labor $	Equipment $	Total $
2" thick	P1@.194	LF	11.70	4.74	—	16.44
4" thick	P1@.228	LF	24.90	5.57	—	30.47
6" thick	P1@.308	LF	37.70	7.53	—	45.23

Jacketed calcium silicate insulation for 10" pipe

Description	Craft@Hrs	Unit	Material $	Labor $	Equipment $	Total $
2" thick	P1@.202	LF	13.70	4.94	—	18.64
4" thick	P1@.239	LF	29.00	5.84	—	34.84
6" thick	P1@.337	LF	45.10	8.24	—	53.34

Jacketed calcium silicate insulation for 12" pipe

Description	Craft@Hrs	Unit	Material $	Labor $	Equipment $	Total $
2" thick	P1@.208	LF	15.90	5.08	—	20.98
4" thick	P1@.261	LF	32.30	6.38	—	38.68
6" thick	P1@.388	LF	51.10	9.48	—	60.58

Pipe fittings and flanges

For each fitting or flange use the cost for 3 LF of pipe of the same pipe size

Valves

Body only: for each valve body use the cost for 5 LF of pipe of the same pipe size

Body and bonnet or yoke: for each valve use the cost for 10 LF of pipe of the same pipe size

Flanged valves: add the cost for of flanges per above

Description	Craft@Hrs	Unit	Material $	Labor $	Equipment $	Total $

Closed Cell Elastomeric Pipe and Tubing Insulation. Semi split, by nominal pipe or tube diameter with insulation wall thickness as shown, no cover. R factor equals 3.58 at 220 degrees F. Manufactured by Rubatex. These costs do not include scaffolding, add for same, if required. Also see fittings, flanges and valves at the end of this section.

Description	Craft@Hrs	Unit	Material $	Labor $	Equipment $	Total $
1/4", 1/2" thick	P1@.039	LF	.31	.95	—	1.26
3/8", 1/2" thick	P1@.039	LF	.33	.95	—	1.28
1/2", 1/2" thick	P1@.039	LF	.37	.95	—	1.32
3/4", 1/2" thick	P1@.039	LF	.42	.95	—	1.37
1", 1/2" thick	P1@.042	LF	.47	1.03	—	1.50
1¼", 1/2" thick	P1@.042	LF	.53	1.03	—	1.56
1½", 1/2" thick	P1@.042	LF	.62	1.03	—	1.65
2", 1/2" thick	P1@.046	LF	.78	1.12	—	1.90
2½", 3/8" thick	P1@.056	LF	.99	1.37	—	2.36
2½", 1/2" thick	P1@.056	LF	1.33	1.37	—	2.70
2½", 3/4" thick	P1@.056	LF	1.73	1.37	—	3.10
3", 3/8" thick	P1@.062	LF	1.14	1.52	—	2.66
3", 1/2" thick	P1@.062	LF	1.50	1.52	—	3.02
3", 3/4" thick	P1@.062	LF	1.95	1.52	—	3.47
4", 3/8" thick	P1@.068	LF	1.33	1.66	—	2.99
4", 1/2" thick	P1@.068	LF	1.76	1.66	—	3.42
4", 3/4" thick	P1@.068	LF	2.29	1.66	—	3.95

Pipe fittings and flanges

For each fitting or flange use the cost for 3 LF of pipe of the same pipe size

Valves

Body only: for each valve body use the cost for 5 LF of pipe of the same pipe size

Body and bonnet or yoke: for each valve use the cost for 10 LF of pipe of the same pipe size

Flanged valves: add the cost for of flanges per above

Description	Craft@Hrs	Unit	Material $	Labor $	Equipment $	Total $
Fiberglass flexible wrap duct insulation with foil facing vapor barrier to 250 degrees F						
1"	S2@.023	SF	.35	.56	—	.91
1½"	S2@.023	SF	.42	.56	—	.98
2"	S2@.025	SF	.55	.60	—	1.15
Deduct foil facing	—	%	30	—	—	—
Fiberglass rigid board duct insulation plain, *without* foil facing vapor barrier to 450 degrees F						
1"	S2@.067	SF	.43	1.62	—	2.05
1½"	S2@.067	SF	.71	1.62	—	2.33
2"	S2@.075	SF	.85	1.81	—	2.66
3"	S2@.085	SF	1.27	2.05	—	3.32
Fiberglass rigid board duct insulation with FSK (foil) facing vapor barrier to 250 degrees F						
1"	S2@.067	SF	.55	1.62	—	2.17
1½"	S2@.067	SF	.68	1.62	—	2.30
2"	S2@.075	SF	.81	1.81	—	2.62
3"	S2@.085	SF	1.17	2.05	—	3.22
Fiberglass flexible blanket duct lining to 250 degrees F						
1/2"	S2@.032	SF	.26	.77	—	1.03
1"	S2@.032	SF	.34	.77	—	1.11
1½"	S2@.035	SF	.48	.85	—	1.33
2"	S2@.035	SF	.62	.85	—	1.47
Fiberglass rigid duct lining to 250 degrees F						
5/8"	S2@.032	SF	.56	.77	—	1.33
1"	S2@.032	SF	.73	.77	—	1.50
1½"	S2@.035	SF	.96	.85	—	1.81
2"	S2@.035	SF	1.23	.85	—	2.08

Note: Material prices include all necessary tape, pins, mastic and 15% waste.

Balancing of HVAC Systems

Description	Craft@Hrs	Unit	Material $	Labor $	Equipment $	Total $

Air Balancing – Air Handling units

Description	Craft@Hrs	Unit	Material $	Labor $	Equipment $	Total $
Central Station	S2@6.00	Ea	—	145.00	—	145.00
Multi-zone A/C	S2@4.00	Ea	—	96.60	—	96.60
Single-zone A/C	S2@3.00	Ea	—	72.40	—	72.40
Packaged A/C	S2@2.50	Ea	—	60.40	—	60.40
Built-up low pressure	S2@4.50	Ea	—	109.00	—	109.00
Built-up high pres.	S2@6.00	Ea	—	145.00	—	145.00
Built-up VAV	S2@6.00	Ea	—	145.00	—	145.00
Built-up Dual Duct	S2@6.00	Ea	—	145.00	—	145.00

Air Balancing – Terminal Boxes

Description	Craft@Hrs	Unit	Material $	Labor $	Equipment $	Total $
Variable air volume	S2@0.40	Ea	—	9.66	—	9.66
Constant volume	S2@0.40	Ea	—	9.66	—	9.66
Dual Duct box	S2@0.70	Ea	—	16.90	—	16.90
Terminal box and all downstream diffusers						
Cost per zone	S2@1.75	Ea	—	42.30	—	42.30

Air Balancing – Diffusers

Description	Craft@Hrs	Unit	Material $	Labor $	Equipment $	Total $
Ceiling diffusers to 500 CFM	S2@.250	Ea	—	6.04	—	6.04
Ceiling diffusers over 500 CFM	S2@.350	Ea	—	8.45	—	8.45
Ceiling diffusers perforated	S2@.370	Ea	—	8.93	—	8.93
Sidewall diffusers double deflection	S2@.250	Ea	—	6.04	—	6.04
Linear slot diffusers 2 slot	S2@.060	LF	—	1.45	—	1.45
Linear slot diffusers 3 slot	S2@.070	LF	—	1.69	—	1.69
Linear slot diffusers 4 slot	S2@.080	LF	—	1.93	—	1.93
Plug-in diffuser	S2@.250	Ea	—	6.04	—	6.04
Light troffers 1 slot	S2@.195	Ea	—	4.71	—	4.71
Light troffers 2 slot	S2@.220	Ea	—	5.31	—	5.31
Laminar flow diff.	S2@.750	Ea	—	18.10	—	18.10
Laminar flow diffusers with HEPA filter	S2@3.00	Ea	—	72.40	—	72.40

Description	Craft@Hrs	Unit	Material $	Labor $	Equipment $	Total $

Air Balancing – Grilles

Description	Craft@Hrs	Unit	Material $	Labor $	Equipment $	Total $
Return air grilles to 800 CFM	S2@.250	Ea	—	6.04	—	6.04
Return air grilles over 800 CFM	S2@.350	Ea	—	8.45	—	8.45
Bar type return air grilles 2 slot	S2@.060	LF	—	1.45	—	1.45
Bar type return air grilles 4 slot	S2@.070	LF	—	1.69	—	1.69
Bar type return air grilles 6 slot	S2@.080	LF	—	1.93	—	1.93

Air Balancing – Centrifugal fans

Description	Craft@Hrs	Unit	Material $	Labor $	Equipment $	Total $
Supply and return air to 3000 CFM	S2@1.00	Ea	—	24.10	—	24.10
Supply and return air to 5000 CFM	S2@1.50	Ea	—	36.20	—	36.20
Supply and return air to 10,000 CFM	S2@2.00	Ea	—	48.30	—	48.30
Supply and return air to 20,000 CFM	S2@2.75	Ea	—	66.40	—	66.40
Roof exhaust to 6000 CFM	S2@.725	Ea	—	17.50	—	17.50
Roof exhaust to 10,000 CFM	S2@.925	Ea	—	22.30	—	22.30
Roof exhaust to 20,000 CFM	S2@1.50	Ea	—	36.20	—	36.20
Range hood exhaust to 10,000 CFM	S2@1.75	Ea	—	42.30	—	42.30
Range hood exhaust to 20,000 CFM	S2@2.50	Ea	—	60.40	—	60.40

Air Balancing – Fan Coil Units

Description	Craft@Hrs	Unit	Material $	Labor $	Equipment $	Total $
Vertical floor model	S2@.500	Ea	—	12.10	—	12.10
Horiz. ceiling mtd	S2@.650	Ea	—	15.70	—	15.70

Air Balancing – Fume Hoods

Description	Craft@Hrs	Unit	Material $	Labor $	Equipment $	Total $
Lab fume hood	S2@2.00	Ea	—	48.30	—	48.30
General purpose	S2@.450	Ea	—	10.90	—	10.90
Dust collectors	S2@1.50	Ea	—	36.20	—	36.20

Wet Balancing

Description	Craft@Hrs	Unit	Material $	Labor $	Equipment $	Total $
Boiler	S2@1.65	Ea	—	39.80	—	39.80
Chiller	S2@3.50	Ea	—	84.50	—	84.50
Cooling Tower	S2@2.00	Ea	—	48.30	—	48.30
Circulating pump	S2@1.65	Ea	—	39.80	—	39.80
Heat exchanger	S2@1.75	Ea	—	42.30	—	42.30
AHU water coil	S2@2.50	Ea	—	60.40	—	60.40
Reheat coil	S2@.450	Ea	—	10.90	—	10.90
Wall fin radiation	S2@.350	Ea	—	8.45	—	8.45
Radiant heat panel	S2@.450	Ea	—	10.90	—	10.90
Fan coil unit, 2 pipe	S2@.350	Ea	—	8.45	—	8.45
Fan coil unit, 4 pipe	S2@.450	Ea	—	10.90	—	10.90
Cabinet heater	S2@.350	Ea	—	8.45	—	8.45
Unit ventilator	S2@.450	Ea	—	10.90	—	10.90
Unit heater	S2@.350	Ea	—	8.45	—	8.45

Ductile Iron, Class 153, Cement-Lined with Mechanical Joints

Cement-lined Class 153 ductile iron pipe with mechanical joints

Description	Craft@Hrs	Unit	Material $	Labor $	Equipment $	Total $
4"	P1@.140	LF	5.83	3.42	—	9.25
6"	ER@.180	LF	8.36	4.69	.90	13.95
8"	ER@.200	LF	11.60	5.22	1.00	17.82
10"	ER@.230	LF	17.20	6.00	1.15	24.35
12"	ER@.270	LF	23.40	7.04	1.35	31.79
14"	ER@.310	LF	35.60	8.09	1.55	45.24
16"	ER@.350	LF	45.80	9.13	1.75	56.68

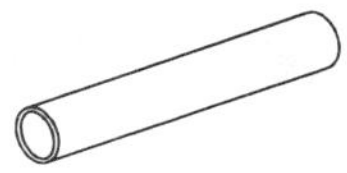

Cement-lined Class 153 ductile iron 1/16 bend with mechanical joints

Description	Craft@Hrs	Unit	Material $	Labor $	Equipment $	Total $
4"	P1@.900	Ea	23.70	22.00	—	45.70
6"	ER@1.30	Ea	36.30	33.90	6.50	76.70
8"	ER@1.70	Ea	57.20	44.30	8.50	110.00
10"	ER@2.15	Ea	78.10	56.10	10.80	145.00
12"	ER@2.40	Ea	107.00	62.60	12.00	181.60
14"	ER@2.70	Ea	238.00	70.40	13.50	321.90
16"	ER@3.00	Ea	268.00	78.20	15.00	361.20

Cement-lined Class 153 ductile iron 1/8 bend with mechanical joints

Description	Craft@Hrs	Unit	Material $	Labor $	Equipment $	Total $
4"	P1@.900	Ea	23.70	22.00	—	45.70
6"	ER@1.30	Ea	35.20	33.90	6.50	75.60
8"	ER@1.70	Ea	56.10	44.30	8.50	108.90
10"	ER@2.15	Ea	77.00	56.10	10.80	143.90
12"	ER@2.40	Ea	106.00	62.60	12.00	180.60
14"	ER@2.70	Ea	230.00	70.40	13.50	313.90
16"	ER@3.00	Ea	268.00	78.20	15.00	361.20

Cement-lined Class 153 ductile iron 1/4 bend with mechanical joints

Description	Craft@Hrs	Unit	Material $	Labor $	Equipment $	Total $
4"	P1@.900	Ea	25.30	22.00	—	47.30
6"	ER@1.30	Ea	41.30	33.90	6.50	81.70
8"	ER@1.70	Ea	86.40	44.30	8.50	139.20
10"	ER@2.15	Ea	129.00	56.10	10.80	195.90
12"	ER@2.40	Ea	177.00	62.60	12.00	251.60
14"	ER@2.70	Ea	455.00	70.40	13.50	538.90
16"	ER@3.00	Ea	478.00	78.20	15.00	571.20

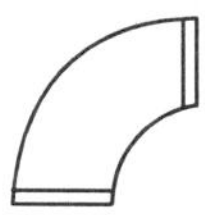

Cement-lined Class 153 ductile iron tee with mechanical joints

Description	Craft@Hrs	Unit	Material $	Labor $	Equipment $	Total $
4"	P1@1.65	Ea	38.50	40.30	—	78.80
6"	ER@2.35	Ea	56.10	61.30	11.80	129.20
8"	ER@3.15	Ea	86.40	82.20	15.80	184.40
10"	ER@4.00	Ea	129.00	104.00	20.00	253.00
12"	ER@4.45	Ea	177.00	116.00	22.30	315.30
14"	ER@4.90	Ea	455.00	128.00	24.50	607.50
16"	ER@5.30	Ea	478.00	138.00	26.50	642.50

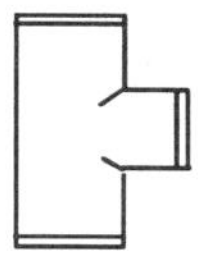

Ductile Iron, Class 153, Cement-Lined with Mechanical Joints

Cement-lined Class 153 ductile iron wye with mechanical joints

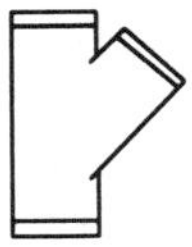

Description	Craft@Hrs	Unit	Material $	Labor $	Equipment $	Total $
4"	P1@1.65	Ea	86.40	40.30	—	126.70
6"	ER@2.35	Ea	107.00	61.30	11.80	180.10
8"	ER@3.15	Ea	160.00	82.20	15.80	258.00
10"	ER@4.00	Ea	301.00	104.00	20.00	425.00
12"	ER@4.45	Ea	452.00	116.00	22.30	590.30
14"	ER@4.90	Ea	927.00	128.00	24.90	1,079.90
16"	ER@5.30	Ea	1,160.00	138.00	26.50	1,324.50

Cement-lined Class 153 ductile iron reducer with mechanical joints

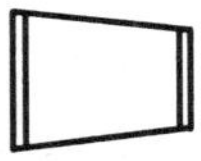

Description	Craft@Hrs	Unit	Material $	Labor $	Equipment $	Total $
6"	ER@1.45	Ea	28.10	37.80	7.25	73.15
8"	ER@2.00	Ea	44.40	52.20	10.00	106.60
10"	ER@2.55	Ea	58.90	66.50	12.80	138.20
12"	ER@3.05	Ea	75.90	79.60	15.30	170.80
14"	ER@3.60	Ea	170.00	93.90	18.00	281.90
16"	ER@4.20	Ea	207.00	110.00	21.00	338.00

Ductile Iron, Class 153, Double Cement-Lined with Mechanical Joints

Double cement-lined Class 153 ductile iron pipe with mechanical joints

Description	Craft@Hrs	Unit	Material $	Labor $	Equipment $	Total $
4"	P1@.140	LF	6.12	3.42	—	9.54
6"	ER@.180	LF	8.78	4.69	.90	14.37
8"	ER@.200	LF	12.20	5.22	1.00	18.42
10"	ER@.230	LF	18.10	6.00	1.15	25.25
12"	ER@.270	LF	24.60	7.04	1.35	32.99
14"	ER@.310	LF	37.40	8.09	1.55	47.04
16"	ER@.350	LF	48.10	9.13	1.75	58.98

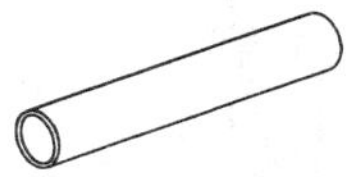

Double cement-lined Class 153 ductile iron 1/16 bend with mechanical joints

Description	Craft@Hrs	Unit	Material $	Labor $	Equipment $	Total $
4"	P1@.900	Ea	27.00	22.00	—	49.00
6"	ER@1.30	Ea	41.80	33.90	6.50	82.20
8"	ER@1.70	Ea	66.00	44.30	8.50	118.80
10"	ER@2.15	Ea	90.00	56.10	10.80	156.90
12"	ER@2.40	Ea	123.00	62.60	12.00	197.60
14"	ER@2.70	Ea	273.00	70.40	13.50	356.90
16"	ER@3.00	Ea	309.00	78.20	15.00	402.20

Double cement-lined Class 153 ductile iron 1/8 bend with mechanical joints

Description	Craft@Hrs	Unit	Material $	Labor $	Equipment $	Total $
4"	P1@.900	Ea	27.00	22.00	—	49.00
6"	ER@1.30	Ea	40.70	33.90	6.50	81.10
8"	ER@1.70	Ea	64.40	44.30	8.50	117.20
10"	ER@2.15	Ea	88.60	56.10	10.80	155.50
12"	ER@2.40	Ea	122.00	62.60	12.00	196.60
14"	ER@2.70	Ea	265.00	70.40	13.50	348.90
16"	ER@3.00	Ea	309.00	78.20	15.00	402.20

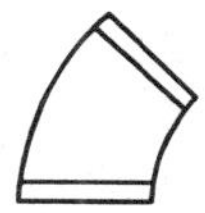

Double cement-lined Class 153 ductile iron 1/4 bend with mechanical joints

Description	Craft@Hrs	Unit	Material $	Labor $	Equipment $	Total $
4"	P1@.900	Ea	29.20	22.00	—	51.20
6"	ER@1.30	Ea	47.30	33.90	6.50	87.70
8"	ER@1.70	Ea	73.20	44.30	8.50	126.00
10"	ER@2.15	Ea	115.00	56.10	10.80	181.90
12"	ER@2.40	Ea	149.00	62.60	12.00	223.60
14"	ER@2.70	Ea	353.00	70.40	13.50	436.90
16"	ER@3.00	Ea	402.00	78.20	15.00	495.20

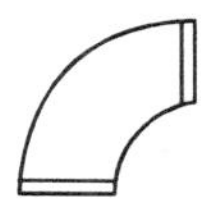

Double cement-lined Class 153 ductile iron tee with mechanical joints

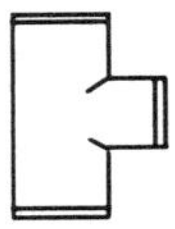

Description	Craft@Hrs	Unit	Material $	Labor $	Equipment $	Total $
4"	P1@1.65	Ea	44.60	40.30	—	84.90
6"	ER@2.35	Ea	64.40	61.30	11.80	137.50
8"	ER@3.15	Ea	99.60	82.20	15.80	197.60
10"	ER@4.00	Ea	149.00	104.00	20.00	273.00
12"	ER@4.45	Ea	203.00	116.00	22.30	341.30
14"	ER@4.90	Ea	523.00	128.00	24.50	675.50
16"	ER@5.30	Ea	549.00	138.00	26.50	713.50

Double cement-lined Class 153 ductile iron wye with mechanical joints

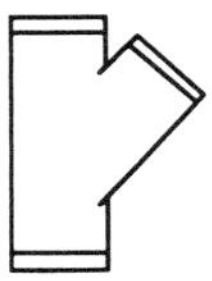

Description	Craft@Hrs	Unit	Material $	Labor $	Equipment $	Total $
4"	P1@1.65	Ea	99.60	40.30	—	139.90
6"	ER@2.35	Ea	127.00	61.30	11.80	200.10
8"	ER@3.15	Ea	186.00	82.20	15.80	284.00
10"	ER@4.00	Ea	350.00	104.00	20.00	474.00
12"	ER@4.45	Ea	520.00	116.00	22.30	658.30
14"	ER@4.90	Ea	1,070.00	128.00	24.50	1,222.50
16"	ER@5.30	Ea	1,330.00	138.00	26.50	1,494.50

Double cement-lined Class 153 ductile iron reducer with mechanical joints

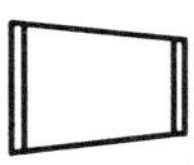

Description	Craft@Hrs	Unit	Material $	Labor $	Equipment $	Total $
6"	ER@1.45	Ea	32.50	37.80	7.25	77.55
8"	ER@2.00	Ea	46.20	52.20	10.00	108.40
10"	ER@2.55	Ea	60.00	66.50	12.80	139.30
12"	ER@3.05	Ea	87.50	79.60	15.30	182.40
14"	ER@3.60	Ea	188.00	93.90	15.00	296.90
16"	ER@4.20	Ea	238.00	110.00	21.00	369.00

Ductile Iron, Class 110, Cement-Lined with Mechanical Joints

Description	Craft@Hrs	Unit	Material $	Labor $	Equipment $	Total $
Cement-lined Class 110 ductile iron pipe with mechanical joints						
4"	P1@.180	LF	6.65	4.40	—	11.05
6"	ER@.220	LF	9.53	5.74	1.10	16.37
8"	ER@.240	LF	13.20	6.26	1.20	20.66
10"	ER@.270	LF	19.60	7.04	1.35	27.99
12"	ER@.320	LF	26.60	8.35	1.60	36.55
16"	ER@.400	LF	52.20	10.40	2.00	64.60
Cement-lined Class 110 ductile iron 1/16 bend with mechanical joints						
4"	P1@1.40	Ea	66.60	34.20	—	100.80
6"	ER@2.00	Ea	89.70	52.20	10.00	151.90
8"	ER@2.50	Ea	124.00	65.20	12.50	201.70
10"	ER@3.20	Ea	187.00	83.50	16.00	286.50
12"	ER@3.60	Ea	240.00	93.90	18.00	351.90
16"	ER@4.50	Ea	620.00	117.00	22.50	759.50
Cement-lined Class 110 ductile iron 1/8 bend with mechanical joints						
4"	P1@1.40	Ea	64.90	34.20	—	99.10
6"	ER@2.00	Ea	88.00	52.20	10.00	150.20
8"	ER@2.50	Ea	124.00	65.20	12.50	201.70
10"	ER@3.20	Ea	180.00	83.50	16.00	279.50
12"	ER@3.60	Ea	239.00	93.90	18.00	350.90
16"	ER@4.50	Ea	608.00	117.00	22.50	747.50
Cement-lined Class 110 ductile iron 1/4 bend with mechanical joints						
4"	P1@1.40	Ea	72.60	34.20	—	106.80
6"	ER@2.00	Ea	97.40	52.20	10.00	159.60
8"	ER@2.50	Ea	141.00	65.20	12.50	218.70
10"	ER@3.20	Ea	216.00	83.50	16.00	315.50
12"	ER@3.60	Ea	326.00	93.90	18.00	437.90
16"	ER@4.50	Ea	787.00	117.00	22.50	926.50
Cement-lined Class 110 ductile iron tee with mechanical joints						
4"	P1@2.50	Ea	109.00	61.10	—	170.10
6"	ER@3.50	Ea	146.00	91.30	17.50	254.80
8"	ER@4.70	Ea	208.00	123.00	23.50	354.50
10"	ER@6.00	Ea	358.00	156.00	30.00	544.00
12"	ER@6.70	Ea	455.00	175.00	33.00	663.00
16"	ER@8.00	Ea	1,190.00	209.00	40.00	1,439.00
Cement-lined Class 110 ductile iron reducer with mechanical joints						
6"	ER@2.20	Ea	72.60	57.40	11.00	141.00
8"	ER@3.00	Ea	100.00	78.20	15.00	193.20
10"	ER@3.80	Ea	164.00	99.10	19.00	282.10
12"	ER@4.60	Ea	194.00	120.00	23.00	337.00
16"	ER@6.30	Ea	532.00	164.00	31.50	727.50

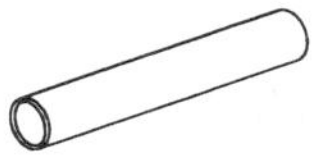

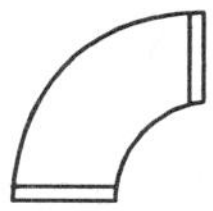
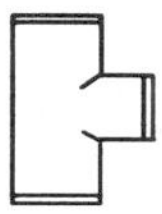
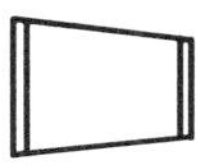

Cast Iron, Class 150 with Mechanical Joints

Class 150 cast iron pipe with mechanical joints

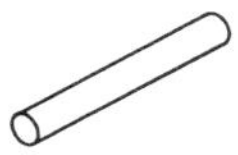

Description	Craft@Hrs	Unit	Material $	Labor $	Equipment $	Total $
1½"	P1@.080	LF	3.24	1.95	—	5.19
2"	P1@.100	LF	3.17	2.44	—	5.61
3"	P1@.120	LF	4.37	2.93	—	7.30
4"	P1@.140	LF	5.68	3.42	—	9.10
6"	ER@.180	LF	9.76	4.69	.90	15.35
8"	ER@.200	LF	15.70	5.22	1.00	21.92

Class 150 cast iron 1/8 bend with mechanical joints

Description	Craft@Hrs	Unit	Material $	Labor $	Equipment $	Total $
1½"	P1@.750	Ea	3.53	18.30	—	21.83
2"	P1@.800	Ea	3.94	19.50	—	23.44
3"	P1@1.00	Ea	5.32	24.40	—	29.72
4"	P1@1.20	Ea	6.74	29.30	—	36.04
6"	ER@1.70	Ea	15.70	44.30	8.50	68.50
8"	ER@2.25	Ea	45.10	58.70	11.30	115.10

Class 150 cast iron 1/4 bend with mechanical joints

Description	Craft@Hrs	Unit	Material $	Labor $	Equipment $	Total $
2"	P1@.800	Ea	5.70	19.50	—	25.20
3"	P1@1.00	Ea	8.13	24.40	—	32.53
4"	P1@1.20	Ea	12.80	29.30	—	42.10
6"	ER@1.70	Ea	26.60	44.30	8.50	79.40
8"	ER@2.25	Ea	71.80	58.70	11.30	141.80

Class 150 cast iron sanitary tee with mechanical joints

Description	Craft@Hrs	Unit	Material $	Labor $	Equipment $	Total $
1½"	P1@.750	Ea	5.89	18.30	—	24.19
2"	P1@1.00	Ea	6.43	24.40	—	30.83
3"	P1@1.20	Ea	7.81	29.30	—	37.11
4"	P1@1.65	Ea	12.10	40.30	—	52.40
6"	ER@2.35	Ea	34.40	61.30	11.80	107.50
8"	ER@3.15	Ea	109.00	82.20	15.80	207.00

Class 150 cast iron wye with mechanical joints

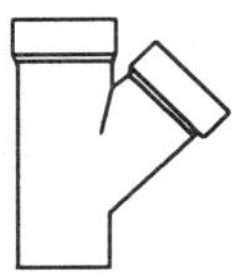

Description	Craft@Hrs	Unit	Material $	Labor $	Equipment $	Total $
1½"	P1@.750	Ea	6.02	18.30	—	24.32
2"	P1@1.00	Ea	6.02	24.40	—	30.42
3"	P1@1.20	Ea	8.54	29.30	—	37.84
4"	P1@1.65	Ea	13.90	40.30	—	54.20
6"	ER@2.35	Ea	37.00	61.30	11.80	110.10
8"	ER@3.15	Ea	87.10	82.20	15.80	185.10

Class 150 cast iron reducer

Description	Craft@Hrs	Unit	Material $	Labor $	Equipment $	Total $
2" x 1½"	P1@.600	Ea	3.24	14.70	—	17.94
3" x 2"	P1@.800	Ea	3.21	19.50	—	22.71
4" x 2"	P1@1.00	Ea	5.01	24.40	—	29.41
6" x 4"	ER@1.45	Ea	13.30	37.80	7.25	58.35
8" x 6"	ER@2.00	Ea	22.90	52.20	10.00	85.10

Asbestos-Cement, Class 2400 or 3000 with Mechanical Joints

Description	Craft@Hrs	Unit	Material $	Labor $	Equipment $	Total $
Asbestos-cement pipe with mechanical joints						
4"	ER@.080	LF	5.03	2.09	.40	7.52
6"	ER@.100	LF	3.18	2.61	.50	6.29
8"	ER@.140	LF	5.03	3.65	.70	9.38
10"	ER@.200	LF	7.81	5.22	1.00	14.03
12"	ER@.260	LF	10.30	6.78	1.30	18.38
14"	ER@.300	LF	14.00	7.82	1.50	23.32
16"	ER@.350	LF	17.30	9.13	1.75	28.18
Asbestos-cement 1/16 bend with mechanical joints						
4"	ER@1.00	Ea	70.70	26.10	5.00	101.80
6"	ER@1.25	Ea	93.70	32.60	6.25	132.55
8"	ER@1.70	Ea	129.00	44.30	8.50	181.80
10"	ER@2.20	Ea	196.00	57.40	11.00	264.40
12"	ER@2.70	Ea	226.00	70.40	13.50	309.90
14"	ER@3.40	Ea	450.00	88.70	17.00	555.70
16"	ER@3.90	Ea	650.00	102.00	19.50	771.50
Asbestos-cement 1/8 bend with mechanical joints						
4"	ER@1.00	Ea	67.90	26.10	5.00	99.00
6"	ER@1.25	Ea	92.00	32.60	6.25	130.85
8"	ER@1.70	Ea	129.00	44.30	8.50	181.80
10"	ER@2.20	Ea	188.00	57.40	11.00	256.40
12"	ER@2.70	Ea	250.00	70.40	13.50	333.90
14"	ER@3.40	Ea	445.00	88.70	17.00	550.70
16"	ER@3.90	Ea	640.00	102.00	19.50	761.50
Asbestos-cement 1/4 bend with mechanical joints						
4"	ER@1.00	Ea	75.90	26.10	5.00	107.00
6"	ER@1.25	Ea	102.00	32.60	6.25	140.85
8"	ER@1.70	Ea	147.00	44.30	8.50	199.80
10"	ER@2.20	Ea	225.00	57.40	11.00	293.40
12"	ER@2.70	Ea	297.00	70.40	13.50	380.90
14"	ER@3.40	Ea	555.00	88.70	17.00	660.70
16"	ER@3.90	Ea	820.00	102.00	19.50	941.50
Asbestos-cement tee or wye with mechanical joints						
4"	ER@1.40	Ea	114.00	36.50	7.00	157.50
6"	ER@1.75	Ea	152.00	45.60	8.75	206.35
8"	ER@2.40	Ea	217.00	62.60	12.00	291.60
10"	ER@3.10	Ea	374.00	80.90	15.50	470.40
12"	ER@3.80	Ea	476.00	99.10	19.00	594.10
14"	ER@4.75	Ea	860.00	124.00	23.80	1,007.80
16"	ER@5.45	Ea	1,250.00	142.00	27.30	1,419.30

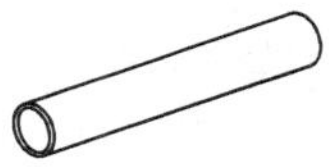

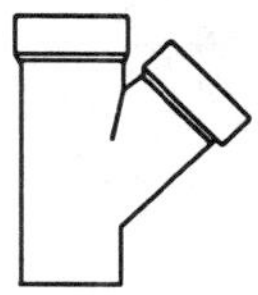

Asbestos-Cement, Class 2400 or 3000 with Mechanical Joints

Description	Craft@Hrs	Unit	Material $	Labor $	Equipment $	Total $

Asbestos-cement reducer with mechanical joints

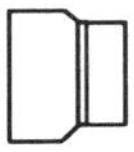

Description	Craft@Hrs	Unit	Material $	Labor $	Equipment $	Total $
6" x 4"	ER@1.20	Ea	80.50	31.30	6.00	117.80
8" x 6"	ER@1.50	Ea	118.00	39.10	7.50	164.60
10" x 8"	ER@1.95	Ea	172.00	50.90	9.75	232.65
12" x 10"	ER@2.45	Ea	239.00	63.90	12.30	315.20

Because of varying soil compositions and densities, trenching costs can sometimes be difficult to accurately estimate. The tables in this section are based on average soil conditions, which are loosely defined as dirt, soft clay or gravel mixed with rocks. The costs for trenching through limestone, sandstone or shale can increase by as much as 1,000 percent. For these conditions, it is best to consult with experts or to subcontract the work to specialists.

The primary consideration in excavations is to ensure the safety of the pipe layers. Current OSHA statutes require all excavations over 5 feet deep to be sloped, shored, sheeted or braced to prevent cave-ins. The use of shoring minimizes trenching costs by eliminating the need to slope the trench walls, but increases pipe-laying costs because of the difficulty in working between cross-bracing.

The estimator should carefully compare the costs for installing and removing shoring, and reduced pipe installation productivity, against the additional excavation costs for sloping the walls of the trench. For average soil conditions, not exceeding 6 to 8 feet in depth, it is usually more economical to slope the trench walls.

The current average cost for a backhoe with operator is $57.10 per hour. Some trenching contractors additionally charge $25.00 move-on and $25.00 move-off costs, particularly for small projects.

Total trenching, backfill and compaction costs are calculated by multiplying the backhoe hours from the following charts by $57.10 per hour, and adding move-on and move-off charges, if applicable.

Backhoe Trenching

Trench walls at a 90 degree angle, based on average soil, including trenching, backfill and compaction. Hours per 100 linear feet of trench

Trench depth (feet)	Trench width (feet)					
	1	2	3	4	5	6
2	2.6	5.1	7.7	10.2	12.8	15.3
3	3.8	7.7	11.5	15.3	19.2	23.0
4	5.1	10.2	15.3	20.4	25.6	30.7
5	6.4	12.8	19.2	25.6	32.0	38.3
6	7.7	15.3	23.0	30.7	38.3	46.0
7	8.9	17.9	26.8	35.8	44.7	53.7
8	10.2	20.4	30.7	40.9	51.1	61.3
9	11.5	23.0	34.5	46.0	57.5	69.0
10	12.8	25.5	38.3	51.1	63.9	76.6

Notes:

1) Compaction is based on the use of a compactor wheel.
2) Additional labor costs must be added if watering-in, stone removal or excess spoil removal is required.

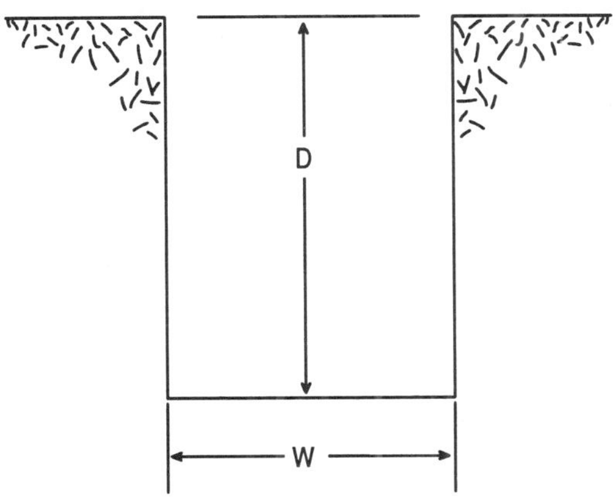

Backhoe Trenching Trench walls at a 45 degree angle, based on average soil including trenching, backfill and compaction. Hours per 100 linear feet of trench						
Trench depth (feet)	Trench width (feet) 1	2	3	4	5	6
2	7.6	10.3	12.6	15.5	17.9	20.3
3	15.2	22.6	22.6	26.6	30.7	34.5
4	25.5	30.7	35.5	40.7	45.9	51.0
5	38.3	44.5	51.0	57.6	63.8	70.0
6	53.4	61.0	69.0	76.6	84.1	91.7
7	71.4	80.3	89.3	98.3	107.2	115.9
8	92.1	102.1	112.4	122.4	132.8	143.1
9	114.8	126.6	137.9	149.3	161.0	172.4
10	140.3	153.1	165.9	178.6	191.4	204.5

Notes:

1) Compaction is based on the use of a compactor wheel.
2) Additional labor costs must be added if watering-in, stone removal excess spoil removal is required.

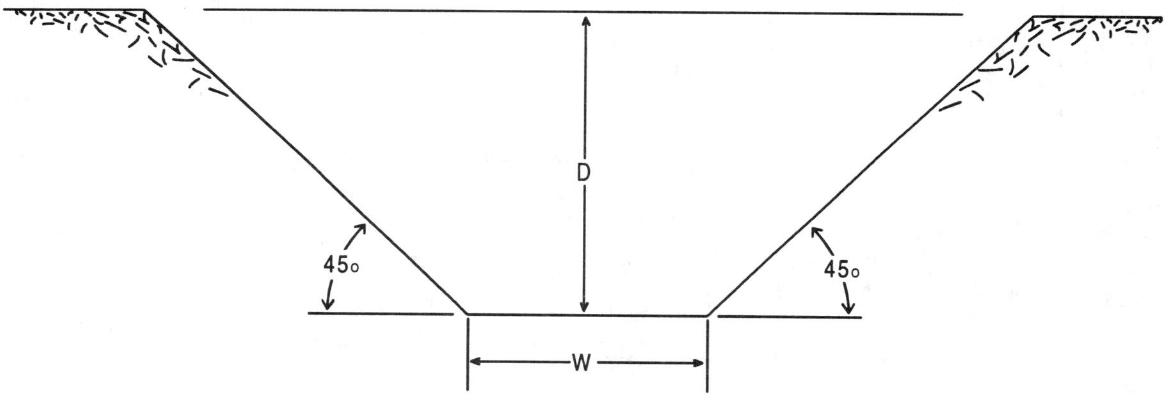

Equipment Rental			
Description	**Per day ($)**	**Per week ($)**	**Per month ($)**
Air compressor, gas or diesel, 100-125 CFM with 100' hose, air tool and 4 bits	81.00	246.00	673.00
Air compressor, gas or diesel, 165-185 CFM with 100' hose, air tool and 4 bits	104.00	315.00	868.00
Air compressor, gas or diesel, 100-125 CFM, no accessories	62.00	195.00	555.00
Air compressor, gas or diesel, 165-185 CFM, no accessories	85.00	264.00	750.00
Backhoe with bucket, 55 HP	230.00	730.00	2,160.00
Backhoe with bucket, 100 HP	285.00	1,120.00	3,280.00
Backhoe hydraulic compactor, 5-10 ton	175.00	600.00	1,730.00
Bender, less than 2.5"	170.00	450.00	1,200.00
Bevel machine with torch	16.00	48.00	140.00
Boom lift, 31' - 40'	243.00	690.00	1,960.00
Boom lift, 51' - 60'	376.00	1,100.00	3,150.00
Chain hoist, to 9.9 tons	35.00	95.00	260.00
Chain-hoist, 10 tons & over	125.00	315.00	850.00
Come-along, 1/2-ton	10.00	20.00	60.00
Come-along, 1-ton	12.00	40.00	85.00
Come-along, 4-ton	24.00	60.00	145.00
Compactor, soil, gas or diesel, 21" x 24" plate	66.00	211.00	592.00
Crane, truck-mounted, 10-ton capacity	380.00	1,090.00	3,200.00
Crane, truck-mounted, 15-ton capacity	340.00	1,100.00	3,300.00
Forklift, 2,000 pound capacity	120.00	360.00	1,100.00
Forklift, 4,000 pound capacity	135.00	400.00	1,200.00
Forklift, 8,000 pound capacity	240.00	725.00	2,200.00
Front-end loader, 1 yard capacity	215.00	755.00	2,220.00
Front-end loader, 2 yard capacity	385.00	1,210.00	3,580.00
Front-end loader, 3¼ yard capacity	625.00	1,960.00	5,620.00
Pipe machine, up to 2" capacity	30.00	110.00	325.00
Pipe machine, 2½" to 4" capacity	50.00	175.00	535.00
Roustabout, small	26.00	94.00	285.00
Roustabout, medium	37.00	125.00	375.00
Roustabout, large	42.00	155.00	460.00
Scissors-lift, to 20', diesel, self-propelled	105.00	285.00	780.00
Scissors-lift, 21' - 30'	135.00	385.00	1,110.00
Skip-loader	150.00	625.00	1,920.00
Trailer, office, 25'	—	—	150.00
Truck, 3/4 ton pickup	55.00	285.00	800.00
Truck, 1 ton pickup	60.00	295.00	820.00
Truck, 1 ton stakebed	70.00	315.00	850.00
Van, storage, 26'	—	—	225.00
Van, storage, 40'	—	—	275.00
Welding machine, gas, 200 amp, trailer-mounted with 50' bare lead	51.00	161.00	450.00
Welding machine, diesel, 400 amp, trailer-mounted with 50' bare lead	60.00	191.00	538.00

Description	Craft@Hrs	Unit	Material $	Labor $	Equipment $	Total $
1½" brass valve tags with black filled numbers						
Typical costs	P1@.150	Ea	1.50	3.67	—	5.17
1½" blank brass valve tags						
Typical costs	P1@.150	Ea	.75	3.67	—	4.42
1½" plastic valve tags with engraved numbers						
Typical costs	P1@.150	Ea	1.25	3.67	—	4.92
Engraved plastic equipment nameplates						
Typical costs	P1@.400	Ea	3.25	9.77	—	13.02
Pipe and duct markers						
Typical costs	P1@.250	Ea	1.85	6.11	—	7.96
As-built drawings						
Typical costs	P1@.500	Ea	—	12.20	—	12.20
Instructing owner's operating personnel						
Typical costs	P1@8.00	Ea	—	195.00	—	195.00

Budget Estimates for Plumbing and HVAC Work

General contractors and owners often ask mechanical subcontractors to prepare budget cost estimates for use in planning projects. A budget estimate isn't a bid for the contract and you probably won't be paid for preparing the estimate. That's why you won't want to spend too much time on these estimates. Accuracy within 8 to 10 percent of the final bid price will usually be acceptable. Normally you'll have only preliminary architectural drawings and possibly a guide specification to work from. The tables that follow will help you prepare budget estimates for nearly any plumbing or HVAC job.

The costs in these tables are averages and will vary considerably with design requirements and by geographic area. Make it clear to the owner or architect requesting the figures that these are budget estimates and will be revised when final plans and specifications are available.

Budget Estimates for Plumbing: First, count the number of plumbing fixtures. Multiply the number of fixtures of each type by the price per fixture shown in the table below. These prices include all the usual labor and material for a complete average plumbing installation. Prices are based on branch piping lengths averaging 35 feet between the fixture and the main. Typical main piping is also included in the prices.

Budget plumbing estimate, cast iron DWV, copper supply pipe

Description	Craft@Hrs	Unit	Material $	Labor $	Equipment $	Total $
Bathtubs	P1@15.0	Ea	1,050.00	367.00	—	1,417.00
Showers	P1@9.00	Ea	575.00	220.00	—	795.00
Lavatories	P1@11.0	Ea	990.00	269.00	—	1,259.00
Kitchen sinks	P1@12.0	Ea	1,030.00	293.00	—	1,323.00
Service sinks	P1@10.0	Ea	895.00	244.00	—	1,139.00
Bar sinks	P1@9.00	Ea	750.00	220.00	—	970.00
Floor sinks	P1@7.00	Ea	465.00	171.00	—	636.00
Water closets						
Tank-type	P1@11.0	Ea	750.00	269.00	—	1,019.00
Flush valve	P1@12.0	Ea	850.00	293.00	—	1,143.00
Urinals	P1@8.00	Ea	1,080.00	195.00	—	1,275.00
Drinking fountains						
(refrigerated)	P1@12.0	Ea	1,250.00	293.00	—	1,543.00
Wash fountains	P1@25.0	Ea	2,700.00	611.00	—	3,311.00
Can washers	P1@18.0	Ea	1,370.00	440.00	—	1,810.00
Floor drains	P1@8.00	Ea	425.00	195.00	—	620.00
Area drains	P1@6.00	Ea	365.00	147.00	—	512.00
Roof drains	P1@18.0	Ea	960.00	440.00	—	1,400.00
Overflow drains	P1@18.0	Ea	960.00	440.00	—	1,400.00
Deck drains	P1@4.00	Ea	300.00	97.70	—	397.70
Cleanouts	P1@5.00	Ea	325.00	122.00	—	447.00
Trap primers	P1@2.00	Ea	165.00	48.90	—	213.90
Sump pumps	P1@6.00	Ea	1,600.00	147.00	—	1,747.00
Water heaters						
Gas, 40 gal.	P1@16.0	Ea	1,900.00	391.00	—	2,291.00
Gas, 80 gal.	P1@17.0	Ea	2,100.00	415.00	—	2,515.00
Gas, 120 gal.	P1@18.0	Ea	2,450.00	440.00	—	2,890.00

Description	Craft@Hrs	Unit	Material $	Labor $	Equipment $	Total $
Budget plumbing estimate, plastic DWV and plastic supply pipe						
Bathtubs	P1@12.0	Ea	950.00	293.00	—	1,243.00
Showers	P1@7.00	Ea	460.00	171.00	—	631.00
Lavatories	P1@9.00	Ea	890.00	220.00	—	1,110.00
Kitchen sinks	P1@10.0	Ea	980.00	244.00	—	1,224.00
Service sinks	P1@8.00	Ea	790.00	195.00	—	985.00
Bar sinks	P1@7.00	Ea	660.00	171.00	—	831.00
Floor sinks	P1@6.00	Ea	400.00	147.00	—	547.00
Water closets						
Tank-type	P1@8.00	Ea	620.00	195.00	—	815.00
Flush valve	P1@9.00	Ea	740.00	220.00	—	960.00
Urinals	P1@6.00	Ea	1,000.00	147.00	—	1,147.00
Drinking fountains						
(refrigerated)	P1@10.0	Ea	1,150.00	244.00	—	1,394.00
Wash fountains	P1@22.0	Ea	2,650.00	538.00	—	3,188.00
Can washers	P1@15.0	Ea	1,300.00	367.00	—	1,667.00
Floor drains	P1@7.00	Ea	330.00	171.00	—	501.00
Area drains	P1@5.00	Ea	270.00	122.00	—	392.00
Roof drains	P1@16.0	Ea	920.00	391.00	—	1,311.00
Overflow drains	P1@16.0	Ea	920.00	391.00	—	1,311.00
Deck drains	P1@3.00	Ea	230.00	73.30	—	303.30
Cleanouts	P1@4.00	Ea	280.00	97.70	—	377.70
Trap primers	P1@2.00	Ea	140.00	48.90	—	188.90
Sump pumps	P1@5.00	Ea	1,600.00	122.00	—	1,722.00
Water heaters						
Gas, 40 gal.	P1@15.0	Ea	1,950.00	367.00	—	2,317.00
Gas, 80 gal.	P1@16.0	Ea	2,150.00	391.00	—	2,541.00
Gas, 120 gal.	P1@17.0	Ea	2,500.00	415.00	—	2,915.00

Budget Estimates for HVAC Work: Budget estimates for heating, ventilating and air conditioning work are figured by the square foot of floor. Multiply the area of conditioned space by the cost per square foot in the table that follows. The result is the estimated cost including a complete heating, ventilating and air conditioning system and 20 percent overhead and profit. The table also shows average heating and cooling requirements for various types of buildings.

	Cooling	Heating	Total Selling Price ($)	
Description	**(Btu/SF)**	**(Btu/SF)**	**Per SF**	**Per Ton**
Apartments and condominiums	25	25	4.90	2,350.00
Auditoriums	40	40	12.50	3,750.00
Banks	45	30	13.80	3,680.00
Barber shops	40	40	12.90	3,870.00
Beauty shops	60	30	15.20	3,040.00
Bowling alleys	40	40	10.00	3,000.00
Churches	35	35	10.80	3,700.00
Classrooms	45	40	19.60	5,230.00
Cocktail lounges	70	30	13.70	2,350.00
Computer rooms	145	20	27.90	2,300.00
Department stores	35	30	8.80	3,010.00
Laboratories	60	45	24.30	4,860.00
Libraries	45	35	14.50	3,860.00
Manufacturing plants	40	35	5.60	1,690.00
Medical buildings	35	35	9.60	3,290.00
Motels	30	30	6.70	2,680.00
Museums	35	40	10.70	3,670.00
Nursing homes	45	35	12.40	3,310.00
Office buildings, low-rise	35	35	9.80	3,350.00
Office buildings, high-rise	40	30	13.20	3,970.00
Residences	20	30	3.00	1,800.00
Retail shops	45	30	10.00	2,660.00
Supermarkets	30	30	5.20	2,080.00
Theaters	40	40	9.80	2,950.00

Plumbing & HVAC Forms and Letters

The following pages contain business forms and letters most plumbing and HVAC contractors can use.

- Change Orders
  Change Estimate Takeoff
  Change Estimate Worksheet
  Change Estimate Summary
  Change Order Log
- Subcontract Forms
  Standard Form Subcontract
  Subcontract Change Order
- Purchase Order
- Construction Schedule
- Letter of Intent
- Submittal Data Cover Sheet
- Submittal Index
- Billing Breakdown Worksheet
- Monthly Progress Billing
- Warranty Letter

An explanation is provided on how the forms function in the contractor's office, how to fill them out, and a sample provided. Where appropriate, a blank copy of the form is provided for your use.

Permission is hereby given by the publisher for the duplication of any of the blank forms on a copy machine or at a printer, provided they are for your own use and not to be sold at a cost exceeding the actual printing and paper cost.

Space has been provided at the top of the forms so you can add your own letterhead or logo.

Change Estimates

Very few projects are completed without some changes to the original plans. Most changes in the scope of work are a result of either errors by the designer or requests by the owner.

Here's the usual sequence in handling changes.

1) The project architect usually initiates a change by sending a Request for Quotation (R.F.Q.) to the general contractor with a detailed description of the work to be modified.
2) The general contractor prepares a Change Estimate (C.E.) if the work is to be done by the general.
3) The subcontractor prepares a change estimate which shows the work to be added to or deleted from the contract and the cost of changes, including markup. The general contractor reviews the change estimate and forwards a summary to the architect for decision.
4) If the decision is to go ahead with the change, the architect issues a Change Order (C.O.) to the general contractor. This change order is authorization to proceed with the change at the price quoted.

Your costs for making changes will normally be higher than your costs for similar items on the original estimate. Here's why:

- Equipment and materials for changes are almost always purchased in smaller quantities. Volume discounts are seldom available.
- Special handling and delivery of small orders will further increase equipment and material costs.
- Scheduled work will be disrupted.
- Overhead costs will be higher because special purchase orders or subcontracts will be required.
- Vendor restocking charges for deleted items will range from 20 to 50 percent of the original cost.

- Large change orders may require additional manpower, which usually results in decreased efficiency.
- The cumulative effect of change orders will be to increase overhead costs if the project completion date is extended.

To compensate for these higher costs, most estimators use a higher markup on change estimates:

- Add an extra 10 percent markup to equipment and material costs to cover increased over-head and handling costs, prior to adding final Overhead and Profit markup.
- Increase competitive labor units (manhours required to perform a specific task) by 15 percent to compensate for inefficiencies such as disruptions in the work schedule, extra handling of equipment and material, and additional manpower.
- Mark up subcontractor prices by 10 percent.
- Figure overhead and profit markup at 15/10 or 10/10.

By following these suggestions, the mathematical gross profit will appear to be 40 to 45 percent, but the true gross will be closer to 25 percent because your costs for change order work are higher, as listed above.

Change Estimate Example

Let's assume that the general contractor on your job has sent a R.F.Q. to your attention. The extra work is a run of hot water piping that will connect to equipment being furnished and set by others. The first step is to list all material and labor required to do the work. Notice the items entered on the Change Estimate Take-off form on page 338.

When the take-off is complete, transfer the manhour total to the Change Estimate Worksheet on page 339. To complete this form:

1) Multiply manhours by the cost per hour.
2) Add the cost of supervision.
3) Total direct labor cost.
4) Add the cost of preparing the change order estimate (C.E.).
5) Add equipment and tool costs.
6) Add subcontract costs.

Then bring totals forward to the Change Estimate Summary on page 340:

1) Enter the material cost total from the Change Estimate Take-off form on page 338.
2) Enter direct and indirect labor cost totals from the Change Estimate Worksheet.
3) Enter equipment, tool and subcontract cost totals from the Change Estimate Worksheet.

Notice the additional bond charge on line M of the Change Estimate Summary. This is the extra premium charged by the bonding company to cover the increase in contract value. The next line, Service Reserve, is the allowance for service work (if any is required) during the warranty period.

A sample Change Order Log, used to record the status of changes, is also included on page 341.

ACME

Mechanical Contractors

7600 Oak Avenue - Anytown, U.S.A. 12345-6789
Voice 1-234-5678 / FAX 1-234-5680

Change Estimate Take-off

Project Chancey Bank Building **C. E. Number** One

Take-off by DP **Job No** 7777

Date 2-2-99

Equipment and Material			Material $		Labor Manhours	
Item	**Description**	**Quantity**	**Unit**	**Total**	**Unit**	**Total**
1	1" Type L hard copper pipe	40 LF	1.50	60.00	.038	1.52
2	1" 90 degree copper ell	6 Ea	2.28	13.68	.23	1.38
3	1" copper coupling	4 Ea	1.39	5.56	.24	.96
4	1" bronze ball valve	2 Ea	8.21	16.42	.36	.72
5	2" x 2" x 1" reducing copper tee	2 Ea	15.77	31.54	.40	.80
6	1" copper union	2 Ea	10.35	20.07	.26	.52
7	1" pipe hanger and inserts	8 Ea	2.60	20.80	.45	3.60
8	Miscellaneous material	1 Ea	LS	15.00	1.78	1.78
		Totals		183.07		11.28

ACME

Mechanical Contractors

7600 Oak Avenue - Anytown, U.S.A. 12345-6789
Voice 1-234-5678 / FAX 1-234-5680

Change Estimate Worksheet

Estimated by DP **C. E. Number** One

Date 2-3-99 **Job No** 7777

Direct Labor Expense (including benefits)

Plumber or pipefitter journeyman	11.28	manhours at $	28.34	per hour =	$	319.68
Plumber or pipefitter supervisor	1.00	manhours at $	38.78	per hour =	$	38.78
Sheet metal journeyman		manhours at $		per hour =	$	
Sheet metal supervisor		manhours at $		per hour =	$	
Laborer journeyman		manhours at $		per hour =	$	
		manhours at $		per hour =	$	
				Total Direct Labor Cost	$	358.46

Indirect Costs

Travel		$	
Subsistance		$	
Labor	Cost Estimate Preparation (1 Hour)	$	33.50
	Total Indirect Cost	$	33.50

Equipment and Tool Costs

Ramset rental	$	23.50
Propane	$	2.50
Total Equipment and Tool Cost	$	26.00

Subcontract Costs

Quikrap Insulation Co.	$	65.00
	$	
Total Subcontract Cost	$	65.00

ACME

Mechanical Contractors

7600 Oak Avenue - Anytown, U.S.A. 12345-6789
Voice 1-234-5678 / FAX 1-234-5680

A C M E

Change Estimate Summary

Quotation to ABC General Contractors, Inc. **Date** 2-3-99
Address 380 First Street **Change Order No.** One
Address ______ **Job Number** 7777
City / State / Zip Anytown, CA 63876
Attention: Mr. J.H. Smith
Job Name Chancey Bank Building
Reference: Your R.F.O. No. 7 Labor and Material to furnish and install additional hot water piping.

A.	Materials and equipment:	$ 183.07	
B.	Sales tax: 7%	$ 12.81	
C.	Direct labor:	$ 358.46	
D.	Indirect costs:	$ 33.50	
E.	Equipment and tools:	$ 26.00	
F.			**Subtotal** $ 613.84
G.	Overhead at 15% of line F:	$ 92.08	
H.	Subcontracts:	$ 65.00	
I.	Overhead at 10% of line H:	$ 6.50	
J.			**Subtotal** $ 777.42
K.	Profit at 10% of line J:	$ 77.74	
L.			**Subtotal** $ 855.16
M.	Bond premium at 1% of line L:	$ 8.55	
N.	Service reserve at 0.5% of line L:	$ 4.28	
O.	**Total cost estimate, lines L thru N:**	☐ Add ☑ Deduct	$ 867.99

P. Exclusions from this estimate: Per Base
Q. ☐ This quotation is valid for 30 days
R. ☐ We require 1 day(s) extension of the contract time.
S. ☐ We are proceeding with this work per your authorization.
T. ☐ Please forward your confirming change order.

Signed by: ______________________
Project Manager

Change Order Log

Job Name ______________________________ **Job Number** ____________

Change Number	Customer Number	Date	Amount	Date Approved	Amount Approved	Remarks

Use this space to insert your company logo and letterhead.

Change Estimate Take-off

Project ____________________ C. E. Number ____________

Take-off by ____________________ Job No ____________

Date ____________

Equipment and Material			Material $		Labor Manhours	
Item	Description	Quantity	Unit	Total	Unit	Total
		Totals				

Use this space to insert your company logo and letterhead.

Change Estimate Worksheet

Estimated by ______________________________ **C. E. Number** ______________

Date __________ **Job No** ________________

Direct Labor Expense (including benefits)

Plumber or pipefitter journeyman	______	manhours at $	______	per hour =	$	______
Plumber or pipefitter supervisor	______	manhours at $	______	per hour =	$	______
Sheet metal journeyman	______	manhours at $	______	per hour =	$	______
Sheet metal supervisor	______	manhours at $	______	per hour =	$	______
Laborer journeyman	______	manhours at $	______	per hour =	$	______
______________	______	manhours at $	______	per hour =	$	______
				Total Direct Labor Cost	$	______

Indirect Costs

Travel ______________________________ $ ______

Subsistance ______________________________ $ ______

Labor ______________________________ $ ______

Total Indirect Cost $ ______

Equipment and Tool Costs

______________________________ $ ______

______________________________ $ ______

Total Equipment and Tool Cost $ ______

Subcontract Costs

______________________________ $ ______

______________________________ $ ______

Total Subcontract Cost $ ______

Use this space to insert your company logo and letterhead.

Change Estimate Summary

Quotation to ______________________ **Date** ____________

Address ______________________ **Change Order No.** ________

Address ______________________ **Job Number** ____________

City / State / Zip ______________________

Attention: ______________________

Job Name ______________________

Reference: ____

____

A.	Materials and equipment:	$ ________	
B.	Sales tax: _______	$ ________	
C.	Direct labor:	$ ________	
D.	Indirect costs:	$ ________	
E.	Equipment and tools:	$ ________	
F.			**Subtotal $** ________
G.	Overhead at _______of line F:	$ ________	
H.	Subcontracts:	$ ________	
I.	Overhead at _______of line H:	$ ________	
J.			**Subtotal $** ________
K.	Profit at _______of line J:	$ ________	
L.			**Subtotal $** ________
M.	Bond premium at 1% of line L:	$ ________	
N.	Service reserve at 0.5% of line L:	$ ________	
O.	**Total cost estimate, lines L thru N:**	☐ Add ☐ Deduct	$ ________

P. Exclusions from this estimate: ______________________________

Q. ☐ This quotation is valid for _______ days

R. ☐ We require ______ day(s) extension of the contract time.

S. ☐ We are proceeding with this work per your authorization.

T. ☐ Please forward your confirming change order.

Signed by: ______________________________

Project Manager

Another term for the general contractor is *prime contractor.* Subcontractors working directly for the prime contractor are called *second-tier subcontractors.* Contractors working for these second-tier subs are usually known as *third-tier subcontractors.* On a large job, many second-tier subs may hire third-tier subs to perform specialized work such as insulation, fireproofing, test and balance, water treatment, instrumentation and trenching.

If a portion of your work will be subcontracted to others, I recommend drawing up a subcontract agreement for that work. Any subcontractors who furnish both material and labor for a job should have a written contract.

Most larger plumbing, heating and A/C subcontractors use a two-part *Standard Form Subcontract* published by the Associated General Contractors of America. This two-part form is used to write the base subcontract. The A.G.C. also publishes a *Subcontract Change Order* form which is used when adding or deleting items after the base subcontract is signed. These forms are sold at modest cost at most A.G.C. offices.

The first section of page one of the Standard Form Subcontract identifies the two contractors entering into the agreement. The party issuing the subcontract is the *contractor.* Recipient of the agreement is the *subcontractor.*

Under *Recitals,* the contractor must affirm the right to issue a subcontract by declaring that a legal contract exists with the *owner* (general contractor).

Section 1 - Entire Contract describes the contract documents: the project plans, specifications and addenda. This list of documents should always be the same as the documents identified in the second-tier subcontractor's contract with the general contractor.

Section 2 - Scope describes the specific scope of work to be performed.

Section 3 - Contract Price. Complete this section by inserting the agreed-upon price for doing the work described.

Section 4 - Payment Schedule on page two should show the same payment terms as your contract with the general contractor.

Section 7 - Special Provisions is used to list any special agreements or exclusions.

When the contract is typed, do not sign it. Instead, forward it to the third-tier subcontractor with instructions to sign it and return it intact. When the signed contract is received, sign it and send a copy to the third-tier subcontractor.

Standard Form Subcontract

Subcontract No. 7777-1

THIS AGREEMENT, made and entered into at Anytown, CA. this 3rd day of May, 1999, by and between Acme Mechanical Contractors hereinafter called **CONTRACTOR,** with principal office at 7600 Oak Avenue, Anytown, California, and Quikrap Insulation Co., 320 Maple Blvd., Anytown, CA hereinafter called a **SUBCONTRACTOR**.

RECITALS

On or about the 2nd day of January, 1999, CONTRACTOR entered into a prime contract with

ABC General Contractors, Inc. hereinafter called OWNER, whose address is 380 First Street, Anytown, California to perform the following construction work: The installation of all mechanical and plumbing systems for the proposed Mile-Hi office building to be located at 2600 Second, North, Anytown, California.

Said work is to be performed in accordance with the prime contract and the plans and specifications. Said plans and specifications have been prepared by or on behalf of Quik-Draw and Associates, Anytown, California, **ARCHITECT**

SECTION 1 - ENTIRE CONTRACT

SUBCONTRACTOR certifies and agrees that he is fully familiar with all of the terms, conditions and obligations of the Contract Documents, as hereinafter defined, the location of the job site, and the conditions under which the work is to be performed, and that he enters into this Agreement based upon his investigation of all of such matters and is in no way relying upon any opinions or representations of CONTRACTOR. It is agreed that this Agreement represents the entire agreement. It is further agreed that the Contract Documents are incorporated in this Agreement by this reference, with the same force and effect as if the same were set forth at length herein, and that SUBCONTRACTOR and his subcontractors will be and are bound by any and all of said Contract Documents insofar as they relate in any part or in any way, directly or indirectly to the work covered by this Agreement. SUBCONTRACTOR agrees to be bound to CONTRACTOR in the same manner and to the same extent as CONTRACTOR is bound to OWNER under the Contract Documents, to the extent of the work provided for in this Agreement, and that where, in the Contract Documents, reference is made to CONTRACTOR and the work or specification therein pertains to SUBCONTRACTOR'S trade, craft, or type of work, then such work or specification shall be interpreted to apply to SUBCONTRACTOR instead of CONTRACTOR. The phrase "Contract Documents" is defined to mean and include:

Drawing Nos. A-1 thru A-17, C-1 thru C-4, S-1 thru S-7, E-1 thru E-6 and M-1 thru M-7, all dated 10-4-1998, Specifications dated 10-2-98 and Addendum No. 1 dated 10-18-98.

SECTION 2 - SCOPE

SUBCONTRACTOR agrees to furnish all labor, services, materials, installation, cartage, hoisting, supplies, insurance, equipment, scaffolding, tools and other facilities of every kind and description required for the prompt and efficient execution of the work described herein and to perform the work necessary or incidental to complete thermal insulation for all plumbing, HHW, CHW and condensate piping systems, including pumps P-1, P-2 and P-3 for the project in strict accordance with the Contract Documents and as more particularly, though not exclusively, specified in:

Division 15, Section 400 entitled "Thermal Insulation systems" and General Conditions, as applicable.

SECTION 3 - CONTRACT PRICE

CONTRACTOR agrees to pay SUBCONTRACTOR for the strict performance of his work, the sum of:

Forty-two thousand, three hundred dollars ($42,300.00), subject to additions and deductions for changes in the work as may be agreed upon, and to make payment in accordance with the Payment Schedule, Section 4.

SECTION 4 - PAYMENT SCHEDULE

CONTRACTOR agrees to pay SUBCONTRACTOR in monthly payments of 90 % of labor and materials which have been placed in position and for which payment has been made by OWNER to CONTRACTOR. The remaining 10 % shall be retained by CONTRACTOR until he receives final payment from OWNER, but not less than thirty-five days after the entire work required by the prime contract has been fully completed in conformity with the Contract Documents and has been delivered and accepted by OWNER, ARCHITECT, and CONTRACTOR. Subject to the provisions of the next sentence, the retained percentage shall be paid SUBCONTRACTOR promptly after CONTRACTOR receives his final payment from OWNER. SUBCONTRACTOR agrees to furnish, if and when required by CONTRACTOR, payroll affidavits, receipts, vouchers, release of claims for labor, material and subcontractors performing work or furnishing materials under this Agreement, all in form satisfactory to CONTRACTOR, and it is agreed that no payment hereunder shall be made, except at CONTRACTOR'S option, until and unless such payroll affidavits, receipts, vouchers or release; or any or all of them, have been furnished. And payment made hereunder prior to completion and acceptance of the work, as referred to above, shall not be construed as evidence of acceptance of any part of SUBCONTRACTOR'S work.

SECTION 5 - GENERAL SUBCONTRACT PROVISIONS

General Subcontract Provisions on back of Pages 1 and 2 are an integral part of this Agreement.

SECTION 6 - GENERAL PROVISIONS

1. *SUBCONTRACTOR* agrees to begin work as soon as instructed by the CONTRACTOR, and shall carry on said work promptly, efficiently and at a speed that will not cause delay in the progress of the CONTRACTOR'S work or work of other subcontractors. If, in the opinion of the CONTRACTOR, the SUBCONTRACTOR falls behind in the progress of the work, the CONTRACTOR may direct the SUBCONTRACTOR to take such steps as the CONTRACTOR deems necessary to improve the rate of progress, including, without limitation, requiring the SUBCONTRACTOR to increase the number of shifts, personnel, overtime operations, days of work, equipment, amount of plant, or other remedies and to submit to CONTRACTOR for CONTRACTOR'S approval an outline schedule demonstrating the manner in which the required rate of progress will be regained, without additional cost to the CONTRACTOR. CONTRACTOR may require SUBCONTRACTOR to prosecute, in preference to other parts of the work, such part or parts of the work as CONTRACTOR may specify.

The SUBCONTRACTOR shall complete the work as required by the progress schedule prepared by the CONTRACTOR, which may be amended from time to time. The progress schedule may be reviewed in the office of the CONTRACTOR and sequence of construction will be as directed by the CONTRACTOR.

The SUBCONTRACTOR agrees to have an acceptable representative (an officer of SUBCONTRACTOR if requested by the CONTRACTOR) present at all job meetings and to submit weekly progress reports in writing if requested by the CONTRACTOR. Any job progress schedules are hereby made a part of and incorporated herein by reference.

2. ***Reserved Gate Usage***

SUBCONTRACTOR shall notify in writing, and assign its employees, material men and suppliers, to such gates or entrances as may be established for their use by CONTRACTOR and in accordance with such conditions and at such times as may be imposed by CONTRACTOR. Strict compliance with CONTRACTOR'S gate usage procedures shall be required by the SUBCONTRACTOR who shall be responsible for such gate usage by its employees, material men, suppliers, subcontractors, and their material men and suppliers.

3. ***Staggered Days and Hours of Work and for Deliveries***

SUBCONTRACTOR shall schedule the work and the presence of its employees at the jobsite and any deliveries of supplies or materials by its material men and suppliers to the jobsite on such days, and at such times and during such hours, as may be directed by CONTRACTOR. SUBCONTRACTOR shall assume responsibility for such schedule compliance not only for its employees but for all its material men, suppliers and subcontractors, and their material men and suppliers.

SECTION 7 - SPECIAL PROVISIONS

1. Pipe hanger inserts will be furnished and installed by Contractor.
2. All trash pickup by Subcontractor; haul-away will be by General Contractor.

Contractors are required by law to be licensed and regulated by the Contractors' State License Board. Any questions concerning a contractor may be referred to the registrar of the board whose address is:

Contractors' State License Board -- P. O. Box 26000, Sacramento, California 95826

IN WITNESS WHEREOF: The parties hereto have executed this Agreement for themselves, their heirs, executors, successors, administrators, and assignees on the day and year first above written.

SUBCONTRACTOR	**CONTRACTOR**
Quikrap Insulation Company ____________	Acme Mechanical Contractors____________
By ____________	By ____________
Name Title	Name Title
☐ Corporation ☐ Partnership ☐ Proprietorship	
(Seal)	
Contractor's State License No. ____________	Contractor's State License No. ____________

Revised 5-81 Published by AGC San Diego

Page 2 of 2

Subcontract Change Order

Number One

Date 6-11-99

To Quickrap Insulation Co.

Subcontract Number 7777-1

Our Job Number

Our Proposal Number

Architect's C.O. Number

Effective Date of Charge 6-17-99

Subject to all the provisions of this Change Order, you are hereby directed to make the following change(s)

Delete HHW and CHW piping insulation located above Room No. 322 per ABC General Contractor's Change Order No. 18, dated 6-5-99.

The following change(s) will alter the price provided in your subcontract by the deduction ***of $*** 465.00

*Surety Consent		
This Change is Approved Date	Adjusted Subcontract Price Through C.O. Number 0	$ 42,300.00
Name:	Amount this C.O. Number:	$ (465.00)
By:		
Title:	Current Adjusted Subcontract Price:	$ 41,835.00

When this Change Order is signed by both parties (and by Subcontractor's surety is subcontract is bonded), it constitutes their agreement:

(A) That the subcontract price is adjusted as shown above and that no further adjustment in that price by reason of the change(s) provided herein shall be made; and (B) That all the terms and conditions of the subcontract, except as modified by this and any previous changes, shall remain in full force and effect and apply to the work as so changed.

Accepted and Agreed:	Date:
By ______ Authorized Signature – Subcontractor	By ______ Authorized Signature – Contractor
Title	Title

* Subcontractor: Sign pink copy and return immediately. If subcontract is bonded, obtain consent of surety endorsed thereon. No payments on account of this Change Order will be made until you have complied with the foregoing.

Published by San Diego Chapter A.G.C.

A purchase order is a legal document used to order equipment and materials from a vendor, with the implied promise to pay for these goods within a specified time after receipt of the items. Purchase orders are sometimes used to write subcontracts whose value does not exceed one or two thousand dollars. Refer to the sample purchase order on page 350.

While most of the purchase order is self-explanatory, certain items need to be emphasized:

“Description” should include all pertinent data such as model numbers, arrangements, performance requirements, applicable standards, electrical characteristics and so on.

“Tag” is used to identify a particular piece of equipment to facilitate unloading it at the proper location at the jobsite. Tag numbers for all major equipment can usually be found on the contract drawings.

“48 hour delivery notice”: This can be very important if the material or equipment being delivered is too heavy to be unloaded by hand. Time will be required to arrange for a crane or forklift to unload the truck.

“Sales tax” is the current state sales tax which must be included unless the goods are to be delivered and installed at a project located out-of-state.

“Terms”: The usual payment terms are “Net 30” which means the entire amount of the purchase order must be paid within thirty days after receipt of all goods, undamaged. If items are received damaged, payment should be withheld until satisfactory repair or replacement is made.

ACME

Mechanical Contractors

7600 Oak Avenue - Anytown, U.S.A. 12345-6789
Voice 1-234-5678 / FAX 1-234-5680

Purchase Order

P.O. Number: 7777-36 **Date:** 1-6-99

Job: Chancey Bank Building

Job No: 7777

Seller:	**Ship to:**
Anytown Industrial Sales Co. 3657 Third Avenue Northwest Anytown, CA 63876	Acme Mechanical Contractors c/o Chancey Bank Building 465 Commercial Street Anytown, CA 63876

Attn: Mr. L.H. Seller By: D.P.

Delivery date: 2-28-99 ❑ **24 Hr. Delivery Notice Required** **Payment terms:** Net 30

Quantity	Description	Unit	Total
Two	Big-Blo Model No. 1JN 86-13 centrifugal	$443.00	$886.00
	fans with 480/3/60 one-half H.P., O.D.P.		
	motor. 3250 CFM @ 0.50 T.S.P.		
	Tag one EF-13 and one EF-23		

Number of copies of operation and maintenance manuals required 2

All shipments are to be F.O.B. jobsite or shop unless otherwise noted.

Total before tax	$886.00
Sales tax	$62.02
Freight	Included
Total Order	$948.02

Buyer: D.P.

Acceptance by seller __________ Date __________

The general contractor's construction schedule may range from a simple bar chart to a complicated computer-generated CPM (Critical Path Method) diagram. More complex jobs require more complete and detailed schedules. On larger jobs, lower-tier subcontractors may be required to prepare a bar chart showing their construction schedule.

If you have to prepare a construction schedule, make conservative estimates of the time required to complete each part of the job. A slower schedule makes it possible to use smaller crews which are usually more efficient.

You can't schedule your own work until the general contractor has supplied a schedule for the balance of the project. When you have that schedule, prepare a list of tasks your crews will perform. Show this list to the general contractor to be certain there are no conflicts with the master schedule.

When the list of tasks to be performed has been approved, begin recording the manhours required for each task. Take these manhours from your estimate. When the duration for each task has been determined, figure the crew sizes required.

For example, suppose the estimate shows that it will take 1,040 manhours to install the underground plumbing for a project. The work must be completed in two months. Assuming one worker averages 173 hours of work in one month, how many workers will be required to meet the schedule? Here's the solution:

$$\frac{\text{1,040 manhours}}{\text{173 MH / Mo. x 2 months}} = \text{3 workers}$$

The first of the two bar charts on the following pages shows the schedule a general contractor might expect to receive from a plumbing and HVAC subcontractor. The chart on page 318 is intended for use by the subcontractor. It includes information a mechanical subcontractor needs to determine proper crew sizes for each task. The *Cumulative Manhours* line shows the budgeted manhours for each month of the project. If labor costs are to be kept within budget, actual hours should not exceed estimated hours. Monitor labor costs each month by comparing the line *Actual MH Used* (actual manhours used) with the line *Cumulative MH* (estimated cumulative manhours).

Construction Schedule

Company: Acme Mechanical Contractors **Project:** Mile-Hi Office Bldg **Job Number:** 7777 **Date:** 6/8/99

	1999						2000					
Activity	**Jul**	**Aug**	**Sep**	**Oct**	**Nov**	**Dec**	**Jan**	**Feb**	**Mar**	**Apr**	**May**	**Crew Sizes**
Submittals												
Purchasing												
U/G Plumbing												
A/G Plumbing												
Fin. Plumbing												
HVAC Piping												
HVAC Ducting												
Equipment												
Insulation												
Temp. Controls												
Start-up												

Construction Schedule

Company: Acme Mechanical Contractors **Project:** Mile-Hi Office Bldg **Job Number:** 7777 **Date:** 6/8/99

			1999					2000			
Activity	**Jul**	**Aug**	**Sep**	**Oct**	**Nov**	**Dec**	**Jan**	**Feb**	**Mar**	**Apr**	**May**	**Crew Sizes**
Submittals												N/A
Purchasing												N/A
U/G Plumbing												1040 MH ÷(173 x 2) = 3
A/G Plumbing												2075 MH ÷ (173 x 4) = 3
Fin. Plumbing												130 MH ÷ (173 x .5) = 1.5
HVAC Piping												2760 MH ÷ (173 x 4) = 4
HVAC Ducting												2790 MH ÷ (173 x 4) = 4
Equipment												690 MH ÷ (173 x 2) = 2
Insulation												Subcontract
Temp. Controls												Subcontract
Start-up												85 MH ÷ (173 x .5) = 1
Crew size/month			3	6	11	11	11	11.5	2	1		
Cumulative MH*			520	1159	3466	5373	7280	9143	9448	9573		MH Estimated = 9573
Acutal MH used												
*Per estimate												
*Rounded												

Letter of Intent

On most large projects the contractor is required to submit drawings and technical data to the owner's representative for approval before work actually begins. Most of these submittals will be prepared by your suppliers and third-tier subs. Vendors and subs may be reluctant to prepare these documents without some assurance that they've been selected to do the work. That's the purpose of a letter of intent. It notifies a proposed vendor or subcontractor that you plan to contract with them when the owner approves the information submitted.

A letter of intent isn't a contract. It's an expressed intention to make a contract, which isn't a contract at all. But a vendor who gets your letter of intent and supplies all the information requested should get the contract. If you place the order with a different vendor, obviously, your letter of intent was worthless. That first vendor may be very reluctant to cooperate in the future.

A sample letter of intent follows this section.

ACME

Mechanical Contractors

7600 Oak Avenue - Anytown, U.S.A. 12345-6789
Voice 1-234-5678 / FAX 1-234-5680

December 20, 1999

Quikrap Insulation Company
320 Maple Boulevard
Anytown, U.S.A. 12345-6789

Attn: Mr. R. S. Quikrap

Dear Mr. Quikrap:

Acme Mechanical Contractors has been issued a subcontract to furnish and install the mechanical and plumbing systems for the proposed Mile-Hi Office Building to be located at 2600 Second Street in Anytown.

It is my intention to award your firm a subcontract to provide the thermal insulation for these systems in accordance with your proposal dated 12-13-99 in the amount of $42,300.00.

Please prepare ten copies of your submittal data and forward them to me no later than January 15, 2000. Upon approval of your submittal, I will mail you the subcontract for your review and signature.

Thank you,

D. Phillips
Project Manager

CC: Project file

Submittal Data

The construction specifications may require that you submit manufacturer's technical data on certain equipment you plan to install. You may have to submit six to ten copies of this technical data for approval before buying the equipment or contracting with third-tier subcontractors.

First, find out how many submittal copies will be required. Request that number of copies plus at least two additional sets for your use. A subcontractor who provides only services (no materials) may have to submit a detailed written explanation of the work to be done.

The second step is to label each submittal with the paragraph number of the specification where that equipment, material or service is described. Type or write this number in the upper right corner of each submittal sheet. From this set of numbered submittals, prepare a submittal index, a list of submittals identified by title and paragraph number. This index should be arranged in specification paragraph order. A blank submittal index form follows this section.

I prefer to use a three-ring binder to hold submittals. That makes it easier to add and delete sections as needed. Be sure to include a cover sheet, usually on company letterhead, which identifies the project, job number and scope of work being submitted. A sample submittal cover sheet follows this section.

Purchase orders and subcontracts should not be written until an approved copy of the submittal brochure has been received.

ACME

Mechanical Contractors

7600 Oak Avenue - Anytown, U.S.A. 12345-6789
Voice 1-234-5678 / FAX 1-234-5680

Division 15

Heating, Ventilating, Air Conditioning and Plumbing Systems

Submittal Data for

Mile-Hi Office Building
2600 Second Street
Anytown, U.S.A.

Submittal Index

Specification Section	Specification Paragraph	Item Description	Proposed Vendor or Sub	As Specified	See Submittal Data Page

The format to be used for monthly progress billings is seldom specified in the contract documents. In any case, your bill has to include enough detail to satisfy the owner and the lender.

Before preparing your bill, ask the general contractor how much detail is required. Some general contractors will want detailed cost breakdowns showing costs for each system on each building floor. Others will require much less detail.

Find out if billings can include the cost of equipment and materials delivered and stored either on site or off site but not yet installed. If billings can include materials stored off site, do these materials have to be stored in a bonded warehouse? Do vendor's invoices have to accompany progress billings?

When you understand what's required for monthly progress billings, prepare a billing breakdown worksheet. A sample worksheet follows this section.

Column one, Activity Lists each work item in the order work will be performed. The first item listed will usually be mobilization. This includes the cost of the job site trailer, electrical and telephone hookups, office furniture and supplies and initial labor costs. Billings for mobilization costs are sometimes denied by the lender. Many contractors minimize these costs to increase the chance of approval. It's better to receive partial payment for these costs than none at all.

Column two, Cost ($) Lists the actual contractor's costs for each activity, including labor, material, equipment and sales taxes. Markup for overhead and profit are not included.

Column three, Factor This column shows the markup assigned to each activity. Activities scheduled to be completed first are usually assigned the highest markups. This is called front loading and is common in the construction industry. Most bills won't be paid for 30 to 60 days. When the bill is paid, the amount received probably won't include the percentage allowed for retention. Generally, retention is 10 percent and isn't released until the project is complete and accepted by the owner.

Front loading helps contractors carry the financial burden of work in progress. Few subcontractors have enough cash to pay all their bills when due and still wait months to collect for work that's been completed. Employees have to be paid in full and on time. Assigning a higher markup to work completed first accelerates the payment schedule and helps spread payments more evenly over the entire project. Front loading is routine with subcontractors and unpopular with owners and lenders. In practice, subcontractors have no choice but to place a higher price or higher markup on work completed first.

Note that figures in the Factor column are less than 1.0 for work done near the end of the project. That's because front loading doesn't change the contract price. It changes only when that money is received.

Column four, Sell ($) Values for this column are the product of the cost and factor columns. Use the figures in this column when submitting monthly progress invoices.

ACME

Mechanical Contractors

7600 Oak Avenue - Anytown, U.S.A. 12345-6789
Voice 1-234-5678 / FAX 1-234-5680

Billing Breakdown Worksheet

(Chancey Bank Building - Job No. 7777)

Activity	Cost ($)	Factor	Sell ($)
Mobilization	2,000	1.50	3,000
Underground chilled and heated water piping	120,000	1.35	162,000
Underground plumbing & piping	87,000	1.35	117,450
Above ground chilled and heating water piping	115,000	1.20	138,000
Above ground plumbing & piping	87,000	1.20	104,400
HVAC ducting	96,000	1.20	115,200
Equipment	312,000	1.10	343,200
Duct and pipe insulation	119,000	1.00	119,000
Temperature controls	53,000	.90	47,700
Plumbing fixtures	43,000	.65	27,950
Water treatment	1,800	.55	990
Test and balance	13,000	.55	7,150
Validation	500	.50	250
	1,049,300		1,186,290

$$\text{Estimated gross profit} = \frac{\$1{,}186{,}290 - \$1{,}049{,}300}{\$1{,}049{,}300} = 13\%$$

When the billing worksheet has been completed, prepare the billing breakdown as shown on the next page.

The general contractor will usually have a cutoff date each month for progress billings. If the invoice has not been received by that date, it won't be considered until the following month. The cutoff date is usually either the 20th or 25th of the month. The general contractor may permit you to include in each monthly billing work that's not yet complete but should be by the last day of the month.

Column one, Activity List each work item on the billing breakdown worksheet.

Column two, Amount ($) List costs from the Sell ($) column of the billing breakdown worksheet.

Column three, % Complete Show the percentage of completion for each activity listed. Obviously, this figure has to be based on estimates. Be prepared to support your estimate of completion if asked by the general contractor or the lender. If you anticipate a dispute on the figures in this column, call the lender or general contractor to explain the percentage you claim before submitting the bill.

Column four, $ Complete Multiply the percentage in the % Complete column by the figure in the Amount ($) column. Enter the result in the $ Complete column.

Total the $ Complete column. Then subtract the retention percentage and progress billings received to date. The new total is the progress billing for the current month.

ACME

Mechanical Contractors

7600 Oak Avenue - Anytown, U.S.A. 12345-6789
Voice 1-234-5678 / FAX 1-234-5680

Monthly Progress Billing

(Chancey Bank Building — Job No. 7777)

Activity	Amount ($)	% Complete	$ Complete
Mobilization	3,000		
Underground chilled and heated water piping	162,000		
Underground plumbing & piping	117,450		
Above ground chilled and heating water piping	138,000		
Above ground plumbing piping	104,400		
HVAC ducting	115,200		
Equipment	343,200		
Duct and pipe insulation	119,000		
Temperature controls	47,700		
Plumbing fixtures	27,950		
Water treatment	990		
Test and balance	7,150		
Validation	250		
Change order numbers _______ through _______			
Contract price to date	$1,186,290		
		Total completed to date	$
		Less ________% retention	$
		Subtotal	$
		Less previous billings	$
		Amount this billing	$

Upon completion of a project, all subcontractors may be required to provide a warranty letter for the work they've completed. These letters are addressed to the general contractor and will probably be forwarded to the owner.

The warranty usually covers defects in labor or material discovered during the first year after the project is turned over to the owner. Occasionally, the owner will request occupancy of a portion of the project before the entire project is completed. This is called a period of beneficial occupancy. If there will be a period of beneficial occupancy, a separate warranty letter may be required covering only the areas of the building occupied by the owner and only the portion of the plumbing and HVAC systems used in those areas.

Before preparing a warranty letter, get warranty letters from third-tier subs such as the environmental control, thermal insulation, test and balance subs. Any exclusions or limitations in these warranty letters should also appear as exclusions and limitations in your warranty letter. Otherwise you may be accepting responsibility for repairs on work you never performed.

A sample warranty letter follows.

ACME

Mechanical Contractors

7600 Oak Avenue - Anytown, U.S.A. 12345-6789
Voice 1-234-5678 / FAX 1-234-5680

A C M E

Date: ____________

To: ______________________________
____
____
____

Project:____

Gentlemen:

In compliance with the project specifications, we submit our guarantee of equipment, materials and workmanship furnished by Acme Mechanical Contractors for a period of one year. The warranty period is as follows:

From ______________________ to ______________________

Equipment and materials furnished by others, but installed by Acme Mechanical Contractors, are not covered by this warranty, except for the installation work performed by Acme Mechanical Contractors.

Ordinary wear is not covered by this warranty. The owner's abuse, neglect or failure to perform recommended maintenance procedures will void this warranty.

Should any problems occur during the specified warranty period, due to faulty equipment, materials or workmanship, Acme Mechanical Contractors will correct the problems, without charge, to the satisfaction of the owner.

Sincerely,

____
Project Manager

National Estimator and Job Cost Wizard Quick Start

National Estimator is a construction cost estimating program with all the cost estimates in this book. You can page through these costs one screen at a time or use the electronic index to search by keyword for exactly the information you need. When you find what you're looking for, split the screen in two so your estimate is on the bottom half of the screen and the database is on the top half. Then copy and paste into your estimate. Type the estimated quantity. The program extends prices and totals columns automatically. Add your overhead and profit as a percentage and then print the estimate.

Construction Estimating program group.

Job Cost Wizard takes your estimates to the next level:

1. Turning estimates into invoices you can send out in a window envelope.
2. Exporting to *QuickBooks* where you can track actual costs against estimates.

The installation program creates a Construction Estimating program group and puts several icons in that group. If you have trouble installing *National Estimator*, call Craftsman tech support at 760-438-7828. If your computer can't use the CD, exchange it for *National Estimator* on 3½" disk. Send the CD to Craftsman, Box 6500, Carlsbad, CA 92018. There's no charge for this exchange.

Click on File. Click on Run in Windows 3.1.

Installing National Estimator and Job Cost Wizard

Put the *National Estimator* disk in your CD drive (such as D:). Start Windows.

In **Windows 3.1 or 3.11** go to the Program Manager.	In **Windows 95, 98** or **NT**
1. Click on **File**	1. Click on **Start**
2. Click on **Run . . .**	2. Click on **Run . . .**
3. Type D:\SETUP	3. Type D:\SETUP
4. Press Enter↵	4. Press Enter↵

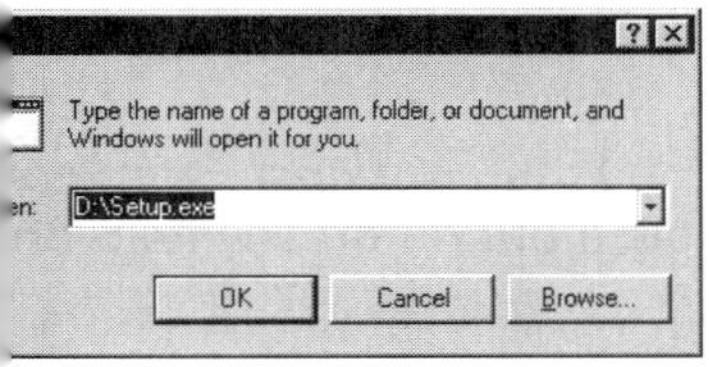

Type D:\SETUP in Windows 95, 98 or NT.

We recommend accepting the installation defaults:

1. National directory
2. "Typical" setup

When installation is complete, click on **Finish**. Note that *Job Cost Wizard* won't be installed if you use Windows 3.1 or 3.11.

Uninstalling National Estimator and Job Cost Wizard

In Windows 3.1 or 3.11: Select the Construction Estimating program group. Then click on the Uninstall icon.

In Windows 95, 98 or NT: Click on **Start, Settings, Control Panel**. Double click on **Add/Remove Programs**. Then click on the name of the program to remove and **Add/Remove**.

Learn the Easy Way!

Estimating with Jay and Julie.

Watch Me icon

The **Watch Me** tutorial is an interactive video guide to *National Estimator*. Let Jay and Julie help you get started using the *National Estimator* program. Sit back and relax an let the video run — or click on buttons and menus to jump from topic to topic inside the video. Exit any time you want. Then go back to Watch Me later to brush up on some topic. Watch Me takes up very little spac on your hard drive and is part of the "Typical" installation. Watch Me requires 8Mb of RAM, a Windows-compatible sound system and at least a 100 MHz 48 computer. Click on the Watch Me icon to begin Watch Me.

Using National Estimator

National Estimator icon

National Estimator begins when you click on the *National Estimator* icon (in Windows 3.1 or 3.11) or click on *Start*, *Programs*, the *Construction Estimating* group and then *National Estimator* (in Windows 95, 98 or NT).

On the title bar at the top of the screen you see the program name, *Nationa Estimator*, and [Estimate1], showing that Estimate1 is on the screen. Let's take closer look at the other information at the top of your screen.

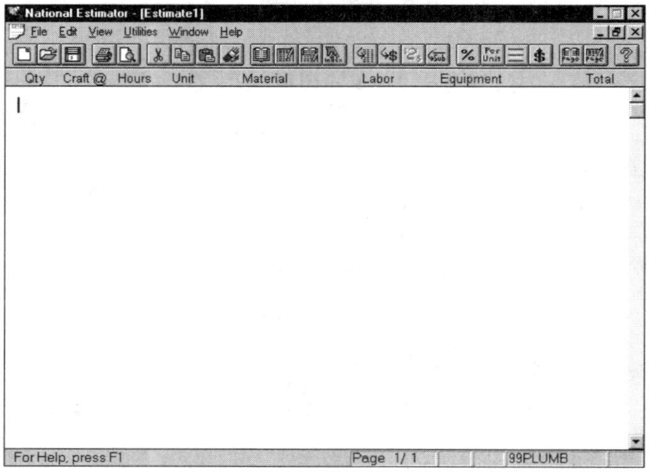

Your estimating form ready to begin the first estimate.

The Menu Bar

Below the title bar you see the menu bar. Every option in *National Estimato* is available on the menu bar. Click with your left mouse button on any item on the menu to open a list of available commands.

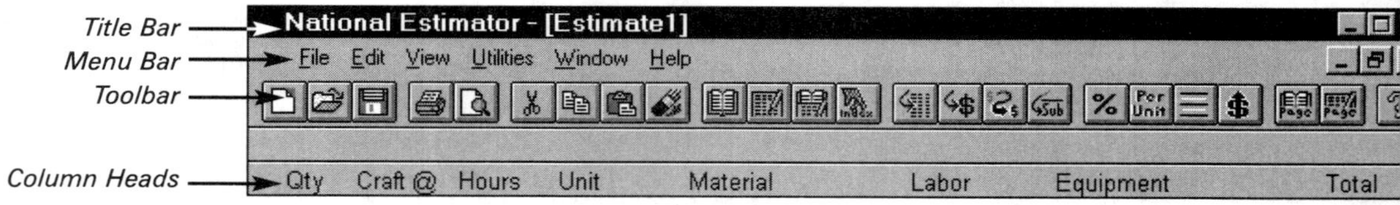

Buttons on the Toolbar

Below the menu bar you see 24 buttons that make up the toolbar. The options you use most in *National Estimator* are only a mouse click away on th toolbar.

Column Headings

Below the toolbar you'll see column headings for your estimate form:

Qty for quantity

Craft@Hours for craft (the crew doing the work) and manhours (to complete the task)

Unit for unit of measure, such as per linear foot or per square foot

Material for material cost

Labor for labor cost

Equipment for equipment cost

Total for the total of all cost columns

The Status Bar

The bottom line on your screen is the status bar. Here you'll find helpful information about the choices available. Notice "Page 1/1" near the right end of the status line. That's a clue that you're looking at page 1 of a one-page estimate. When you see PICT at the right of the status bar, a picture is available in the costbook. Click on PICT to show the picture. Click on PICT again to hide the picture.

Check the status bar occasionally for helpful tips and explanations of what you see on screen.

Beginning an Estimate

| The Blinking Cursor (insert point)

Mouse Pointer

I Mouse Pointer

Let's start by putting a heading on this estimate.

1. Press [Enter] once to space down one line.
2. Press [Tab] four times (or hold the space bar down) to move the Blinking Cursor (the insert point) near the middle of the line.
3. Type "Estimate One" and press [Enter]. That's the title of this estimate, "Estimate One."
4. Press [Enter] again to move the cursor down a line. That opens up a little space below the title.

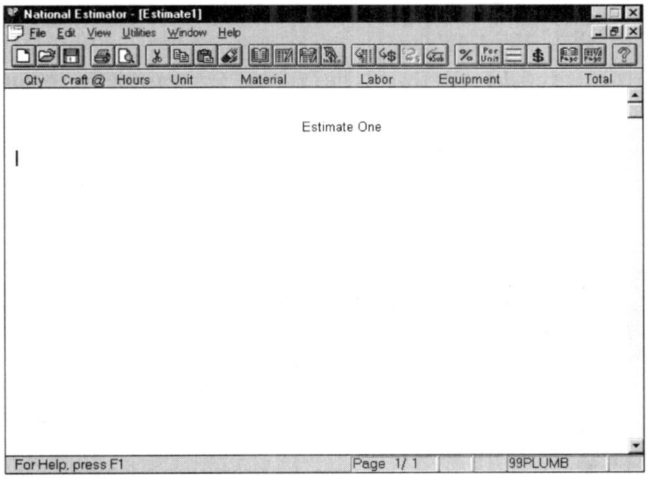

Begin by putting a title on your estimate, such as "Estimate One."

The Costbook

Let's leave your estimating form for a moment to take a look at estimates stored in the costbook. To switch to the costbook, either:

- Click on the (Costbook Window) button, -or-
- Click on **View** on the menu bar. Then click on **Costbook Window**, -or-

- Press [Esc]. Press the [↓] arrow key to highlight Costbook Window. Then press [Enter↵].

The entire *1999 National Plumbing & HVAC Estimator* is available in the Costbook Window. Notice the words *Page 15 Electric domestic hot water heater, (residential)* at the left side of the screen just below the toolbar. That's your clue that the hot water heater section of page 15 is on the screen. To turn to the next page, either:

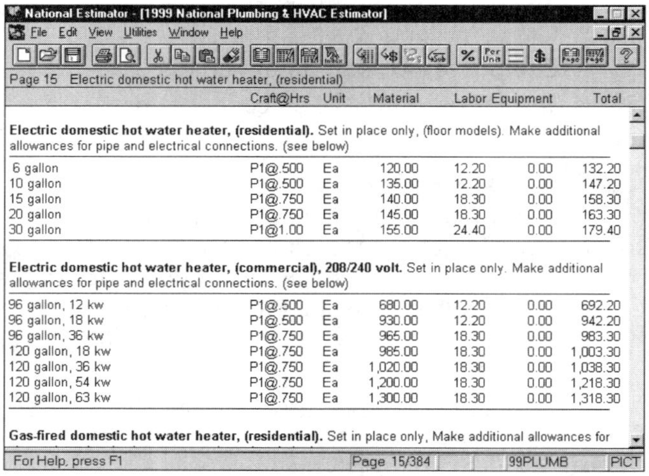

Electric domestic hot water heater, (residential). Set in place only, (floor models). Make additional allowances for pipe and electrical connections. (see below)

	Craft@Hrs	Unit	Material	Labor	Equipment	Total
6 gallon	P1@.500	Ea	120.00	12.20	0.00	132.20
10 gallon	P1@.500	Ea	135.00	12.20	0.00	147.20
15 gallon	P1@.750	Ea	140.00	18.30	0.00	158.30
20 gallon	P1@.750	Ea	145.00	18.30	0.00	163.30
30 gallon	P1@1.00	Ea	155.00	24.40	0.00	179.40

Electric domestic hot water heater, (commercial), 208/240 volt. Set in place only. Make additional allowances for pipe and electrical connections. (see below)

	Craft@Hrs	Unit	Material	Labor	Equipment	Total
96 gallon, 12 kw	P1@.500	Ea	680.00	12.20	0.00	692.20
96 gallon, 18 kw	P1@.500	Ea	930.00	12.20	0.00	942.20
96 gallon, 36 kw	P1@.750	Ea	965.00	18.30	0.00	983.30
120 gallon, 18 kw	P1@.750	Ea	985.00	18.30	0.00	1,003.30
120 gallon, 36 kw	P1@.750	Ea	1,020.00	18.30	0.00	1,038.30
120 gallon, 54 kw	P1@.750	Ea	1,200.00	18.30	0.00	1,218.30
120 gallon, 63 kw	P1@.750	Ea	1,300.00	18.30	0.00	1,318.30

Gas-fired domestic hot water heater, (residential). Set in place only, Make additional allowances for

The costbook window has the entire National Plumbing & HVAC Estimator.

- Press [PgDn] (with Num Lock off), -or-
- Click on the lower half of the scroll bar at the right edge of the screen.

To move down one line at a time, either:

- Press the [↓] arrow key (with Num Lock off), -or-
- Click on the arrow on the down scroll bar at the lower right corner of the screen.

Press [PgDn] about 1,000 times and you'll page through the entire *National Plumbing & HVAC Estimator*. Obviously, there's a better way. To turn quickly to any page, either:

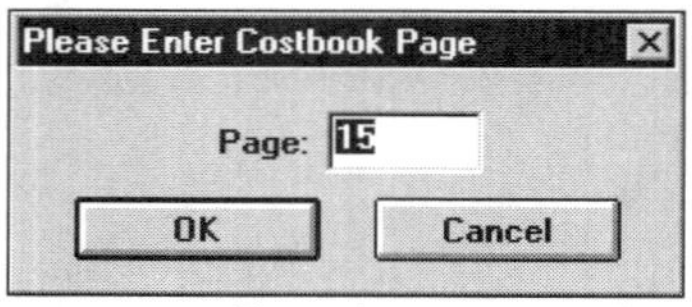

Type the page number you want to see.

- Click on the (Turn to Costbook Page) button near the right end of the toolbar, -or-
- Click on **View** on the menu bar. Then click on **Turn to Costbook Page**, -or-
- Press [Esc]. Press the [↓] arrow key to highlight Turn to Costbook Page. Then press [Enter↵].

Type the number of the page you want to see and press Enter. *National Estimator* will turn to the top of the page you requested.

An Even Better Way

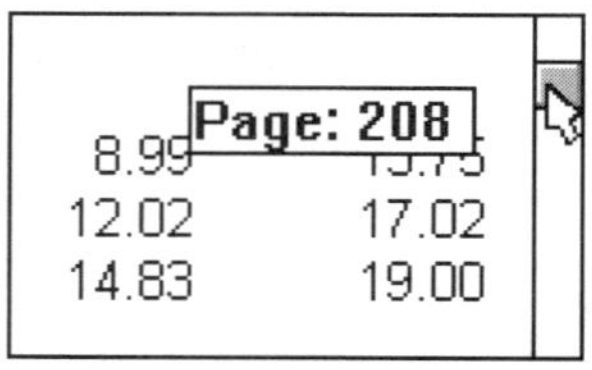

Drag the square to see any page.

Find the small square in the slide bar at the right side of the Costbook Window. Click and hold on that square while rolling the mouse up or down. Keep dragging the square until you see the page you want in the Page: box. Release the mouse button to turn to the top of that page.

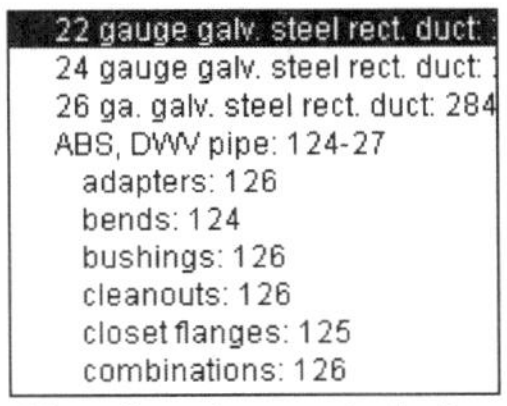

Use the electronic index to find cost estimates for any item.

A Still Better Way: Keyword Search

To find any cost estimate in seconds, search by keyword in the index. To go to the index, either:

- Click on the (Index) button near the center of the toolbar, -or-
- Press [Esc]. Press [Enter↵], -or-
- Click on **View** on the menu bar. Then press [Enter↵].

Notice that the cursor is blinking in the Enter Keyword to Locate box at the right of the screen. Obviously, the index is ready to begin a search.

Enter keyword to locate.

Your First Estimate

Suppose we're estimating the cost of copper DWV pipe with soldered joints. Let's put the index to work with a search for copper DWV pipe. In the box under Enter Keyword to Locate, type *copper*. The index jumps to the heading *Copper pipe*.

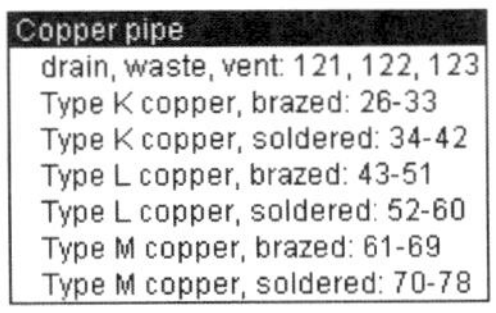

The index jumps to Copper pipe.

The first item under Copper pipe is *drain, waste, vent: 121,122,123*. Either:

- Click once on that line and press [Enter↵], -or-
- Double click on that line, -or-
- Press [Tab] and the [↓] arrow key to move the highlight to *drain, waste, vent: 121,122,123*. Then press [Enter↵].

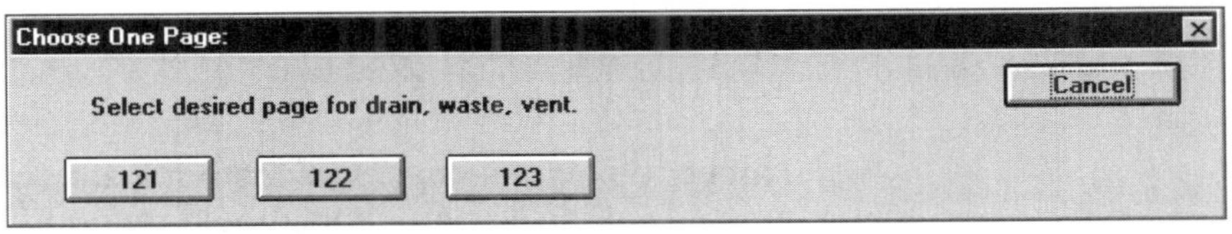

If costs appear on several pages, click on the page you prefer.

To select the page you want to see (page 121 in this case), either:

- Click on the number 121, -or-
- Press [Tab] to highlight 121. Then press [Enter↵].

National Estimator turns to the top of page 121. See the example below. Press the [↓] arrow key until "DWV copper pipe" is at the top of your screen.

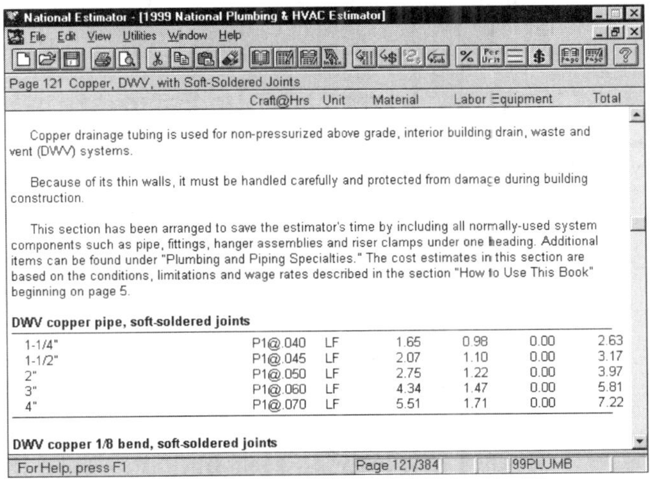

Costs for DWV copper pipe on page 121.

Splitting the Screen

Most of the time you'll want to see what's in both the costbook and your estimate. To split the screen into two halves, either:

- Click on the (Split Window) button near the center of the toolbar, -or-
- Press Esc and the ↓ arrow key to move the selection bar to Split Window, -or-
- Click on **View** on the menu bar. Then click on **Split Window** and your screen should look like the example below.

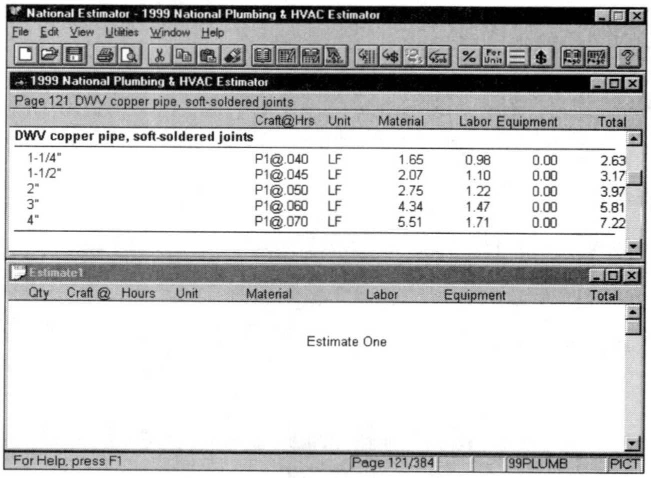

The split window: Costbook above and estimate below.

Notice the eight lines of the costbook are at the top of the screen and your estimate is at the bottom. You should recognize "Estimate One." It's your title f the estimate. Column headings are at the top of the costbook and across the middle of the screen (for your estimate).

To Switch from Window to Window

- Click in the window of your choice, -or-
- Hold the Ctrl key down and press Tab.

Notice that a window title bar turns dark when that window is selected. The selected window is where keystrokes appear as you type.

Copying Costs to Your Estimate

Next, we'll estimate the cost of 128 feet of 1-1/2" copper DWV pipe. Click on the (Split Window) button on the toolbar to be sure you're in the split windo Click anywhere in the costbook (the top half of your screen). Then press the ↓ arrow key until the cursor is on the line:

1-1/2"	P1@.045	LF	2.07	1.10	0.00	3.1

To copy this line to your estimate:

1. Click on the line.
2. Click the (Copy) button.
3. Click on the (Paste) button to open the Enter Cost Information dialog box.

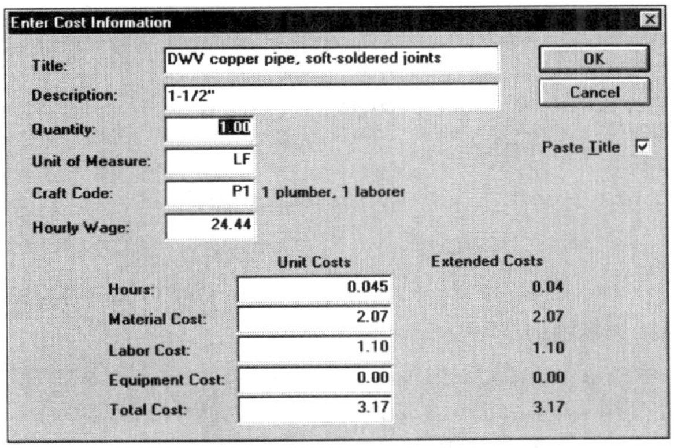

Use the Enter Cost Information dialog box to copy or change costs.

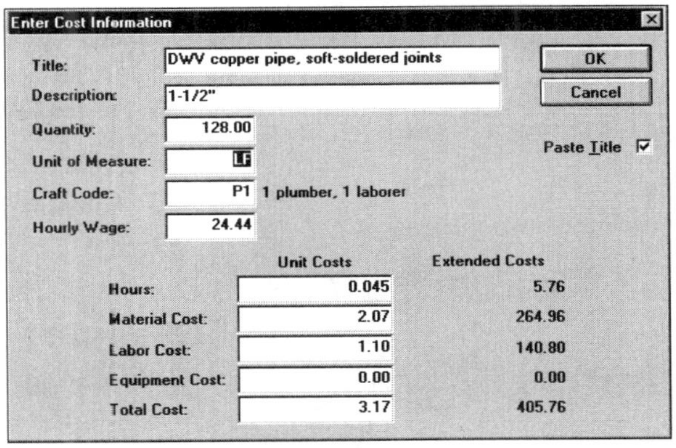

Costs for 128 linear feet (extended costs) are on the right.

Notice that the blinking cursor is in the Quantity box:

1. Type a quantity of 128 because the job has 128 feet of copper DWV pipe.
2. Press [Tab] and check the estimate for accuracy.
3. Notice that the column headed Unit Costs shows costs per unit, per "LF" (linear foot) in this case.
4. The column headed Extended Costs shows costs for 128 linear feet.
5. The lines opposite Title and Description show what's getting installed. You can change the words in either of these boxes. Just click on what you want to change and start typing or deleting.
6. You can also change any numbers in the Unit Cost column. Just click and start typing.
7. When the words and costs are exactly right, press [Enter] or click on OK to copy these figures to the end of your estimate.

The new line at the bottom of your estimate shows:

Extended costs for 128 LF of 1½" copper pipe as they appear on your estimate form.

DWV copper pipe, soft-soldered joints						
1-1/2"						
128.00	P1@5.760	LF	264.96	140.80	0.00	405.76

128.00 is the quantity of copper DWV pipe

P1 is the recommended crew, a plumber and a laborer

@5.760 shows the manhours required for the work

LF is the unit of measure, linear feet in this case

264.96 is the material cost (the pipe)

140.80 is labor cost for the job

0.00 shows there is no equipment cost

405.76 is the total of material, labor and equipment columns

Copy Anything to Anywhere in Your Estimate

Anything in the costbook can be copied to your estimate. Just click on the line (or select the words) you want to copy and press the [F8] key. It's copied to the last line of your estimating form. If your selection includes costs, you'll hav chance to enter the quantity. To copy to the *middle* of your estimate:

1. Select what you want to copy.
2. Click on the (Copy) button.
3. Click in the estimate where you want to paste.
4. Click on the (Paste) button.

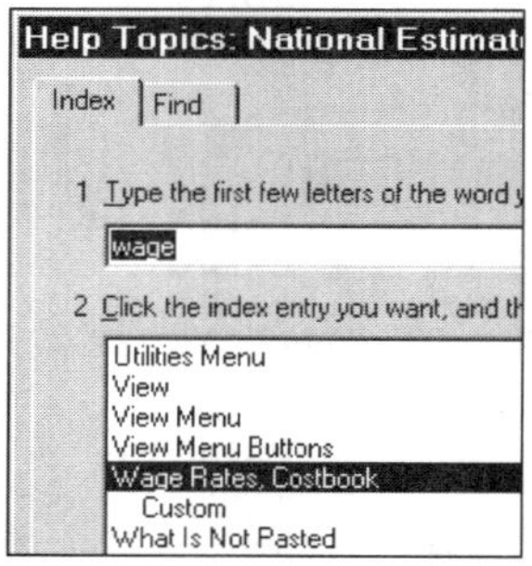

Search for information on setting wage rates.

Changing Wage Rates

The labor cost in the example above is based on a plumber and a laborer working at a cost of $24.44 per hour. (See page 7 in the *National Plumbing & HVAC Estimator* for crew rates used in the costbook.) Suppose $24.44 per hou isn't right for your estimate. What then? No problem! It's easy to use your own wage rate for any crew or even make up your own crew codes. To get more information on setting wage rates, press F1. At *National Estimator* Help Contents, click on the **Search** button. Type *wage* then double click on **Custom** under Wage Rates. To return to your estimate, click on File on the *National Estimator* Help menu bar. Then click on **Exit**.

Changing Cost Estimates

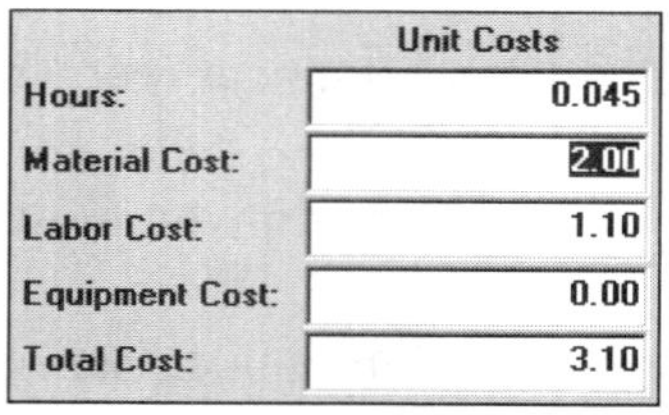

Change the material cost to two dollars.

With Num Lock off, use the ↑ or ↓ arrow key to move the cursor to the lin you want to change (or click on that line). In this case, move to the line that begins with a quantity of 128. To open the Enter Cost Information Dialog box, either:

- Press Enter↵, -or-
- Click on the (Change Cost) button on the toolbar.

To make a change, either:

- Click on what you want to change, -or-
- Press Tab until the cursor advances to what you want to change.

Then type the correct figure. In this case, change the material cost to 2.00 (two dollars).

Press Tab and check the Extended Costs column. If it looks OK, press Enter and the change is made on your estimating form.

Changing Text (Descriptions)

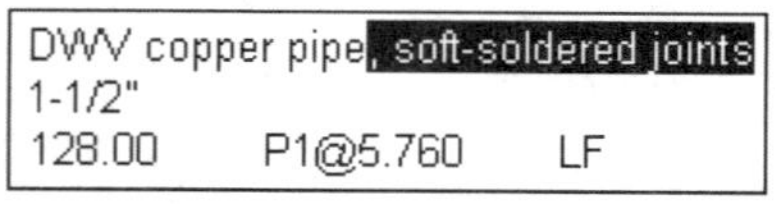

To select, click and hold the mouse button while dragging the mouse.

Click on the (Estimate Window) button on the toolbar to be sure you're i the estimate. With Num Lock off, use the ↑ or ↓ arrow key or click the mouse button to put the cursor where you want to make a change. In this case, we're going to make a change on the line that begins "DWV copper pipe . . ."

To make a change, click where the change is needed. Then either:

- Press the Del or ←Bksp key to erase what needs deleting, -or-
- Select what needs deleting and click on the (Cut) button on the toolba
- Type what needs to be added.

In this case, click just after "DWV copper pipe." Then hold the left mouse but down and drag the mouse to the right until you've put a dark background behin ", soft-soldered joints." The dark background shows that these words are select and ready for editing.

Press the Del key, or click on the (Cut) button on the toolbar, and the selection is cut from the estimate. If that's not what you wanted, click on the (Undo) button and "soft-soldered joints" is back again.

DWV copper pipe
1-1/2"
128.00 P1@5.760
Angle stops

Adding "Angle stops."

Adding Text (Descriptions)

Some of your estimates will require descriptions (text) and costs that can't be found in the *National Plumbing & HVAC Estimator*. What then? With *National Estimator* it's easy to add descriptions and costs of your choice anywhere in the estimate. For practice, let's add an estimate for four angle stops to Estimate One.

Click on the (Estimate Window) button to be sure the estimate window is maximized. We can add lines anywhere on the estimate. But in this case, let's make the addition at the end. Press the ↓ arrow key to move the cursor down until it's just above the horizontal line that separates estimate detail lines from estimate totals. To open a blank line, either:

- Press Enter↵, -or-
- Click on the (Insert Text) button on the toolbar, -or-
- Click on **Edit** on the menu bar. Then click on **Insert a Text Line.**

Type "Angle stops" and press Enter↵.

Adding a Cost Estimate Line

Now let's add a cost for "Angle stops" to your estimate. Begin by opening the Enter Cost Information dialog box. Either:

- Click on the (Insert Cost) button on the toolbar, -or-
- Click on **Edit** on the menu bar. Then click on **Insert a Cost Line.**

1. The cursor is in the Quantity box. Type the number of units (4 in this case) and press Tab.
2. The cursor moves to the next box, Unit of Measure.
3. In the Unit of Measure box, type *Each* and press Tab.
4. Press Tab twice to leave the Craft Code blank and Hourly Wage at zero.
5. Since these angle stops will be installed by the supplier, there's no material, labor or equipment cost. So press Tab four times to skip over the Hours, Material Cost, Labor Cost and Equipment Cost boxes.
6. In the Total Cost box, type 20.00. That's the cost per angle quoted by your supplier.
7. Press Tab once more to advance to OK.
8. Press Enter↵ and the cost of four angle stops is written to your estimate.

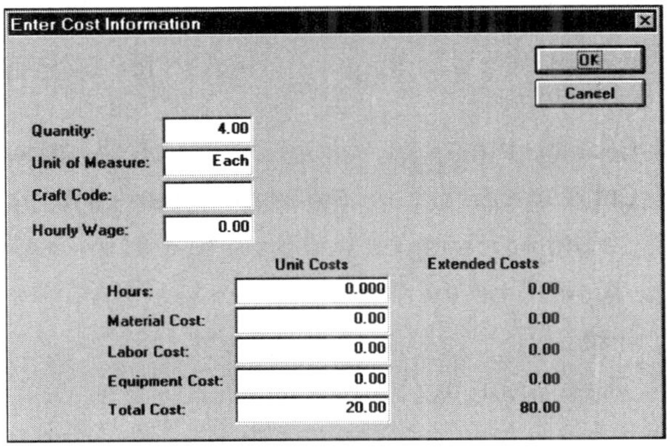

Unit and extended costs for four angle stops.

Note: The sum of material, labor and equipment costs appears automatically in the Total Cost box. If there's no cost entered in the Material Cost, Labor Cost or Equipment Cost boxes (such as for a subcontracted item), you can enter any figure in the Total Cost box.

Adding Lines to the Costbook

Add lines or make changes in the costbook the same way you add lines or make changes in an estimate. The additions and changes you make become p of the user costbook. For more information on user costbooks, press F1. Click **Search**. Type "user" and press [Enter].

Subtotals Become QuickBooks Cost Categories

At the end of each section in your estimate, insert a subtotal. For a general contractor, estimate sections might be Demolition, Excavation, Foundation, Framing, etc. Estimate sections for a plumbing and HVAC contractor might include Piping, DWV, Fixtures and Ducting. Section subtotals help organize you estimates and make them easier to read and understand. Insert section subtot wherever they make the most sense to you. These subtotals become cost cate gories when printing bids and invoices. The first ten characters of subtotal names become *QuickBooks* cost category names when exporting to *QuickBoo*

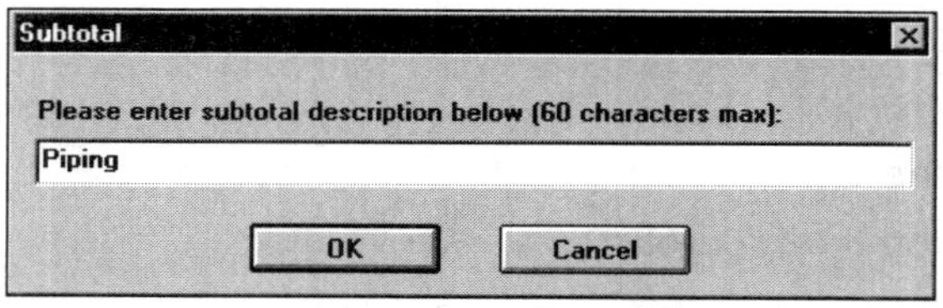

Insert a subtotal.

To insert a subtotal:

1. Click on the last cost line of the section (or on any blank line below the section).
2. Click on the (Subtotal) button on the toolbar (or click on **Edit** and **Insert Subtotal**).
3. Type a name or description for the section (such as "Piping"
4. Press [Enter].

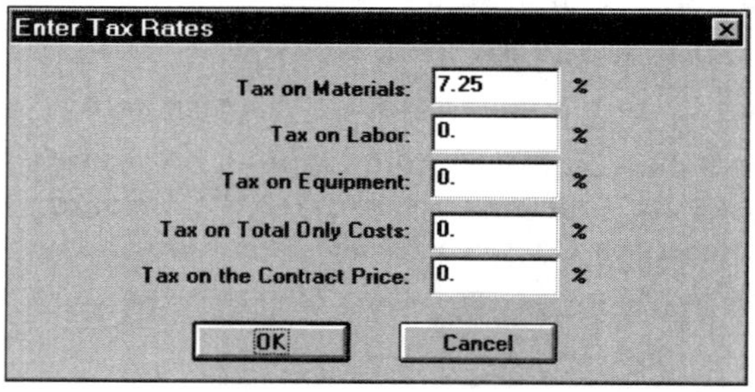

Type the tax rate that applies.

Adding Tax

To include sales tax in your estimate:

1. Click on **Edit**.
2. Click on **Tax Rates**.
3. Type the tax rate in the appropriate box.
4. Press [Tab] to advance to the next box.
5. Press [Enter] or click on **OK** when done.

In this case, the tax rate is 7.25% on materials only. Tax will appear as the last line of the estimate.

Adding Overhead and Profit

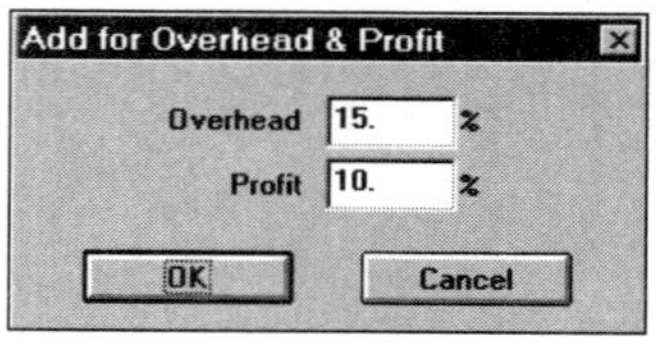

Adding overhead & profit.

Set markup percentages in the Add for Overhead & Profit dialog box. To ope the box, either:

- Click on the (Markup) button on the toolbar, -or-
- Click on **Edit** on the menu bar. Then click on **Markup**.

Type the percentages you want to add for overhead. For this estimate:

1. Type 15 on the Overhead line.
2. Press [Tab] to advance to Profit.
3. Type 10 on the Profit line.
4. Press [Enter].

Markup percentages can be changed at any time. Just reopen the Add for Overhead & Profit dialog box and type the correct figure.

Preview Your Estimate

You can display an estimate on screen just the way it will look when printed on paper. To preview your estimate, either:

- Click on the (Print Preview) button on the toolbar, -or-
- Click on **File** on the menu bar. Then click on **Print Preview**.

Use buttons on Print Preview to see your estimate as it will look when printed.

In Print Preview:

- Click on **Next Page** or **Prev Page** to turn pages.
- Click on **Two Page** to see two estimate pages side by side.
- Click on **Zoom In** to get a closer look.
- Click on **Close** when you've seen enough.

Printing Your Estimate

When you're ready to print the estimate, either:

- Click on the (Print) button on the toolbar, -or-
- Click on File on the menu bar. Then click on **Print**, -or-
- Hold the Ctrl key down and type the letter P.
- Press Enter↵ or click on **OK** to begin printing.

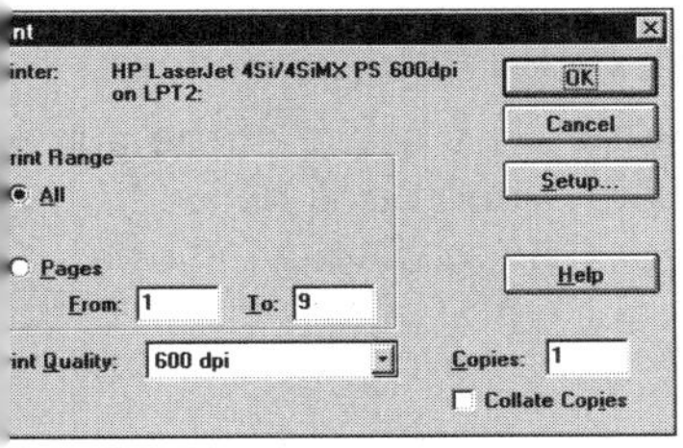

Options available depend on the printer you're using.

Save Your Estimate to Disk

To store your estimate on the hard disk where it can be re-opened and changed at any time, either:

- Click on the (Save) button on the toolbar, -or-
- Click on **File** on the menu bar. Then click on **Save**, -or-
- Hold the Ctrl key down and type the letter S.

The cursor is in the File Name box. Type the name you want to give this estimate, such as *First*. The name can be up to eight letters and numbers, but don't use symbols or spaces. Press Enter↵ or click on **OK** and the estimate is written to disk.

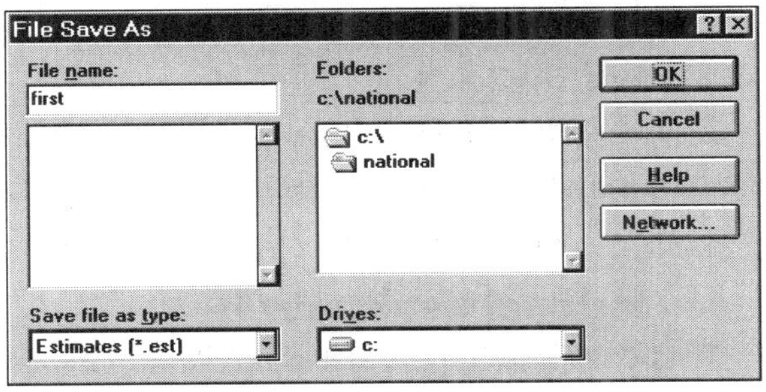

Type the estimate name in the File Name box to assign a file name.

Opening Other Costbooks

Many construction cost estimating databases are available for the *National Estimator* program. The order form at the back of this manual has more information on these costbooks.

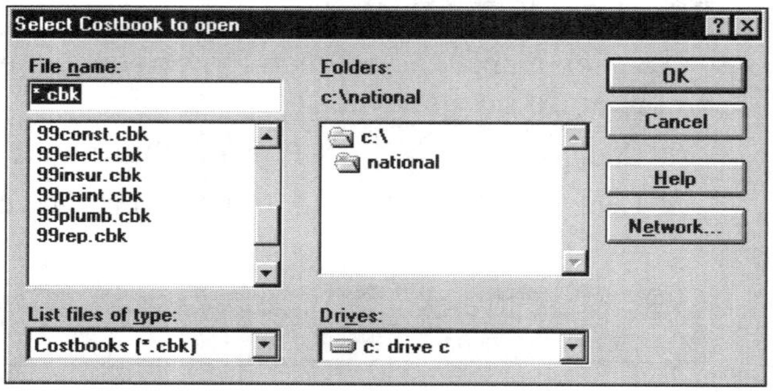

Open the costbook of your choice.

If you own several of these manuals and have installed the database that comes with each, you can open several costbooks at the same time. To open another costbook:

- Click on **File**.
- Click on **Open Costbook**.
- Be sure the drive and directory are correct, usually *c:\national*.
- Double click on the costbook of your choice.

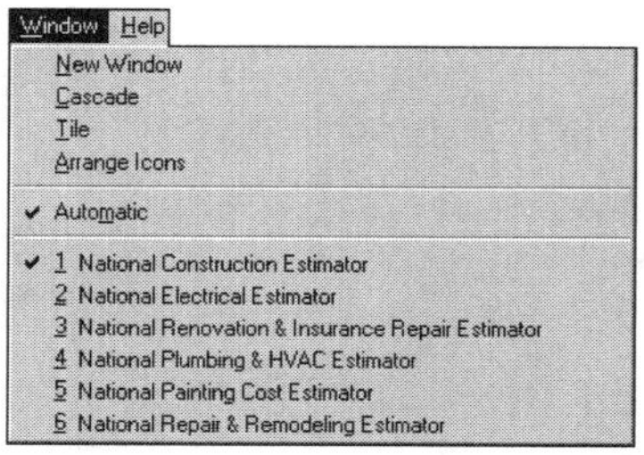

Click to switch costbooks.

To see a list of the costbooks open, click on **Window**. The name of the curre estimate or costbook will be checked. Click on any other costbook name to display that costbook. Click on **Window**, then click on **Tile** to display all open costbooks and estimates.

Select Your Default Costbook

Your default costbook is the last costbook installed. It opens automatically every time you begin using *National Estimator*. Save time by making the defau costbook the one you use most.

To change your default costbook, click on **Utilities** on the menu bar. Then click on **Options**. Next, click on **Select Default Costbook**. Click on the costboo of your choice. Click on **OK**. Then click on **OK** again.

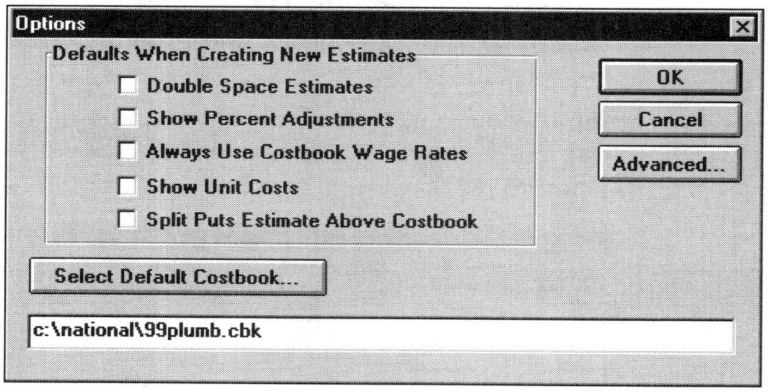

Select the default costbook.

Use National Estimator Help

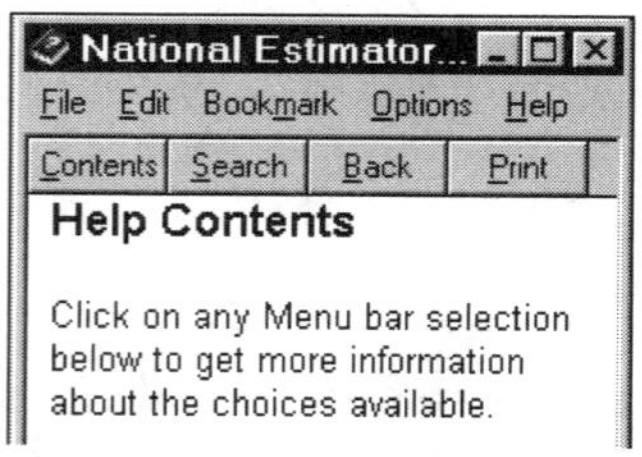

Click on File and then Print Topic.

That completes the basics of *National Estimator*. You've learned enough to complete most estimates. When you need more information about the fine points, use *National Estimator* Help. Click on the [?] (Help) button to see Help Contents. Then click on the menu selection of your choice. To print 28 pages o instructions for *National Estimator*, go to **Help Contents**. Click on **Print All Topics** (at the bottom of Help Contents). Click on **File** on the Help menu bar. Then click on **Print Topic**. Click on **OK**.

Converting Estimates with Job Cost Wizard

To start *Job Cost Wizard*, click on the Job Cost Wizard icon. (Be sure to close the estimate in *National Estimator* before trying to open it in *Job Cost Wizard*.)

Job Cost Wizard icon

Job Cost Wizard starts by asking what estimate you want to open. The blinking cursor will be in the File Name box. To open the estimate of your choice:

- Double click on the estimate name, -or-
- Click on the estimate name and click on **Open**.

For practice, try opening Stillwel.est in the National folder.

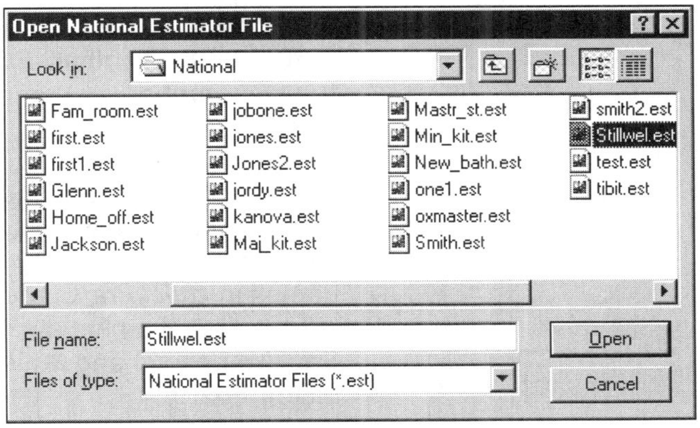

Open a National Estimator file.

The Company Information dialog box will open the first time you use *Job Cost Wizard*. Type your company name and address. This will appear at the top of every estimate and invoice.

Zoom, Scroll and Turn Pages

If the estimate doesn't fit your screen, set the percentage of zoom. For 640x480 resolution, type 81% in the zoom window and press [Enter↵].

Click and drag the vertical slide bar at the right of your screen to scroll down the page. Turn pages by clicking on the Previous Page or Next Page buttons.

Enter Job Information

Job Cost Wizard needs some information about the job to create a nice-looking bid or invoice. For practice, enter job information for the Stillwel estimate:

1. **Customer name and address.** Click on the  (Customer Info) button on the toolbar to enter information about the customer. Only the customer first name and last name are required. All other information is optional. When done, click on **OK**.

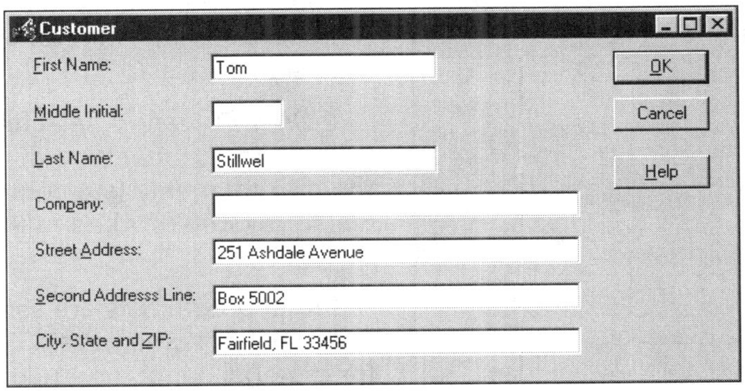

Fill in Customer Information.

Enter a Job Name.

2. **Job Name.** Click on the (Job Name) button on the toolbar. Then type the name of the job, such as "Room Addition." When done, click on OK.

   For transfers to QuickBooks You can change the custome job name after the file has been imported into *QuickBooks* In *QuickBooks*, click on **Lists**. Click on **Customers:Jobs**. Right-click on customer name or job name. Click on **Edit**. Then click on the tab of your choice.

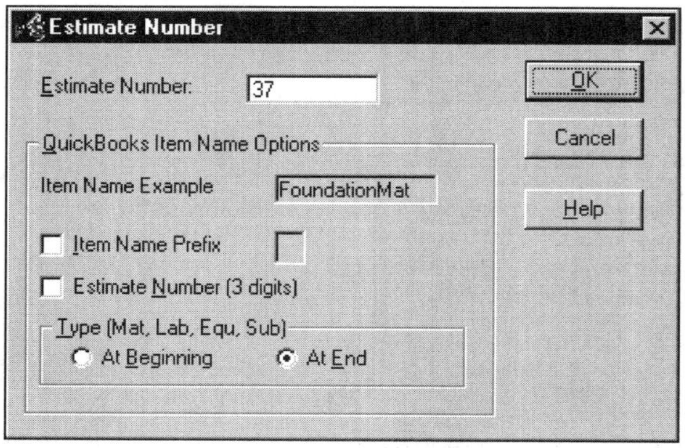

Check the Estimate Number.

3. **Estimate Number.** Click on the (Estimate Number) button on the toolbar. *Job Cost Wizard* keeps track of the last number used and recommends using the next number in sequence.

   For transfers to QuickBooks When *QuickBooks* imports an estimate or invoice, subtotals in your estimate become co categories ("items") in *QuickBooks*. By default, cost catego names in *QuickBooks* are the first 10 characters of estimat subtotal names plus the work type, either Mat, Lab, Equ, c Sub. You can change this default in the Estimate Number dialog box.

Job Cost Wizard Prints Invoices Your Way

Your estimates should cover every cost in a job. But your bids and invoice don't have to show all the details and reveal your markup. So *Job Cost Wizar* gives you choices about showing or hiding the details and markup.

Amount of Detail

To set the amount of detail, click on the (Detail) button.

If *Show All* is selected, every item in your bid or invoice will show a cost a each will become a cost category on the QuickBooks Items list. *Subtotals Onl* is the default and will usually be a better choice.

If *Subtotals Only* is selected, subtotals will be the only costs in your bid or invoice. Each subtotal in your estimate becomes a cost category on the QuickBooks Items list. That's usually the best choice. If Subtotals Only is selec ed for all four cost categories, click on Omit Work Descriptio to show subtotal categories but hide all work descriptions.

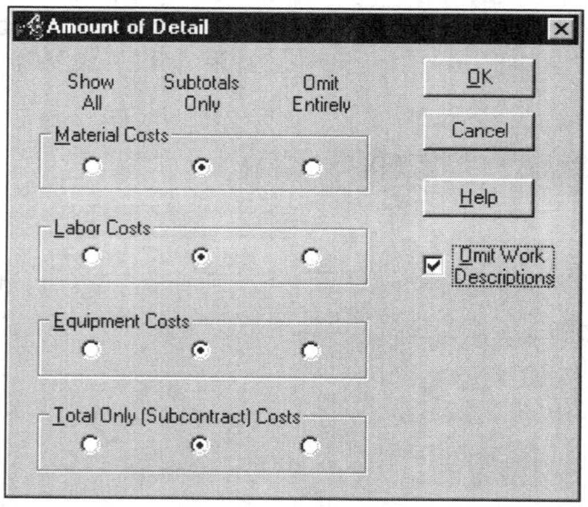

How much detail do you want to show?

The names you give to subtotals in *National Estimator* become cost category names in *QuickBooks*. Cost lines in *National Estimator* not followed by a subtotal become the "Project" subtotal.

If *Omit Entirely* is selected, neither costs nor descriptions will appear for that type of cost — either material, labor, equi ment or total only (subcontract). Use Omit Entirely for materi als, for example, when materials are being furnished by the owner.

"Total only" costs are assumed to be subcontract items. Subcontract items have a cost in the total column but no cos for material, labor, or equipment.

Markup

To control how markup appears, click on the (Markup) button.

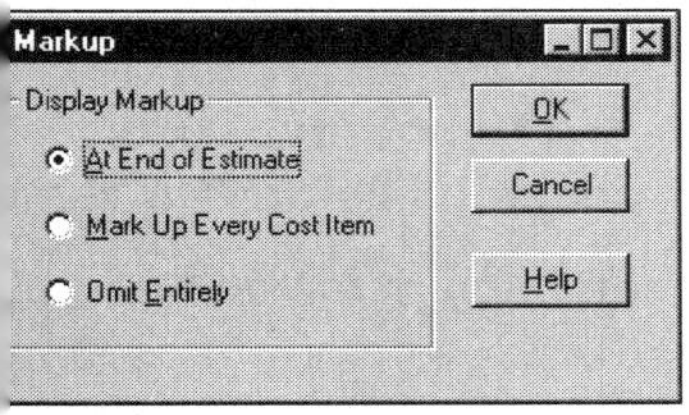

Where do you show markup?

At End of Estimate puts overhead and profit at the end of the estimate, as in *National Estimator*.

Mark Up Every Cost Item distributes overhead and profit proportionately throughout the estimate. There's no mention of overhead, profit or markup anywhere in the estimate or invoice.

Omit Entirely omits overhead and profit from the *QuickBooks* estimate or invoice. Use this option if you prefer to add markup to each cost item in *QuickBooks*.

QuickBooks Account Names

Estimates and invoices imported into *QuickBooks* include expense and income account names. If the imported accounts do not exist already in your *QuickBooks* company, *QuickBooks* will create new accounts. You can control the names of these accounts by making changes in the QuickBooks Account Names box.

- Click on the (Account Names) button on the toolbar, -or-
- Click on **Options** on the menu bar. Then click on **QuickBooks Account Names.**

Enter the names you prefer for construction income and construction expense accounts. Change "ST Sales Tax" to "FL Sales Tax", for example, if the job is taxable under Florida law. *QuickBooks* will keep track of tax due in each state where you do business.

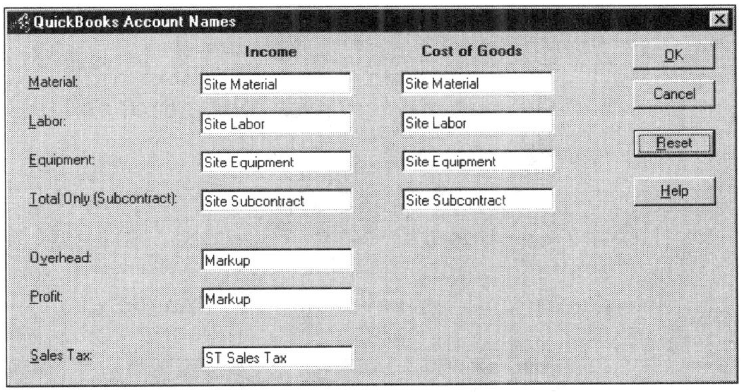

Check the QuickBooks account names.

Automatic Mode

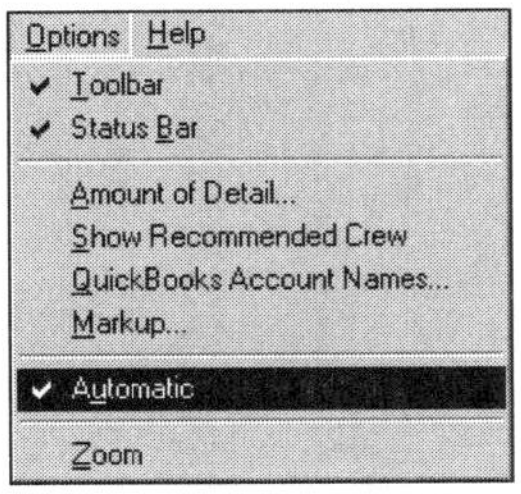

Set up for automatic operation.

Job Cost Wizard always requires customer information, a job name and a job number before saving an estimate. In automatic mode, *Job Cost Wizard* opens the Customer Info, Job Name and Estimate Number dialog boxes automatically after opening any estimate.

Job Cost Wizard runs in automatic mode when there is a check mark beside Automatic on the Options menu. To change to automatic mode, click on **Options** on the menu bar. Then click on **Automatic.**

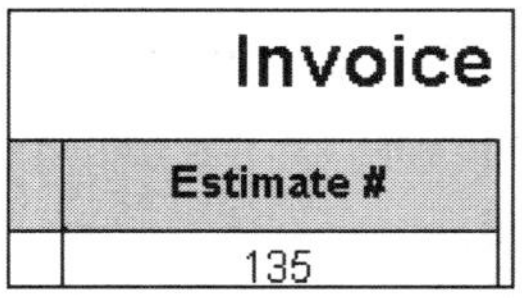

"Invoice" is the document title

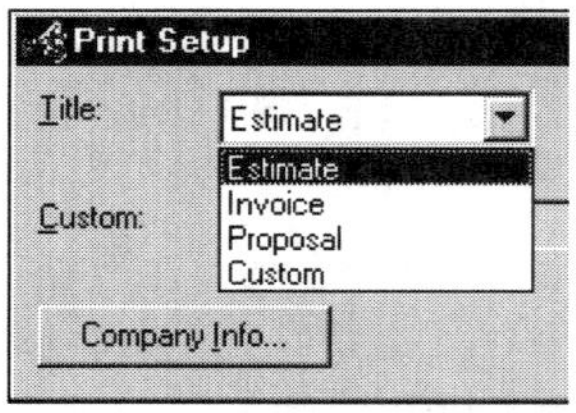

Change the title of a document

Printing a Bid or Invoice

To print one copy of the entire document, click on the [print] button on the toolbar. Then press [Enter] or click on **OK**.

To change the title of the document, click on **File** and **Print Setup**. Select **Estimate, Invoice, Proposal**, or click on **Custom** and enter a title of your choi

Click on **Company Info** to enter or change your company name and addre to be printed on documents.

Exporting to QuickBooks

QuickBooks can not use estimates created by *National Estimator* until *Job Cost Wizard* converts them into the *iif* (Intuit Interface Format) type. This conv sion is done automatically in *Job Cost Wizard* when you open and then save any estimate.

You can save to *QuickBooks* any estimate that is open and has been given customer name and job name.

1. Open the *Save QuickBooks File* dialog box:
   - Click on the [Save] (Save) button on the toolbar, -or-
   - Click on **File** on the menu bar. Then click on **Save**, -or-
   - Hold the [Ctrl] key down and tap the letter S.
2. Change the drive or the folder if the *QuickBooks* folder is not listed at the right of *Save in*.
3. Check the file name to be sure it is what you want.
4. If you use *QuickBooks Pro*, click on **Estimate**.
5. If you use regular *QuickBooks*, click on **Invoice**.
6. When complete, click on **Save**.

Saving an estimate in *Job Cost Wizard* does not affect the original estimate any way. *Job Cost Wizard* can open and save an estimate as many times as y want. To create a second copy of the same estimate with different *Job Cost Wizard* options, save with a slightly different file name.

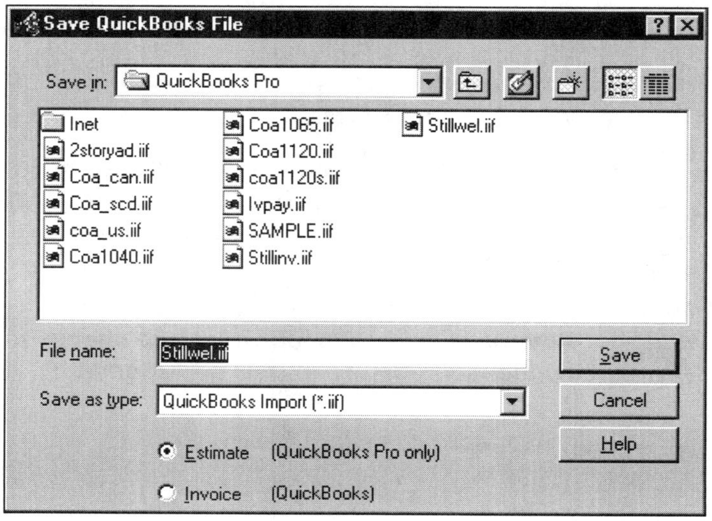

Saving as an estimate to QuickBooks Pro.

Job Cost Tracking with QuickBooks

If you're using *QuickBooks* for the first time, begin by setting up a new *QuickBooks* company. Click on **File** and **EasyStep Interview**. Allow a day or two to set up income and expense accounts, enter existing account balances, format reports to your liking and set up payroll processing. *QuickBooks Pro* "Qcards" will coach you through the setup process. A book titled *Contractor's Guide to QuickBooks Pro* (available at http://costbook.com or 800-829-8123) may also be helpful.

To track job costs, you'll want to have *QuickBooks* write company checks and prepare the company payroll. To turn on *QuickBooks* payroll, click on **File, Preferences, Payroll & Employees**, and **Full payroll features**. You'll need the current tax table for your state. Click on **Help** and **About Tax Table**. Your first tax table is free on the Internet. Thereafter the cost is about $60 per year.

When company files have been created in *QuickBooks*, click on **File** and **Back Up** to preserve the work you've done. *Backups are important.* Intuit, the developer of *QuickBooks*, recommends making a backup before opening import files.

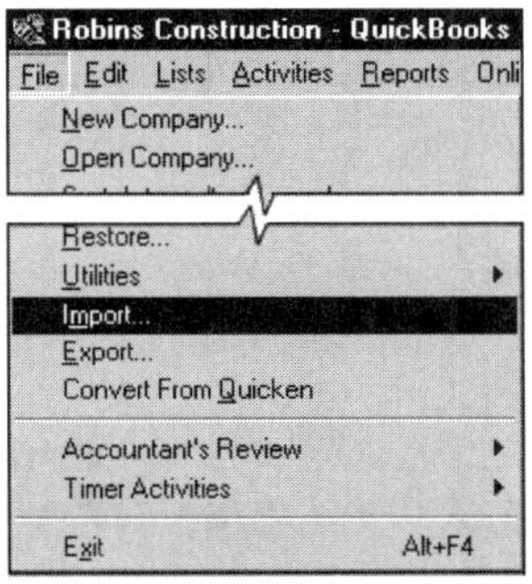

Importing into QuickBooks.

Opening Import Files in QuickBooks

Once an estimate or invoice has been saved with *Job Cost Wizard*, start *QuickBooks*:

1. Click on **File**.
2. Click on **Import**.
3. Double click on the name of the estimate or invoice you want.
4. Click on **OK** when the import is complete.
5. Click on **Activities**.
6. Click on **Create Estimates** if you saved the file as an estimate.
7. Click on **Create Invoices** if you saved the file as an invoice.
8. Click on **Prev** to see the file just imported.

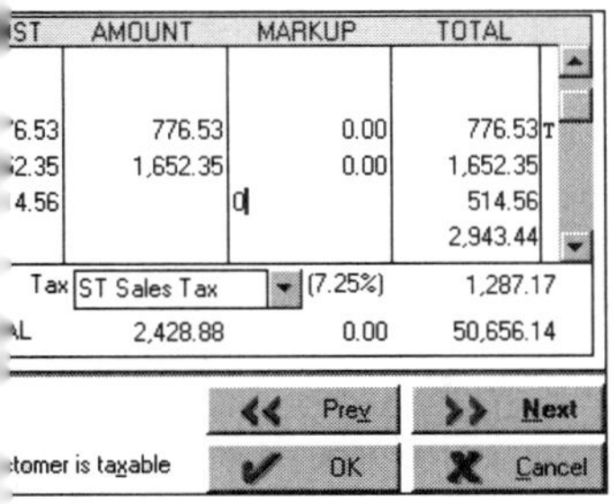

ering zero markup in an estimate.

Entering Markup in QuickBooks Estimates

An imported estimate is not complete until some figure, either a percentage or number, is typed in the Markup column of each cost line. So long as the Markup column is blank, *QuickBooks* will consider the estimated cost for that line to be zero (even when numbers appear in the Total column). If the Markup column for an entire estimate is left blank, *QuickBooks* reports will show the estimated cost for that job to be zero.

The fastest way to fill in the Markup column for a *QuickBooks* estimate is to click in the Markup column on a cost line. Type a zero. Then click on the next cost line in the Markup column and type another zero. Continue clicking and entering zeros until zero markup has been entered on each cost line.

Turn an Estimate into an Invoice in QuickBooks

First, decide if you want an invoice for the whole job or for just part of the job (progress billing). If you prefer progress billing:

1. Click on **File**.
2. Click on **Preferences**.
3. Click on **Jobs & Estimates**.

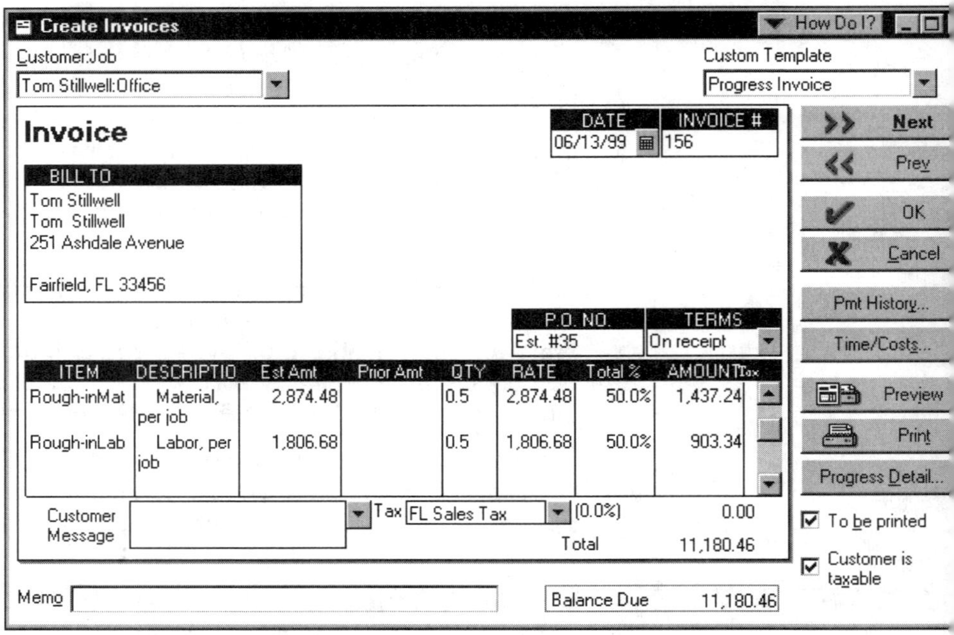

An invoice for the first half of the plumbing rough-in.

4. Click on the **Company Preferences** tab.
5. Click on **Yes** under **Do You Do Progress Invoicing?**
6. Click on **OK.**

To create the invoice:

1. Click on the **Create Invoice** button at the bottom of the *QuickBooks* estima screen.
2. Click on **Yes** to record changes to the estimate.
3. If you selected progress billing, enter a percentage or select items to be invoiced.
4. Click on **OK** and *QuickBooks* creates the invoice.
5. Make changes to the invoice if you want.
6. Click to print one copy for your file and another for your customer.
7. Click on **OK** when done. (Processing the file may take a little time.) *QuickBooks* reports will now include totals from the job just invoiced.

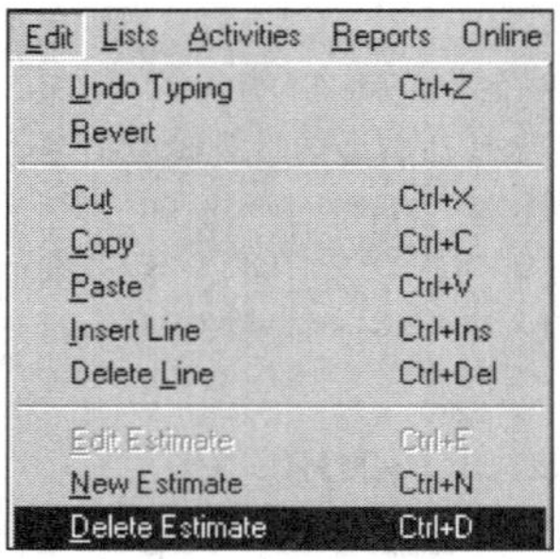

It's easy to delete an estimate.

Don't Worry About Making a Mistake

QuickBooks is very forgiving. Practice all you want. Experiment any way y want. Then delete any estimate or invoice to remove every trace of it from *QuickBooks*. With the offending estimate or invoice displayed:

1. Click on **Edit.**
2. Click on **Delete Estimate** (or Invoice).
3. Click on **OK**. The estimate (or invoice) is deleted.

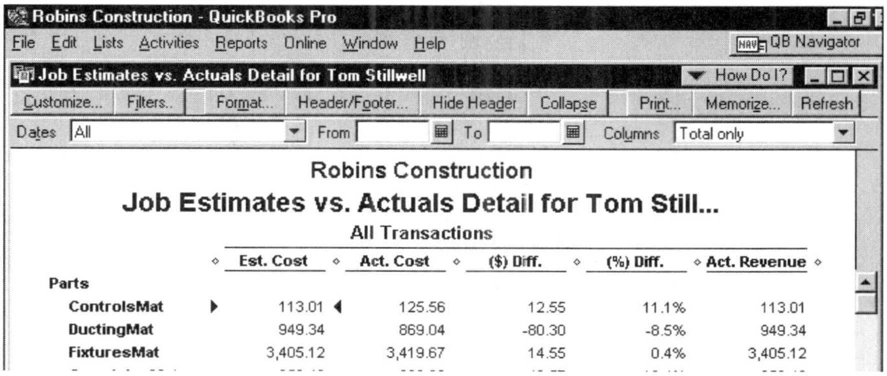

Estimates vs. actuals detail.

Your Jobs in QuickBooks

- Click on **Reports, Profit & Loss, Standard** to see job income and expense.
- Click on **Reports, Balance Sheet, Standard** to see the new receivables total.
- Click on **Reports, Project Reports, Job Estimates vs. Actuals Detail,** select the customer and job to see a detailed cost comparison for the job. Until you start paying bills, the Actual Cost column will be all zeros.

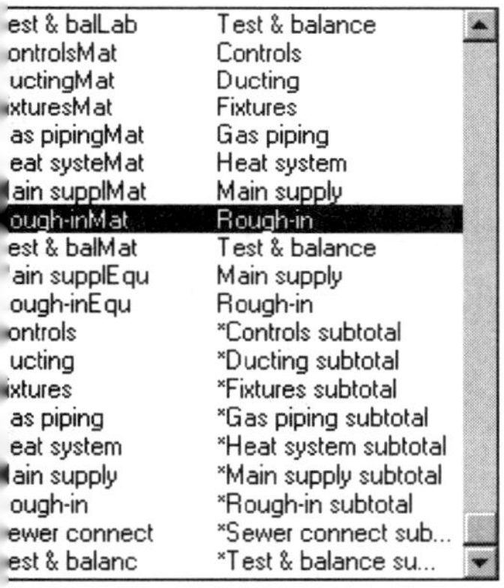

Select a cost category from the Item list.

Paying Bills by Cost Category

When you pay vendors and subcontractors, the amount paid is charged to the job and deducted from your bank balance.

1. Click on **Activities**.
2. Click on **Write Checks**.
3. Fill in the **Pay to the Order of** line.
4. Click on the **Items** tab.
5. Click on the down triangle in the Item column of the check stub to open the Item list.
6. Select a *Mat, Sub or Equ* cost category from the Item list. These are the material, subcontract and equipment subtotal costs from your estimate.

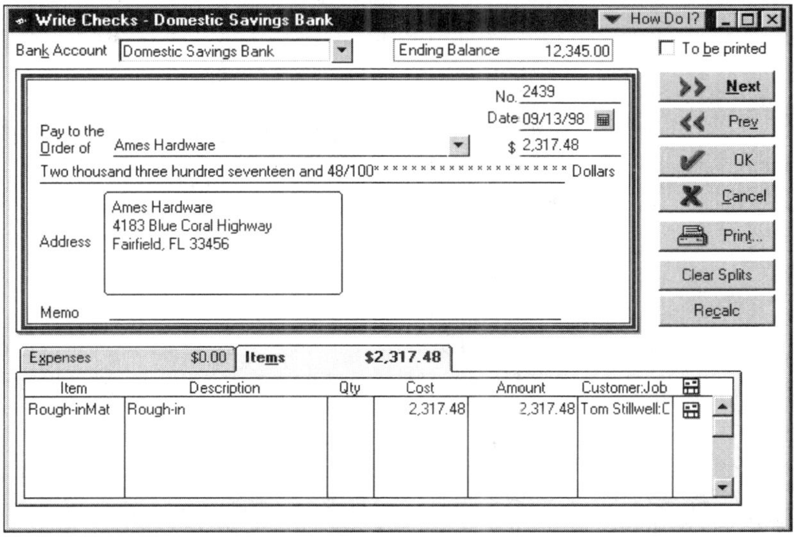

Paying a bill for the Stillwel rough-in.

7. Click in the amount column and enter the amount paid for the item described.
8. Click on the down triangle under **Customer:Job** and select the correct c tomer and job.
9. This amount is not billable to your customer. So click on the icon representing an invoice to put a red X over the icon.
10. One check can cover items in several cost categories and even costs or several jobs. Click again on the next line down in the Item column and f the next cost item.

Creating Payroll by Cost Category

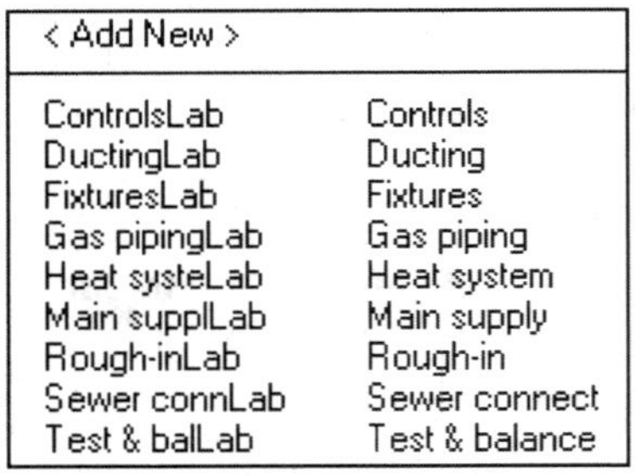

Under Service Item, select a labor cost category from the list.

When you write payroll checks, the amount paid is charged to the job anc deducted from your bank balance.

1. Click on **Activities**.
2. Click on **Time Tracking**.
3. Click on **Use Weekly Time Sheets**.
4. Click on the down triangle opposite **Name** and select the employee to be paid.
5. On the **Timesheet,** click on the down arrow under **Customer:Job** and se the first job where the employee worked.
6. Under **Service Item**, select a *Lab* cost category from the list. These are labor subtotal costs from your estimate.
7. Under **Payroll Item**, select the type of pay, such as hourly.
8. Enter the number of hours worked for that cost category.
9. Any timesheet can cover work done on several jobs and many service items.
10. When finished recording time for an employee, click on **New Sheet**.
11. When finished recording time for all employees, click on **OK**.

To actually produce paychecks:

1. Click on **Activities** and then **Payroll**.
2. Click on **Pay Employees**.
3. Check the names of the employees to be paid.
4. When done, click on **Create**.

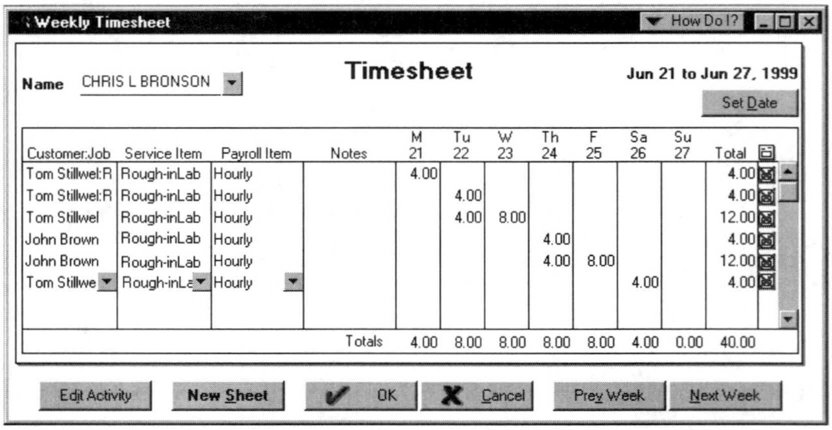

Fill out the timesheet by Customer and Service Item.

Index

Practical References for Builders

c Plumbing with Illustrations, Revised

This completely-revised edition brings this comprehensive manual fully up-to-date with all the latest plumbing codes. It is the journeyman's and apprentice's guide to installing plumbing, piping, and fixtures in residential and light commercial buildings: how to select the right materials, lay out the job and do professional-quality plumbing work, use essential tools and materials, make repairs, maintain plumbing systems, install fixtures, and add to existing systems. Includes extensive study ques- at the end of each chapter, and a section with all the correct answers.
ages, 8½ x 11, $33.00

onal Construction Estimator

nt building costs for residential, commercial, and industrial construc- stimated prices for every common building material. Provides man- , recommended crew, and gives the labor cost for installation. les a CD-ROM with an electronic version of the book with *National ator*, a stand-alone *Windows*™ estimating program, plus an interac- nultimedia video that shows how to use the disk to compile con- ion cost estimates. **592 pages, 8½ x 11, $47.50. Revised annually**

bing & HVAC Manhour Estimates

reds of tested and proven manhours for installing just about any bing and HVAC component you're likely to use in residential, com- al, and industrial work. You'll find manhours for installing piping sys- specialties, fixtures and accessories, ducting systems, and HVAC ment. If you estimate the price of plumbing, you shouldn't be with- e reliable, proven manhours in this unique book.
ages, 5½ x 8½, $28.25

Pocket Guide to HVAC Electricity

andy guide will be a constant source of useful informa- or anyone working with electrical systems for heating, ating, refrigeration, and air conditioning. Includes essen- bles and diagrams for calculating and installing electri- stems for powering and controlling motors, fans, heating nts, compressors, transformers and every electrical part HVAC system. **304 pages, 3½ x 5½, $18.00**

truction Forms & Contracts

rms you can copy and use — or load into your computer (from the disk enclosed). Then you can customize the forms to fit your compa- them out, and print. Loads into *Word* for *Windows*™, *Lotus 1-2-3*, *erfect*, or *Excel* programs. You'll find forms covering accounting, esti- g, fieldwork, contracts, and general office. Each form comes with ete instructions on when to use it and how to fill it out. These forms designed, tested and used by contractors, and will help keep your ss organized, profitable and out of legal, accounting and collection es. Includes a CD-ROM for *Windows*™ and Macintosh.
ages, 8½ x 11, $39.75

truction Estimating Reference Data

les the 300 most useful manhour tables for practically every item of uction. Labor requirements are listed for sitework, concrete work, ry, steel, carpentry, thermal and moisture protection, door and win- finishes, mechanical and electrical. Each section details the work estimated and gives appropriate crew size and equipment needed. es a CD-ROM with an electronic version of the book with *National tor*, a stand-alone *Windows*™ estimating program, plus an interac- ultimedia video that shows how to use the disk to compile con- on cost estimates. **432 pages, 8½ x 11, $39.50**

ber's Exam Preparation Guide

eds of questions and answers to help you pass the apprentice, jour- an, or master plumber's exam. Questions are in the style of the actu- m. Gives answers for both the Standard and Uniform plumbing Includes tips on studying for the exam and the best way to prepare lf for examination day. **320 pages, 8½ x 11, $29.00**

Contractor's Guide to QuickBooks Pro

This user-friendly manual walks you through QuickBooks Pro's detailed setup procedure and explains step-by-step how to create a first-rate accounting system. You'll learn in days, rather than weeks, how to use QuickBooks Pro to get your contracting business organized, with simple, fast accounting procedures. On the CD included with the book you'll find a full version of QuickBooks Pro, good for 25 uses, with a QuickBooks Pro file preconfigured for a construction company (you drag it over onto your computer and plug in your own company's data). You'll also get a complete estimating program, including a database, and a job costing program that lets you export your estimates to QuickBooks Pro. It even includes many useful construction forms to use in your business.
288 pages, 8-1/2 x 11, $39.75

CD Estimator

If your computer has *Windows*™ and a CD-ROM drive, CD Estimator puts at your fingertips 85,000 construction costs for new construction, remodeling, renovation & insurance repair, electrical, plumbing, HVAC and painting. You'll also have the *National Estimator* program — a stand-alone estimating program for *Windows*™ that *Remodeling* magazine called a "computer wiz." Quarterly cost updates are available at no charge on the Internet. To help you create professional-looking estimates, the disk includes over 40 construction estimating and bidding forms in a format that's perfect for nearly any word processing or spreadsheet program for *Windows*™. And to top it off, a 70-minute interactive video teaches you how to use this CD-ROM to estimate construction costs. **CD Estimator is $68.50**

The Contractor's Legal Kit

Stop "eating" the costs of bad designs, hidden conditions, and job surprises. Set ground rules that assign those costs to the rightful party ahead of time. And it's all in plain English, not "legalese." For less than the cost of an hour with a lawyer you'll learn the exclusions to put in your agreements, why your insurance company may pay for your legal defense, how to avoid liability for injuries to your sub and his employees or damages they cause, how to collect on lawsuits you win, and much more. It also includes a FREE computer disk with contracts and forms you can customize for your own use. **352 pages, 8½ x 11, $59.95**

National Building Cost Manual

Square foot costs for residential, commercial, industrial, and farm buildings. Quickly work up a reliable budget estimate based on actual materials and design features, area, shape, wall height, number of floors, and support requirements. Includes all the important variables that can make any building unique from a cost standpoint.
240 pages, 8½ x 11, $23.00. Revised annually

Rough Framing Carpentry

If you'd like to make good money working outdoors as a framer, this is the book for you. Here you'll find shortcuts to laying out studs; speed cutting blocks, trimmers and plates by eye; quickly building and blocking rake walls; installing ceiling backing, ceiling joists, and truss joists; cutting and assembling hip trusses and California fills; arches and drop ceilings — all with production line procedures that save you time and help you make more money. Over 100 on-the-job photos of how to do it right and what can go wrong. **304 pages, 8½ x 11, $26.50**

Contractor's Guide to the Building Code Revised

This new edition was written in collaboration with the International Conference of Building Officials, writers of the code. It explains in plain English exactly what the latest edition of the *Uniform Building Code* requires. Based on the 1997 code, it explains the changes and what they mean for the builder. Also covers the *Uniform Mechanical Code* and the *Uniform Plumbing Code*. Shows how to design and construct residential and light commercial buildings that'll pass inspection the first time. Suggests how to work with an inspector to minimize construction costs, what common building shortcuts are likely to be cited, and where exceptions may be granted. **320 pages, 8½ x 11, $39.00**

Builder's Guide to Accounting Revised

Step-by-step, easy-to-follow guidelines for setting up and maintaining records for your building business. This practical, newly-revised guide to all accounting methods shows how to meet state and federal accounting requirements, explains the new depreciation rules, and describes how the Tax Reform Act can affect the way you keep records. Full of charts, diagrams, simple directions and examples, to help you keep track of where your money is going. Recommended reading for many state contractor's exams. **320 pages, 8½ x 11, $26.50**

Planning Drain, Waste & Vent Systems

How to design plumbing systems in residential, commercial, and industrial buildings. Covers designing systems that meet code requirements for homes, commercial buildings, private sewage disposal systems, and even mobile home parks. Includes relevant code sections and many illustrations to guide you through what the code requires in designing drainage, waste, and vent systems. **192 pages, 8½ x 11, $19.25**

Contractor's Index to the 1997 *Uniform Building Code*

Finally, there's a common-sense index that helps you quickly and easily find the section you're looking for in the *UBC*. It lists topics under the names builders actually use in construction. Best of all, it gives the full section number and the actual page in the *UBC* where you'll find it. If you need to know the requirements for windows in exit access corridor walls, just look under *Windows*™. You'll find the requirements you need are in Section 1004.3.4.3.2.2 in the *UBC* — on page 115. This practical index was written by a former builder and building inspector who knows the from both perspectives. If you hate to spend valuable time hunting through pages of fine print for the information you need, this is the book for you.
192 pages, 8½ x 11, paperback edition, $26.00
192 pages, 8½ x 11, loose-leaf edition, $29.00

Plumber's Handbook Revised

This new edition shows what will and won't pass inspection in drai vent, and waste piping, septic tanks, water supply, fire protection, anc piping systems. All tables, standards, and specifications completely u date with recent plumbing code changes. Covers common layouts fo idential work, how to size piping, selecting and hanging fixtures, prac recommendations, and trade tips. The approved reference for the pl ing contractor's exam in many states. **352 pages, 8½ x 11, $32.00**

Renovating & Restyling Older Homes

Any builder can turn a run-down old house into a showcase of perfection — if the customer has unlimited funds to spend. Unfortunately, most customers are on a tight budget. They usually want more improvements than they can afford — and they expect you to deliver. This book shows how to add economical improvements that can increase the property value by two, five or even ten times the cost of the remodel. Sound impossible? Here you'll find the secrets of a builder who has been putting these techniques to work on Victoriar Craftsman-style houses for twenty years. You'll see what to repair, wh replace and what to leave, so you can remodel or restyle older home the least amount of money and the greatest increase in value.
416 pages, 8½ x 11, $33.50

Professional Kitchen Design

Remodeling kitchens requires a "special" touch — one that blends ar flair with function to create a kitchen with charm and personality as as one that is easy to work in. Here you'll find how to make the best u the space available in any kitchen design job, as well as tips and lesso how to design one-wall, two-wall, L-shaped, U-shaped, peninsula island kitchens. Also includes what you need to know to run a profi kitchen design business. **176 pages, 8½ x 11, $24.50**

Craftsman Book Company
6058 Corte del Cedro
P.O. Box 6500
Carlsbad, CA 92018

☎ 24 hour order line
1-800-829-8123
Fax (760) 438-0398

Name ______________________

Company ______________________

Address ______________________

City/State/Zip ______________________

❍ This is a residence

Total enclosed______________________(In California add 7.25% tax)

We pay shipping when your check covers your order in full.

In A Hurry?

We accept phone orders charged to your
❍ Visa, ❍ MasterCard, ❍ Discover or ❍ American Express

Card#______________________

Exp. date______________Initials______________

Order online http://www.craftsman-book.com
Free on the Internet! Download any of Craftsman's estimating costbooks for a 30-day free trial! http://costbook.com

Tax Deductible: Treasury regulations make these references tax deductible when used in your work. Save the canceled check or charge card statement as your receipt.

10-Day Money Back Guarantee

- ❍ 33.00 Basic Plumbing with Illustrations
- ❍ 26.50 Builder's Guide to Accounting Revised
- ❍ 68.50 CD Estimator
- ❍ 39.50 Construction Estimating Reference Data with FREE *National Estimator* on a CD-ROM.
- ❍ 39.75 Construction Forms & Contracts with a CD-ROM for *Windows*™ and Macintosh.
- ❍ 39.75 Contractor's Guide to QuickBooks Pro
- ❍ 39.00 Contractor's Guide to the Building Code Revised
- ❍ 26.00 Contractor's Index to the *UBC*— *Paperback*
- ❍ 29.00 Contractor's Index to the *UBC*— *Loose-leaf*
- ❍ 59.95 Contractor's Legal Kit
- ❍ 23.00 National Building Cost Manual
- ❍ 47.50 National Construction Estimator with FREE *National Estimator* on a CD-ROM.
- ❍ 19.25 Planning Drain, Waste & Vent Systems
- ❍ 29.00 Plumber's Exam Preparation Guide
- ❍ 32.00 Plumber's Handbook Revised
- ❍ 28.25 Plumbing & HVAC Manhour Estimates
- ❍ 24.50 Professional Kitchen Design
- ❍ 33.50 Renovating & Restyling Older Homes
- ❍ 26.50 Rough Framing Carpentry
- ❍ 18.00 Vest Pocket Guide to HVAC Electricity
- ❍ 48.25 National Plumbing & HVAC Estimator with FREE *National Estimator* on a CD-ROM.
- ❍ FREE Full Color Catalog

Prices subject to change without notice

Craftsman Book Company
6058 Corte del Cedro
P.O. Box 6500
Carlsbad, CA 92018

24 hour order line
1-800-829-8123 Fax (760) 438-0398
Order online: http://www.craftsman-book.com
In A Hurry? We accept phone orders charged to your Visa, MasterCard, Discover or American Express

ame ______
ompany ______
ddress ______
ity/State/Zip ______
ital enclosed ______ (In California add 7.25% tax)

Send check or money order and we pay postage, or use your Visa ☐ MasterCard ☐ Discover or ☐ Amex

ard# ______
piration Date ______ Initials ______

10-Day Money Back Guarantee

- ☐ 34.00 Basic Engineering for Builders for Builders
- ☐ 38.00 Basic Lumber Engineering
- ☐ 35.50 Carpentry Estimating with FREE estimating program on disk.
- ☐ 68.50 CD Estimator
- ☐ 69.00 CD Estimator — Heavy
- ☐ 25.00 Concrete Construction & Estimating
- ☐ 36.00 Contracting in All 50 States
- ☐ 39.50 Construction Estimating Reference Data with FREE stand-alone *Windows* estimating program on a CD-ROM.
- ☐ 39.75 Construction Forms & Contracts with a CD-ROM for both *Windows* and *Mac*.
- ☐ 39.00 Contractor's Guide to Building Code Revised
- ☐ 39.75 Contractor's Guide to QuickBooks Pro with FREE CD-ROM
- ☐ 26.00 Contractor's Index to the 1997 UBC (paperback)
- ☐ 29.00 Contractor's Index to the 1997 UBC (loose-leaf)
- ☐ 22.25 Contractor's Survival Manual
- ☐ 27.50 Drafting House Plans
- ☐ 17.00 Estimating Home Building Costs
- ☐ 21.50 Estimating Tables for Home Building
- ☐ 22.75 Excavation & Grading Handbook Revised
- ☐ 22.50 Finish Carpenter's Manual
- ☐ 23.00 National Building Cost Manual
- ☐ 47.75 National Electrical Estimator with FREE stand-alone *Windows* estimating program on CD ROM.
- ☐ 48.00 National Painting Cost Estimator with FREE stand-alone *Windows* estimating program on CD ROM.
- ☐ 48.25 National Plumbing & HVAC Estimator with FREE stand-alone *Windows* estimating program on CD ROM.
- ☐ 49.50 National Renovation & Insurance Repair Estimator with FREE stand-alone *Windows* estimating program on CD ROM.
- ☐ 48.50 National Repair & Remodeling Estimator with FREE stand-alone *Windows* estimating program on CD ROM.
- ☐ 28.25 Plumbing & HVAC Manhour Estimates
- ☐ 29.00 Residential Electrical Estimating
- ☐ 38.80 Residential Steel Framing Guide
- ☐ 35.00 Roofing Construction & Estimating
- ☐ 26.50 Rough Framing Carpentry
- ☐ 47.50 National Construction Estimator with FREE stand-alone *Windows* estimating program on CD ROM

Receive Fresh Cost Data Every Year – Automatically

Join Craftsman's Standing Order Club and automatically receive special membership discounts on annual cost books!

Qty. ___ National Building Cost Manual
Standing Order price $19.55

Qty. ___ National Construction Estimator
Standing Order price $40.38

Qty. ___ National Electrical Estimator
Standing Order price $40.59

Qty. ___ National Plumbing & HVAC Estimator
Standing Order price $41.01

Qty. ___ National Painting Cost Estimator
Standing Order price $40.80

Qty. ___ National Renovation & Insurance Repair Est.
Standing Order price $42.08

Qty. ___ National Repair & Remodeling Est.
Standing Order price $41.23

How many times have you missed one of Craftsman's pre-publication discounts, or procrastinated about buying the updated book and ended up bidding your jobs using obsolete cost data that ou "updated" on your own? As a Standing Order Member you never have to worry about ordering every year. Instead, you will receive a confirmation of your standing order each September. If the rder is correct, do nothing, and your order will be shipped and billed as listed on the confirmation card. If you wish to change your address, or your order, you can do so at that time. Your standing rder for the books you selected will be shipped as soon as the publications are printed. You'll automatically receive any pre-publication discount price.

Bill to:
Name ______
ompany ______
ddress ______
ity/State/Zip ______

Ship to:
Name ______
Company ______
Address ______
City/State/Zip ______

Purchase order# ______
Phone # (   ) ______
Signature ______

Mail This Card Today for a Free Full Color Catalog

Over 100 construction references at your fingertips with information that can save ou time and money. Here you'll find information on carpentry, contracting, estimating, emodeling, electrical work and plumbing.

All items come with an unconditional 10-day money-back guarantee. If they don't save ou money, mail them back for a full refund.

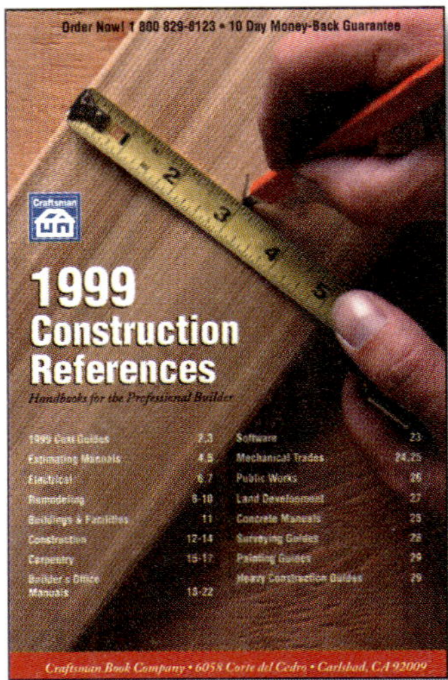

Name ______
Company ______
Address ______
City/State/Zip ______

Craftsman Book Company
6058 Corte del Cedro
P.O. Box 6500 • Carlsbad, CA 92018

NO POSTAGE NECESSARY IF MAILED IN THE UNITED STATES

BUSINESS REPLY MAIL
FIRST CLASS MAIL PERMIT NO. 271 CARLSBAD, CA

POSTAGE WILL BE PAID BY ADDRESSEE

Craftsman Book Company
6058 Corte del Cedro
P.O. Box 6500
Carlsbad, CA 92018-9974

NO POSTAGE NECESSARY IF MAILED IN THE UNITED STATES

BUSINESS REPLY MAIL
FIRST CLASS MAIL PERMIT NO. 271 CARLSBAD, CA

POSTAGE WILL BE PAID BY ADDRESSEE

Craftsman Book Company
6058 Corte del Cedro
P.O. Box 6500
Carlsbad, CA 92018-9974

NO POSTAGE NECESSARY IF MAILED IN THE UNITED STATES

BUSINESS REPLY MAIL
FIRST CLASS MAIL PERMIT NO. 271 CARLSBAD, CA

POSTAGE WILL BE PAID BY ADDRESSEE

Craftsman Book Company
6058 Corte del Cedro
P.O. Box 6500
Carlsbad, CA 92018-9974